DIGITAL RADIO SYSTEMS ON A CHIP

A Systems Approach

DIGITAL RADIO SYSTEMS ON A CHIP

A Systems Approach

Charles Chien
Rockwell Science Center

KLUWER ACADEMIC PUBLISHERS
Boston / Dordrecht / London

Distributors for North, Central and South America:
Kluwer Academic Publishers
101 Philip Drive
Assinippi Park
Norwell, Massachusetts 02061 USA
Telephone (781) 871-6600
Fax (781) 871-6528
E-Mail <kluwer@wkap.com>

Distributors for all other countries:
Kluwer Academic Publishers Group
Distribution Centre
Post Office Box 322
3300 AH Dordrecht, THE NETHERLANDS
Telephone 31 78 6392 392
Fax 31 78 6546 474
E-Mail <orderdept@wkap.nl>
Electronic Services <http://www.wkap.nl>

Library of Congress Cataloging-in-Publication Data

Chien, Charles.
Digital radio systems on a chip : a systems approach / Charles Chien.
p. cm.
Includes bibliographical references and index.
ISBN 0-7923-7260-3 (alk. paper)
1. Wireless communication systems--Equipment and supplies. 2. Digital integrated circuits. 3. Radio--Transmitter-receivers. I. Title.

TK7874.65 .C45 2000
621.384'12--dc21 00-048784

Printed on acid-free paper.
Printed in the United States of America

To Sheri, Daniel, and Samuel, God's greatest blessing in my life.

Table of Contents

Preface

This book focuses on a specific engineering problem that is and will continue to be important in the forth-coming information age: namely, the need for highly integrated radio systems that can be embedded in wireless devices for various applications, including portable mobile multimedia wireless communications, wireless appliances, digital cellular, and digital cordless. Traditionally, the design of radio IC's involves a team of engineers trained in a wide range of fields that include networking, communication systems, radio propagation, digital/analog circuits, RF circuits, and process technology. However as radio IC's become more integrated, the need for a diverse skill set and knowledge becomes essential for professionals as well as students to broaden beyond their trained area of expertise and to become proficient in related areas. The key to designing an optimized, economical solution for radio systems on a chip hinges on the designer's thorough understanding of the complex trade-offs from communication systems down to circuits.

To acquire the insight and understanding of the complex system and circuit trade-offs, a designer must digest volumes of books covering diverse topics, such as communications theory, radio propagation, and digital/analog/RF circuits. While books are available today that cover the individual areas, they tend to be narrowly focused and do not provide the necessary insight in the specific problem of integrating a complete radio system on a chip. The purpose of this book is to provide a focused, top-down treatment of radio system IC design with an emphasis on digital radio systems that will play a dominant role in wireless communications in the 21st century.

The intended readers of this book include both engineers as well as students whose main area of interest or line of work lies in the design and integrated circuits implementation of wireless communication IC's, with an emphasis on the integration of an entire radio system on a chip.

This book is organized into the following chapters and is intended to provide the reader a system perspective on design insights and trade-offs in integrating digital radio systems on a chip.

Chapter 2 gives an overview of key terminologies and techniques for digital communication systems. In particular, commonly used terminologies are explained in context of implementation. These terminologies include bandwidth efficiency, energy efficiency, channel capacity, bit-error-rate, packet/frame error rate, and coding efficiency. Commonly encountered degradations in a wireless environment are also discussed in context of their effect on design trade-offs. These degradations include noise, interference, and multipath fading. Key techniques are then described for digital modulation, coding, and signal diversity. This chapter also covers practical aspects of implementing a digital system, such as phase and timing recovery and frequency acquisition.

Chapter 3 covers spread-spectrum communications, which has become important over the past decade in the commercial sector. Systems based on it are being deployed worldwide. These include code-division multiple access (CDMA) systems and unlicensed transmission based on the IEEE 802.11 WLAN standard. The chapter again explains commonly used terminologies and techniques in the context of their impact on implementation. Terminologies discussed include processing gain, chip rate, and hop rate. Key techniques that are discussed include frequency-hop spread-spectrum, direct-sequence spread-spectrum, Rake receiver, and frequency diversity.

Chapter 4 presents an overview of system constraints and trade-offs for the RF front-end. Issues that are covered include the impact of RF noise and distortion on system performance. Important metrics such as noise figure and intercept points are thoroughly defined in this chapter.

Chapter 5 gives an overview of radio systems and motivates the tradeoffs that exist for a radio system IC design, driven top-down. Issues that are covered include the impact of quality-of-service, system capacity, and link quality on radio implementation.

Chapter 6 deals with digital radio architecture. In particular, the chapter focuses on the partition of digital and analog processing and discusses the relative merits of different partitioning schemes.

Chapter 7 deals with implementation details of the RF transceiver and its associated components, such as low noise amplifier, mixer, voltage-controlled oscillator, synthesizer, and power amplifiers.

Chapter 8 deals with implementation details of the digital transceiver and the associated signal processing, such as modulation, maximum likelihood equalization, and diversity reception. The chapter also discusses how system requirements impact the digital transceiver design parameters such as gate complexity, power, and the operations per second.

Chapter 9 illustrates the design principles and radio components covered in specific case studies, including a complete top-down design example for the GSM digital cellular system. This chapter also discusses future trends in single-chip radios and challenges designers will face in integrating highly sensitive radio-frequency circuits along with fast switching digital circuits on the same substrate.

The completion of this book would not have been possible without the supports from many of my friends and colleagues. In particular, I would like to thank Dr. Shanthi Pavan at Texas Instruments and Professor Robert Brodersen for their helpful and constructive suggestions. I would especially like to thank the support of Dr. Derek Cheung, Dr. Hank Marcy, Dr. Jon Rode, and Dr. Marius Vassiliou for their strong support of my research at the Rockwell Science Center. I would also like to thank my former colleague Brian Schiffer, Christopher Deng at UCLA, Robby Gupta at University of Michigan, Dr. Trudy Stetzler at Texas Instrument, Dr. Maryjo Nettles at

Mobilian, Dr. Thomas Cho at Intel, and Dr. Victor Lin at Rockwell for their careful review of various chapters in this book. Finally, I would like to acknowledge my colleagues at the Rockwell Science Center in the many valuable and stimulating discussions in the area of RF circuit design, IC processing technology, and wireless communications. In particular, I would like to thank Drs. Bobby Brar, Aiden Higgins, Myung-Jun Choe, and James Li. I would also like to thank Dr. John Walley at Conexant Systems for many interesting discussions on system IC design. Also, I would like to express my appreciation to the staff at Kluwer Publishers for their patience and assistance in preparing for the final manuscript. I would like to acknowledge the love and support of my wife Sheri and my sons Daniel and Samuel, who waited patiently as I worked to complete this manuscript. Finally, I am indebted to my Lord and Savior Jesus Christ who has been and always will be my source of inspiration and strength.

Charles Chien
September 2000

CHAPTER 1

Introduction

The telecommunication industry has advanced rapidly over the past century. The invention of telephone has sparked the beginning of the telecommunications era. Today, an immense Public Switching Telephone Network (PSTN) exists to provide services to millions of users worldwide. Though for many years our communication needs have been satisfied by the telephone networks, in the past few decades computer and integrated circuits (IC) technology have reshaped our communication needs to include not only voice but data and video as well. Conveying data information worldwide, computer networks, such as the Internet, have become prevalent. This growth has been fueled by advances in IC technology that have enabled affordable consumer products, such as high speed modems which have made computer accesses as easy as making a phone call. Sophisticated image and speech processing are also becoming available in integrated circuits to enable embedded multimedia capabilities in portable laptops.

Currently, high quality affordable multimedia content is becoming available through emerging technologies, such as digital cable and digital subscriber lines (DSL). Communication speed on twisted pair using sophisticated echo cancellation and equalization is enabling data rates up to several megabits per second on twisted pairs and hundreds of megabits per sec on coaxial cable. In this new millennium, abundant bandwidth will be supplied through optical fiber backbones, enabling the delivery of high quality multimedia services to homes via cable, twisted pair, or even directly through fibers.

Concurrent with rapid advances in wireline-based technology, there is a surging demand for wireless communications. Although most of our communication needs are being met by the current wireline systems, our communications are limited to areas with pre-installed wireline access to information services. We are, in a sense, tethered to the wire or fiber that carries our communications. When we are on the move, it is desirable to have untethered communications such that we may have ubiquitous access to the available information services. The key to this ubiquitous communication lies in the ability to communicate via wireless links.

1.1 A Brief History of Wireless

In the past century, wireless communications have been largely furnished by analog systems. In 1894, Marconi experimented with radio wave transmissions for telegraphy and was able to demonstrate transmission over a mile. Later in 1897, the range was extended to 18 miles, then to 200 miles in 1901, and eventually across the Atlantic [1]. Soon thereafter wireless grew rapidly: from radio broadcast in 1920's to television broadcast in 1930's. Wireless has then been predominantly used for broadcasting until 1977 when Advanced Mobile Phone System (AMPS) was introduced. AMPS is the result of research efforts by AT&T since the late 1940's and is designed to provide mobile telephone services. After its introduction, however, the public was slow in adopting its use.

In the late 1980's, AMPS begin to show signs of growth and by early 1990's it had a substantial subscriber base of 20-30 million users [2]. Not surprisingly, the growth coincides with the growth in personal computers, which was fueled by the reduction in cost of microprocessor chips of increasing processing capability. Similarly, due to recent reduction in the cost of cellular handsets with increased level of integration in communications integrated circuits, the number of cellular subscribers has risen dramatically, stressing the capacity limit of the current analog systems (See Figure 1.1).

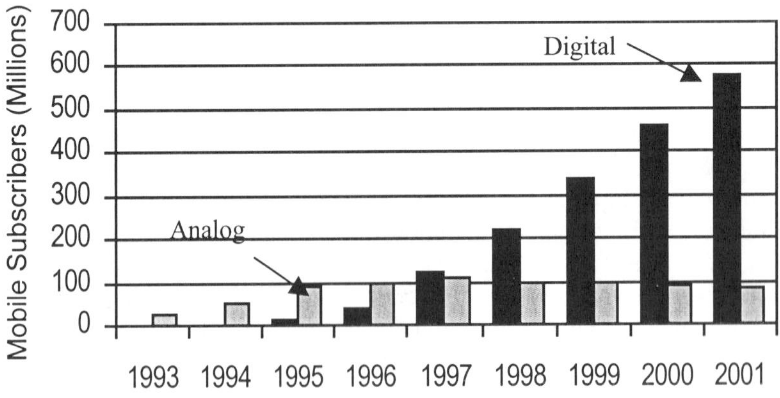

Figure 1.1 Worldwide mobile subscriber growth.

To meet the increasing demands for high capacity, second-generation digital cellular systems have emerged worldwide. For instance, in the U.S. several systems have been deployed. These include the Interim Standard (IS)-136 [3], IS-95 [4][5], and a large number of personal communication systems (PCS) [6]. In Europe, Digital European Cordless Telephone (DECT) [7] and Global System for Mobile Communication (GSM) [8] have been successfully deployed and are showing worldwide adoption. The second-generation cellular systems use digital rather than analog modulation to improve capacity and for an efficient integration of data services. Currently, there is a major push for digital technology even further to support high capacity, truly ubiquitous multimedia services in a new third-generation digital standard, known as IMT-2000 (international mobile telecommunications by the year 2000) [9]. The International Telecommunications Union (ITU) has spearheaded the standardization process for IMT-2000 and has recently converged on a digital standard that covers multiple air-interfaces, with wideband code division multiple access (WCDMA) [10] being the most popular. The projected number of cellular subscribers worldwide is projected to reach over 600 million in the next few years and total worldwide revenue in wireless will exceed $100 billion. Most of the growth will be in digital technology while analog is on a decline as shown in Figure 1.1.

Besides digital cellular, digital radio technology is also finding its way into emerging areas, such as wireless local area networks (WLAN), wireless local loops (WLL), and personal area networks (PAN). WLAN was first specified in the early 1990's as part of the IEEE 802.11 standard [11] for wireless network cards that can support up to 1-2 Mbps throughput. IEEE 802.11 specifies RF transmission in the unlicensed bands approved by the FCC. The frequency bands are at 900 MHz, 2.4 GHz, and 5.7 GHz. New standards are now emerging for broadband wireless networking (10 to 155 Mbps): e.g. Broadband Radio Access Network (BRAN) and IEEE 802.11a,b. WLAN products and services are projected to reach multi-billion dollars in revenue within the next few years.

WLL, also known as fixed wireless access (FWA) enables broadband services via fixed radio connectivity to homes, businesses, or public facilities. The technology offers wireless alternative to wireline hubs. FWA offers ease of deployment of broadband services for the last mile connection to fixed subscribers. In the US, fixed wireless access is known as local multipoint distribution system (LMDS) or multi-channel multipoint distribution system (MMDS). For LMDS, licensed bands are available in the 28 GHz range with over 1 GHz of bandwidth available. For MMDS, licensed bands are available in the 2.5 GHz range with 200 MHz of bandwidth. Multiple OC1-rates can be supported.

PAN is the most recent result of the digital revolution. It includes Bluetooth [12] and HomeRF. Both systems are targeted for extremely low cost short-range wireless networks. Bluetooth supported by an industry consortium has a charter to develop miniature embeddable radio chips for laptops, handheld computers, personal digital assistants (PDA), and cellular phones. The radio chips must be extremely low power (less than a few mW) and low cost (less than a couple of dollars). The miniature radios will enable various personal devices to communicate with one another in an autonomous fashion; for instance, the ability to download address books from a laptop to a cellular phone automatically. HomeRF has a similar goal but is applied to home appliances, such as personal computers and digital set-top boxes.

In summary, we have witnessed a long history in wireless and are currently in midst of the digital revolution in creating wireless technology that will enable ubiquitous connectivity for future services and information access, such as Internet access, E-commerce, video-conference, and mobile data networking. Figure 1.2 maps current commercial systems in terms of their throughput and operation environment. In the coming years, we will see a convergence into three main areas: 1) broadband wireless networks, 2) third-generation digital cellular, and 3) personal area networks.

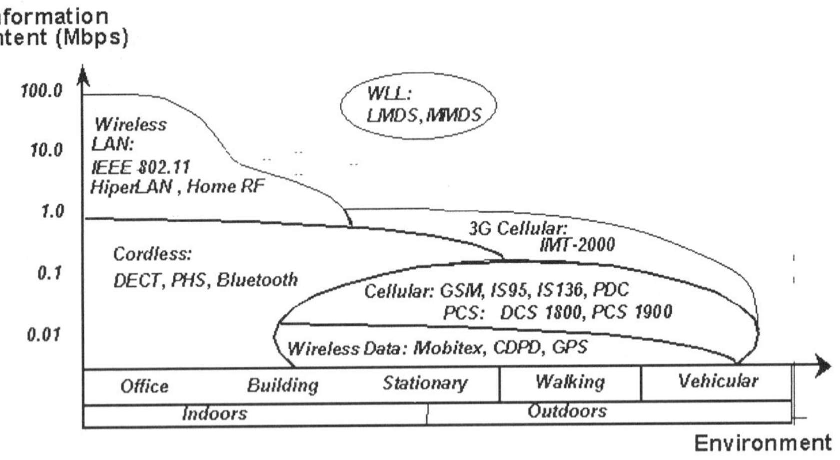

Figure 1.2 Current and emerging digital communication systems.

1.2 The Challenges

As systems advance, consumers are able to obtain new wireless services and more sophisticated wireless devices, all at a lower price due to economy of scale. On the other hand, engineers must bear with all the challenges that must be overcome to implement radio devices that can support these new services, such high-speed wireless Internet access and mobile data networking, and still be able to maintain low cost, size, and power. Some specific challenges are listed below.

- **High Capacity:** The rising demand for wireless services and appliances stresses the need for higher capacity. The challenge lies in accommodating more users and appliances in the spectrally limited wireless environment. To make the problem even more difficult, new broadband multimedia services impose a high demand on the per-user bandwidth. To accommodate more users in broadband services, the communication device must be designed to operate at substantially higher interference levels. This requires implementation of computationally intense signal processing to suppress interference, such as multiuser detection, space-time processing, coding/modulation, and spread-spectrum.

- **Quality of service (QoS):** Unlike wire-line transmissions, serious degradations such as multipath fading and interference occur in the wireless channel. These degradations are not as predictable as in the wireline environment and are detrimental to the system capacity and QoS. As new services come on line that demand QoS for different information contents, such as voice, data, and/or video, the underlying radio device must implement complex signal processing that can adapt to the changing channel environment and service requirements. The signal processing may include adaptive decision-feedback equalizers, adaptive Rake receivers, or adaptive space-time processing.

- **Portable:** Since most wireless applications require portability, wireless devices must be lightweight and small enough to fit in a shirt pocket. For instance, a StarTac phone made by Motorola weighs less than 4 oz. and 50 cc in volume. To maintain their appeal to consumers, new devices must be just as small and be able to accommodate complex signal processing at substantially higher sampling rate. The new devices must also operate over multiple RF bands. The latter requirement results from the need to interoperate with multiple air-interface standards.

- **Low Power:** Related to portability is low power. A portable wireless device must live on battery for substantial amount of time (e.g. several weeks or even months) yet not adding unacceptable weight to the user. In other words, users should not feel like they are being loaded with a brick inside their shirt pockets. Current technology allows about 3-8 hours talk time and 200-300 hours of standby time with a Lithium Ion battery. The power dissipation must either be maintained or improved further even when more complex high-speed signal processing and high frequency RF circuits are needed.

- **Affordable:** To maintain competitive advantage, cost must be factored into the design process. In particular, low cost is critical for high volume products such as digital handsets, which are sold at a volume of 100's millions units per year. A few dollar savings translate to $100's million increase in profit. More importantly, consumer's expectation on low price continues to squeeze the profit margin from wireless vendors so any additional cost saving becomes an essential factor for staying in business. Today, a handset has less than a dozen of integrated circuit chips and cost per handset is down to $100-200 dollars. The cost pressure will remain, forcing designers to integrate more functions onto a single chip for low system cost. Furthermore, for some embedded applications, such as Bluetooth, a fully integrated solution is unavoidable. The designers face substantial challenge in integrating complex radio systems on a chip, which consists of complex high-speed digital signal processing circuits along with highly sensitive radio-frequency circuits on the same substrate.

As systems become more complex and capable, requiring advanced digital signal processing, radio-frequency circuits, and signal conditioning circuits, the design of radio IC's can no longer be treated in isolation from the larger system since these devices must eventually work within the overall system. Moreover, as cost factor continues to drive toward a fully integrated wireless device, i.e. radio systems on a chip, radios can no longer be treated as a dumb device which one can design, implement, put it to a logic analyzer and expect it work properly. With a major portion of the digital system implemented on a chip and with digital circuits that must operate at high speed together with highly sensitive radio-frequency circuits on the same substrate, a thorough system understanding and circuit savvy are needed to design a working radio system IC. This book attempts to address the design challenges associated with radio system IC design from a top-down system perspective, offering the reader a systems understanding and how it constrains the circuit design and requirements.

1.3 Organization of the Text

The focus of this book is on integrated circuits design for digital radio systems and addresses the many challenges discussed earlier needed to obtain a fully integrated radio system on a chip. A digital radio system is shown in Figure 1.3 and consists of a user interface, digital transceiver, analog-to-digital interface, and an RF transceiver. For asymmetric systems, the wireless device is different at the user-end as compared to at the infrastructure. For instance, in a digital cellular system, the infrastructure consists of basestations which service a large number of user-end equipment, also known as handsets. The basestation is therefore much more capable then the user handset and demands proportionally more complex signal processing and high-powered RF circuits. On the other, hand, a symmetric system such as WLAN, the complexity of radio equipment at the user-end and the infrastructure is comparable. This book focuses mainly on system issues and radio circuits that are needed to implement the user-end equipment, although whenever necessary, relevant infrastructure issues will also be covered.

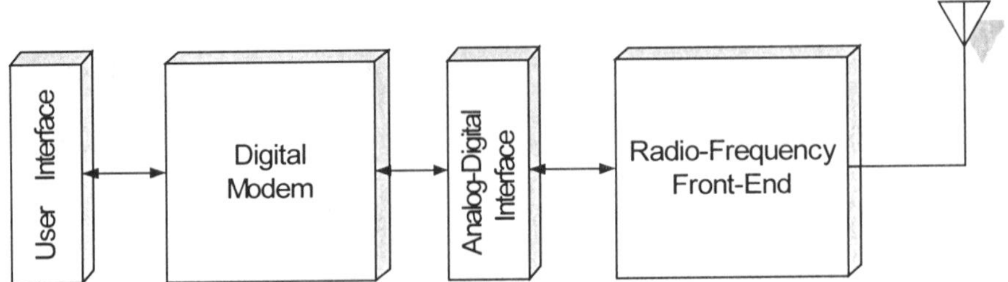

Figure 1.3 Block diagram of a digital radio.

The book is organized into the following chapters.

Chapter 2 gives an overview of key terminologies and techniques for digital communication systems. In particular, commonly used terminologies are explained in context of implementation. These terminologies include bandwidth efficiency, energy efficiency, channel capacity, bit-error-rate, packet/frame error rate, and coding efficiency. Commonly encountered degradations in a wireless environment are also discussed in context of their effect on design trade-offs. These degradations include noise, interference, and multipath fading. Key techniques are then described for digital modulation, coding, and signal diversity. This chapter also covers practical aspects of implementing a digital system, such as phase and timing recovery and frequency acquisition.

Chapter 3 covers spread-spectrum communications, which has become important over the past decade in the commercial sector. Systems based on it are being deployed worldwide. These include code-division multiple access (CDMA) systems and unlicensed transmission based on the IEEE 802.11 WLAN standard. The chapter again explains commonly used terminologies and techniques in the context of their impact on implementation. Terminologies discussed include processing gain, chip rate, and hop rate. Key techniques that are discussed include frequency-hop spread-spectrum, direct-sequence spread-spectrum, Rake receiver, and frequency diversity.

Chapter 4 presents an overview of system constraints and trade-offs for the RF front-end. Issues that are covered include the impact of RF noise and distortion on system performance. Important metrics such as noise figure and intercept points are thoroughly defined in this chapter.

Chapter 5 gives an overview of radio systems and motivates the tradeoffs that exist for a radio system IC design, driven top-down. Issues that are covered include the impact of quality-of-service, system capacity, and link quality on radio implementation.

Chapter 6 deals with digital radio architecture. In particular, the chapter focuses on the partition of digital and analog processing and discusses the relative merits of different partitioning schemes.

Chapter 7 deals with implementation details of the RF transceiver and its associated components, such as low noise amplifier, mixer, voltage-controlled oscillator, synthesizer, and power amplifiers.

Chapter 8 deals with implementation details of the digital transceiver and the associated signal processing, such as modulation, maximum likelihood equalization, and diversity reception. The chapter also discusses how system requirements impact the digital transceiver design parameters such as gate complexity, power, and the operations per second.

Chapter 9 illustrates the design principles and radio components covered in specific case studies, including a complete top-down design example for the GSM digital cellular system. This chapter also discusses future trends in single-chip radios and challenges designers will face in integrating highly sensitive radio-frequency circuits along with fast switching digital circuits on the same substrate.

CHAPTER 2

Digital Communication Systems

A basic digital communication system consists of a transmitter and a receiver. The transmitter takes digitized data at the information source and preprocesses the data for reliable transmission over the communication channel. To efficiently utilize the available bandwidth, the transmitter must accomplish this with minimum bandwidth usage per information source. The transmitted waveform experiences numerous degradations in a communications channel before arriving at the receiver. These degradations include noise, interference, and fading. To achieve adequate performance when such degradations occur, the receiver must process the received signal appropriately to recover the transmitted information. The challenge in the design of a digital communication system lies in the complex design trade-offs that must be considered to achieve adequate performance in the presence of an imperfect communication channel. The design problem is particularly difficult for the wireless environment due to rapid time variation in the channel characteristics and vulnerability to interference from nearby transmissions.

This chapter is devoted to providing an overview of basic techniques and design parameters for the design of a wireless digital communication system. The chapter is organized as follows. In Section 2.1, a basic digital transceiver system and its associated design parameters are described. The purpose is to provide a common set of terminologies needed to understand the design trade-offs. In Section 2.2, degradations experienced in a wireless channel are described. In Section 2.3, an overview of different modulation schemes is presented. In Section 2.4, the performance of selected modulation schemes is analyzed. Section 2.5, describes synchronization techniques that are essential in a practical system. Section 2.6 describes coding techniques required for reliable communications in noise-limited environments and Section 2.7 discusses the diversity techniques used to combat degradations due to fading and interference.

2.1 Digital System Basics

In general, a digital communication system includes a source coder/decoder (codec), a channel codec, a modulator/demodulator (modem), and diversity processing, as shown in Figure 2.1. The source coder takes sampled digital data from a data source and compresses the information to reduce the required bitrate, which minimizes the amount of bandwidth used per information source. At the receiver, the corresponding source decoder decompresses the data and reconstructs the original source information. In general, the source codec is *lossy*, that is the reconstruction exhibits distortion due to the loss in information during the compression process. Such loss is considered worthwhile given the increase in bandwidth utilization. While the source coder strips out redundant information from the source and thereby reduces the required transmission bitrate, the channel coder adds redundant parity check bits that improve the reliability of the compressed source data over noisy transmission channels. Though the channel coder imposes a bandwidth penalty with the parity bits, the improvement in reliability can be significant in poor channel conditions. Since all physical channels are analog in nature, the digital bit stream must be converted into appropriate analog waveforms for transmission. The modulator performs this translation from digital to analog. In contrast to analog modulation, digital modulation translates to discrete analog values rather than a continuum. At the receiver the demodulator performs the inverse operation to obtain the transmitted digital bit stream.

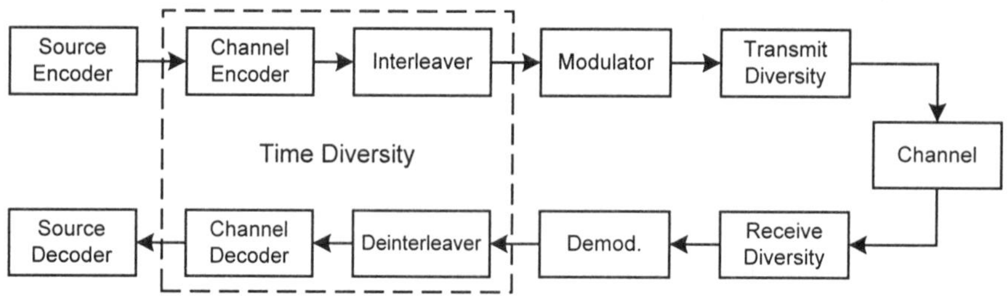

Figure 2.1 A digital communication system.

The baseline communication blocks consisting of a channel codec and modem can be designed to work in a benign environment, degraded mainly by noise. However, in severe multipath fading and interference dominant conditions, additional processing for diversity is required. Multipath fading and interference introduce correlated error statistics in the received bit stream that degrade the performance of the channel decoder and demodulator. Diversity processing attempts to decorrelate the effects of fading and interference before demodulation and channel decoding. Figure 2.1 shows three blocks for diversity processing: namely, FEC codec with interleaving/de-interleaving, diversity transmitter, and diversity receiver. The interleaver permutes the transmitted bit stream such that by inverting the permutation at the receiver, bursty errors due to time-selective fading appear with random statistics at the output of the de-interleaver, effectively converting the channel to a noisy channel with uncorrelated error statistics. The diversity transmitter can improve the performance by preprocessing the modulated waveforms in both space and time domains. Transmit diversity can be accomplished in two ways:

1) It can generate a waveform with the property that after passing through the physical channel, the error statistics appear uncorrelated. This differs from the interleaver in which case errors appear uncorrelated only after the de-interleaver.
2) It can also do the exact opposite by introducing fading in a controlled fashion such that with a diversity receiver, fading can be optimally combined for increased performance.

To aid the understanding of design trade-offs and underlying algorithms for each of the blocks shown in Figure 2.1, we first describe the basic terminologies and parameters needed to characterize the basic attributes of a digital system.

2.1.1 Signal Representation

In analog communications, the source information denoted by $i(t)$ is either amplitude modulated (AM) or frequency modulated (FM) onto a radio-frequency (RF) carrier. As shown in Figure 2.2a, for AM, $i(t)$ resides in the amplitude of the carrier and for FM, $i(t)$ resides in the frequency. In both cases, the constant m is the modulation index and $i(t)$ is modulated as a continuous level of amplitude or frequency.

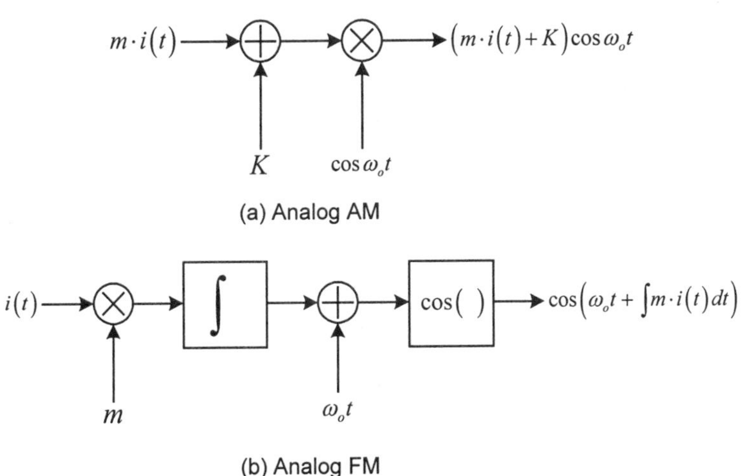

Figure 2.2 Signal representations for analog transmission.

In contrast, a digital system first converts $i(t)$ into digital samples that are then processed digitally by the source coder, interleaver, and the channel coder to form an input $d(t)$ to the modulator. The digital bit stream in $d(t)$ is grouped into symbols, each having $log_2 M$ bits. A symbol has a duration $T_s = T_b \log_2 M$ $T_s = T_b$ and runs at a symbol rate of $R_s = R_b/\log_2 M$ in symbols/sec or baud. T_b is the bit duration and $R_b = 1/T_b$ is the bitrate in bits/sec or bps. Each symbol is mapped to one of M possible combinations of amplitude $\{a_k\}$, phase $\{\phi_k(t)\}$, and/or frequency $\{\omega_k\}$ as shown below:

$$s(t) = \sum_{k=-\infty}^{\infty} a_k \cos[\omega_0 t + \omega_k t + \phi_k(t)]u(t - kT_s) \qquad (2.1)$$

where $u(t)$ is a unit-amplitude pulse with duration T_s, and a_k, ω_k, and $\phi_k(t)$ are respectively the amplitude, frequency, and phase trajectory for the kth symbol time, $kT_s < t \le (k+1)T_s$.

In general, digital modulation that uses amplitude, phase, or frequency is referred to as amplitude shift keying (ASK), phase-shift keying (PSK), and frequency-shift keying (FSK), respectively. Modulations that use both amplitude and phase are referred to as quadrature-amplitude modulation (QAM). In the case of PSK, the phase trajectory stays constant during the symbol duration while in the more general case of continuous phase modulation, the phase trajectory varies with time. Figure 2.3 shows an example of a digital modulation that uses a four-phase symbol where each symbol has four possible binary values *{00, 01, 10, 11}* that are mapped to $\{0.25\pi, 0.75\pi, 1.25\pi, 1.75\pi\}$, respectively. In the four-phase example above, the modulation is also known as quadrature phase shift keying (QPSK).

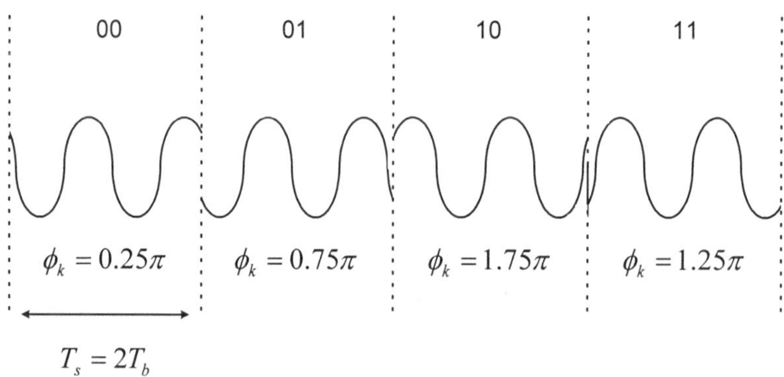

Figure 2.3 Modulator waveforms for QPSK.

To find a more general signal representation for digitally modulated waveforms, the modulator output $s(t)$ is expanded using standard trigonometric identities:

$$s(t) = \sum_{k=-\infty}^{\infty} [I_k \cos(\omega_0 t) - Q_k \sin(\omega_0 t)]u(t - kT_s) \qquad (2.2)$$

where $I_k = a_k \cos(\omega_k t + \phi_k(t))$ and $Q_k = a_k \sin(\omega_k t + \phi_k(t))$. I_k and Q_k are known, respectively, as the in-phase and quadrature-phase components. For simplicity of notation, the time variable is implicit in I_k and Q_k. If $\xi_k = I_k + jQ_k$ then (2.2) can also be expressed in terms of a complex number as shown below

$$s(t) = \Re\{\tilde{s}(t)e^{j\omega_o t}\} \tag{2.3}$$

$$\begin{aligned}\tilde{s}(t) &= \sum_{k=-\infty}^{\infty} \xi_k u(t-kT_s) \\ &= \sum_{k=-\infty}^{\infty} \sqrt{I_k^2+Q_k^2}\, e^{j\arctan\left(\frac{Q_k}{I_k}\right)} u(t-kT_s) \\ &= \sum_{k=-\infty}^{\infty} a_k e^{j(\omega_k t+\phi_k(t))} u(t-kT_s)\end{aligned} \tag{2.4}$$

where without a carrier component $\tilde{s}(t)$ is the baseband representation of the modulated waveform. Expression (2.4) directly relates the amplitude, phase, and frequency to the in-phase and quadrature components as shown below:

$$\begin{aligned}a_k &= \sqrt{I_k^2+Q_k^2} \\ \omega_k t+\phi_k(t) &= \arctan\left(\frac{Q_k}{I_k}\right).\end{aligned} \tag{2.5}$$

In general, the baseband signal is a more compact representation and is therefore used throughout the book.

A digital modulator is illustrated in Figure 2.4a. A digital bitstream $d(t)$ enters into the modulator and gets mapped into symbols I_k and Q_k. The signal lines which carry I_k and Q_k are also referred to as the I-channel and Q-channel. The I and Q channels are upconverted with a carrier in quadrature to generate $s(t)$. A more compact but somewhat abstract representation of the modulator is shown in Figure 2.4b. The symbols and carrier are now represented as complex variables ξ_k and $e^{j\omega_o t}$, respectively.

By plotting (2.4) on the complex *I-Q* plane, one can obtain a graphical repensentation (i.e. phase diagram) of the baseband signal that often proves useful in assessing the performance of different modulation schemes. Igorning $\{\omega_k\}$ for the moment, a scatter plot of (a_k, ϕ_k) for all k reveals the relative distance among the different amplitude-phase pairs. The plot of (a_k, ϕ_k) is also generally known as a *constellation diagram*. Since each point on the constellation diagram corresponds to a specifc symbol transmitted, an error is made at the receiver only when the receiver mis-identifies the constellation points. The likelihood that such an error occurs depends on the proximity of one constellation point to another. For instance, take the previous example of QPSK with the constellation diagram shown in Figure 2.5. A decision rule at the receiver might be based on the location of the received baseband I/Q component relative to the I-Q axes on the constellation diagram.

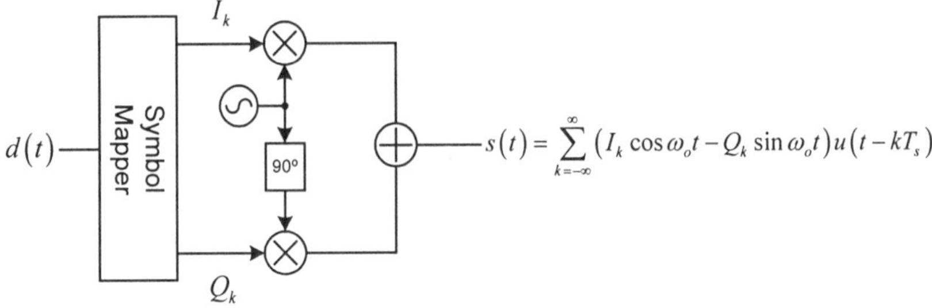

(a) I/Q Representation

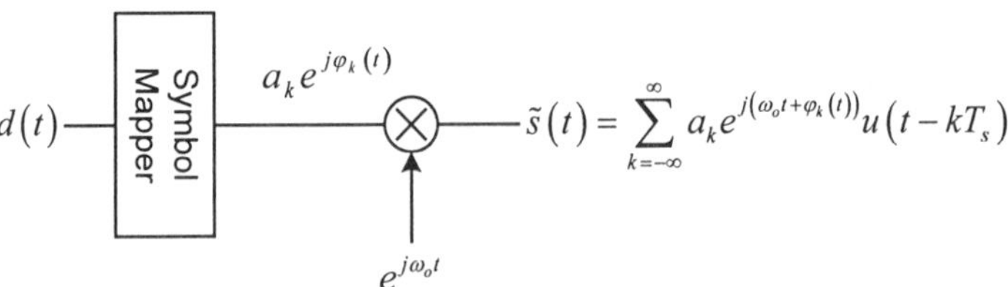

(b) Complex Representation

Figure 2.4 A digital modulator.

Each symbol are associated with a decision region such that any received signal that falls within the decision region is declared the corresponding symbol value. For instance, in this example the decision region could be the quadrant that each transmitted symbol is in. That is quadrant I, II, III, and IV are associated with symbols 00, 01, 11, and 10, respectively. If symbol 00 has been transmitted but due to noise or other types of degradations encountered in the channel, the received baseband signal falls in quadrant II rather than quadrant I, an incorrect decision will be made such that symbol 01 rather than 00 is declared as the received symbol. Section 2.3 further details the analytical relationship between the distance and reliability of different modulation schemes.

Constellation diagrams based on I-Q components are not applicable to FSK where the digital data is mapped to $\{\omega_k\}$ since for each k the constellation is a circle with the same radius. Therefore, the distance between different symbols cannot be determined. Instead, to obtain a constellation plot that is useful for FSK, a new set of axese is chosen and for ease of analysis the axese should be composed of an orthonormal basis since any signal can be expressed in terms of its projection onto the set of orthonormal basis functions. Analysis based on well known techniques in vector space can be applied.

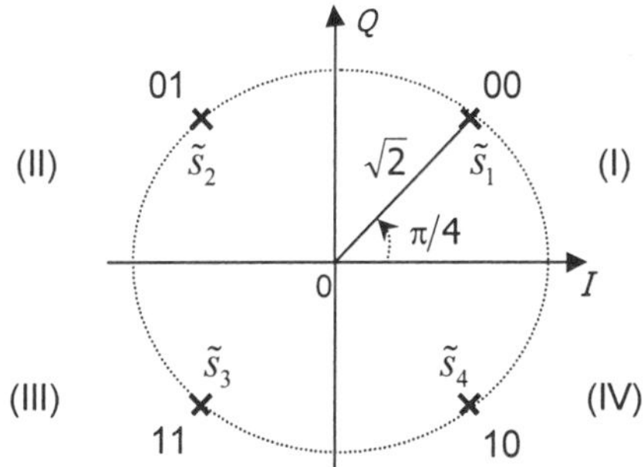

Figure 2.5 Constellation diagram of QPSK.

Taking a vector space approach, the inner product is used to represent the modulated signals. Given two arbitrary signals x and y with finite energy, the inner product is defined by

$$(x, y) = \int_{-\infty}^{\infty} x(t) y^*(t) dt \tag{2.6}$$

An orthonormal basis $\{\varphi_k\}$ in the vector space satisfies the following property

$$(\varphi_i, \varphi_j) = \begin{cases} 1, & i = j \\ 0, & i \neq j \end{cases} \tag{2.7}$$

Any signal $x(t)$ that lies within the vector space spanned by $\{\varphi_k\}$ can be represented by its projections onto $\{\varphi_k\}$, where the ith element of the projection can be readily obtained with (2.4)as shown below:

$$x_i = (x, \varphi_i) = \int_{-\infty}^{\infty} x(t) \varphi_i^*(t) dt \tag{2.8}$$

The corresponding signal vector is $\boldsymbol{x} = [x_1, x_2, \cdots, x_{N_d}]$, where N_d is the dimension of the signal vector and $x(t)$ can be expressed in terms of $\boldsymbol{x}$ and the basis functions.

$$x(t) = \sum_{i=1}^{N_d} x_i \varphi_i(t) \tag{2.9}$$

For FSK, a convenient set of basis functions to chose is a set of sinusoids, i.e. $\{e^{j\omega_k}\}$, which satisfies (2.7). FSK then has a signal set that forms an orthonormal basis. In general, such a modulation is known as orthongonal modulation, of which FSK is a special case. A constellation diagram based on the orthormal basis is shown in the Figure 2.6 for 3-ary FSK. Note that the constellation diagram of FSK cannot be visualized for a constallation size greater than three. It should also be noted that the I-Q representation for amplitude-phase modulations can be seen as a special case of the orthonormal representation. In Section 2.3, the use of orthonormal representation to analyze the performance of various modulation schemes will be described.

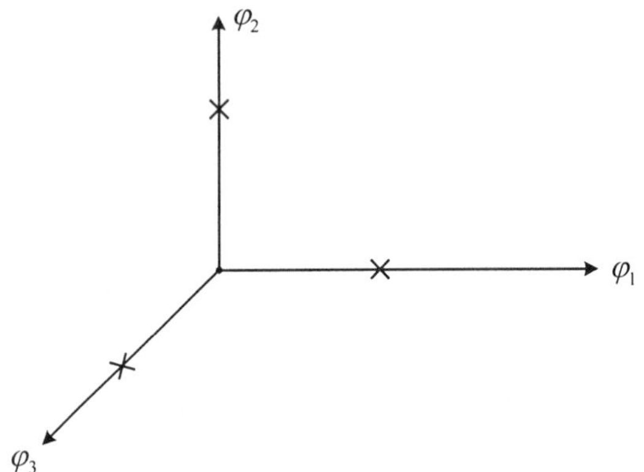

Figure 2.6 Constellation diagram of FSK with M = 3.

2.1.2 Bandwidth Efficiency

In an analog system, the bandwidth required for transmission is in general determined directly by the information bandwidth W_i. For example, audio information occupies 4 kHz to 20 kHz. The high end is required for high fidelity audio and the low end is adequate for speech. In a digital system, on the other hand, transmission bandwidth W_T is determined by five other components as shown below

$$W_T = \frac{cW_i}{rS_c\eta_W\eta_c} \tag{2.10}$$

where c is the conversion rate of the analog-to-digital converter in bps/Hz for the information source, S_c is the compression ratio of the source coder, r is the rate expansion of the channel coder (i.e. code rate), η_W is the bandwidth efficiency of the modulation in bps/Hz, and η_c is the channel efficiency which measures the amount of overhead required to establish the communication channel (See (5.2) in Chapter 5). Channel efficiency is essentially defined as the ratio of the time to transmit user data to the total over-the-air transmission time.

This section focuses on bandwidth efficiency, defined as the bitrate R_b that can be transmitted within W_T. Since most practical signals are not band limited, the signal bandwidth is infinite if not appropriately constrained through carefully selected filtering. This is especially important in the wireless environment where the spectrum is a scarce resource. Figure 2.7 shows the filtering that is commonly used in a digital system. The baseband symbols ξ_k are shaped by $\widetilde{P}(f)$ before transmission while the received signal $\widetilde{r}(t)$ is filtered by $\widetilde{G}(f)$, also referred to as a matched filter. The output of the matched filter is sampled at the symbol interval T_s to obtain the received symbol $\tilde{x}_n$ which has been corrupted by the channel. For the time being, the channel is modeled by a filter with response $\widetilde{C}(f)$ plus a noise source $\widetilde{n}(t)$. The tilde notation indicates that the signals are in their baseband complex representation.

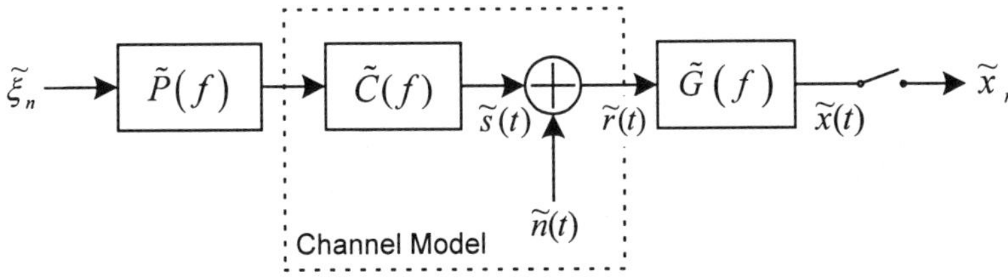

Figure 2.7 A digital system with transmit and receive filters.

To account for a complex pulse shaping function $\widetilde{p}(t) = p(t) + jq(t)$, expression (2.4) can be generalized to

$$\widetilde{s}(t) = \sum_{k=-\infty}^{\infty} \xi_k \widetilde{p}(t - kT_s) \tag{2.11}$$

Let the bandwidth of $\widetilde{P}(f)$ be denoted by W_s and note that $W_T = 2W_s$ for double sideband modulation (DSB) while $W_T = W_s$ for single sideband modulation (SSB) where only the upper or the lower sideband is transmitted. Then, it is desirable to minimize W_s such that the bandwidth efficiency is maximized. However, due to the well known Nyquist criteria [13], W_s cannot be made smaller than one half the symbol rate R_s without substantial increase in intersymbol interference (ISI):

$$W_s \geq 0.5R_s \tag{2.12}$$

ISI occurs when a symbol interferes with other adjacent symbols due to correlation introduced by the transmit filter, the channel, and the receive filter. To see this, let $\widetilde{h}(t)$ be the impulse response of the combined response $\widetilde{P}(f)\widetilde{C}(f)\widetilde{G}(f)$. Then, $\widetilde{x}(t)$ can be represented by

$$\widetilde{x}(t) = \sum_{k=-\infty}^{\infty} \xi_k \widetilde{h}(t - kT_s) + \widetilde{v}(t) \tag{2.13}$$

where $\widetilde{v}(t)$ is a complex noise process. At the nth sampling instance, the output of the sampler consists of the nth received symbol plus ISI and a sampled noise term $\widetilde{v}_n$:

$$\widetilde{x}_n = \widetilde{h}(0)\xi_n + \underbrace{\sum_{k \neq n} \xi_k \widetilde{h}(nT_s - kT_s) + \widetilde{v}_n}_{ISI} \tag{2.14}$$

To have zero ISI, the following constraint must be met for $\widetilde{h}(t)$:

$$\widetilde{h}_n = \begin{cases} 1, & n = 0 \\ 0, & n \neq 0 \end{cases} \tag{2.15}$$

where $\widetilde{h}_n$ is a simplified notation for $\widetilde{h}(t = nT_s)$ and without loss of generality, $\widetilde{h}(0) = \widetilde{h}_0$ is assumed to be 1. Equivalently, (2.15) can be expressed in the frequency domain

$$\sum_{k=-\infty}^{\infty} H(f - kR_s) = T_s. \tag{2.16}$$

For a band-limited system, it can be shown that to meet (2.15), the Nyquist Criteria (2.12) must be satisfied. To see this, let us first assume that (2.12) is not met. Then, for an arbitrary $\widetilde{H}(f)$, the sampled frequency response (2.16) will exhibit dips as shown in Figure 2.8 and therefore will not meet the constant amplitude requirement.

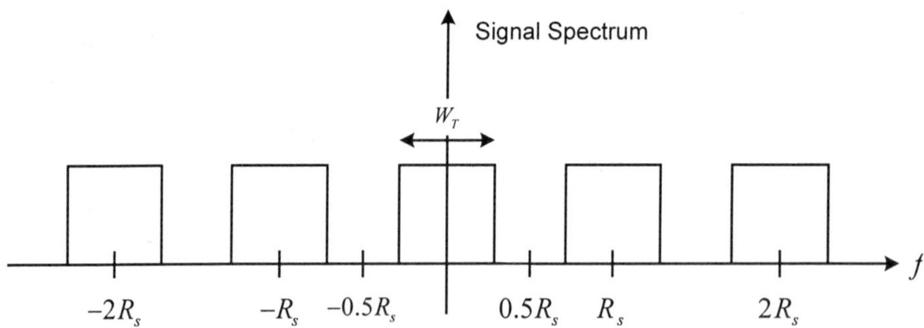

Figure 2.8 Received signal spectrum after the sampler with $W_T < 0.5R_s$.

To satisfy (2.12), many possible responses exist for $\widetilde{H}(f)$ that meet the constant amplitude requirement for the sampled response. One such example is a brick-wall rectangular frequency response with cutoff at $\pm 0.5R_s$. Such a filter is also referred to as the Nyquist filter. The bandwidth efficiency shown in (2.17) can then be achieved with zero ISI using DSB modulation. For SSB modulation, the bandwidth efficiency is twice that of (2.17).

DSB:
$$\eta_W = \frac{R_b}{R_s} \tag{2.17}$$

In practice, however, the rectangular response is not physically realizable due to the sharp cutoff. Instead, a tapered response is generally designed by selecting the pulse shaping functions $\tilde{P}(f)$ and $\tilde{G}(f)$ at the transmitter and receiver respectively such that (2.16) can be met. One such response is the raised-cosine with the following transfer function.

$$\tilde{H}(f) = \begin{cases} T_s & 0 \le |f| \le \dfrac{1-\beta}{2T_s} \\ \dfrac{T_s}{2}\left[1 - \sin\dfrac{\pi T_s}{\beta}\left(|f| - \dfrac{1}{2T_s}\right)\right] & \dfrac{1-\beta}{2T_s} \le |f| \le \dfrac{1+\beta}{2T_s} \\ 0 & |f| > \dfrac{1+\beta}{2T_s} \end{cases} \tag{2.18}$$

$$\tilde{h}(t) = \frac{\sin \pi t/T_s}{\pi t/T_s} \frac{\cos \beta\pi t/T_s}{1 - 4\beta^2 t^2/T_s^2} \tag{2.19}$$

Figure 2.9 shows the filter response in the frequency domain for various β which is also referred to as the excess bandwidth; i.e. the amount of bandwidth in excess of the minimum Nyquist rate of $0.5R_s$. A 100% excess bandwidth or $\beta = 1$ indicates a filter bandwidth of twice the Nyquist rate.

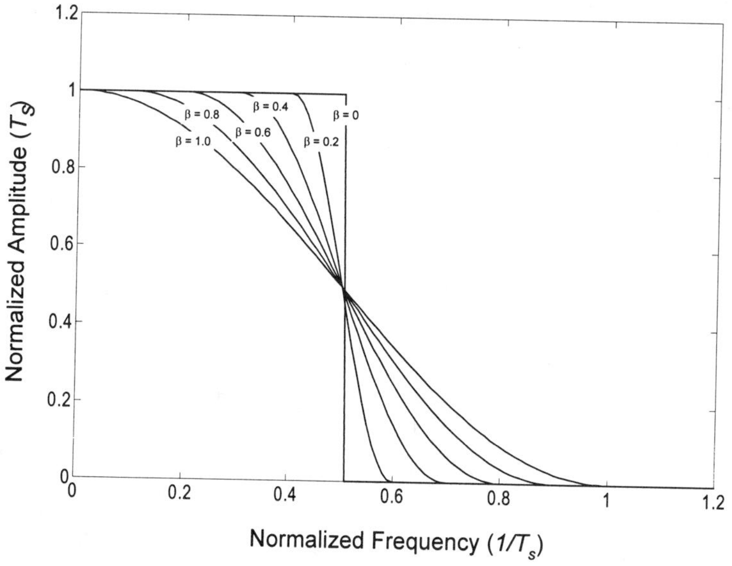

Figure 2.9 Frequency response of raised cosine pulse shaping.

The shaping functions $\tilde{p}(t)$ and $\tilde{g}(t)$ that produce the overall raised response as shown in Figure 2.9 can be determined by constraining $\tilde{G}(f)$ to be the complex conjugate of $\tilde{P}(f)$ as shown below:

$$\tilde{P}(f)\tilde{C}(f)\tilde{P}^*(f) = \tilde{H}(f) \tag{2.20}$$

$$\left|\tilde{P}(f)\right| = \sqrt{\frac{\tilde{H}(f)}{\tilde{C}(f)}} \tag{2.21}$$

Since the channel response in general is not deterministic, a fixed pulse shaping function cannot be designed based on (2.21). Instead, for now it is assumed that the channel is perfect (i.e. $\tilde{C}(f) = 1$), a reasonable assumption if equalization is used to minimize multipath fading. Equalization is discussed in Section 2.7.2. With the ideal channel assumption, the pulse-shaping function then becomes a square-root raised cosine filter with the following response:

$$\tilde{p}(t) = \frac{\sin[\pi(1-\beta)t] + 4\beta t\cos[\pi(1+\beta)t]}{\pi[1-(4\beta t)^2]t} \tag{2.22}$$

Since $\tilde{P}(f)$ is the square root of $\tilde{H}(f)$, a transmitted signal shaped by $\tilde{P}(f)$ will have the same bandwidth as $\tilde{H}(f)$. In the case of a square-root raised cosine pulse shape, the signal bandwidth is then $(1+\beta)R_s/2$, which results in the bandwidth efficiency shown below:

DSB:
$$\eta_W = \frac{R_b}{(1+\beta)R_s} \tag{2.23}$$

Equations (2.17) and (2.23) can be applied for QAM, PSK, and ASK but for FSK the efficiency is M times lower. The poor bandwidth efficiency of FSK results from the fact that each symbol is represented by a tone with a minimum separation of $0.5R_s$ from other tones. *M* symbols then results in an *M*-fold bandwidth expansion in the denominator of (2.17) and (2.23). Bandwidth efficiency of various modulations is described in more detail in Section 2.3.

2.1.2.1 Power Spectral Density

The bandwidth efficiency thus derived is based on the bandwidth constrained by the Nyquist Criteria. This section introduces other measures of bandwidth that are important for a wireless system. One of the major criteria in a wireless system is to limit the amount of interference introduced to other wireless links for increased spectral efficiency; i.e. getting as many concurrent links in a given available bandwidth. A meaningful measure for signal bandwidth would then be based on the bandwidth needed to constrain most of the signal power. Some examples of bandwidths defined in this way are given below:

- 3-dB bandwidth W_{3dB} is the bandwidth within which the signal maintains a power greater than half its peak power per Hz.
- Null-to-null bandwidth W_{NN} is the bandwidth measured between the first nulls from the zero frequency.
- Equivalent bandwidth W_{eq} is the effective bandwidth that contains all of the signal energy:

$$W_{eq} = 0.5\int_{-\infty}^{\infty} G_{\xi}(f)df \Big/ G_{\xi}(0) \tag{2.24}$$

- Fractional power containment bandwidth W_{pc} is the bandwidth that contains a fraction $1-\varepsilon$ of the total signal energy:

$$\int_{-W_{pc}}^{+W_{pc}} G_{\xi}(f)df \Big/ \int_{-\infty}^{+\infty} G_{\xi}(f)df = 1-\varepsilon \tag{2.25}$$

All of the above definitions require one to know the behavior of the signal power as a function of frequency; i.e. $G_{\xi}(f)$ which is also known as the power spectral density (PSD) of the baseband signal $\tilde{s}(t)$ as defined in (2.4). Since the transmitted baseband sequence ξ_k is a random process, the power spectral density depends not only on the pulse shape $\tilde{P}(f)$ but also on the autocorrelation of ξ_k shown below [14]:

$$G_{\xi}(f) = \frac{\left|\tilde{P}(f)\right|^2}{T_s} \sum_{m=-\infty}^{\infty} R_{\xi}(mT_s)e^{-j2\pi f mT_s} \tag{2.26}$$

where $R_{\xi}(mT_s)$ is defined as the autocorrelation of ξ_k

$$R_{\xi}(mT_s) = E\left\{\xi_{kT_s}\xi_{(k+m)T_s}\right\}. \tag{2.27}$$

A random process refers to the behavior of random variables as a function of time. An example is a plot of the outcome of someone flipping a coin (heads or tails) as a function of time. Equation (2.26) is valid only for wide-sense stationary process where the 1st order and 2nd order statistics of the process is invariant to time shifts. The 1st and 2nd order statistics are also known as the average and autocorrelation of the random process. For a thorough discussion of this subject the reader is referred to [15].

The average power of the transmitted baseband signal (P) can then be obtained by integrating (2.26):

$$P_{\tilde{s}} = \int_{-\infty}^{+\infty} G_{\xi}(f)df \tag{2.28}$$

From (2.4), it can also be shown that the average power of the passband signal (P_s) is simply one half of (2.28) and that its PSD is related to (2.26) by

$$P_{PB}(f) = 0.25P_{\xi}(f - f_0) + 0.25P_{\xi}(f + f_0) \tag{2.29}$$

As an example, let $\tilde{p}(t) = u(t)$ and let ξ_k be a sequence of independent, uniformly distributed random variables with unit variance and zero mean. Applying (2.26), the PSD in this case is simply

$$T_s \left| \frac{\sin(\pi f T_s)}{\pi f T_s} \right|^2 . \tag{2.30}$$

Figure 2.10 shows a plot of (2.30) and illustrates the four types of bandwidth described at the beginning of this subsection.

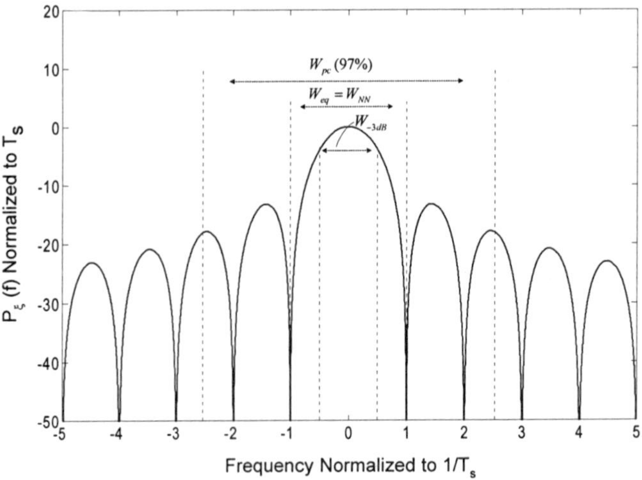

Figure 2.10 Power spectral density of a digital signal with rectangular-pulse shaping.

2.1.3 Error Rate

The reliability of a digital system is measured in terms of the error rate in the transmission link. There are several types of error rates, namely: bit error rate (BER), symbol error rate (SER), frame error rate (FER), or packet error rate (PER). Throughout this book, $\overline{P}_b$, $\overline{P}_S$, $\overline{P}_F$, $\overline{P}_P$ are used to denote the average BER, SER, FER, and PER, respectively. At the network level, typically FER and PER are the meaningful measures. The SER characterizes the performance of modulators. BER is measured at the bit level in terms of the number of bits that are received erroneously.

For instance, a BER of 0.001% indicates that on average 1 out of 10,000 bits are in error.

On the other hand, SER is measured at the symbol level in terms of the number of symbols that are in error. As mentioned earlier, a symbol error is made when the received signal falls outside of its decision region. A symbol error leads to bit errors as the symbol is erroneously mapped to an incorrect bit pattern. Let n be the number of bits per symbol. Then, SER may be bounded in terms of BER as shown below

$$\overline{P}_b \leq \overline{P}_S \leq n \cdot \overline{P}_b. \tag{2.31}$$

At the source level, error rate is measured by FER. A frame is the basic unit that represents the digitized source data. The duration of the frame depends on the type of source. In general, a voice source uses a frame with a duration of 20 ms while a video source uses a frame with a duration of 16.7 ms to 66.7 ms. These values for the frame duration tend to give good perceptual quality in the reconstructed voice and video information. For example, in a digital cellular system such as GSM, the voice source information is first filtered to band limit the signal to 3.4 kHz. The filtered signal is then digitized at a sample rate of 8 kSps. With 13 bits per sample, the digitized source generates a bit stream at 104 kbps, stored in 20 ms blocks. Each block of 2080 bits is compressed by a factor of 8 using a voice encoder (vocoder) based on regular pulse excitation long-term prediction (RPE-LTP) [16]. Each coded speech frame consists of 260 bits that are channel coded and interleaved for transmission. In GSM, the required FER for the voice channel is 0.1-11.2 % [255].

FER is also a function of BER and in the case of random bit error statistics, the FER can be expressed in terms of the BER and the number of bits in a frame L_f.

$$\overline{P}_F = 1 - (1 - \overline{P}_b)^{L_f} \tag{2.32}$$

Note that random bit error statistics is a reasonable assumption with channel coding with interleaving that compensates for the channel fluctuations in a mobile fading channel described in Section 2.7.4.1.

For data transmissions, the measure of error rate is the PER. In general, in data transmission, the bit stream from the data source is formatted into a packet that contains two general fields: a control field and a data field. The control field contains a preamble and a header followed by cyclic redundancy check (CRC) bits. The data field contains the user data of size L_p bits followed by another set of CRC parity check bits. The preamble contains special bit patterns needed for the radio to synchronize and/or to train the equalizer to properly recover the data. The header contains system level information that indicates how the data should be processed by the system. For example, the header can contain source and destination addresses that tell the system where the data came from and where it is going. Since the header contains essential information needed for the system to determine what to do with the data payload, CRC is computed for the header to ensure data integrity. With a CRC, the system can determine whether an error has occurred in the header and if so to take the appropriate action, such as request retransmission of the packet. The PER is a function of L_p and for the case of random error statistics can be expressed as

$$\overline{P}_P = 1-\left(1-\overline{P}_b\right)^{L_P} \tag{2.33}$$

In Internet applications, the packet size can range from a few hundred bytes to 1500 bytes. Depending on the type of channel the BER can range from 0.1 % to 0.001%. For acceptable data rate, a low BER of 0.001% is generally needed.

2.1.4 E_b/N_0

Since SER, FER, and PER all depend on BER, a basic measure for digital systems is based on BER which can be expressed in terms of E_b/N_0, where E_b is the energy per bit and N_0 is the equivalent noise spectral density over the signal bandwidth. The variation of BER with E_b/N_0 depends on the channel condition and the type of demodulator. In general, in an additive white Gaussian noise (AWGN) channel, BER is exponentially related to E_b/N_0 while in a fading channel, BER is inversely related to E_b/N_0. For example, the BER for binary-phase shift keying (BPSK) in AWGN and fading channels are shown below:

AWGN:
$$\overline{P}_b = Q\left(\sqrt{2E_b/N_0}\right) \tag{2.34}$$

where

$$Q(x) = \frac{1}{\sqrt{2\pi}}\int_x^{\infty} e^{-t^2/2}\,dt \tag{2.35}$$

and

Rayleigh Fading:
$$\overline{P}_b = \frac{1}{2}\left(1-\sqrt{\frac{E_b/N_0}{1+E_b/N_0}}\right) \approx \frac{1}{4E_b/N_0}. \tag{2.36}$$

These channels are described in Sections 2.2.1.1 and 2.2.5.1.

For a given BER, a digital system with lower E_b/N_0 requires lower transmission power, which can improve battery lifetime of the communication device and the system capacity. The improvement in system capacity results from reduced levels of interference introduced by collocated transmissions. Therefore, more users can be accommodated in a given area.

E_b/N_0 can be computed as an average value or a peak value. To see this, we first express E_b in terms of the symbol energy:

$$E_b = \frac{E_s}{\log_2 M}. \tag{2.37}$$

Since the distribution of the symbol in a signal constellation has a random distribution, an average E_b can be expressed in terms of the average symbol energy:

$$\overline{E}_b = \frac{\sum_i p_i E_s^{(i)}}{\log_2 M} \tag{2.38}$$

where p_i is the probability for the occurrence of the ith symbol with energy $E_s^{(i)}$. From (2.38), the peak E_b is simply expressed in terms of the peak symbol energy in the signal constellation:

$$\max E_b = \frac{\max_i E_s^{(i)}}{\log_2 M} \tag{2.39}$$

For simplicity of notation, the average and peak E_b/N_0 corresponding to the average and peak E_b will henceforth be denoted as $\overline{\gamma}$ and γ_p, respectively.

Since the symbol energy is proportional to the square of the signal amplitude, then according to (2.38) and (2.39) any modulation that has constellation points with constant amplitude has equal peak and average E_b. For example, QPSK shown in Figure 2.5 satisfies this condition whereas QAM does not. In fact, as shown in Figure 2.42 QAM displays a much higher peak E_b than the average E_b. A large peak to average energy (PAE) ratio or equivalently $\gamma_p/\overline{\gamma}$ has a significant impact on the power efficiency of the power amplifier circuits as discussed in Section 7.7 of Chapter 7.

2.1.5 Channel Capacity

In 1948, Claude Shannon derived an analytical bound for channel capacity for a digital system. Shannon showed that in an AWGN channel, the maximum bitrate C that can be achieved with arbitrarily low error rate over a given transmission bandwidth W_T is bounded by the expression below [17]

$$\frac{C}{W_T} \le \log_2\left(1 + \frac{P}{N_0 W_T}\right) = \log_2\left(1 + \overline{\gamma}\frac{C}{W_T}\right), \tag{2.40}$$

where $P/N_0 W_T$ is the SNR, C/W_T is the maximum achievable bandwidth efficiency in bps/Hz, and $\overline{\gamma}$ is the average E_b/N_0 defined by (2.17) and (2.38), respectively.

The objective in a digital system design is to approach the Shannon Capacity as closely as possible for a given $\overline{\gamma}$ constraint. It is instructive to observe the asymptotic behavior of the Shannon capacity. When bandwidth efficiency approaches infinity, the required $\bar{\gamma}$ also approaches infinity. This implies that to achieve high bandwidth efficiency, the transmission system must pay in terms of energy dissipation. On the other hand, when bandwidth efficiency approaches zero, $\bar{\gamma}$ approaches ln(2) which is -1.6 dB. Therefore, even when an infinitely small amount of information is transmitted or when an infinite amount of bandwidth is used for transmission, some finite amount of energy is still required to offset the degradation due to noise. In this case, the energy per bit must be -1.6 dB above the noise spectral density.

Figure 2.11 plots the Shannon capacity as a function of $\bar{\gamma}$. Area above the plot indicates the physically unrealizable region; i.e. not meeting the condition in (2.40). The area under the plot that satisfies (2.40) can be classified into a bandwidth limited region and an energy limited region. The bandwidth-limited region can be defined as the region with bandwidth efficiency greater than 1 bps/Hz. In this case, the transmission system has plenty of energy resource but lacks available bandwidth. Applications such as fixed point-to-point microwave communications fall under this category. QAM is particularly suitable for bandwidth limited systems since it can approach within 10 dB of the Shannon Capacity at a SER of 0.001%.

Energy-limited systems, on the other hand, do not have unlimited energy resource but instead have plenty of available bandwidth. This type of system includes low-data rate pager networks. Such systems fall in the region where bandwidth efficiency is less than 1 since bandwidth is not the limiting factor with a low data rate. Orthogonal modulation schemes such as FSK are most suitable for energy limited systems and can asymptotically achieve the Shannon Capacity at a SER of 0.001%. However, with practical constellation size, uncoded FSK achieves within 8 dB of the Shannon Capacity. With advanced coding such as trellis codes and turbo codes, the gap narrows to a few tenths of a dB in an AWGN channel.

2.1.6 Energy Efficiency

While bandwidth and energy limited systems cover some of the wireless applications, many other systems fall in the middle where both bandwidth and energy resources are constrained. Digital cellular systems are an excellent example where the handset battery lifetime is a critical performance differentiator and high bandwidth efficiency is also required to allow many users to share a given RF spectrum. For such systems, the efficiency at which the system uses up energy for a given bandwidth becomes an important measure. A possible definition for energy efficiency might be based on the actual E_b/N_0 required by the system, denoted by $\bar{\gamma}_{req}$, versus the E_b/N_0 required by the Shannon Capacity bound, denoted by $\bar{\gamma}_{Shannon}$ as shown below

$$\frac{\bar{\gamma}_{Shannon}}{\bar{\gamma}_{req}} = \frac{2^{C/W_T} - 1}{\bar{\gamma}_{req} C / W_T}. \tag{2.41}$$

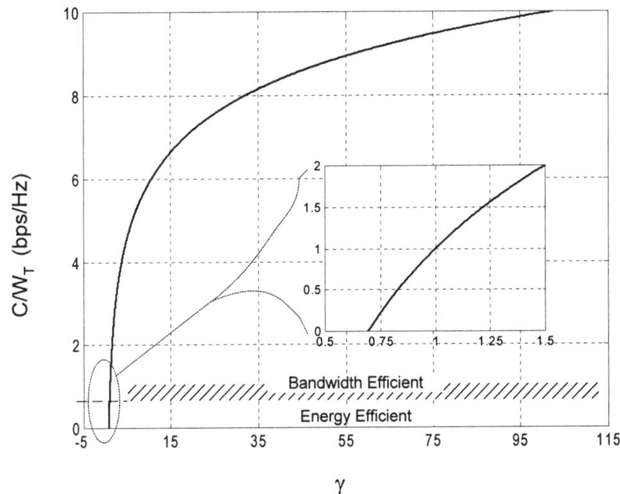

Figure 2.11 Shannon channel capacity for the AWGN channel.

However, Table 2.1 shows the problem associated with this definition where systems that require a high E_b/N_0 can still have a high energy efficiency according to (2.41). This is because modulations such as QAM that require high energy also have high bandwidth efficiency. The total effect is that (2.41) remains high. Such systems occupy the upper part of the Shannon Capacity curve (i.e. bandwidth-limited region). Furthermore, systems that occupy the lower part of the Shannon Capacity curve (i.e. energy-limited region) can also have a high efficiency rating. In this case, (2.41) correctly indicates the low energy achieved but hides the poor bandwidth efficiency.

Table 2.1 Energy efficiency of PSK, QAM, and FSK according to (2.41)

Modulation	$\bar{\gamma}$ **at SER = 0.001%**	η_w (bps/Hz)	**Eff. (%)**
BPSK	9.09	1	11.0
GMSK	10.8	1.35	11.5
QPSK	9.09	2	16.5
8-PSK	19.82	3	11.8
16-PSK	55.41	4	6.8
32-PSK	171.2	5	3.6
8-QAM	13.93	3	16.8
16-QAM	22.05	4	17.0
32-QAM	34.67	5	17.9
BFSK	17.78	1	5.6
4-FSK	9.77	1	10.2
8-FSK	7.08	0.75	12.8
16-FSK	5.62	0.5	14.7
32-FSK	4.52	0.3125	17.1

For systems that require both low energy and high bandwidth efficiency, a more accurate definition for energy efficiency is needed. To take into account both energy and bandwidth, we define an efficiency function $f_E(x)$ of a system to be the amount of E_b/N_0 required for a given bandwidth efficiency:

$$f_E(C/W_T) = \frac{\overline{\gamma}_{Shannon}}{C/W_T} = \frac{2^{C/W_T} - 1}{\left(C/W_T\right)^2}. \tag{2.42}$$

The unit for (2.42) is in (cycle/bit)^2 which represents the minimum energy needed to transmit data above the noise power density at a given bandwidth efficiency. Figure 2.12 shows a plot of (2.42). Note that a minimum exists at $\eta_W = 2.3$ bps/Hz which implies that to achieve high energy efficiency, systems should be designed around this optimum.

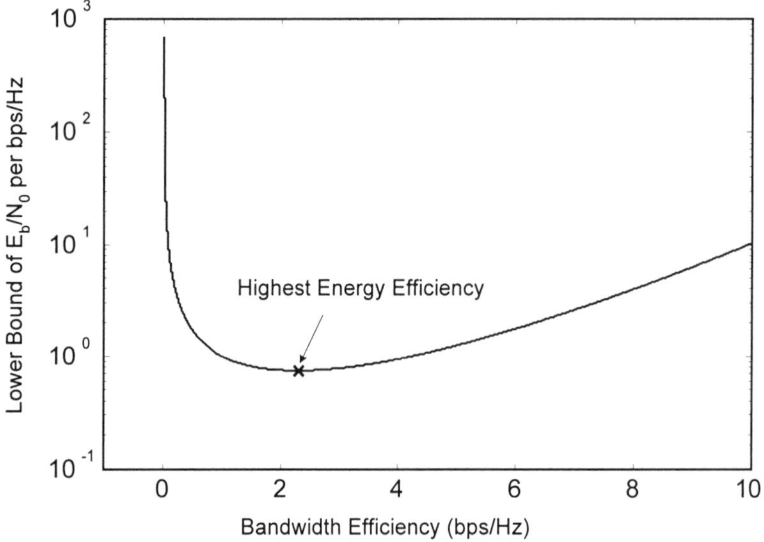

Figure 2.12 Energy efficiency (2.42) versus bandwidth efficiency.

Energy efficiency can now be defined based on (2.42) for a given system using $\overline{\gamma}_{req}$ and bandwidth efficiency (η_W). To obtain the efficiency as a percent, the minimum of (2.42), $f_E(2.3)$, is used as the reference:

$$\eta_E = \frac{f_E(2.3)}{\dfrac{\overline{\gamma}_{req}}{\eta_W}} \tag{2.43}$$

Figure 2.13 shows the energy efficiency as defined by (2.43) for QAM, PSK, and FSK. The high energy requirement of PSK and QAM with large constellation size and poor bandwidth efficiency of FSK both result in low energy efficiency. Note, the highest energy efficiency that can be achieved is with QPSK where $\eta_W = 2$.

Later in Section 2.4.1, it will be shown that while QPSK shows good energy efficiency according to (2.43), when its interaction with the power amplifier circuit is taken into account, QPSK in fact requires significantly higher energy. Due to abrupt phase changes, non-linearity in the power amplifier tends to restore the sidebands in filtered QPSK signals, a phenomenon also referred to as spectral regrowth that will be discussed in more detail in Section 2.4.1. To reduce spectral regrowth, the output power of the amplifier must be backed off from the operating point at which it is most efficient. Therefore, while η_E provides a good measure for energy efficiency when considering modulation alone, other factors related to implementation must be considered when evaluating the energy performance of the overall system.

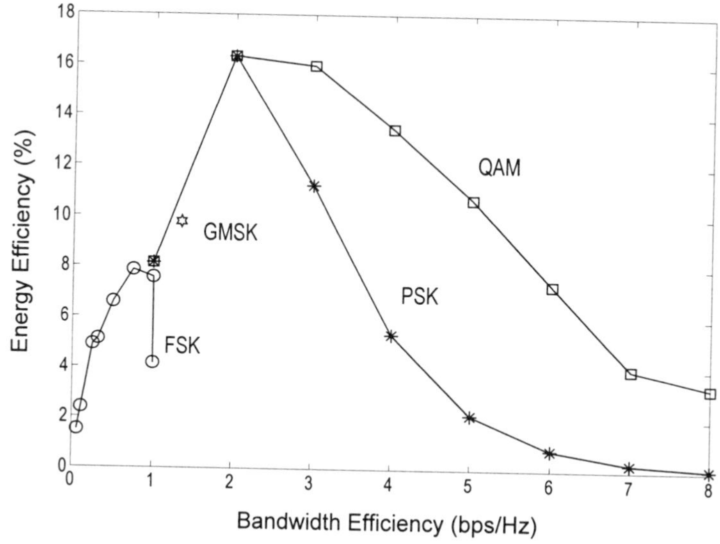

Figure 2.13 Energy efficiency (2.43) versus bandwidth efficiency for BER = 0.001%.

To model the spectral regrowth in closed form is difficult. However, a more stringent constraint can be derived for η_E that accounts for the power amplifier backoff. The efficiency of the power amplifier decreases proportionally to the amount of power that must be backed off, denoted by γ_{OBO}. Expression (2.43) can then be modified to take into account output backoff as shown below

$$\eta_E = \frac{f_E(2.3)}{\dfrac{\bar{\gamma}_{\text{Required}}}{\eta_W}} \cdot \frac{1}{\gamma_{OBO}}. \tag{2.44}$$

It should be noted that γ_{OBO} can be lower bounded by the peak-to-average energy ratio of the modulation. However, for certain modulations such as PSK the actual OBO can be substantially off since for PSK the peak-to-average ratio is one.

2.1.7 Coding Gain and Code Rate

Forward error correction (FEC) can be performed to decrease the required E_b/N_0. This is necessary when the channel condition is poor due to noise or fading. Different sources of degradation are described in Section 2.2. A simple coding system takes k bits and forms an N-bit codeword with $n-k$ parity bits used in the decoding process to increase the chance of error-free decoding even in the presence of significant channel degradation. Code rate r is defined as the amount of information transmitted per codeword:

$$r = k / n \tag{2.45}$$

Coding gain G_c is defined as the amount of E_b/N_0 reduction compared with an uncoded system for a given BER when the channel degradation is due to AWGN. Specifically, G_c is a function of the minimum distance $d_{\min}$ and the code rate r [18] of a given code

$$G_c \approx rd_{\min}. \tag{2.46}$$

$d_{\min}$ is a distance measure for error correction codes and is described in Section 2.6. Equation (2.46) is an approximation and becomes more accurate at higher E_b/N_0. It also assumes soft-decision decoding whereby the unquantized decision signal is used for decoding. In contrast, hard-decision decoding uses the quantized decision signal for decoding and results in approximately 3 dB lower coding gain. Coding gain helps to reduce the required E_b/N_0 and therefore enables a system to achieve higher energy efficiency. However, to obtain the coding gain, the system must pay in reduced bandwidth efficiency by a factor equal to the code rate as shown in (2.10).

2.1.8 Diversity Gain

When channel degradations introduce correlation to the received signal, the error performance degrades with respect to that of the AWGN channel. Correlation may be introduced by multipath fading and correlated interference. Figure 2.14, shows the BER performance of a BPSK transmission system over a Rayleigh fading and an AWGN channel, obtained by plotting (2.34) and (2.36). At a BER of 10^{-5}, the increase in required E_b/N_0 can be as much as 44 dB for the Rayleigh fading channel as compared to that of the AWGN channel. To compensate for the degradation in BER, digital systems employ diversity techniques described later in Section 2.7.

The optimum error performance that can be achieved for systems using diversity is a function of the diversity order N_D available through the channel; i.e. the number of independent fading components created by the channel. Diversity techniques use K_D diversity branches, where each branch carries a part of the digital message, which can be a symbol or part of a packet. If fading corrupts one branch, other branches still have a high probability of being received correctly; thus, error performance is improved. Strictly speaking, if each diversity branch is designed to have independent fading statistics then increasing K_D beyond N_D results in no additional improvement. However, depending on the diversity scheme, it is likely that not all branches are independent and therefore K_D may exceed the diversity order of the channel with additional improvement in performance gain. For instance, such a situation occurs when there is ISI or interference in the channel.

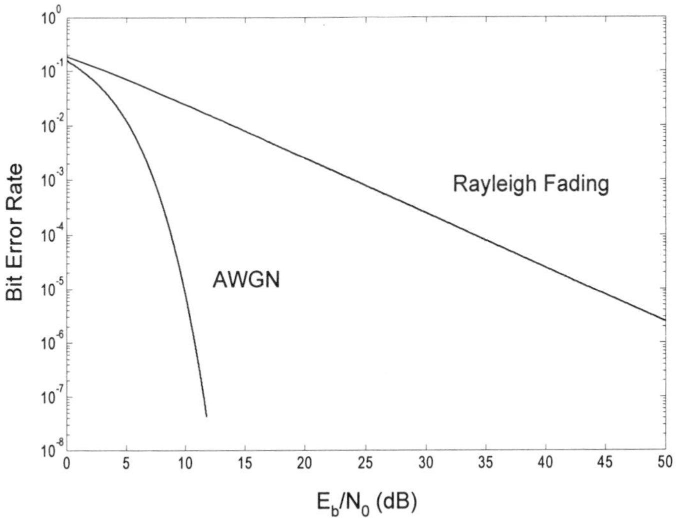

Figure 2.14 BER curve for BPSK in AWGN and Rayleigh fading channels.

Diversity gain measures the reduction in the required E_b/N_0 for a given error performance when diversity is used as compared to that without diversity. For example, with BPSK transmissions, the average BER with ideal diversity combining has the form shown below [19]

MRC
$$\bar{P}_b \le \left(\frac{1}{1+\bar{\gamma}/K_D}\right)^{K_D}, \qquad K_D \le N_D. \tag{2.47}$$

Ideal combining refers to the case when each diversity branch captures an independent fading component of the channel and weights each component according to its SNR. It is also often referred to as maximum ratio combining (MRC) [20], which will be discussed in greater detail later in Section 2.7.1.

Comparing (2.47) with (2.36), the performance improvement is exponentially related to K_D. More specifically, the diversity gain at a given BER in this case can be approximated by

$$G_d \approx 10\log_{10}\left(e^{\overline{\gamma}_n}-1\right)-10\log_{10}K_D\left(e^{\overline{\gamma}_n/K_D}-1\right). \tag{2.48}$$

where $\overline{\gamma}_n$ is the average E_b/N_0 required for an AWGN channel. The unit for (2.48) is in dB. Figure 2.15 plots the diversity gain of a BPSK system for different values of K_D and $\overline{\gamma}_n$. As expected, the diversity gain increases with increasing K_D and $\overline{\gamma}_n$.

Example 2.1 To achieve a BER of 0.001% using BPSK, the required E_b/N_0 the AWGN channel is 9.6 dB. With $K_D = 2$ and applying (2.48), the diversity gain is 16.84 dB. Therefore, even with two-branch diversity, a substantial diversity gain is achieved. □

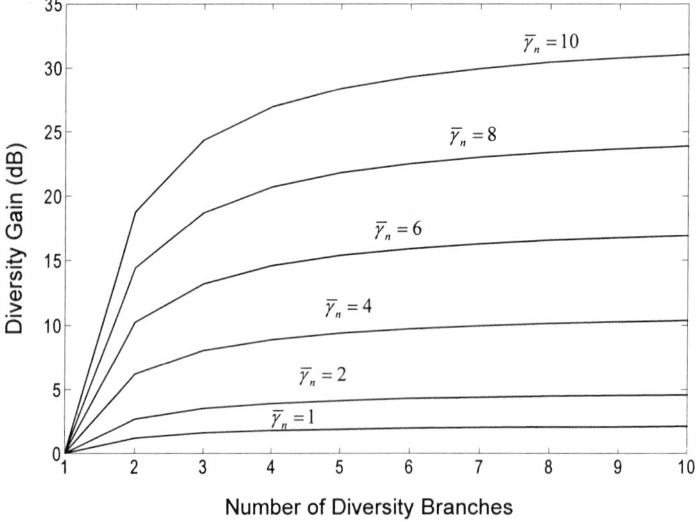

Figure 2.15 Diversity gain of a BPSK system.

It is instructive to derive for a given K_D and BER, the loss in E_b/N_0 for a diversity receiver relative to the E_b/N_0 needed if the channel were AWGN. Applying (2.47) and (2.36), this loss denoted by $L_{d/n}$ can be approximated by

$$L_{d/n} \approx 10\log_{10}K_D\left(e^{\overline{\gamma}_n/K_D}-1\right)-10\log_{10}\overline{\gamma}_n. \tag{2.49}$$

Note that when the diversity order of the channel becomes large, the number of diversity branches that a receiver can use also increases. With an increasing number of diversity branches according to (2.49), it becomes possible to asymptotically approach the performance achievable in an AWGN channel; i.e. $L_{d/n}$ approaches zero as K_D becomes infinitely large as shown in Figure 2.16.

Example 2.2 Given the same conditions as in Example 2.1 and applying (2.49), the loss in E_b / N_0 relative to that required for an AWGN channel is 13.2 dB for BPSK with two-branch diversity and operating at a BER of 0.001%. □

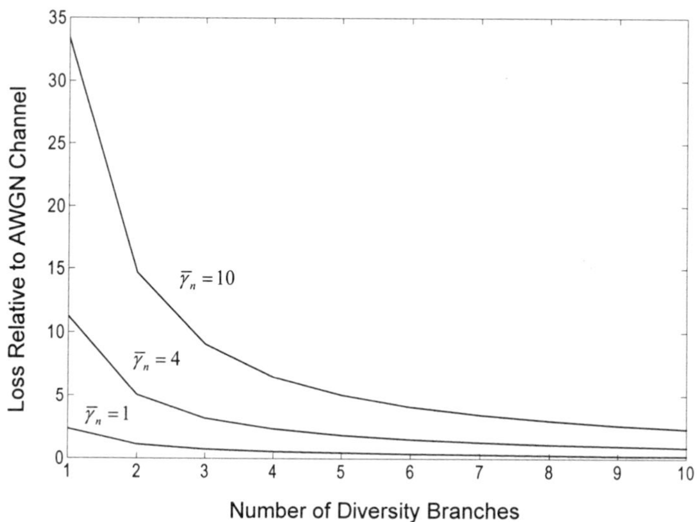

Figure 2.16 Loss in E_b/N_0 relative to the AWGN channel for K_D – branch diversity.

2.2 Sources of Degradation in Mobile Radio Link

A mobile wireless channel suffers from two general types of degradation. The first type of degradation results from signal distortion due to other undesired signal sources such as noise and interference. The second type of degradation results from decreased power in the desired signal itself due to path loss and multipath fading that arise from the propagation of radio waves in the physical medium. An understanding of the underlying mechanism for each type of degradation is important in developing techniques to achieve adequate radio performance for mobile wireless channels.

2.2.1 Noise

The process that generates noise can be either naturally occurring or man made which cannot be described in a deterministic fashion. For instance, it is nearly impossible to describe the precise motion of individual air molecules in an enclosure or the amount of electromagnetic radiation generated by someone turning on a car in a given area and at a particular time. Either of these two processes appears random with no apparent rules to predict precisely its behavior. Such processes are also called random processes. As another example, the transmitted symbol sequence ξ_k can be represented as a random process. In general, we will be dealing with random processes that are stationary; i.e. the statistical behavior is independent of time. An example of a stationary random process is one with a statistical behavior that depends only on the difference in time. That is the process is widesense stationary (WS).

Let $X(t)$ be a WS random process. Then its statistical behavior at a given time is described by a probability density function (PDF), denoted by $P_X(x)$ which measures the probability for a given value x of the random variable X. From the PDF, important statistical parameters can be determined about $X(t)$, such as its mean $\bar{x}$, variance σ_x^2, and autocorrelation function

$$R_{xx}(\tau) = E\{(x(t) - \bar{x})(x^*(t+\tau) - \bar{x}^*)\} \tag{2.50}$$

where $E\{\}$ denotes expectation. Section 2.1.2.1 showed that the Fourier Transform of a autocorrelation function is the PSD of the associated random process. If the autocorrelation is a delta function (i.e. exhibits an infinite peak at $\tau = 0$ and zero for all other values of τ) then the random process is white and the corresponding noise is also known as *white noise*. The PSD of *white noise* is flat for all frequency. If the PSD is not a constant function of frequency then the random process is colored. A noise process with non-constant PSD is also referred to as *colored noise*. White noise has the unique property that the noise behavior is completely uncorrelated in time since the autocorrelation function is non-zero only when the time difference is zero. On the other hand, samples of colored noise at different times tend to be correlated. In general, channels that display correlation require diversity processing at the receiver to achieve adquate link performance as shown in Section 2.7.

2.2.1.1 Thermal Noise

Thermal noise [21][22] has a relatively flat PSD up to about 1000 GHz and has been found as the ambient noise in most physical devices and environments. Its PSD is given by

$$G_X(f) = N_0/2 = kT/2 \tag{2.51}$$

where $k = 1.38x10^{-23} Ws/K$ is Boltzmann's constant and T is the noise temperature in degrees kelvin. The total termal noise power in a given bandwidth W is then N_0W.

Thermal noise is thought to be formed by a large collection of independent random processes. Therefore, by applying the well-known central-limite theorem [15], the PDF of thermal noise can be approximated by a Gaussian distribution. Since its PDF is relatively flat it can also be approximated as white noise. Hence, the white Gaussian noise distribution is often used to describe the statistical behavior of thermal noise. A channel degraded by thermal noise is typically know as an additive white Gaussian noise (AWGN) channel and has the following noise PDF:

$$P_X(x) = \frac{1}{\sqrt{2\pi}\sigma} e^{-(x-\bar{x})^2/2\sigma^2} \tag{2.52}$$

In practice, the mean of thermal noise is zero and its variance is equal to $N_0/2$. The latter fact follows from $R_X(0) = \sigma^2$ and that for white noise $R_X(0)$ also equals $N_0/2$. The probability that the random variable X exceeds x can be found with the Q-function as defined in (2.35)

$$P(X > x) = Q\left(\frac{x - \bar{x}}{\sigma}\right) \tag{2.53}$$

An AWGN channel is in practice valid only for very few systems, such as deep space communications, but due to its analytical tractability, it is often used to estimate the link performance of most digital systems.

2.2.1.2 Man-made Noise

In practice, most systems encounter colored noise rather than white noise. For instance, if a filter is used in the receiver as shown in Figure 2.7, the received noise becomes correlated and therefore behaves as colored noise. Another example is man-made noise encountered in most communication systems located in populated areas, such as an urban environment. Man-made noise results from random electromagnetic radiation from various electronic devices, such as microwave ovens, car ignitions, and emissions from power lines. The noise power is particularly high at low frequencies due to a lower propagation loss and numerous devices that emit low frequency disturbances [23].

Colored noise tends to make analysis intractable. Typically, simulations are performed instead to estimate system performance. A colored noise channel is modeled by using a digital filter that has a frequency response $H_n(f)$ with the following property:

$$|H_n(f)|^2 = G_X(f) \tag{2.54}$$

To generate the colored noise samples, the digital filter is excited with an AWGN input as shown in Figure 2.17 below:

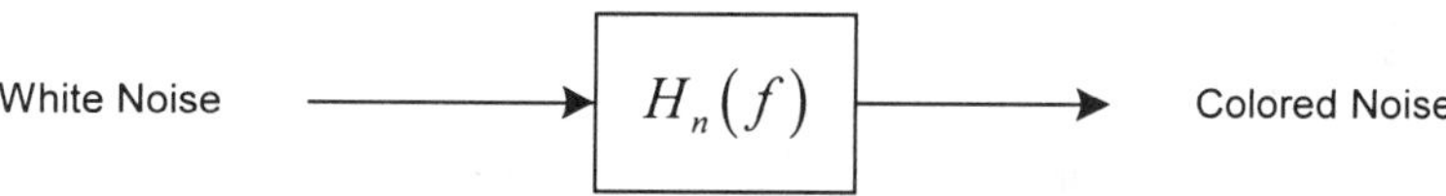

Figure 2.17 Colored noise generation.

As a first order approximation on system performance, one can model the man-made noise with an equivalent rectangular PSD that has a bandwidth equal to the signal bandwidth W and an equivalent noise power density described below:

$$N_{mm} = \frac{1}{W}\int_{-\infty}^{\infty} G_X(f)df \tag{2.55}$$

The total noise density experienced by the receiver can then be expressed as the sum of (2.51) and (2.55).

2.2.2 Interference

Interference can also be treated as noise. For clear distinction, interference is defined as unwanted disturbance from spurious RF transmissions. In the following subsections, two types of interference are examined: 1) co-channel interference and 2) adjacent-channel interference.

2.2.2.1 Co-Channel Interference

Ideally, if all transmissions were on different frequencies, then there would be no RF interference. However, in practice, the RF spectrum is a limited resource and therefore it is not possible to dedicate a frequency to each communication link. For example, in a GSM system, the bandwidth efficiency η_W is 1.35 bps/Hz, the code rate r is 0.57, the compressed voice bitrate (full rate) is 13 kbps, and the channel efficiency η_c is 67%. Substituting these numbers into (2.10) results in a transmission bandwidth of 25 kHz. Given that the total available bandwidth is 25 MHz, only 1000 users can be supported simultaneously, far less than what the system can actually support.

The key in providing many simultaneous channels for a given available bandwidth is frequency reuse. The idea is to reuse frequency in local areas separated by a distance D large enough such that the locally transmitted signal is sufficiently attenuated by the path loss and therefore does not cause significant degradation in other localities. Figure 2.18 illustrates the concept of frequency reuse. R in Figure 2.18 denotes the maximum communication range within each locality and the signal attenuation (i.e. path loss) is approximated by cd^{-n}, where d is the communication range, and c and n are constants determined by the propagation environment. Path loss is described in more detail in Sections 2.2.3 and 2.2.4.

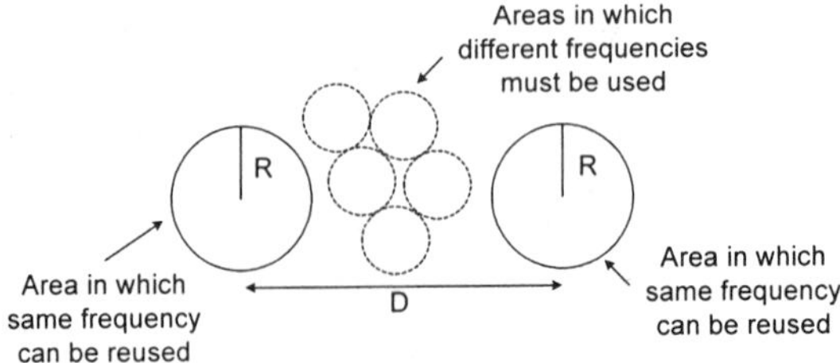

Figure 2.18 Frequency reuse.

While the system in Figure 2.18 provides more usable channels (i.e. higher system capacity), it inevitably introduces unwanted interference. Since the interference is associated with the same frequency channels being reused at different locations, the interference is also known as co-channel interference (CCI). The amount of tolerable CCI depends on parameters D, R, n, K_I, and the required signal-to-interference ratio (SIR) as shown below:

$$SIR = 10\log_{10}\left[\left(\frac{D}{R}\right)^n K_I^{-1}\right]. \tag{2.56}$$

where K_I is the number of co-channel interferers at a distance D from the receiver. In general, to improve system capacity, D/R must be minimized so that more frequencies can be reused over shorter distances. However, to decrease D/R, the required SIR must also decrease, implying that the radio equipment must be less sensitive to interference. This is particularly important for code-division multiple access systems since many users transmit in the same frequency band with different spreading codes, effectively raising K_I. Interference results from imperfect isolation among the different codes.

To aid the design of radios that are less susceptible to increased interference levels, the effect of interference on BER needs to be characterized. Unfortunately, for most systems, the number of co-channel interferers is small. Thus, the central-limit theorem cannot be applied to obtain the Gaussian distribution as in the case of thermal noise. Other more involved computation methods can be used though they provide little insight on the behavior of the system. In this text, the Gaussian approximation is still used to provide intuitive design guidelines. Given the Gaussian assumption, the effect of interference is shown to reduce the average E_b/N_0 by α_I,

$$\alpha_I = \left(1 - \frac{\left|\overline{I}\right|^2}{\overline{\gamma} N_0}\right)\left(1 + \frac{\eta_W \overline{\gamma}}{SIR}\right)^{-1}. \tag{2.57}$$

where $\overline{I}$ is the mean of the interference level. SIR is defined as the signal power over the interference power I_0.

Example 2.3 Consider BPSK modulation. Then, applying (2.57), the BER performance in presence of interference would be $Q\left(\sqrt{2\overline{\gamma}\alpha_I}\right)$. □

2.2.2.2 Adjacent-Channel Interference

In contrast to CCI, adjacent channel interference (ACI) results from RF transmissions at frequency channels adjacent to the desired frequency channel, as shown in Figure 2.19a. Such a situation occurs in systems that allow multiple transmissions in a local area using FDMA. Ideally, transmission from nearby channels should be completely isolated and not cause any loss in performance on the desired channel. However, in practice the ideal situation cannot be achieved due to non-zero out-of-band emission and the near-far problem.

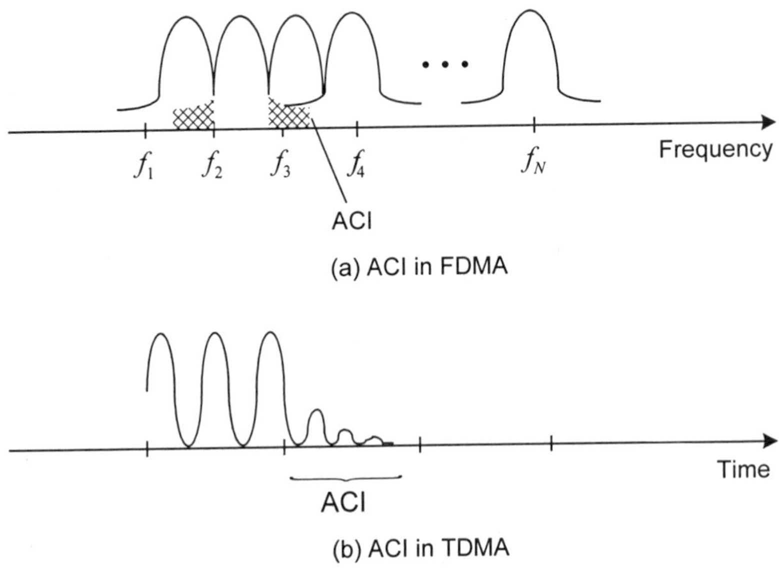

Figure 2.19 Adjacent channel interference.

The out-of-band emission results mainly from the spectral regrowth of the transmitted baseband signal as discussed in Section 2.4.1.3. The near-far problem arises from the possibility of having a strong interferer nearby a receiver which is receiving weak signals from a transmitter far away. Denote the transmission power and distance of the nearby interferer by $P_{TX}^{'}$ and $d^{'}$, respectively. Let the corresponding quantities for the weaker link be P_{TX} and d, respectively. The SIR for this near-far situation can then be described by the following relationship

$$SIR = C\frac{P_{TX}}{d^n}\left(\frac{P_{TX}^{'}\int_W G_\xi(f)df}{\left(d^{'}\right)^m}\right)^{-1} \tag{2.58}$$

where n and m are the path loss exponentials for the nearby and the distant transmission paths, respectively. $G_\xi(f)$ is the normalized power spectral density of the transmitted waveform that includes the effect of spectral regrowth, and C is a constant that depends on path loss (See Section 2.2.4). The bandwidth of each channel is assumed to be W Hz wide. The term in brackets is the interference power resulting from ACI. When m is less than n and d much greater than $d^{'}$, the integral term gets amplified and can cause significant reduction in SIR.

Example 2.4 Assume that a system has total out-of-band emission in the adjacent channel that is –60 dBc, normalized to its transmit power. Let's assume a nearby transmitter with transmit power P'_{TX} that is 100 m from a receiver. A nearby receiver trying to receive from a distant transmitter 10 km away will experience an adjacent channel interference of $10^{-6} P'_{TX} / (d')^3$. Assuming that both transmitters are transmitting with equal power and that m = 3 and n = 4, the SIR computed based on (2.58) is

$$SIR = 10 \frac{P'_{TX}}{10^4} \left(\frac{P'_{TX}}{10^6} \frac{1}{0.1^3} \right)^{-1} = 1$$

where the interferer and the desired signal are assumed to have the same transmit power, the distances are measured in km, and the constant factor C is 10. Assume that the system requires a $\bar{\gamma}$ of 10 dB with a bandwidth efficiency of 1.35 bps/Hz and that the mean of the ACI is zero. With the above assumptions, substituting the computed SIR into (2.57) results in an 11.6 dB reduction in E_b/N_0, rendering the system inoperable. If the transmit power on the strong link is decreased by 20 dB through power control, the performance loss is reduced to about 0.5 dB. □

To improve the SIR, the radio must be designed to minimize the effects of the ACI, the bracketed term in (2.58). In general, power control is employed to keep P'_{TX} at the minimum required to establish the link for the nearby transmitters. To improve the performance even further, interference diversity (Section 2.7.6.1) is employed to reduce the interference power via signal processing in the radio transceiver. For instance, spread-spectrum can be employed to mitigate the effects of interference.

ACI can also degrade the receiver performance via intermodulation distortion, discussed in Chapter 4. Moreover, when a system allows multiple transmissions to occur in a local frequency channel by allocating non-overlapping time slots (e.g. time-division multiple access), ACI exists in the time domain as well. In this case as shown in Figure 2.19b, transmission from a previous time slot may *leak* into the next time slot due to the finite amount of time needed for the transmitter to shut down. If local channels are created using different spreading codes as in CDMA, then ACI also occurs among the different codes due to the fact that perfect orthogonality cannot be achieved in practice. Since different code channels typically share the same frequency channel, no frequency isolation exists and therefore extremely accurate power control is required for CDMA systems. The different multiple access schemes are described in more detail in Chapter 5.

2.2.3 Free-Space Path Loss

The baseband signal modulates an RF carrier for transmission. The transmitted RF signal is subject to signal attenuation that varies as a function of the carrier wavelength λ and the distance d between the transmitter and the receiver. The propagation environment is free space if the transmitter acts like a point source with no obstruction between it and the receiver. In general, free space propagation occurs only in space, such as inter-satellite links where there are no obstructions between the links. In practice, certain line-of-sight (LOS) communication links may closely approximate free space propagation. Such links include satellite links and terrestrial microwave links (e.g. wireless cable).

For free-space propagation, the received power P_{RX} can be expressed in terms of d, λ, the transmitted power P_{TX}, the transmitter antenna gain G_{TX}, and the receiver antenna gain G_{RX} [24].

$$P_{RX} = \frac{P_{TX} G_{TX} G_{RX}}{16\pi^2 d^2 / \lambda^2} \quad (2.59)$$

The antenna gains G_{TX} and G_{RX} depend on the physical characteristics of the antenna such as dielectric constant and geometry. The denominator of the expression in (2.59) represents the free-space path loss L_s that the transmitted signal experiences before reaching the receiver radio frequency (RF) front-end as shown below:

$$L_s = \frac{16\pi^2 d^2}{\lambda^2} \quad (2.60)$$

At one meter, (2.60) becomes

$$L_{s,1m} = \frac{16\pi^2}{\lambda^2} \quad (2.61)$$

which is the path loss at 1 meter in free space. These expressions provide a design guideline to choose adequate transmission power and antenna gains to achieve a given received power required at the RF front-end. The P_{RX} required to achieve a given BER is also known as the receiver sensitivity. A high sensitivity implies a low P_{RX}. The use of (2.59) in the link budget analysis will be examined in Chapter 5.

Example 2.5 Given a carrier frequency of 1 GHz, the wavelength λ is 0.3 m. The one-meter loss is 32.4 dB, according to (2.61). At a distance of 1 km, the free-space path loss becomes $20\log(1000) + 32.4 = 92.4$ dB. Let both the transmitter and receiver antenna gains be 0 dB. If the receiver requires a signal power of -102.4 dBm to properly decode the data, then according to (2.59), the transmitter must transmit at least -10 dBm of power or 100 μW. □

2.2.4 Large-scale Multipath Fading

Propagation in wireless applications exhibits multipath fading. Some applications such as satellite and LOS microwave links approximate free-space propagation most of the time. However, even in these links multipath fading may occur for a fraction of time. The multipath fading could be due to blockage resulting from high-rise when the satellite is seen at a low elevation angle or due to attenuation from rain in wireless cable systems operating at a high 20-30 GHz carrier frequency.

Multipath fading can be classified into large-scale and small-scale fading. This section discusses large-scale fading resulting from reflection and diffraction of the RF transmission from obstructions in the propagation medium. It is large-scale since the obstructions being considered are at a macroscopic scale - i.e. representative of the overall topological characteristic of the environment, such as hilly terrain, dense foliage, urban, or suburban environments. RF signals reflect off obstructions that are much larger than the carrier wavelength and diffract off obstructions that are a few carrier wavelengths in dimension. For instance, at 1 GHz the RF signal reflects off buildings and diffracts off signposts and trees. Every time a signal reflects or diffracts a fraction of the power is re-radiated and the rest is absorbed. The transmitted RF signal arrives at the receiver through multiple paths where at each path the RF signal would have encountered many obstructions resulting in multiple re-radiations. The vector sum of these arriving paths results in multipath fading that can result in a worse signal attenuation (e.g. 40 dB) than the path loss predicted for free-space propagation.

Let the fraction of the re-radiated power by the jth path be represented by Γ_j^2. If the propagation between obstructions is assumed to be free space then the fraction of signal amplitude re-radiated is simply Γ_j / d_j, where d_j is the accumulated distance between the transmitter and the receiver for the jth path. The effective signal amplitude r_j for the jth path at the receiver can then be represented by

$$r_j = r_{TX} \frac{\lambda}{4\pi d_j} \prod_{i=1}^{K_j} \Gamma_{j,i} = r_{TX} \frac{\lambda}{4\pi d_j} \Gamma_j \tag{2.62}$$

where r_{TX} is the un-attenuated signal amplitude at the transmitter, $\Gamma_{j,i}$ is the reflection coefficient for the jth path at the ith obstruction, K_i is the number of obstructions encountered by the jth path, and d_j is the cumulative distance from the transmitter to the receiver, traversing through the K_i obstructions [25]. The received signal $re^{j\phi}$ can then be represented by

$$re^{j\phi} = r_{TX} \frac{\lambda}{4\pi} \sum_{j=1}^{L} \frac{\Gamma_j}{d_j} e^{j\phi_j} \tag{2.63}$$

where L is the total number of paths observed by the receiver, Γ_j is the effective reflection coefficient as defined in (2.62), and ϕ_j is the phase of the jth path, which can also be represented by $2\pi d_j / \lambda$. Given (2.63), one can readily express the received power as

$$P_{RX} = P_{TX} \left| \frac{\lambda}{4\pi} \sum_{j=1}^{L} \frac{\Gamma_j}{d_j} e^{j\phi_j} \right|^2 G_{TX} G_{RX} \tag{2.64}$$

Comparing to (2.59), the path loss is simply expressed by the reciprocal of the square magnitude term in (2.64). In general, the path loss is proportional to d^n where n

depends on the terrain characteristics. In Section 2.2.3, free-space propagation has been shown to have an $n = 2$.

2.2.4.1 Planar-Earth Model

Figure 2.20 shows a two-ray propagation model where there is a LOS path and a reflected path off the ground. The two-ray model is simple but effective in illustrating the effects of large-scale fading. In Figure 2.20, h_{TX}, h_{RX}, and d denote, respectively, the transmitter antenna height, the receiver antenna height, and the distance between the transmitter and receiver. Let d_1 and d_2 represent respectively the distance traveled by the LOS path and the reflected path. Using (2.64), the received power can be expressed as

$$P_{RX} = P_{TX}\left(\frac{\lambda}{4\pi}\right)^2\left|\frac{\Gamma_1}{d_1}e^{j\phi_1} + \frac{\Gamma_2}{d_2}e^{j\phi_2}\right|^2 G_{TX}G_{RX} \tag{2.65}$$

where Γ_1 and Γ_2 are the reflection coefficient for the LOS path and the reflected path, respectively.

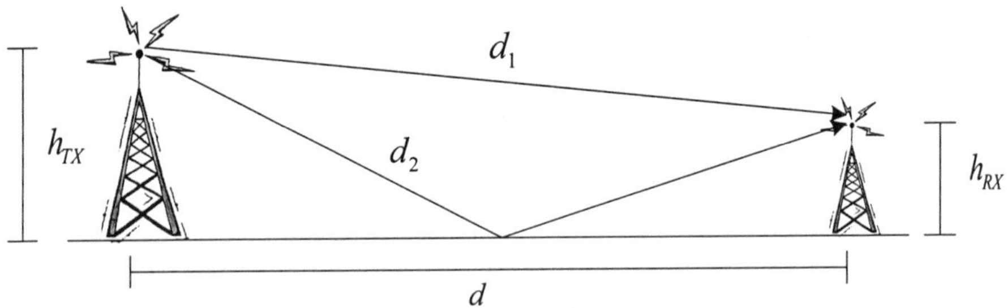

Figure 2.20 Two-ray propagation model.

For the LOS path, $\Gamma_1 = 1$ since the signal experiences no loss due to obstructions. For the reflected path, on the other hand, $\Gamma_2 = -1$ since the signal is reflected off the ground, which can be approximated as a perfect reflector. Expression (2.65) can be simplified if d is assumed to be much greater than h_{TX} and h_{RX}, such that, d_1 and d_2 can be approximated by d. The simplified expression is shown below:

$$P_{RX} \approx P_{TX}G_{TX}G_{RX}\frac{h_{TX}^2 h_{RX}^2}{d^4}. \tag{2.66}$$

Expression (2.66) indicates that the path loss is independent of the carrier frequency, though in practice this is not true. Moreover, the path loss is inversely proportional to d^4 and proportional to the square of the antenna heights. Therefore, to reduce loss, the antenna should be raised as high as possible above ground.

2.2.4.2 Partitioned Model

While the two-ray model gives a simple measure of path loss as a function of distance and antenna heights, it is of limited use to practical channels where usually there exist more than two paths. The partitioned model [26] offers more flexibility in predicting the path loss through a piecewise approximation using different path loss exponentials for different distance segments. Let d_{i-1} be the starting point and d_i the end point of the ith segment, $[d_{i-1}, d_i)$. Then, the path loss for the kth segment can be expressed as

$$L_k(d) = 10\log L_{s,1m} + 10\sum_{i=1}^{k} n_i \log\frac{d}{d_{i-1}}. \tag{2.67}$$

where n_i is the path loss exponential for the ith segment and $L_{s,1m}$ is the 1-m path loss in free space, (2.61).

Example 2.6 Let the partition be of four segments with d_0 = 1 m, d_1 = 100 m, d_2 = 1 km, d_3 = 5 km, and d_4 = 10 km. Also, let the path loss exponentials be n_1 =2, n_2 = 3, n_3 = 3.5, and n_4 = 4. The path loss becomes

$$L_k(d) = 10\log L_{s,1m} + \begin{cases} 20\log d, & 1 \le d < 100 \\ 40 + 30\log\dfrac{d}{100}, & 100 \le d < 1000 \\ 70 + 35\log\dfrac{d}{1000}, & 1000 \le d < 5000 \\ 94.5 + 40\log\dfrac{d}{5000}, & 5000 \le d < 10000 \end{cases} \tag{2.68}$$

where d is in meters. □

2.2.4.3 Lognormal Fading Distribution

Thus far, the multipath fading model describes propagation characteristics of radio transmissions for a given terrain topology where Γ_j, ϕ_j, and d_j are fixed. Sections 2.2.4.1 and 2.2.4.2 discussed two examples where the path loss is computed using (2.64), assuming that the terrain does not change for a given transmitter-receiver separation *d*. However, the assumption of a uniform terrain characteristic does not hold. For instance, the terrain may consist of rolling hills with different spacing, size, and shape. Even at a constant *d*, the link may experience variability in path loss due to the transmitter and/or receiver traversing the non-uniform terrain. Therefore, in practice, large-scale fading shows slow amplitude fluctuations. If the fluctuations are caused by a large number of rays arriving at the receiver with different reflecting coefficients and phases, then the central-limit theorem [15] may be applied to the fading amplitude in dB [27]. The resulting fading can then be approximated by a lognormal distribution, shown below:

$$P_R(r) = \frac{10}{\ln 10\sqrt{2\pi}\sigma r^2} e^{-(20\log r-\mu)/2\sigma^2} \tag{2.69}$$

where and μ and σ are respectively the mean and standard deviation of the fading amplitude r in dB. In this case, the fading amplitude is normalized to the transmitted power level. The mean path loss μ can be determined by (2.64) with appropriately chosen values for Γ_j, ϕ_j, and d_j.

2.2.5 Small-Scale Multipath Fading

Near the receiver the arriving rays tend to experience additional multipath fading due to local scatters. Such fading is known as small-scale fading since it is caused by obstructions nearby the receiver. The fluctuations due to small-scale fading tend to be on the order of a wavelength because of the proximity of the scatterers whereas large-scale fading shows path loss variations on the order of several wavelengths as shown in Figure 2.21. Also, the fading behavior exhibits rapid amplitude fluctuations in contrast to the slow fluctuations typical in large-scale fading. Therefore, sometimes small-scale fading may be referred to as fast fading. In general, small-scale fading can be characterized by either a Rayleigh distribution or a Rician distribution [28].

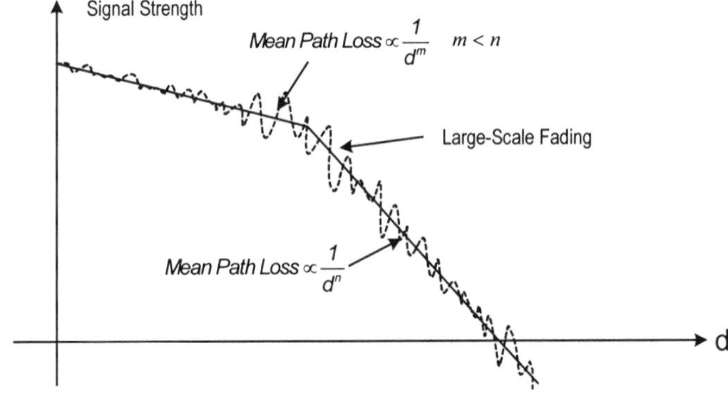

Figure 2.21 Typical path loss with large scale fading.

2.2.5.1 Rayleigh Fading Distribution

When the link is non-LOS (NLOS), the small-scale fading tends to be Rayleigh distributed. A NLOS link has no direct path from the transmitter to the receiver. Intuitively, the Rayleigh distribution arises from the large number of indirect paths when the link is NLOS. The amplitude statistics of the in-phase I_k and quadrature Q_k signal components tend to be Gaussian distributed. It can be shown that $I_k^2 + Q_k^2$, i.e. amplitude squared r^2, is then exponential distributed and $\sqrt{I_k^2 + Q_k^2}$, i.e. amplitude r, is Rayleigh distributed.

The corresponding exponential distribution is shown below:

$$P_{R^2}(\delta) = \frac{1}{\overline{\delta}} e^{-\delta/\overline{\delta}} \tag{2.70}$$

where $\delta = r^2$, $\overline{\delta}$ denotes the mean of r^2, which determines the variance shown in (2.71) below

$$\overline{\delta} = \sigma_{R^2}^2 = 2\sigma^2. \tag{2.71}$$

σ^2 is the variance of the underlying Gaussian process.

The Rayleigh distribution is shown below

$$P_R(r) = \frac{2r}{\overline{\delta}} e^{-r^2/\overline{\delta}} \tag{2.72}$$

And its mean $\overline{r}$ and variance σ_R^2 are shown in (2.73) and (2.74) below.

$$\overline{r} = \sqrt{\pi\overline{\delta}}/2 \tag{2.73}$$

$$\sigma_R^2 = (1 - \pi/4)\overline{\delta} \tag{2.74}$$

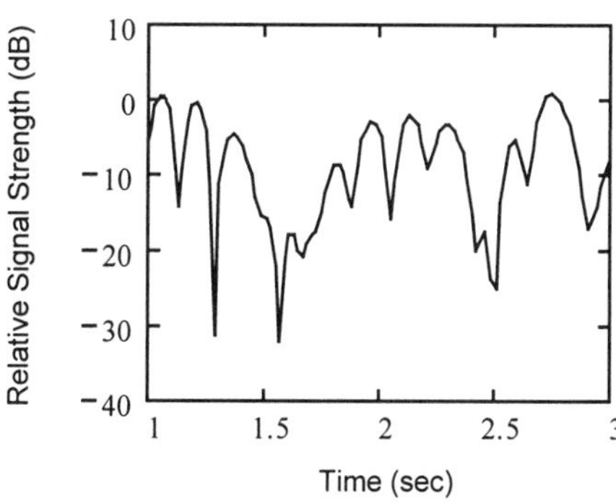

Figure 2.22 Rayleigh fading in an indoor channel with obstructed LOS and a Doppler spread of 5 Hz.

The above distributions are important to the design of digital systems and RF front-ends in the presence of fading. A typical mobile wireless channel can exhibit fading fluctuations up to 30-40 dB as shown in Figure 2.22 for an indoor channel.

Appropriate diversity processing should be used to accommodate the bursty errors due to fading that attenuates the signal below the sensitivity of the RF front-end. Also, sufficient dynamic range should be designed in the RF front-end to accommodate the slow envelope variations due to fading.

2.2.5.2 Rician Fading Distribution

When a direct path exists, the received amplitude has a Rician distribution shown below

$$P_R(r) = \frac{2r}{\overline{\delta}} e^{-(r^2-\upsilon^2)/\overline{\delta}} I_0\left(\frac{2r\upsilon}{\overline{\delta}}\right) \qquad (2.75)$$

where υ is the amplitude of the LOS component. Let $\kappa = \upsilon^2 / \overline{\delta}$ which indicates the amount of power in the LOS component to that of the NLOS components. A typical value of κ for the indoor environment is around 6 dB [29]. The amount of amplitude variation is much less than that of Rayleigh fading and presents a best-case scenario for benign environments such as open terrains and open office areas as shown in Figure 2.23. . For outdoor environments where there is no direct LOS, the value of κ becomes much less.

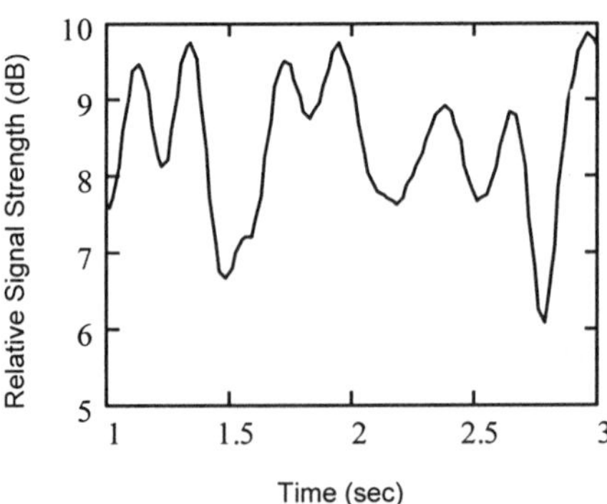

Figure 2.23 Rician fading in a LOS indoor environment (κ = 6 dB).

2.2.6 Selectivity of Fading

So far we have discussed the effects of fading on signal levels, which allows one to set constraints on the transmitter power, the amount of power control, and dynamic range on the receiver. We now consider the selectivity that multipath fading has in time, frequency, and space [30]. An understanding of the fading selectivity forms the basis for the design of diversity techniques to improve the BER performance.

2.2.6.1 Frequency-Selective Fading

Let the transmitted signal be an impulse $\delta(t)$. Then, as the impulse is dispersed by obstructions, it arrives at the receiver as multiple reflected impulses of different time delays, phases, and amplitudes, as shown in Figure 2.24.

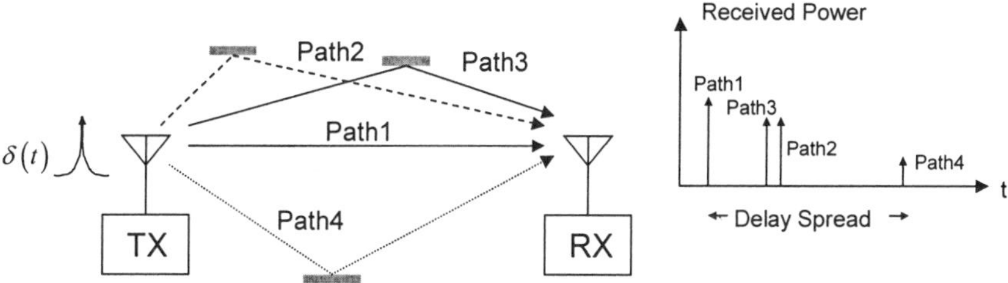

Figure 2.24 Frequency-selective fading.

A channel that disperses the transmitted signal in time as shown in Figure 2.24 can be characterized by its impulse response $h(\tau,t)$, where τ represents the time dispersion and t represents the time variation of the impulse response [31][32]. For instance, the impulse response corresponding to the channel in Figure 2.24 can be represented by

$$h(\tau,t)=\sum_{i=1}^{l}\beta_i(t)e^{\phi_i(t)}\delta\left(t-\tau_i(t)\right) \tag{2.76}$$

where $\beta_i(t)$, $\phi_i(t)$, and $\tau_i(t)$ are the amplitude, phase, and path delay of the ith path.

Figure 2.24 shows the impulse response for a given time $t = t_0$. In general, the impulse response varies with time due to motion of the transmitter, receiver, or the obstructions. Assuming that each path experiences scattering local to the receiver, the amplitude will then have either a Rayleigh or Rician distribution depending on the value of κ, as described in Section 2.2.5.2. The phase generally has a uniform distribution and the path delay has an exponential distribution. Note that in general each impulse experiences independent amplitude fading.

Given (2.76) and that a baseband signal $\tilde{s}(t)$ is transmitted through the channel, the received signal can be represented by

$$r(t)=\sum_{i=1}^{l}\beta_i(t)e^{\phi_i(t)}\tilde{s}\left(t-\tau_i(t)\right). \tag{2.77}$$

The path delays cause the baseband signals in each path to shift in time by $\tau_i(t)$. If the differential path delay between the paths is greater than a symbol duration then symbols from one path will interfere with neighboring symbols from other paths. This is also referred to as intersymbol interference which can severely degrade the performance of the receiver.

Since the channel behaves just like a time varying filter, it is instructive to determine its frequency behavior. Since the impulse response contains random variables, the frequency behavior is determined based on its autocorrelation function shown below

$$R_{hh}(\tau_1,\tau_2,t_1,t_2) = E\left\{h(\tau_1,t_1)h^*(\tau_2,t_2)\right\}. \tag{2.78}$$

To simplify the above expression, a widesense stationary uncorrelated scattering (WSSUS) channel is assumed with the following properties:

1. The channel statistics at times t_1 and t_2 depends only on the time difference $\Delta t = t_1 - t_2$.
2. The scattering in each path is uncorrelated with that of other paths.

Given the WSSUS channel, (2.78) becomes

$$R_{hh}(\tau_1,\tau_2,t_1,t_2) = R_{hh}(\tau,\Delta t). \tag{2.79}$$

Furthermore, if the channel is observed at an instance in time i.e. $\Delta t = 0$, then the autocorrelation function is strictly a function of the path delay τ, i.e. $Q(\tau) = R_{hh}(\tau,0)$ where $Q(\tau)$ is also known as the *delay power spectrum* of the channel. The RMS delay spread of the channel can be expressed in terms of the delay power spectrum as shown below

$$\tau_{rms} = \sqrt{\frac{\int_{-\infty}^{\infty}(\tau-\bar{\tau})^2 Q(\tau)d\tau}{\int_{-\infty}^{\infty}Q(\tau)d\tau}} \tag{2.80}$$

where $\bar{\tau}$ is the average path delay. Also, it is often important to know the maximum delay spread of the channel, which is denoted by $\tau_{\max}$. When the symbol duration is a small fraction of the RMS delay spread, equalization is required to compensate for the intersymbol interference [33]. Equalization is a form of path diversity described in Section 2.7.2.

By taking the Fourier transform of (2.79), the autocorrelation function of the impulse response can be obtained in the frequency domain which measures the correlation between two frequencies, f_1 and f_2. The frequency domain autocorrelation function is shown below

$$R_{Hh}(f_1,f_2,\Delta t) = R_{Hh}(\Delta f,\Delta t) = \int_{-\infty}^{\infty} R_{hh}(\tau,\Delta t)e^{-j2\pi\Delta f\tau}d\tau. \tag{2.81}$$

where $\Delta f = f_1 - f$. Therefore, for a WSSUS channel, the correlation between two frequency components depends only on their separation.

The coherence bandwidth B_c is defined as the frequency separation at which the correlation begins to drop below 50%. B_c is inversely proportional to τ_{rms}. For example, B_c of a typical mobile channel is inversely related to τ_{rms} [28]:

$$B_c = \frac{1}{2\pi\tau_{rms}} \tag{2.82}$$

The physical interpretation of B_c is that signal frequency components separated greater than B_c are likely to experience independent fading statistics while the frequency components separated by less than B_c are likely to fade together due to the greater correlation. If the signal bandwidth is greater than B_c, then the signal will exhibit frequency-selective fading since only a portion of the signal band experiences independent fading. On the other hand, if the signal bandwidth is less than B_c, then the signal fading will be non-frequency-selective or flat. Equivalently, a frequency-selective channel corresponds to a channel that induces ISI and a flat fading channel corresponds to a channel with no ISI. The frequency selectivity of the channel can be determined by the following criteria:

$$\begin{array}{ll} \text{Frequency selective:} & B_c T_s \ll 1 \\ \text{Frequency non-selective (flat):} & B_c T_s \gg 1 \end{array} \tag{2.83}$$

The type of channel impacts the design of diversity technique in the digital receiver to compensate for the BER degradation due to fading. In general, coding with interleaving is used for a flat fading channel while equalization is used for a frequency-selective channel.

2.2.6.2 Time-Selective Fading

So far, the behavior of the channel in time has been largely ignored. Time independence might be a fair assumption for static or quasi-static systems such as a microwave point-to-point link or indoor environments where motion is due to people walking. However, in a mobile channel, the assumption breaks down. Mobility induces a Doppler frequency shift f_d :

$$f_d = f_c \frac{v}{c} \cos\theta \tag{2.84}$$

where c is the velocity of light, θ is the angle of arrival, v is the velocity of the mobile, and f_c is the carrier frequency. When $\theta = 0$ the Doppler shift is at its maximum and is denoted by f_m.

Due to the local scattering at the receiver where each scattered path has a different angle of arrival, the Doppler frequency tends to spread the carrier frequency. The Doppler spread causes signal amplitude to fade whenever the phase becomes 180 degrees out of phase with respect to other signal components. The time between fades is known as the coherence time of the channel τ_c and is inversely proportional to the Doppler spread. In particular, with a Rayleigh fading amplitude, the average downward crossings for a given level A per second and the duration of each fade can be expressed by the following expressions [28]

$$N(\rho) = \sqrt{2\pi} f_m \rho e^{-\rho^2} \tag{2.85}$$

$$\tau(\rho) = \frac{e^{\rho^2} - 1}{\rho f_m \sqrt{2\pi}} \tag{2.86}$$

where ρ is defined as A/A_{rms} and A_{rms} is the RMS signal amplitude. The fade rate is defined to be $1.475 f_m$ when $A = \sqrt{2 \ln 2} A_{rms}$. Figure 2.25 shows the signal envelope as a function of time for a Doppler frequency of 200 Hz. ρ is set to 1 and θ is set to 0. Clearly the high Doppler frequency results in fast amplitude fluctuations and short fade duration. In other words, the channel becomes bursty as the Doppler frequency increases and may cause a large block of data to be lost if fades occur during the transmission of a frame or packet. Later, it will be shown that the definition of time-selectivity of the channel is related to the burstiness of the fade. Finally, it is interesting to see that if the channel also exhibits frequency selective fading as shown in Figure 2.25, the different paths experience independent fading and thus tend to exhibit fades at different times that further increase the burstiness of the channel.

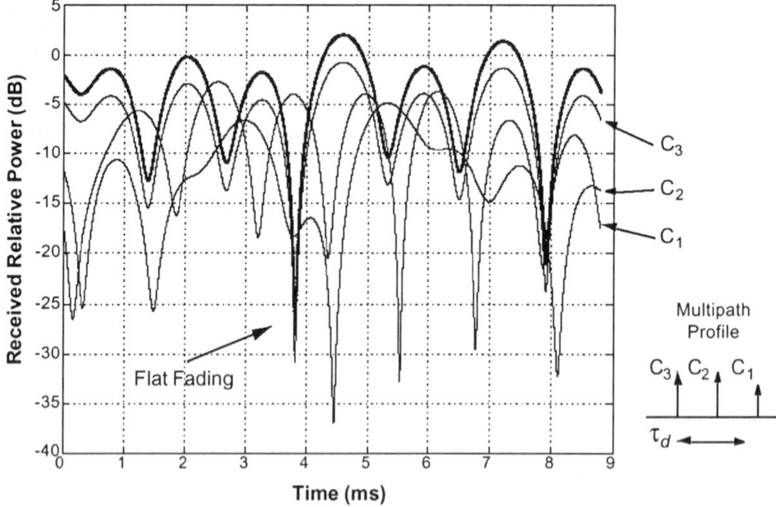

Figure 2.25 Time and frequency selective fading with a Doppler shift of 444 Hz and a delay spread of 3 μs; Chien [163], © 1998 IEEE.

The Doppler spread can be found by taking the Fourier transform of (2.81) with respect to the time variable Δt. The resulting function $R_{HH}(\Delta f, \lambda)$ measures the power at different Doppler shifts denoted by λ:

$$R_{HH}(\Delta f, \lambda) = \int_{-\infty}^{\infty} R_{Hh}(\Delta f, \Delta t) e^{-j2\pi\lambda\Delta t} d(\Delta t). \tag{2.87}$$

Let $\Delta f = 0$, i.e. looking at a specific frequency. Then, (2.87) becomes independent of frequency, i.e. $R_{HH}(0, \lambda) = D(\lambda)$. $D(\lambda)$ is also known as the *Doppler power spectrum*. For a mobile channel with dense local scatterers, the Doppler power spectrum can be expressed by

$$D(\lambda) = 1/2\pi f_m \left[1 - (\lambda / f_m)^2\right]^{-1/2}. \tag{2.88}$$

[28]. For a slow time varying indoor channel, $D(\lambda)$ tends to be flat extending from $-f_m$ to f_m. Given $D(\lambda)$, the RMS Doppler spread $B_{d\text{-}rms}$ can be expressed similarly as in (2.80), except τ and $Q(\tau)$ are replaced by λ and $D(\lambda)$, respectively.

A useful measure of the dynamics of the channel is the coherence time τ_c that measures the time separation over which the correlation of the signal amplitude drops below 50%. The coherence time is inversely proportional to the maximum Doppler spread [28]

$$\tau_c = \frac{9}{16\pi f_m}. \tag{2.89}$$

The physical interpretation of τ_c is that signal amplitudes separated greater than τ_c in time are likely to experience independent fading statistics while amplitudes separated by less than τ_c are likely to fade together due to the higher correlation value. If the frame or packet duration is greater than τ_c, then the signal will exhibit time-selective fading since only a portion of the frame or packet experiences independent fading. On the other hand, if the frame or packet duration is less than τ_c, then the signal fading becomes non-time-selective. Equivalently, a time-selective channel corresponds to a channel where a frame or packet experiences burst errors due to sharp attenuations in signal amplitude over the duration of time, i.e. (2.86). For a non-time-selective channel, the signal envelope tends to stay fairly constant over time without major fluctuations. The following can be used to determine the time selectivity of the channel

$$\begin{aligned} &\text{Time selective:} && L_x T_s \tau_c^{-1} >> 1 \\ &\text{Time non-selective (flat):} && L_x T_s \tau_c^{-1} << 1 \end{aligned} \tag{2.90}$$

where x may either be f or p to denote a frame or packet and L_x is the duration of the frame or packet. In general, coding with interleaving is used for a time-selective channel where the interleaver is designed according to a given fade duration and τ_c. An interleaver randomizes the bursty error induced by the time-selective channel to improve the E_b/N_0 performance through coding as described in Section 2.7.4.

2.2.6.3 Space-Selective Fading

Space-selective fading occurs in systems with multiple antennas at the transmitter and/or receiver. Multiple antennas are also often referred to as antenna arrays [34]. Radio systems use antenna arrays to improve the gain of the antenna and thereby help to reduce the total RF transmission power for a given path loss. See Sections 2.2.3 and 2.2.4. The additional gain achieved by the array over that of a single antenna is also known as the array gain G_{array}, which depends on the configuration of the array. In general, array gain cannot be larger than the number of antenna elements N in the array [34]:

$$G_{array} \leq 20\log_{10} N. \tag{2.91}$$

The simplest array configuration is the uniform linear array shown in Figure 2.26, where the antennas are spaced out evenly on a line with spacing d_a. The received signal at each antenna element along the array experience a different phase shift given the particular angle of arrival θ. The amount of phase shift between two elements depends on the frequency of the carrier as well as the difference in distance Δd traveled by the received signal. Assuming that the RF signal arriving at the antenna array is approximately planar, then for a linear array $\Delta d = d_a \sin\theta$ as shown in Figure 2.27. The phase difference $\Delta\theta$ is then

$$\Delta\theta = \omega_c \frac{\Delta d}{c} = \frac{2\pi f_c d_a \sin\theta}{c}. \tag{2.92}$$

For a linear array, it is clear that the phase difference is the same for any two elements adjacent to each other. Thus, the phase shift for any antenna element in a uniform linear array can be computed with respect to a reference element in the array. Let the first element be the reference. The array response $a(\theta)$ is defined as

$$a(\theta) = \sum_{k=1}^{N} w_k e^{j(k-1)\Delta\theta} \tag{2.93}$$

where w_k is the weight for the output of the kth antenna element. The array response determines the antenna gain as a function of θ, d_a, and $\mathbf{w} = [w_1 w_2 \cdots w_N]$. By appropriately selecting the antenna spacing d_a, the antenna response can be steered using **w** to peak at a given θ. In general, d_a is chosen to be $\lambda/2$ since if it is much more than half a wavelength, grating occurs where undesired sidelobe levels rise as shown in Figure 2.27. Also, the beamwidth decreases, requiring high precision in pointing the antenna at the right direction.

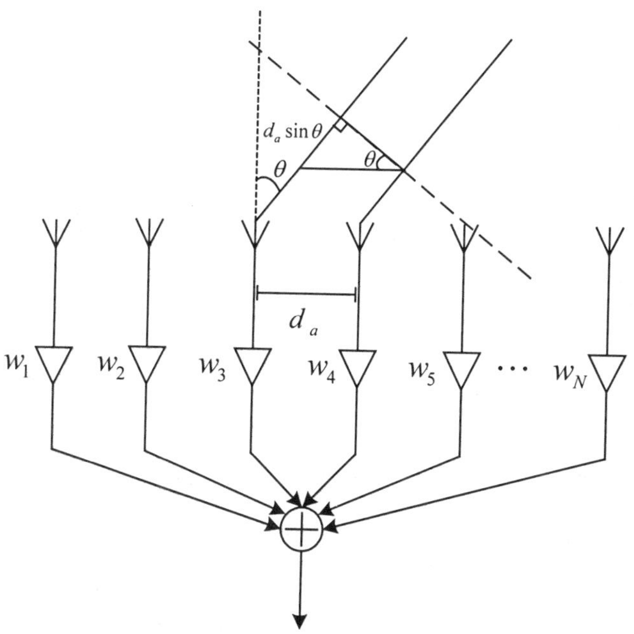

Figure 2.26 A uniform linear antenna array.

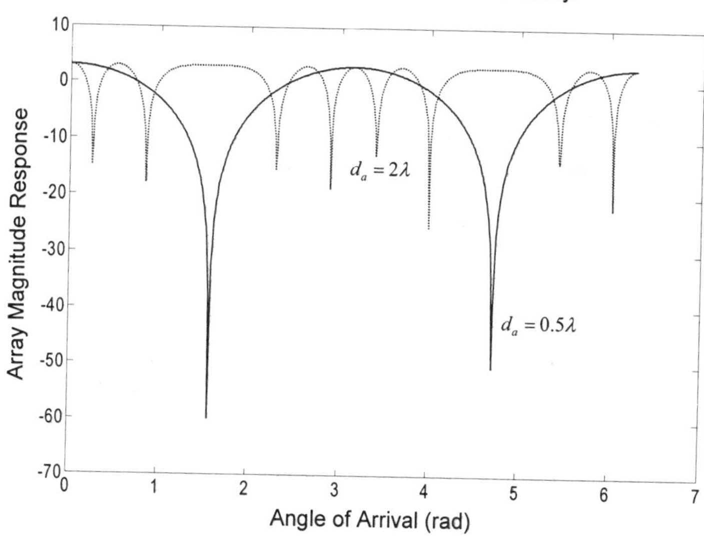

Figure 2.27 Array response of a two-element array with equal weights.

The array response gives some insight on the space-selective behavior of the channel with relation to the spacing of the antenna elements. Given any two elements in an array, the magnitude response is

$$|a(\theta)| = \left| 1 + e^{j\frac{2\pi d_a \sin\theta}{\lambda}} \right|, \tag{2.94}$$

Assuming equal weights. The array response becomes small when $d_a \approx 0.5\lambda/\sin\theta$. If θ is high (e.g. 90°) then $d_a \approx \lambda/2$. On the other hand, if θ is low (e.g. 10°) then $d_a \approx 3\lambda$. In general, the distance between the antenna elements at which the fading becomes independent is known as the coherence distance d_c. Given the above example, it can be seen that signal amplitude is affected by the relationship between the antenna spacing and the angle of arrival. To obtain independent fading, the required antenna spacing tends to increase with decreasing angle of arrival.

In practice, signals arrive at the array through multiple paths and with different angles of arrival. The angular spread θ_a defines the amount of variations in the angle of arrival. It can be shown that the coherence distance d_c is inversely proportional to the RMS angular spread of the channel θ_{rms} [35]. In general, the angular spread at a basestation is small due to multipaths originating from remote mobile terminals. The height advantage of the basestation also contributes to the small angular spread. On the other hand, at the mobile terminal, the angular spread is larger due to local scatter around the mobile terminal. Therefore, to achieve spatial diversity antenna spacing at the basestation should be large on the order of several wavelengths and the antenna spacing at the mobile terminal should be at least one half the wavelength.

Space-selectivity of a channel is determined by the following constraints:

$$\begin{array}{ll} \text{Space selective} & d_a d_c^{-1} >> 1 \\ \text{Space non-selective (flat)} & d_a d_c^{-1} << 1. \end{array} \tag{2.95}$$

While one of the main purposes for using an antenna array is to increase the gain of the antenna, it can also serve to improve the system's performance against fading if the antenna elements are chosen to induce space-selectivity. Diversity schemes that leverage the space-selectivity of the channel are described in Section 2.7.5.

2.2.6.4 Generalized Channel Model

In practice, multipath fading induces some finite amount of frequency, time, as well as space selectivity on the received signal. The scattering function shown below provides insight on the relationship among the different types of fading:

$$S(\tau,\lambda) = \int_{-\infty}^{\infty} R_{hh}(\tau,\Delta t) e^{-j2\pi\lambda\Delta t} d(\Delta t). \tag{2.96}$$

In particular, it relates the Doppler spread to a given path delay τ. Since the scattering is assumed to be uncorrelated, each signal component delayed by a different value of τ will have an independent Doppler spread. A discrete channel model that captures the essence of (2.96) is shown in Figure 2.28.

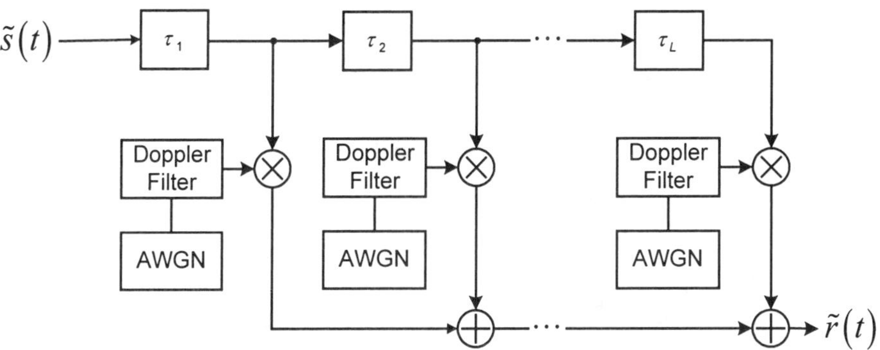

Figure 2.28 A frequency and time selective channel model.

The tapped delay line represents the delay experienced by the signal in each of the L paths. The amplitude fading statistics of each path has an independent Rayleigh distribution. To generate the Rayleigh distributions, L independent complex filtered Gaussian sources are used. The noise-shaping filter is derived by taking the square root of $D(\lambda)$ [36]. The filter response can be designed to generate the desired Doppler spread on each path. The output of the Doppler filter for the ith tap corresponds to $\beta_i(t)$ in (2.76).

The model in Figure 2.28 captures both time and frequency selectivity of the fading process. Finally, to incorporate space selectivity into this model, the impulse response for each of the M antenna elements is multiplied by the antenna array response at a given angel of arrival θ as shown in Figure 2.29. $h_j(\tau,t)$ denotes the channel response from the transmitter to the jth receiver antenna and $a(\theta_j)$ denotes the array response with an angle of arrival θ_j at the jth antenna. Each $h_j(\tau,t)$ can be modeled as shown in Figure 2.28. If an antenna array is used at the transmitter as well, then the channel can be modeled as a matrix whose i,j^{th} element is $h_{ij}(\tau,t)$, which denotes the impulse response from the ith transmit antenna to the jth receive antenna. Each $h_{ij}(\tau,t)$ is multiplied by the transmitter array response $a_T(\theta_i)$ and the receiver array response $a_R(\theta_j)$.

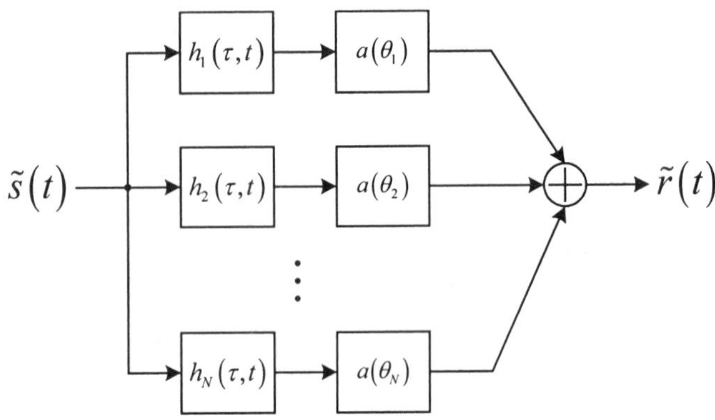

Figure 2.29 A channel model that includes the effect of space selectivity.

Table 2.2 summarizes the conditions required for each type of selective fading. In practice, any combination of the three types of fading can occur and appropriate diversity methods must be used to compensate for the loss due to fading. In Section 2.7, the trade-offs associated with different diversity methods in each fading scenario are discussed.

Table 2.2 Conditions for selective fading.

Fading Type	Selective	Non-Selective
Time	$L_x T_s \tau_c^{-1} >> 1$	$L_x T_s \tau_c^{-1} << 1$
Frequency	$B_c T_s << 1$	$B_c T_s >> 1$
Space	$d_a d_c^{-1} >> 1$	$d_a d_c^{-1} << 1$

2.2.6.5 Summary of Fading

Table 2.3 shows empirical results on the path loss exponential n, lognormal fading variance σ^2, RMS delay spread τ_{rms}, and maximum Doppler shift f_m for different environments at a 1-2 GHz transmission frequency. The path loss exponential n impacts the power amplifier design in terms of the nominal power needed to establish a link at a given range and E_b/N_0. The standard deviation of path loss, i.e. σ in (2.69), determines the dynamic range of power control and the link margin required to meet a given coverage area. The path loss exponential varies from 1.5 to 6 and the lognormal fading varies from 4-14 dB [25][37][38]. Although the transmission range is often shorter indoors, the path loss exponential and lognormal fading loss can still be high due to obstructions, such as floors and walls. The RMS delay spread varies from 25-150 ns indoors and 50 – 15000 ns outdoors [39][40][41][42]. The Doppler spread for different environments varies from 0 to 500 Hz in the 1-2 GHz range

[38][43]. The RMS angular spread depends on the antenna height and the scatters in the environment. In general, a large angular spread is observed at the user terminal with a low antenna height due to a large number of scatters while it is only a few degrees at the basestation [44].

Table 2.3 Channel parameters for different environments at 1-2 GHz.

Channel Parameters	Warehouse	Partitioned Office	Suburban	Urban	Rural (open)
n	1.5-3	2-6	2-3	3-4	2-3
σ (dB)	5-8	7-14	4-8	4-10	4-8
τ_{rms} (ns)	50-150	25-50	50-1000	300-15000	100-500
f_m (Hz)	0-5	0-5	0-50	0-200	0-200

2.3 Performance Analysis

Section 2.1.4 defined the energy per bit over noise density ratio (E_b/N_0) and illustrated some examples of its influence on the error rate performance of a digital system. This section focuses on the basic techniques used to derive error performance of digital systems in an AWGN and fading channels. In particular, a general class of optimal receivers based on maximum likelihood (ML) criteria is derived for symbol detection in the AWGN channel [14][15]. While in practice a pure AWGN channel is not encountered, assuming such a channel simplifies the analysis and provides insight on the design of more sophisticated receiver architectures that combat multipath fading.

2.3.1 Maximum Likelihood Detection

Let the received signal be $\tilde{y}(t)$ which consists of the transmitted signal $\tilde{s}_i(t)$ corrupted by an AWGN source $\tilde{n}(t)$ as shown below:

$$\tilde{y}(t) = \tilde{s}_i(t) + \tilde{n}(t) \tag{2.97}$$

where i has a value between 1 and M, corresponding to the M possible transmitted symbols. Based on the orthonormal expansion described in Section 2.1.1, (2.97) can be expressed in terms of expansion coefficients corresponding to a set of orthonormal basis functions $\{\varphi_l(t)\}$ of N_d elements. In other words, each signal in (2.97) can be expressed as a vector with N_d elements corresponding to the expansion coefficients derived using (2.8). The vector representation is shown below:

$$\tilde{\mathbf{y}} = \tilde{\mathbf{s}}_i + \tilde{\mathbf{n}}. \tag{2.98}$$

A digital receiver should determine the transmitted signal in the presence of noise with minimum error rate. A measure of error rate is the symbol error rate defined as the percentage of incorrect symbol decisions made at the receiver. The symbol error rate can be expressed as follows

$$P_s = 1 - \sum_{i=1}^{M} p_i \int_{R_i} p_{\mathbf{Y}|\mathbf{S}_i}(\tilde{\mathbf{y}} \mid \tilde{\mathbf{s}}_\mathrm{i})\, d\tilde{\mathbf{y}} \tag{2.99}$$

where

$$p_{\mathbf{Y}|\mathbf{S}_\mathrm{i}}(\tilde{\mathbf{y}} \mid \tilde{\mathbf{s}}_\mathrm{i}) = \frac{p(\tilde{\mathbf{y}}, \tilde{\mathbf{s}}_\mathrm{i})}{p(\tilde{\mathbf{s}}_\mathrm{i})} \tag{2.100}$$

is the probability density function (pdf) of the received vector $\tilde{\mathbf{y}}$ conditioned on the transmitted symbol $\tilde{\mathbf{s}}_\mathrm{i}$ and p_i is the probability that $\tilde{\mathbf{s}}_\mathrm{i}$ has been transmitted. The integral in (2.99) is integrated over the ith decision region R_i to obtain the probability that $\tilde{\mathbf{s}}_\mathrm{i}$ has been detected correctly. The weighted sum of this integral represents the average probability P_c that an arbitrary symbol has been detected correctly. The average symbol error rate is simply $1 - P_c$.

The key to minimizing P_s lies in choosing an appropriate R_i for all i such that P_c is maximized. The selection criterion for R_i requires that if $\mathbf{y}$ is a vector in R_i, it must then satisfy the following constraint:

$$p_i p_{\mathbf{Y}|\mathbf{S}_\mathrm{i}}(\mathbf{y} \mid \tilde{\mathbf{s}}_i) = \max_l p_l p_{\mathbf{Y}|\mathbf{S}_l}(\mathbf{y} \mid \tilde{\mathbf{s}}_l). \tag{2.101}$$

That is the weighted conditional pdf of the vector $\boldsymbol{y}$ given that symbol $\tilde{\mathbf{s}}_i$ is transmitted is larger than that of all other transmitted symbols $\tilde{\mathbf{s}}_{l \neq i}$.

In general, since the distribution of the transmitted symbol is not known in advance, it is assumed to be uniformly distributed, i.e. $p_i = M^{-1}$. Given this assumption, (2.101) simplifies to

$$p_{\mathbf{Y}|\mathbf{S}_\mathrm{i}}(\mathbf{y} \mid \tilde{\mathbf{s}}_i) = \max_l p_{\mathbf{Y}|\mathbf{S}_l}(\mathbf{y} \mid \tilde{\mathbf{s}}_l). \tag{2.102}$$

The constraint in (2.102) is also known as the maximum likelihood (ML) decision rule and the region R_i that satisfies this decision rule is known as the ML decision region. It is maximum likelihood since decisions based on it maximize the average probability of making the correct decision on a given transmitted symbol $\tilde{\mathbf{s}}_i$. Any vector z that falls within a decision region R_i indicates that $\tilde{\mathbf{s}}_i$ has been transmitted.

The ML decision rule (2.102) cannot always be easily analyzed in a form that provides instructive insight on the receiver architecture. However, if the disturbance is due to AWGN, then a simple receiver structure emerges. A signal $\tilde{\mathbf{s}}_i$ transmitted over the AWGN channel has a conditional density that can be expressed as follows

$$p_{\mathbf{Y}|\mathbf{S}_j}(\tilde{\mathbf{y}} \mid \tilde{\mathbf{s}}_j) = \frac{1}{(2\pi)^{N_d/2}\sqrt{N_d N_0/2}} e^{-\frac{|\tilde{\mathbf{y}}-\tilde{\mathbf{s}}_j|^2}{N_0}}. \tag{2.103}$$

To satisfy (2.102), it is clear that (2.103) is maximized if the term $|\tilde{\mathbf{y}}-\tilde{\mathbf{s}}_j|^2$ is minimized. Since the square magnitude of the received vector $\mathbf{y}$ is a fixed term, the condition can be further simplified to

$$\max\left\{ 2\Re(\tilde{\mathbf{y}}\cdot\tilde{\mathbf{s}}_j^{\mathbf{H}}) - |\tilde{\mathbf{s}}_j|^2 \right\} \tag{2.104}$$

Based on the definition of inner product in (2.8), (2.104) can be expressed in terms of the associated time functions as shown below

$$\max\left\{ 2\Re\int_0^T \tilde{\mathbf{y}}(\mathbf{t})\, \tilde{\mathbf{s}}_j^*(t)\,dt - \int_0^T |\tilde{\mathbf{s}}_j|^2\, dt \right\}. \tag{2.105}$$

The first term is simply a time correlation of the received signal with $\tilde{\mathbf{s}}_j$ and the second term is the energy of $\tilde{\mathbf{s}}_j$. The physical interpretation of (2.105) is that $\tilde{\mathbf{s}}_j$ is decided as the transmitted signal if $\tilde{\mathbf{s}}_j$ results in the largest cross-correlation with the received signal among all transmitted signal $\tilde{\mathbf{s}}_{l\neq j}$. The energy of each transmitted signal biases the correlation in the decision process.

The correlator receiver structure is optimum only when the noise is AWG and when the channel is perfect (i.e. $\tilde{c}(t) = \delta(t)$). For a fading channel where this assumption no longer holds, diversity must be applied to mitigate the effects of fading to achieve a performance comparable to that of an AWGN channel. Also, (2.105) assumes that the receiver has precise knowledge of the timing and phase of the received signal, i.e. the receiver is coherent. In practice, the coherent assumption does not hold and the receiver must be synchronized prior to making a decision as discussed in Section 2.5. Receivers which make decisions without phase coherency show degradations in E_b/N_0, though non-coherent receivers tend to have lower complexity.

2.3.2 Geometric Interpretation

The ML receiver based on (2.105) has a geometric interpretation. Expression (2.105) is derived from minimizing $|\tilde{\mathbf{y}}-\tilde{\mathbf{s}}_j|^2$ which measures the Euclidean distance between the received vector and the symbol constellation point $\tilde{\mathbf{s}}_i$. The ML receiver can therefore be viewed as a minimum distance receiver where a symbol decision is based on comparing the Euclidean distances of the received vector with all possible constellation points transmitted. The symbol constellation that has the minimum distance from the received vector is decided as the received symbol. Geometrically, the decision regions that satisfy the minimum distance constraint consist of bisectors between adjacent constellation points as shown in Figure 2.30.

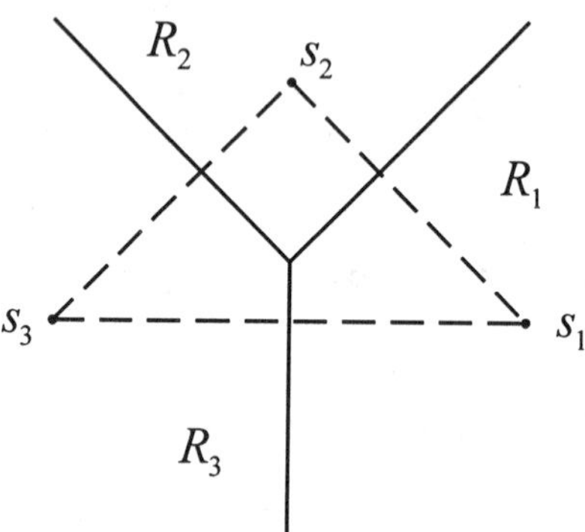

Figure 2.30 ML decision regions derived from bisectors of the signal constellation.

The error performance is a function of the amount of noise needed to push the received vector out of the correct decision region for a given transmitted symbol. Consider two constellation points $\tilde{\mathbf{s}}_i$ and $\tilde{\mathbf{s}}_j$ separated by an angle θ_{ij} and with energies E_{si} and E_{sj}, respectively. The distance between $\tilde{\mathbf{s}}_i$ and $\tilde{\mathbf{s}}_j$ denoted by d_{ij} can be represented by

$$d_{ij} = \sqrt{E_{si} + E_{sj} - 2\sqrt{E_{si}E_{sj}}\cos\theta_{ij}}\,. \tag{2.106}$$

The amount of noise required to cause an error then must exceed $d_{ij}/2$. Applying (2.53), the probability of error $p(e_{ij})$ can then be expressed in terms of (2.106):

$$P(e_{ij}) = P(n > \frac{d_{ij}}{2}) = Q\left(\sqrt{\frac{E_{si} + E_{sj} - 2\sqrt{E_{si}E_{sj}}\cos\theta_{ij}}{2N_0}}\right). \tag{2.107}$$

Expression (2.107) is reduced when the argument in $Q(\cdot)$ is increased. Therefore, to design a robust digital system requires that the pairwise distance be maximized for protection against noise, which implies that symbol energies need to be increased. An increase in symbol energies leads to an increase in the average energy per bit (2.38), degrading the energy efficiency of the overall system. The design challenge lies in finding methods to increase d_{ij} while minimizing the average energy per bit.

2.3.3 Symbol Error Rate

When the constellation is large, the derivation of exact error performance can be quite intricate. It is instructive to obtain an upper bound on error performance that can be more easily derived while still providing insights on the design of digital systems. The simplest bound is the union bound [14] which states that the probability of the union of all likely error events is upper bounded by the sum of the probabilities of the individual events. For example, the symbol error that occurs when $\tilde{\mathbf{s}}_i$ is incorrectly received depends on the union of all error events corresponding to the received vector being located in decision regions other than its own. Denote the probability that a transmitted $\tilde{\mathbf{s}}_i$ is received incorrectly by P_{si}. Then, by using the union bound, P_{si} can be expressed as

$$P_{si} \le \sum_{\substack{j=1 \\ j \ne i}}^{M} P(e_{ij}) = \sum_{\substack{j=1 \\ j \ne i}}^{M} Q\left(\frac{d_{ij}}{\sqrt{2N_0}}\right) \tag{2.108}$$

where $p(e_{ij})$, defined by (2.107), is the error probability that symbol $\tilde{\mathbf{s}}_i$ has been transmitted but incorrectly received as symbol $\tilde{\mathbf{s}}_j$ [14]. To obtain the average symbol error rate $\overline{P}_s$, (2.108) is averaged over all possible transmitted symbols as shown below:

$$\overline{P}_s \le \frac{1}{M} \sum_{i=1}^{M} \sum_{\substack{j=1 \\ j \ne i}}^{M} Q\left(\frac{d_{ij}}{\sqrt{2N_0}}\right), \tag{2.109}$$

where a uniform density is assumed for the transmitted symbols.

At high E_b/N_0, error performance is largely dependent on the minimum distance $d_{\min}$ of all possible d_{ij}. In this case, a simpler upper bound can be obtained by replacing d_{ij} with $d_{\min}$ as shown below

$$\overline{P}_s \le (M-1) Q\left(\frac{d_{\min}}{\sqrt{2N_0}}\right). \tag{2.110}$$

Example 2.7 Consider QPSK with $M = 4$ and constellation points $\tilde{\mathbf{s}}_1 = \sqrt{E_s} e^{j\pi/4}$, $\tilde{\mathbf{s}}_2 = \sqrt{E_s} e^{j3\pi/4}$, $\tilde{\mathbf{s}}_3 = \sqrt{E_s} e^{-j3\pi/4}$, and $\tilde{\mathbf{s}}_4 = \sqrt{E_s} e^{-j\pi/4}$ as shown in Figure 2.5. Because of the symmetry in the constellation, only one constellation point needs to be considered in computing the union bound, i.e. $\overline{P}_s = P_{si}$. Let $j = 1$. Then, $d_{12} = d_{14} = \sqrt{2E_s} = 2\sqrt{E_b}$ and $d_{13} = 2\sqrt{2E_b}$. By applying (2.109), $\overline{P}_s$ is upper bounded by $2Q\left(\sqrt{2\gamma}\right)+Q\left(2\sqrt{\gamma}\right)$, where γ denotes E_b/N_0. A looser bound of $3Q\left(\sqrt{2\gamma}\right)$ can be obtained with (2.110).

The exact $\overline{P}_s$ can be computed based on the observation that the probability of making a correct decision is to stay within the quadrant corresponding to the transmitted constellation point. A correct decision occurs when the noise components for the I and Q channels are less than $d_{12}/2$ and $d_{14}/2$, respectively. The probability when this occurs is simply $P(n_I < d_{12}/2)P\left(n_Q < d_{14}/2\right)=\left(1-Q\left(\sqrt{2\gamma}\right)\right)^2$. The exact probability of error can then be expressed as

$$\begin{aligned}\overline{P}_s &= 1-\left(1-Q\left(\sqrt{2\gamma}\right)\right)^2 \\ &= 2Q\left(\sqrt{2\gamma}\right)-Q^2\left(\sqrt{2\gamma}\right)\end{aligned} \tag{2.111}$$

Note that the bound based on (2.109) closely approximates the exact value at high γ while the bound based on (2.110) is off by 50%. □

2.3.4 Bit Error Rate

Thus far the error rate has been expressed in terms of the percentage of symbols in error. However, as discussed in Section 2.1.3, the performance of most digital systems is ultimately measured by the BER. Analytically, it is more involved to find BER since BER depends on the manner in which bits are allocated to each symbol. Each symbol is allocated a label with log_2M bits, such that the M possible symbols can each have a unique label. For example, in Figure 2.5, the labels for QPSK correspond to 00, 01, 11, and 10 for symbols $\tilde{\mathbf{s}}_1$, $\tilde{\mathbf{s}}_2$, $\tilde{\mathbf{s}}_3$, and $\tilde{\mathbf{s}}_4$, respectively.

Let $\mathbf{x}_t$ and $\mathbf{x}_r$ be the labels corresponding to the transmitted symbol and the decoded symbol, respectively. A symbol error then leads to k bits in error, where k depends on the number of bit positions that differ between $\mathbf{x}_t$ and $\mathbf{x}_r$. A measure of k is the Hamming distance d_H which is defined as the Exclusive Or (XOR) of $\mathbf{x}_t$ and $\mathbf{x}_r$.

$$d_H(\mathbf{x}_r,\mathbf{x}_t)=\mathbf{x}_r \otimes \mathbf{x}_t \tag{2.112}$$

Intuitively, BER is minimized by assigning bits such that symbols, which are closer in Euclidean distance, have a lower Hamming distance. At high SNR, when a symbol error occurs, the decoded symbol will have a low Hamming distance from the correct symbol, implying a low BER.

BER can be upper bounded in a similar fashion using the Union Bound as in (2.109) except that the pairwise probabilities $p\left(e_{ij}\right)$ are weighted by the Hamming distance between the labels corresponding to $\tilde{\mathbf{s}}_i$ and $\tilde{\mathbf{s}}_j$

$$\overline{P}_b \leq \frac{1}{M}\sum_{i=1}^{M}\sum_{\substack{j=1\\j\neq i}}^{M}\frac{d_H(\mathbf{x}_i,\mathbf{x}_j)}{\log_2 M}Q\left(\frac{d_{ij}}{\sqrt{2N_0}}\right). \tag{2.113}$$

Therefore, the design trade-offs with respect to symbol energy and the relative distance between constellation points discussed earlier for SER also applies to BER with the additional constraint that $d_H(\mathbf{x}_i,\mathbf{x}_j)$ should be taken into account.

Example 2.8 Consider the QPSK modulation in Example 2.7, with labels $\mathbf{x}_1 = 00$, $\mathbf{x}_2 = 01$, $\mathbf{x}_3 = 11$, and $\mathbf{x}_4 = 10$ for symbols $\tilde{\mathbf{s}}_1$, $\tilde{\mathbf{s}}_2$, $\tilde{\mathbf{s}}_3$, and $\tilde{\mathbf{s}}_4$, respectively. Because of the symmetry in the constellation, only one constellation point needs to be considered in computing the union bound. Let $j = 1$. Then, $d_{12} = d_{14} = \sqrt{2E_s} = 2\sqrt{E_b}$ and $d_{13} = 2\sqrt{2E_b}$. Also, $d_H(\mathbf{x}_1,\mathbf{x}_2) = d_H(\mathbf{x}_1,\mathbf{x}_4) = 1$ and $d_H(\mathbf{x}_1,\mathbf{x}_3) = 2$. By applying (2.31) and (2.111), $\overline{P}_b$ is then lower bounded by $Q(\sqrt{2\gamma}) - 0.5Q^2(2\sqrt{\gamma})$. The exact $\overline{P}_b$ can be computed based on the observation that the probability of making an error for $\tilde{\mathbf{s}}_1$ consists of the union of three possible events: 1) the decision is in quadrant II, 2) the decision is in quadrant III, and 3) the decision is in quadrant IV. The corresponding probabilities for the three events are respectively:

1) $P(n_I \geq d_{12}/2)P(n_Q \leq d_{14}/2)$,
2) $P(n_I \geq d_{12}/2)P(n_Q \geq d_{14}/2)$,
3) $P(n_I \leq d_{12}/2)P(n_Q \geq d_{14}/2)$.

Given that $P(X \leq a) = 1 - P(X > a)$, the above three probabilities simplify to $Q(\sqrt{2\gamma}) - Q^2(\sqrt{2\gamma})$, $Q^2(\sqrt{2\gamma})$, and $Q(\sqrt{2\gamma}) - Q^2(\sqrt{2\gamma})$. Now weight each of these probabilities by the corresponding weights given by $0.5d_H(\mathbf{x}_i,\mathbf{x}_j)$ and take the sum of their weighted values. The exact probability can then be expressed as follows

$$\overline{P}_b = Q(\sqrt{2\gamma}) \tag{2.114}$$

which is the same as that of BPSK. Note that bounds based on (2.31) and (2.111) closely approximate the exact value at high γ. Also, the BER depends strongly on how the symbols are labeled. The reader can verify that if the labels for $\mathbf{x}_3$ and $\mathbf{x}_4$ were switched, the BER would have increased by 100% at high SNR. The bit allocation (i.e. labeling of each symbol) shown in Figure 2.5 is based on Gray encoding, which is a common bit allocation used to improve the BER performance of a digital system. □

2.3.5 Effects of Amplitude Fading and Interference

In the presence of frequency-flat fading, the received symbol energy is perturbed by a factor r^2, where r is the fading envelope. Since $d_{\min}$ is a function of the square root of the transmitted symbol energy, it varies with the fading envelope r in the presence of frequency-flat fading. Therefore, the error probability becomes a function of $rd_{\min}$.

The average SER is readily obtained by averaging (2.110) over r using the probability density function $P_R(r)$ as shown below

$$\overline{P}_s \leq (M-1)\int_0^\infty Q\left(\frac{rd_{\min}}{\sqrt{2N_0}}\right)P_R(r)dr \tag{2.115}$$

where r may be either Rayleigh distributed (2.72) or Rician distributed (2.75). For a Rayleigh amplitude distribution, (2.115) becomes

$$\overline{P}_s \leq \frac{(M-1)}{d_{\min}^2/N_0} \tag{2.116}$$

The expression for Rician distribution is much more complex. The average BER in the presence of flat fading can be similarly determined by averaging (2.113) over r using $P_R(r)$, with d_{ij} replaced by rd_{ij}.

Example 2.9 Consider BPSK with $M = 2$. Then the minimum distance is $2\sqrt{E_s}$. Applying (2.116), expression (2.36) is obtained. □

In the presence of interference, (2.57) can be used along with (2.110) to obtain the average SER:

$$\overline{P}_s \leq (M-1)Q\left(d_{\min}\sqrt{\frac{\alpha_I}{2N_0}}\right). \tag{2.117}$$

2.4 Digital Modulations

In Section 2.1, we studied the parameters and metrics that are important to determine the design trade-offs for a digital system. In Section 2.2, we described the different types of degradation that are possible in a wireless environment. In Section 2.3, we introduced the optimal ML receiver and discussed its performance in the presence of AWGN, fading, and interference. In reference to these earlier discussions, this section summarizes the performance of four main types of digital modulation commonly used in many wireless systems.

2.4.1 Phase Shift Keying

Phase-shift keying (PSK) refers to modulations where information modulates the phase of the carrier and not its amplitude or frequency. The following general expression describes the constellation points of a PSK system:

$$s_i = \sqrt{E_s}e^{j\frac{2\pi}{M}i}. \tag{2.118}$$

The symbol energy is the same for all constellation points since only phase is modulated and not the amplitude. The constellation diagram, as shown in Figure 2.31, consists of a circle with constellation points evenly distributed around the circle. The constellation points are evenly distributed to maximize $d_{\min}$ for improved error performance.

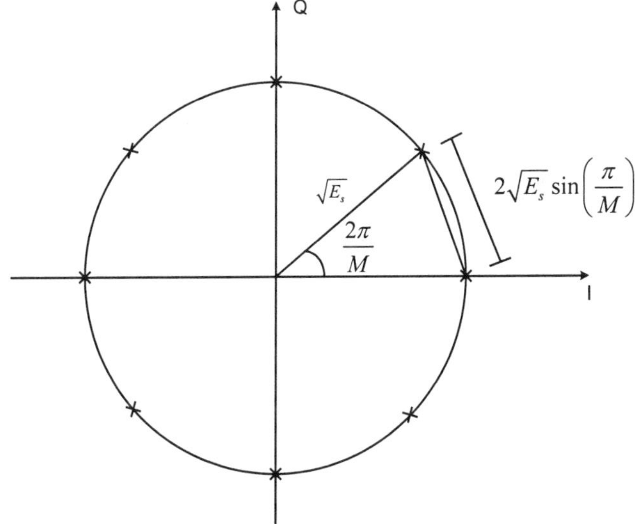

Figure 2.31 Signal Constellation of PSK

Table 2.4 summarizes the performance metrics for a PSK system. PSK is bandwidth efficient. Doubling the constellation size results in an increase of 1 bps/Hz in bandwidth efficiency. However, PSK is not energy efficient as discussed in Section 2.1.6 Figure 2.13. The poor energy efficiency can be inferred from the fact that the constellation points are all equally spaced on a circle of radius $\sqrt{E_s}$. If the energy is kept constant, then as *M* increases, the distance between the points becomes smaller, degrading the BER. To maintain a constant BER as *M* increases requires that the radius of the circle be increased, which is equivalent to increasing the symbol energy.

In particular, consider the effect of doubling the constellation size on the required E_b/N_0. Denote by γ_M and γ_{2M} the required E_b/N_0 for constellation sizes of *M* and *2M* at a fixed BER, respectively. Then, by using the expression for error rate in Table 2.4, it can be shown that for a large M

$$\frac{\gamma_{2M}}{\gamma_M} = \left(\frac{\sin\dfrac{\pi}{M}}{\sin\dfrac{\pi}{2M}}\right)^2 \frac{\log_2 M}{\log_2 M+1} \approx 4\frac{\log_2 M}{\log_2 M+1}.$$

Therefore, for large *M*, the required E_b/N_0 increases by 6 dB as the size of the constellation doubles. Similarly, it can be shown that the energy efficiency decreases by $\left(\frac{\log_2 M+1}{2\log_2 M}\right)^2$ for low BER. For large *M*, the energy efficiency decreases by a factor of four as size of the constellation doubles.

Table 2.4 Performance metrics for phase-shift keying.

Metric	Expression	Derived from
η_W, DSB (bps/Hz)	$\frac{\log_2 M}{1+\beta}$	(2.23)
η_E (%) Fixed BER AWGN	$\le \frac{f_E(2.3)(\sin(\pi/M)\log_2 M)^2}{(1+\beta)\ln\left(\frac{1}{\overline{P}_b \log_2 M}\right)}$	(2.43)
$\gamma_p/\overline{\gamma}$	1.0	(2.37)-(2.39) $p_i = 1/M$
$\overline{P}_s$ AWGN*	$\le 2Q\left(\sqrt{2\overline{\gamma}\log_2 M}\sin\frac{\pi}{M}\right)$	(2.109), [14]
$\overline{P}_s$ Rayleigh	$\approx \frac{1}{2\overline{\gamma}\log_2 M\left(\sin\frac{\pi}{M}\right)^2}$	(2.116)

2.4.1.1 Differentially Encoded Coherent PSK (DECPSK)

The BER performance shown in Table 2.4 does not take into account the noise floor that occurs due to phase ambiguity inherent in phase modulation. The ambiguity arises when the channel introduces a constant phase shift ϕ_c that is greater than 45° or less than -45° for the constellation shown in Figure 2.5. When that occurs, the error rate will be high. For instance, if ϕ_c = 50°, then symbols 00, 01, 11, 10 will be incorrectly identified as 01, 11, 10, 00, resulting in a BER of 50%.

To overcome the phase ambiguity problem, differential encoding is used where the information is encoded in the phase transition rather than the absolute phase value.

* BER can be approximated by $P_s/\log_2 M$, (2.31).

To see this, let $\hat{\phi}_k$ be the running sum of the phases of all previous symbols up to the current k^{th} symbol. $\hat{\phi}_k$ can be represented as

$$\hat{\phi}_k = \phi_k + \hat{\phi}_{k-1} \tag{2.119}$$

where ϕ_k is the absolute phase value of the kth symbol. If the transmitter were to encode the phase according to (2.119), then the absolute phase value can be recovered by taking the difference between the encoded phase values at time k and k-1. Any constant phase introduced by the channel is therefore removed.

However, differential encoding does negatively impact the ideal system performance of Table 2.4, notably the error rate. Since the decision process requires the previous encoded phase, intuitively, the SER is doubled as compared to the ideal case when $\phi_c = 0$. The doubling of SER is due to the fact that an error can result when either the kth or k-1th symbol is incorrectly received.

Example 2.10 Given a QPSK constellation as discussed in Example 2.7, the SER can be approximated by

$$\overline{P}_s \approx 2Q(\sqrt{2\gamma}) \tag{2.120}$$

When compared to the SER of a differentially encoded QPSK shown below, it can be seen that the SER doubles at high E_b/N_0.

$$P_s = 4Q(\sqrt{2\gamma}) - 8Q^2(\sqrt{2\gamma}) + 8Q^3(\sqrt{2\gamma}) - 4Q^4(\sqrt{2\gamma}) \tag{2.121}$$

□

The additional E_b/N_0 required to compensate for the factor of two increase in error rate is typically small. For instance, with a 10-dB E_b/N_0, the loss is only 0.3 dB. Therefore, differential encoding incurs minimal loss in the energy efficiency of the system. The negligible performance loss compared to an ideal system has made differential encoding an essential element in any practical system that uses phase modulation.

2.4.1.2 Differentially Coherent PSK (DCPSK)

In Chapter 8, we will see that differential encoding and decoding can be easily implemented with simple digital logic. This section discusses an alternate approach to differential decoding that combines the decoding along with the decision process. Let the received symbol be $s_i = \sqrt{E_s} e^{j\phi i}$ and the differentially encoded phase be $\hat{\phi}_k = \phi_k + \hat{\phi}_{k-1}$, as shown in (2.119). Then, ϕ_k can be extracted by multiplying the received signal by its complex conjugate delayed by one symbol time:

$$\tilde{z}_k = \tilde{r}_k \tilde{r}^*_{k-1} = E_s e^{j\phi_k}. \tag{2.122}$$

Recall the relationship in (2.5), ϕ_k can then be recovered by computing the phase of $\tilde{z}_k$ via the following expression

$$\phi_k = \tan^{-1}\left(\Im(\tilde{z}_k)/\Re(\tilde{z}_k)\right) \tag{2.123}$$

A receiver that implements (2.123) is differentially coherent since the receiver phase reference is taken to be the previous received symbol. A differentially coherent receiver is much less complex than a fully coherent receiver since a fixed frequency reference is no longer needed. In Chapter 8, the complexity trade-off will be described in more detail.

The reduction in complexity, however, comes with a price in terms of E_b/N_0. Since, the receiver uses the previously received symbol as the phase reference, any received noise is amplified due to the multiplication operation. Essentially, a DCPSK receiver can be viewed as a coherent receiver with a noisy reference. At high signal to noise ratio and $M > 4$, the loss in E_b/N_0 is 3 dB. Intuitively, the 3-dB loss arises from the same noise process being injected at the receiver input and at the phase reference. The noise energy effectively has been doubled. Note, however, for the binary and quaternary cases (DCBPSK and DCQPSK), the loss tends to be smaller. At a 10^{-5} BER, the loss is about 0.9 dB for DCBPSK and 2.3 dB for DCQPSK [45].

DCPSK has the same bandwidth efficiency as a DECPSK. Its energy efficiency degrades by 100% for $M > 4$. For DCBPSK and DCQPSK, the loss in energy efficiency tends to be lower – 23% and 70%, respectively. Therefore, for portable wireless systems, only binary and quaternary differentially coherent receivers have been used.

2.4.1.3 Offset QPSK (OQPSK)

Bandwidth efficiency is constrained by ISI as well as the out-of-band emission requirements in the radio transmission system, as discussed in Section 2.1.2. The bandwidth efficiency η_W in Table 2.4 is achieved for PSK by designing a transmit filter which meets the ISI constraint and also provides sufficient filtering to meet the out-of-band emission requirement. However, the filtered digital signal must be converted to analog levels, up-converted to the carrier frequency, and amplified for transmission. Nonlinear elements in the RF transmitter, in particular the power amplifier, tend to restore and even enhance the filtered sidebands, a phenomenon also known as spectral regrowth.

Spectral regrowth arises from abrupt phase changes in the signal constellation which result in variations in the signal envelope. When the signal passes through a non-linear device, the variation in the signal envelope enhances the distortion induced by the non-linear element and thereby broadens the spectral content, giving rise to increased spectral power in the sidebands. To minimize the spectral regrowth, the nonlinear device must be driven at an operation region which is fairly linear. For a power amplifier, this implies that the input signal power must be backed off substantially from saturation. It will be seen in Chapter 7 that the efficiency of the amplifier is proportional to the level of input drive. Thus, backing off the amplifier to minimize spectral growth results in increased power dissipation in the radio for a given transmit power.

PSK described thus far is prone to spectral regrowth and must be typically backed off by 6-9 dB. The problem arises from the large amplitude variation due to abrupt phase changes, especially, those that are 180 degrees. For example, if all the possible phase transitions are plotted in the QPSK constellation diagram shown in Figure 2.32a, then it is clear that when a symbol makes a direct transition between quadrants I and III or II and IV, its amplitude goes from a maximum of $\sqrt{E_s}$ to a minimum of zero. Let the maximum amplitude change be defined as the ratio between the maximum and minimum amplitude. Then QPSK has an infinitely large amplitude change. To reduce this amplitude change, offset QPSK (OQPSK) [46] is used with a signal constellation shown in Figure 2.32b. The two diagonal phase transitions are then eliminated. Therefore, the maximum change is reduced to $\sqrt{2}$, a dramatic improvement over QPSK.

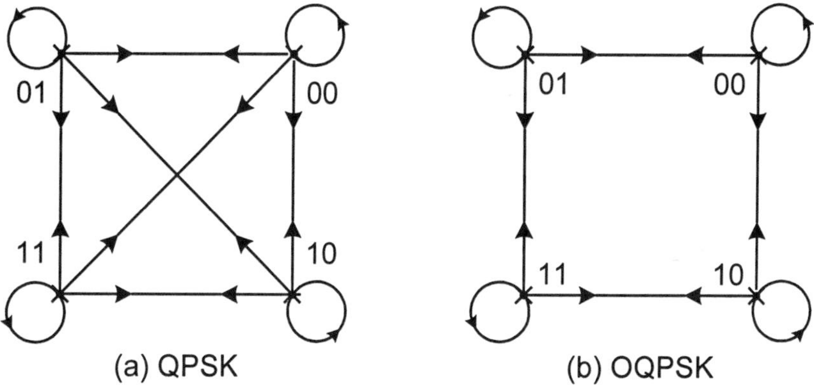

Figure 2.32 Phase transition of QPSK and OQPSK.

OQPSK can be generated by delaying the Q-channel by half a symbol period. To see how this offset eliminates the diagonal transitions, consider the following symbol sequence {00, 11}. With QPSK, both the I and Q channel values immediately switch signs, corresponding to a transition from quadrant I to III. With OQPSK, only the I channel immediately changes sign since Q-channel is offset by half a symbol time. Therefore, for half a symbol time prior to the transition from 00 to 11, symbol 00 transits to 10 located in quadrant IV. In the remaining half of the symbol time, the Q-channel switches sign, making the transition to 11. In general, OQPSK eliminates the diagonal phase transitions by making transitions through intermediate symbols, 01 or 10.

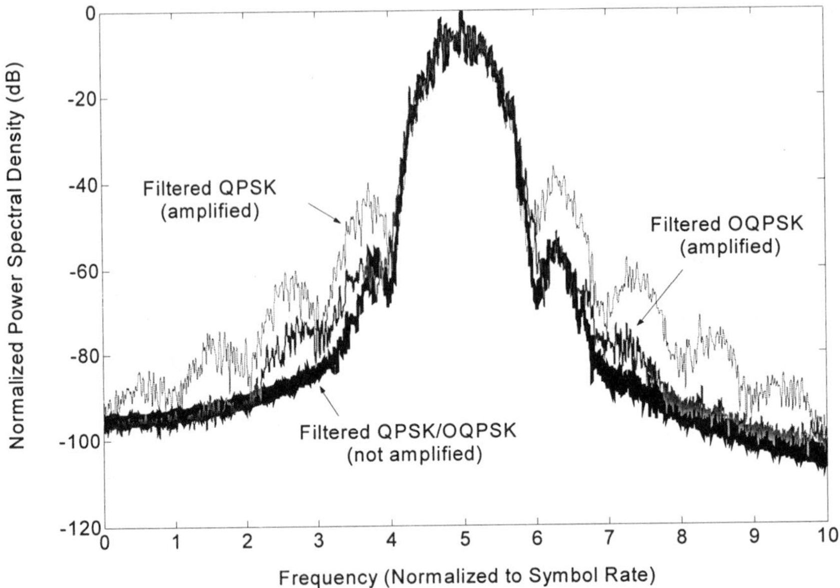

Figure 2.33 PSD of QPSK and OQPSK with linear power amplification.

Figure 2.33 shows the PSD of filtered QPSK and OQPSK when transmitted through a linear power amplifier driven near its compression point, i.e. close to saturation. See Chapter 7 for the definition of compression points. With amplification, the benefit of OQPSK becomes apparent when the amplifier is driven near saturation and if filtering is applied to the modulated signals. For comparison purposes, the PSD curves in Figure 2.33 are normalized to have the same DC power. Note that the PSD of OQPSK is equivalent to that of QPSK with no amplification. Also, the BER performance is the same as that of QPSK since $d_{\min}$ has not changed.

2.4.1.4 π/4 QPSK

OQPSK requires a coherent detector. If a differentially coherent detector is used, then performance degradation results from intersymbol interference from the offset Q-channel. Instead π/4-QPSK [47] should be employed when differentially coherent detection is used. π/4-QPSK has the constellation and phase transitions shown in Figure 2.34. Though there are eight points in the constellation, only four are used for any symbol at a given time. The symbols are interleaved into an even and odd stream. The even stream uses the constellation points $\{\pi/4, \;\; 3\pi/4, \;\; -3\pi/4, \;\; -\pi/4\}$ while the odd stream uses constellation points that are offset by $\pi/4$: namely, $\{\pi/2, \;\; \pi, \;\; 3\pi/2, \;\; 0\}$. The minimum amplitude occurs when the phase changes by $\pm 3\pi/4$. The maximum amplitude change is $1/\sin(0.125\pi)$, which is 85% greater than that of OQPSK. As expected, π/4-QPSK shows a degraded spectral efficiency as compared to OQPSK when the amplifier is driven near saturation as shown in Figure 2.35.

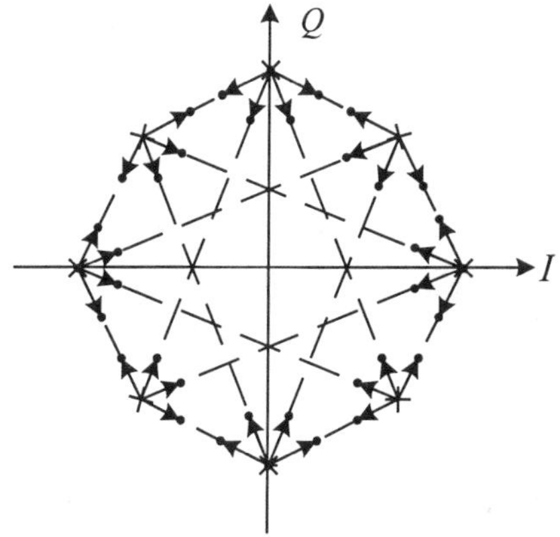

Figure 2.34 Constellation and phase transition of π/4-QPSK.

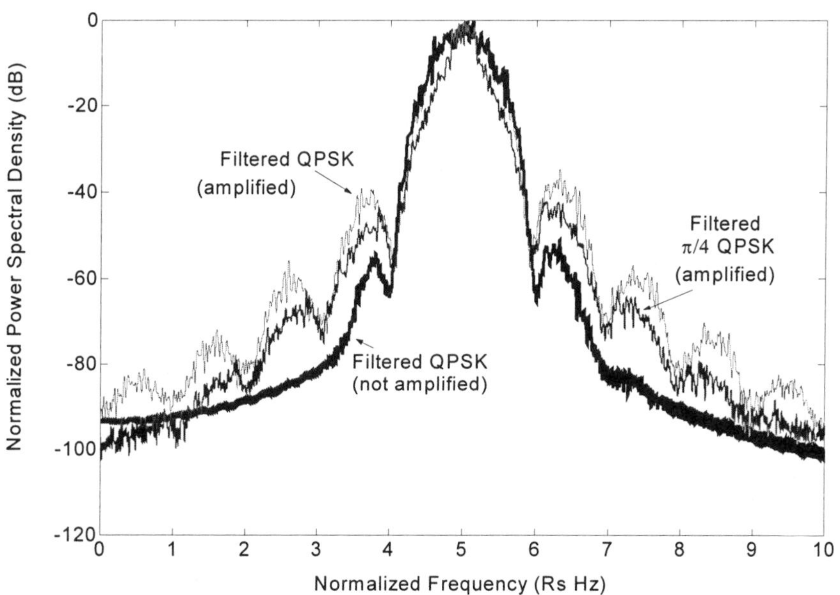

Figure 2.35 PSD of π/4-QPSK with linear power amplification.

2.4.2 Continuous Phase Modulation

Continuous phase modulation (CPM) [48][49] has the advantage that there are no abrupt phase transitions. Even when compared with OQPSK and $\pi/4$ QPSK, CPM shows much smoother phase transitions. In fact, the phase transitions form a circle.

The circular phase trajectory is obtained by modulating the phase in a continuous fashion as shown below

$$\phi(t) = 2\pi h \int_0^t \sum_k \xi_k g(\tau - kT_s) d\tau + \Phi_0 \qquad (2.124)$$

where ξ_k is the symbol at time kT_s, h is the modulation index, Φ_0 is a phase constant, and $g(t)$ is a frequency pulse shape with duration LT_s. The frequency pulse $g(t)$ can have discontinuities and still maintain a smooth phase trajectory because of the integration. Expression (2.124) resembles that of analog FM modulation with the exception that the modulated waveform is constrained to the pulse shape $g(t)$.

It is instructive to expand (2.124) in terms of a phase pulse shape $q(t) = \int g(\tau) d\tau$ to reveal the memory of the modulation, i.e. its dependence on previous symbols, as shown below:

$$\phi(t;\xi_k,\sigma_k) = 2\pi h \xi_k q(t - kT_s) + 2\pi h \sum_{i=k-L+1}^{k-1} \xi_i q(t - iT_s) + \pi h \sum_{i=0}^{k-L} \xi_i + \Phi_0 \qquad (2.125)$$

where ξ_k is the current k^{th} symbol, $q(t)$ is constrained by

$$q(t) = \begin{cases} 0, & t \le 0 \\ \int_0^t g(\tau) d\tau, & 0 \le t \le LT_s \\ 0.5, & t > LT_s \end{cases} \qquad (2.126)$$

and σ_k is the modulator state at time kT_s.

The modulator state σ_k denotes a set of phase values the CPM signal traversed up to time kT_s. σ_k consists of the correlative states represented by the second term of (2.125) and the phase states represented by the third term in (2.125). The correlative states are functions of the *L –1* previous symbols while the phase state φ_k is a function of all transmitted symbols up to the Lth previous symbol. For M-ary modulation, the number of correlative states is M^{L-1}. The number of phase states *p* depends on the modulation index *h*. If *h* is an irrational number then *p* is infinite, which makes it impractical to demodulate. When *h* is a rational number (*m/n*) where *m* and *n* are integers then $p = n$ when m is even and $p = 2n$ when *m* is odd. The total number of modulator states N_σ is then pM^{L-1} and each state can be labeled by the *L*-tuple $(\varphi_k, \xi_{k-1}, \xi_{k-2}, \cdots, \xi_{k-L+1})$.

Because of the memory in the modulation, correlation is introduced in the transmitted symbols. The correlation broadens the 3dB bandwidth of the signal but lowers the side lobes. Therefore, in general, CPM shows worse bandwidth efficiency than PSK.

However, when considering the overall power requirement on the system, CPM has an advantage in meeting a given out-of-band emission requirement with a much more efficient power amplifier in the front-end. This is because the continuous phase transition has eliminated amplitude variation in the transmitted signal, making CPM extremely robust to non-linear power amplifications. Therefore, in contrast to PSK or QAM, efficient non-linear amplifiers can be used for CPM.

Another advantage is that CPM can require a lower E_b/N_0 for a given BER. The lower E_b/N_0 improves the energy efficiency of the system as discussed in Section 2.1.6. The improvement in E_b/N_0 results from the memory of the modulation, which effectively increases the minimum distance of the transmitted signal. Intuitively, the increased distance property can be explained by the correlation introduced among the transmitted symbols, allowing the receiver to make a decision based on a sequence of symbols.

However, the improved performance comes at the price of increased complexity. To see the increase in complexity, let us first apply the ML detector to a sequence of length N. Then, the following expression can be obtained:

$$\hat{\xi}_N = \max_{\xi_N}{}^{-1}\{\Re\int_0^{(N+1)T_s} r(t)\tilde{s}^*(t;\xi_N,\sigma_N)dt\} \tag{2.127}$$

where $\xi_N = \xi_0, \xi_1, \cdots, \xi_N$ denotes all possible symbol sequences of length N and the $\max^{-1}\{\ \}$ function returns the optimally detected sequence $\hat{\xi}_N$ that maximizes its argument. The complexity of the ML sequence detector increases exponentially with the length of the sequence. Since each symbol can have M possible values, a direct implementation of the ML detector has a complexity of M^N. For a large M and N this can become computationally expensive. To reduce the complexity and still achieve the optimal performance, the Viterbi algorithm [50] is typically used to compute (2.127).

2.4.2.1 Viterbi Algorithm

The integral in (2.127) can be partitioned into a sum of integrals over the individual symbol duration as shown below:

$$\hat{\xi}_{N-D} = \max_{\xi_N}{}^{-1}\left\{\sum_{i=0}^{N}\lambda_i(\xi_i,\sigma_i)\right\} \tag{2.128}$$

where $\lambda_i(\xi_i,\sigma_i)$ is the correlation from time iT_s to $(i+1)T_s$ and is also known as the branch metric for the Viterbi Algorithm. N is the total number of symbols processed and D indicates that the algorithm can have a decoding delay DT_s. The branch metric can be expressed as follows:

$$\lambda_i(\xi_i,\sigma_i) = \Re\int_{(i-1)T_s}^{iT_s} r(t)\tilde{s}^*(t;\xi_i,\sigma_i)dt. \tag{2.129}$$

Referring to (2.128), it is clear that the next modulator state σ_{k+1} is a function of only the current symbol ξ_k and the current state $\sigma_{k.}$ Therefore, the branch metrics can be expressed in terms of the state transition from σ_k to σ_{k+1}. Expression (2.128) can then be expressed as

$$\hat{\xi}_{N-D} = \max_{\sigma_k,\sigma_{k+1}}{}^{-1}\left\{\sum_{i=0}^{N}\lambda_i(\sigma_k,\sigma_{k+1})\right\}. \tag{2.130}$$

Because the current transition (σ_k,σ_{k+1}) depends only on the previous transition (σ_{k-1},σ_k), the maximization in (2.130) can be computed iteratively as follows

$$\begin{aligned}\mu_0 &= 0\\ \mu_l(\sigma_l) &= \max_{\xi_l,\sigma_{l-1}}\{\mu_{l-1}(\sigma_{l-1})+\lambda_l(\sigma_{l-1},\sigma_l)\}\\ \hat{\xi}_N &= \max_{\sigma_N}{}^{-1}\{\mu_N(\sigma_N)\}\end{aligned} \tag{2.131}$$

where $\mu_l(\sigma_l)$ is the path metric for a given state at time lT_s and is obtained by taking the maximum of the sum of the previous path metric at $(l-1)T_s$ and the transition metric from state σ_{l-1} to σ_l. The maximum is taken over all possible transitions from σ_{l-1} to σ_l.

The iterative solution shown in (2.131) is known as the Viterbi algorithm which has a complexity equal to the number of states. For CPM, the complexity is ρM^{L-1}. For each state, there is the addition operation with a compare to select the maximum path metric to store for the next iteration. The operation is also known as add-compare select (ACS). Two storage elements are needed, one to store the path metric for the next iteration and one to store the path that leads to the maximum path metric for each state. This path is also known as the survivor path $\Gamma_l(\sigma_l)$ and contains information on the optimum sequence.

If the survivor path is plotted against time, it can be seen that the paths converge to a single path L_d iterations in the past as shown in Figure 2.36. The vertices in Figure 2.36 represent the different possible states, σ_l, while the branches connecting the vertices represent the transition metrics. If all possible transitions are plotted rather than just the surviving transitions, then the resulting graph is known as the trellis diagram. The converged path from time 0 to $(N-L_d)T_s$ contains the optimum sequence, $\tilde{\xi}_{n-D}$. Based on the above observation, decoding of the optimum sequence requires a delay equal to at least L_dT_s. In general, L_d is not deterministic. However, a good rule of thumb is to decode after N_d iterations, where N_d is the decoding depth and should be 5-10 times the correlation length L.

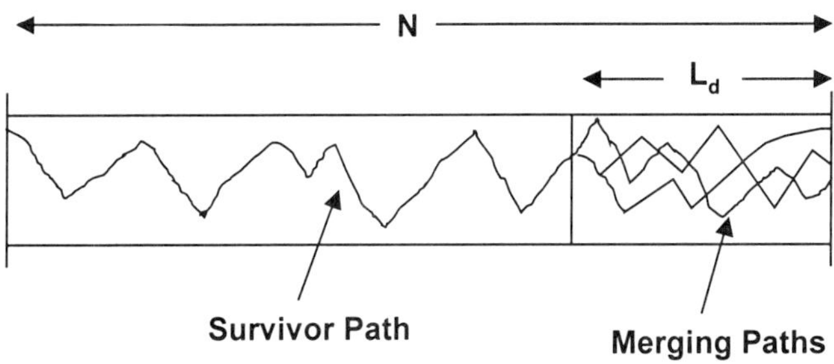

Figure 2.36 Survivor path as a function of time.

Three key methods exist to decode the optimum sequence. The first method, known as sliding window decoding, decodes the symbol N_d iterations in the past, i.e. symbol ξ_{n-N_d} at the nth iteration. In this method, the decoding delay is $D = N_d$ and N_d symbols are required to decode. The second method, known as block decoding, decodes a block of B symbols all at once. In general, B is greater than N_d; therefore, paths may have converged only for part of the block. To maximize the chance of selecting the optimum path, the transmitter can transmit a sequence of symbols to force the decoder into a known state. Block decoding then begins from this known state. This approach incurs an overhead of L-1 symbols, needed to force the decoder into a known state. The decoding delay is $D = B$ and $B+L-1$ symbols are needed to decode a block of B symbols. The third method is a hybrid of the first two methods, where the first $B-N_d$ symbols are decoded using a sliding window and the last N_d symbols are decoded using the block method. In this case, the decoding delay is $D = N_d$ and $B+L-1$ symbols are needed to decode a block of B symbols if L–1 symbols are used to force the Viterbi decoder into a known state.

2.4.2.2 Minimum Distance

In Section 2.3.2, the minimum distance was defined as the minimum distance between the symbol constellation points. This definition can be extended to sequences, where the distance is now measured between two sequences that diverge from a common state S_d and then merge again some time later into another state, as shown in Figure 2.37. The minimum distance for a sequence is the minimum of the distance between paths of equal length that diverge from a common state and then remerge at a later time at another state. Since the number of possible paths that must be compared grows exponentially with the length of the sequence, it is computationally expensive to determine $d_{\min}$ exhaustively. Instead, it is more convenient to measure the minimum distance for a finite sequence of length N, denoted by d_N. This distance is always greater than $d_{\min}$ and provides a rough measure on the error performance of the system. By applying the union bound (2.109) and substituting d_N for d_{ij}, the SER for CPM is obtained:

$$\overline{P}_s \le (M-1)M^{N-1}Q\left(\frac{d_N}{\sqrt{2N_0}}\right). \tag{2.132}$$

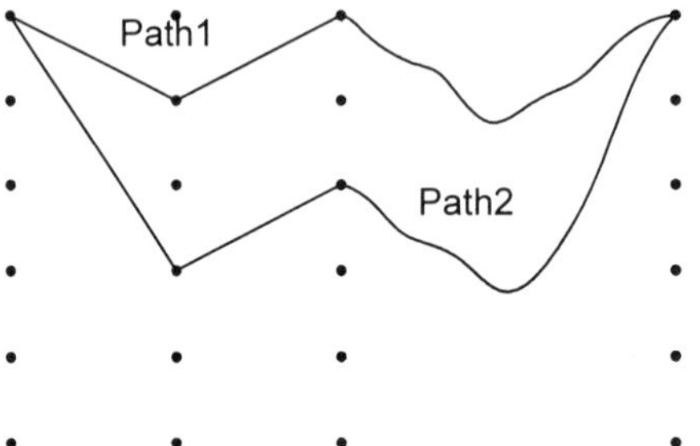

Figure 2.37 Distance measure in a trellis.

MSK Example: Consider $g(t)$ as a raised cosine pulse over a symbol duration, i.e. L = 1 as shown below:

$$g(t) = \begin{cases} \cos(\pi t / T_s), & 0 \le t \le T_s \\ 0, & \text{otherwise} \end{cases} \tag{2.133}$$

Assume a modulation index of 0.5 and binary transmitted symbols. This modulation is also known as minimum shift keying (MSK) [51]. In this case, the number of phase states is four and there are no correlative states. The total number of modulator states is four. The constellation diagram has four constellation points, each corresponding to one of the phase states. The phase transitions form a circle.

The spectral efficiency of MSK can be derived from (2.26) and is plotted in Figure 2.38 along with that of OQPSK. It can be seen that MSK has a broader main lobe but lower side lobes than that of OQPSK. When using the null-to-null bandwidth, the spectral efficiency of MSK and OQPSK are respectively 0.67 bps/Hz and 1 bps/Hz[1]. By applying (2.25) for 99.5% power containment, the bandwidth efficiency is 0.71 bps/Hz for MSK versus 0.15 bps/Hz for unfiltered OQPSK. Therefore, if the containment bandwidth metric is used, MSK has higher bandwidth efficiency than OQPSK. Filtering is often applied to the modulated signals to reduce side lobes that degrade bandwidth efficiency. However, when the filtered signal experiences non-linear amplification, spectral regrowth occurs and negates the attenuation introduced by the filtering. In general, MSK introduces less envelope fluctuations in the filtered waveform and thus reduces the amount of spectral regrowth as compared to OQPSK as shown in Figure 2.46.

---

[1] The 1 bps/Hz of OQPSK assumes 100% excess bandwidth. With 0% excess bandwidth, the theoretically achievable bandwidth efficiency is actually 2 bps/Hz.

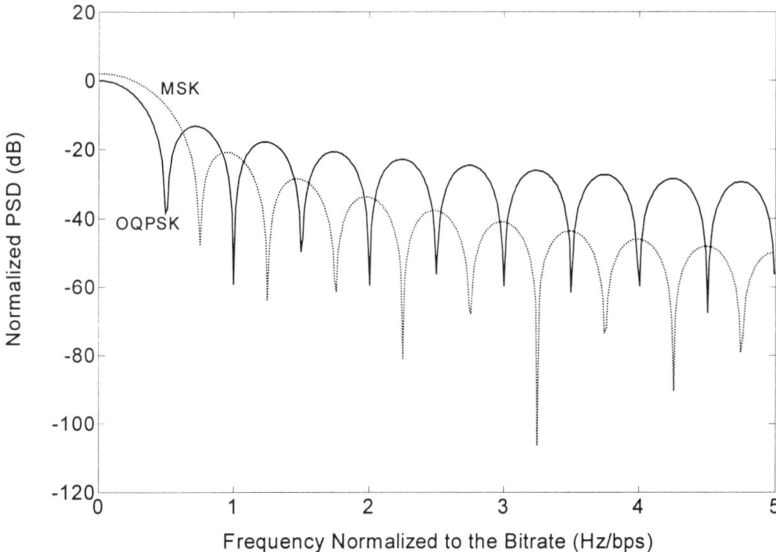

Figure 2.38 Power spectral density of MSK and OQPSK.

The trellis diagram of MSK consists of four modulator states, corresponding to the four phase states $\{0.25\pi, 0.75\pi, 1.25\pi, 1.75\pi\}$. The transition from one phase state to the next can be determined recursively using the third term in (2.125). Let φ_n be the phase state at time index *n*, then the next phase state $\varphi_{n+1} = \varphi_n + \pi h \xi_n$ where $\xi_n = \pm 1$. There are two transitions from each state to the next state, where each transition corresponds to a current symbol value of *+1* or *−1*. The phase functions, also known as phase branches, $\tilde{s}(t-(i-1)T_s;\xi_i,\sigma_i)$ along each transition in the time interval $[(i-1)T_s, T_s]$ are shown in Figure 2.39a and are used to obtain the transition metrics, according to (2.129). The resulting trellis diagram is shown in Figure 2.39b.

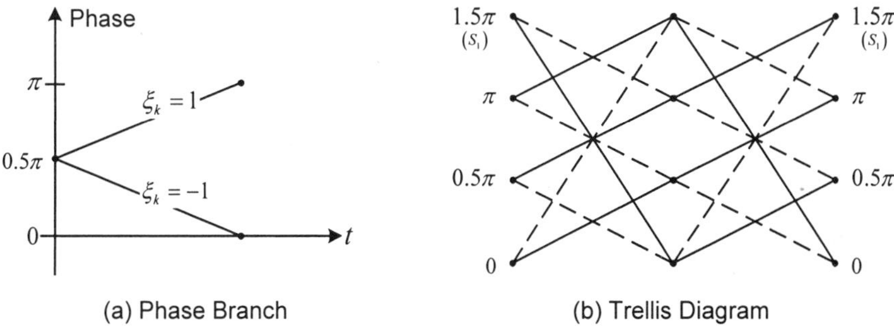

Figure 2.39 Phase branch and trellis diagram for MSK with h = 0.5. Dashed transitions = -1 and solid transitions = 1.

The minimum distance is upper bounded by d_N, as defined in Section 2.4.2.2. By inspecting the trellis diagram, d_N can be found for a given sequence length N, as the minimum distance between paths which diverge from a state and then remerge into another state at a later time. For example, let $N = 2$. Then, starting from state S_l, the two paths which remerge to state S_l after two symbols are {-1, 1}, and {1, -1}, where the numbers in brackets are the symbol values at $k = 0$ and $k = 1$. d_2 can be expressed as the distance between these two paths:

$$d_2^2 = \frac{2E_s}{T}\left|e^{j2\pi h[-q(t)+q(t-T_s)]} - e^{j2\pi h[q(t)-q(t-T_s)]}\right|^2 \tag{2.134}$$

The square magnitude term $|x|^2$ is simply the inner product of x with itself. Following the definition of inner product (2.6), (2.134) simplifies to

$$\begin{aligned} d_2^2 &= \frac{4E_s}{T_S}\left[T_S - \int_0^{2T_s} \cos\left\{4\pi h\left[q(t) - q(t-T_s)\right]\right\} dt\right] \\ &= 4E_s\left(1 - \frac{\sin 2\pi h}{2\pi h}\right) \end{aligned} \tag{2.135}$$

Given that $h = 0.5$, $d_2 = 2\sqrt{E_s}$, which is the same as that of BPSK. Using (2.132), the SER of MSK behaves according to $2Q(\sqrt{2E_s / N_0})$. MSK therefore has a similar error performance as that of BPSK and QPSK.

GMSK Example: Consider a Gaussian pulse shape for $g(t)$ as shown below:

$$g(t) = \begin{cases} \frac{1}{2T_b}[Q(2\pi BT_b \frac{t/T_b - 0.5 - 0.5L}{\sqrt{\ln 2}}) - Q(2\pi BT \frac{t/T_b + 0.5 - 0.5L}{\sqrt{\ln 2}})], & 0 \le t \le LT_b \\ 0, & \text{otherwise} \end{cases} \tag{2.136}$$

The parameter B is the 3-dB bandwidth of the Gaussian pulse shape. This modulation is also known as Gaussian minimum shift keying (GMSK) [52].

Assume a modulation index of 0.5, binary transmitted symbols (i.e. $M = 2$), and $L = 3$. In this case, the number for both the phase states and correlative states is four. Strictly speaking, the total number of modulator states should be sixteen. However, since the phase pulse $q(t)$ levels off when $L = 2$, the modulation can be approximated by 8 states with four phase states and two correlative states. The constellation diagram still has four constellation points, each corresponding to one of the phase states and the phase transition still forms a circle. The main difference between MSK and GMSK is that the phase pulse extends two symbols for GMSK whereas for MSK the phase pulse extends only one symbol. Therefore, GMSK inherently introduces ISI and is a special case of partial response CPM where $L > 1$.

The spectral efficiency of GMSK signal is shown in Figure 2.40 along with that of OQPSK and MSK. It can be seen that for $BT = 0.3$ GMSK has a narrower main lobe than MSK and also has lower side lobes than that of MSK. The null-to-null bandwidth of GMSK is not easily defined due to a lack of distinct nulls in the PSD. However, a bounded bandwidth could be defined above which the PSD is constrained to be less than a certain power density. In this case, the bound could be defined as the –38-dB point at which the bandwidth corresponds to the null-to-null bandwidth of MSK. Due to the lower sidelobe levels, the 99.5% containment bandwidth of GMSK is 1.1 bps/Hz compared to 0.71 bps/Hz for MSK. When subject to non-linear amplification, GMSK also outperforms both MSK and OQPSK in terms of lower spectral regrowth as shown in Figure 2.46.

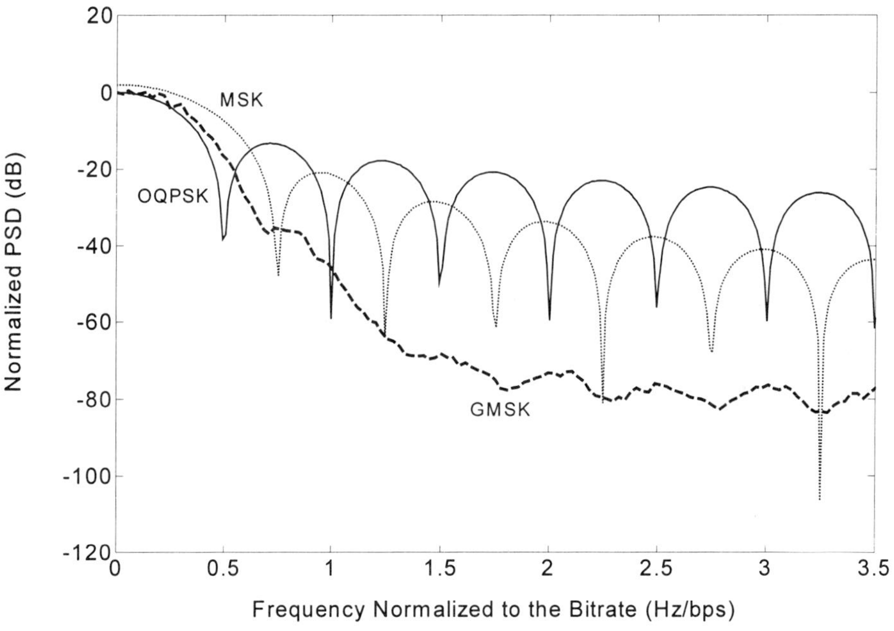

Figure 2.40 Power spectral density of GMSK ($BT = 0.3$), MSK, and OQPSK.

The trellis diagram of MSK consists of eight modulator states, corresponding to the four phase states $\{0.25\pi, 0.75\pi, 1.25\pi, 1.75\pi\}$ and two correlative states {-1,1} determined by the previous symbol. There are two transitions from each state to the next, where each transition corresponds to a current symbol value of *+1* or *–1*. The transition phase branches are shown in Figure 2.41a. There are four such phase branches depending on the current and previous symbols, for $L = 2$. The resulting trellis diagram is shown in Figure 2.41b.

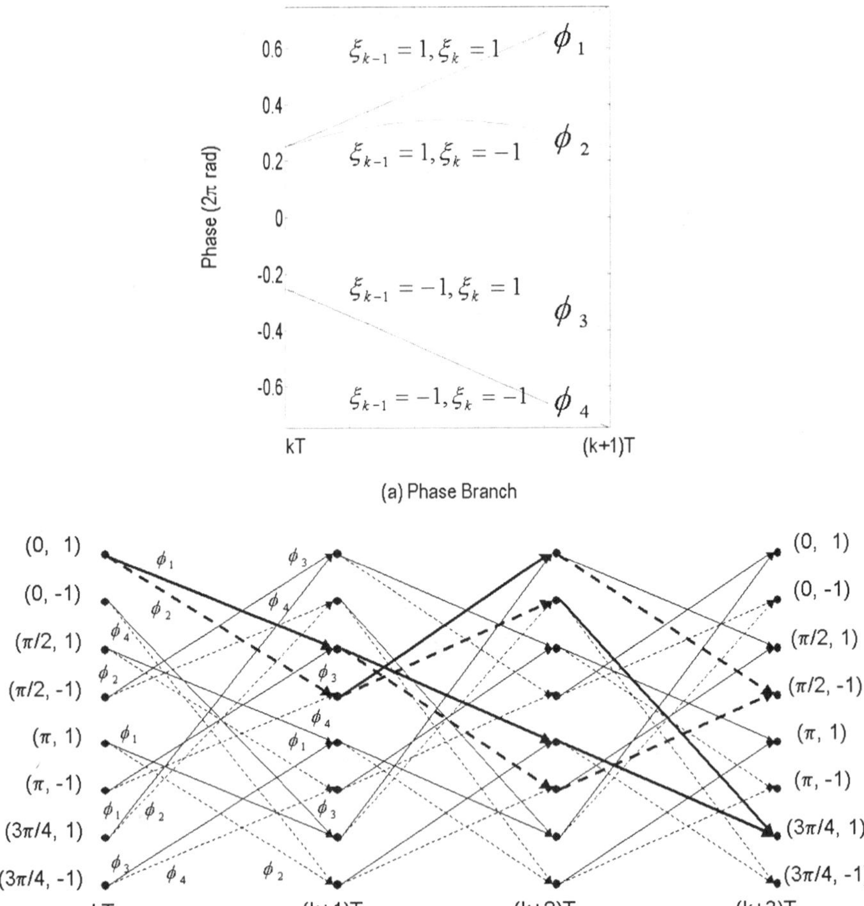

Figure 2.41 Phase branch and trellis of GMSK with h = 0.5 and BT = 0.3. Dashed transitions = -1 and solid transitions = +1.

By inspecting the trellis diagram for a fixed length $N = 3$, two paths diverge from the $(0,1)$ state and then later remerge into two other states $(0.5\pi,-1)$ and $(0.75\pi,1)$ in three symbol times. These paths are highlighted in Figure 2.41a. The minimum distance between the highlighted paths can be determined similar to the procedure illustrated for MSK and has the form shown below:

$$d_3^2 = \frac{2E_s}{T_S}\left[3T_S - 2\int_0^{3T_s} \cos\left\{4\pi h\left[q(t) - q(t - T_s)\right]\right\} dt\right] \tag{2.137}$$

An exact solution requires numerical integration. Instead, $q(t)$ is approximated by a ramp that extends over two symbol periods. In this case, d_3^2 can be approximated by [14]

$$d_3^2 = 2E_s\left[2\left(1-\frac{\sin\pi h}{\pi h}\right)+1-\cos\pi h\right]. \tag{2.138}$$

Applying (2.132), the error performance behaves according to $4Q(\sqrt{1.73E_s / N_0})$, which shows a loss of 0.62 dB in E_b/N_0 as compared to MSK. However, GMSK shows better spectral efficiency. In particular, with an appropriate choice of *h*, *M*, *N*, and $g(t)$, more spectrally and energy efficient CPM may be obtained. Usually, the higher efficiency comes with higher cost in complexity.

2.4.3 Quadrature Amplitude Modulation

In PSK or CPM, information modulates the phase of the carrier and not its amplitude. Quadrature amplitude modulation, on the other hand, encodes the information in both phase and amplitude. By exploiting both amplitude and phase, QAM achieves much higher energy efficiency than PSK while maintaining the same bandwidth efficiency. The following general expression describes the constellation points of a QAM system:

$$s_i = \sqrt{E_{si}}\,e^{j\phi_i}. \tag{2.139}$$

The symbol energy is now different for individual constellation points since the amplitude of the baseband signal varies with different symbols.

The signal constellation for QAM can have a variety of geometric forms. A commonly used form is the rectilinear constellation shown in Figure 2.42. Other geometries exist and are all designed to maximize $d_{\min}$ for improved error performance while keeping the average symbol energy as low as possible. For the purpose of illustration, the following discussions focus only on symmetric rectilinear QAM constellations with M_I and M_Q constellation points along the I and Q signal axes, respectively. To maintain symmetry about the origin of the I-Q axes, the total constellation size $M_I M_Q$ should be constrained to $4k$, where *k* is an integer.

Table 2.5 summarizes the performance metrics for QAM. In terms of bandwidth efficiency, QAM has the same efficiency as PSK. However, unlike PSK, it is more energy efficient as illustrated in Figure 2.13. In particular, it can be shown that for large *M* the required E_b/N_0 for a square constellation (i.e. $M_I = M_Q$) increases by 3 dB as *M* doubles. Likewise, it can be shown that the energy efficiency decreases by a factor of two as *M* doubles. Therefore, the energy efficiency of QAM degrades much more slowly as *M* increases as compared to PSK. Note that the above observation holds only for large *M*. For instance, the energy efficiency remains the same for binary and 4-QAM.

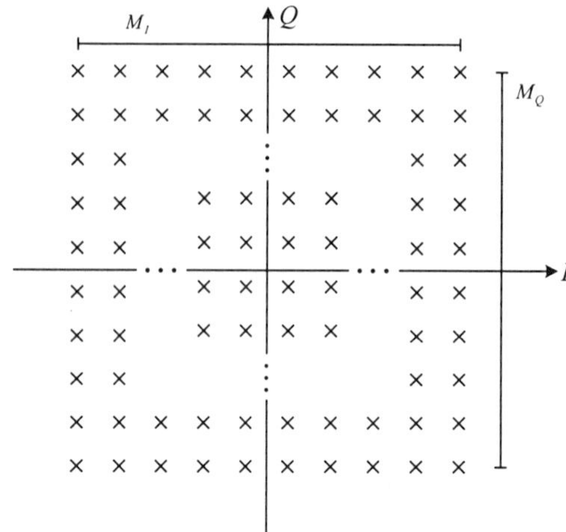

Figure 2.42 Signal Constellation of rectilinear QAM.

Table 2.5 Performance metrics for quadrature-amplitude modulation.

Metric	Expression	Derived From
η_W (bps/Hz) DSB	$\frac{\log_2(M_I M_Q)}{1+\beta}$	(2.23)
η_E (%) Fixed BER, AWGN	$\le \frac{3 f_E(2.3)\left(\log_2(M_I M_Q)\right)^2}{(1+\beta)\ln\left(\frac{2}{\overline{P}_b \log_2\left(M_I M_Q\right)}\right)\left(M_I^2+M_Q^2-2\right)}$	(2.43)
$\gamma_p / \overline{\gamma}$	$3\frac{(M_I-1)^2+(M_Q-1)^2}{(M_I{}^2+M_Q{}^2-2)}$	(2.37)-(2.39) $p_i=\left(M_I M_Q\right)^{-1}$
$\overline{P}_s$ & $\overline{P}_b$ AWGN, Gray code	$\le\left(4-\frac{M_I+M_Q}{M_I M_Q}\right) Q\left(\sqrt{\frac{6\gamma \log_2\left(M_I M_Q\right)}{(M_I^2+M_Q^2-2)}}\right)$	(2.31) (2.109)
$\overline{P}_s$ & $\overline{P}_b$ Rayleigh	$\approx \frac{M_I^2+M_Q^2-2}{3\gamma \log_2\left(M_I M_Q\right)}$	(2.116)

2.4.4 Frequency Shift Keying

PSK and QAM modulate the phase and/or the amplitude of the carrier. FSK, on the other hand, modulates the carrier frequency. The following general expression describes the constellation points of a FSK system:

$$s_i = \sqrt{E_s} e^{j2\pi f_i t}. \tag{2.140}$$

By exploiting frequency, FSK can maintain orthogonality among the transmitted symbols. The minimum requirement for orthogonality can be derived by applying (2.6) and (2.7) with $x(t) = \sqrt{E_s} e^{j2\pi f_i t}$ and $y(t) = \sqrt{E_s} e^{j2\pi f_j t}$. It can be shown that orthogonality is achieved as long as the frequency separation f_d between each tone is at least $R_s / 2$ for coherent detection and R_s for non-coherent detection. Therefore, the set of frequencies used for coherent and non-coherent detections are $f_i = \pm(1 + 0.5i)R_s$ and $f_i = \pm(i+1)R_s$, respectively. Coherent detection refers to a detection process that tracks the phase whereas non-coherent detection does not track the phase. The signal constellation for $M = 3$ is shown in Figure 2.6. Orthogonality constrains the angle between any two signal vectors to 90 degrees.

Table 2.6 summarizes the performance metrics for FSK. The main advantage of FSK is that the energy required to maintain a certain error rate is much less than that of QAM, PSK, and CPM, for M > 2. The energy advantage of FSK arises from the orthogonality of the signal constellation. Since the constellation is orthogonal, d_{ij} is no longer a function of the constellation size. In fact, by letting $\theta_{ij} = 90^o$, the distance between any two symbols becomes $\sqrt{2E_s}$, which is also the minimum distance. By applying the union bound in (2.110), the resulting SER is shown in Table 2.6. Comparing this expression with those of QAM and PSK, it can be seen that the symbol energy required for FSK is roughly M times lower as compared to QAM and M^2 times lower as compared to PSK. However, the energy saving comes with a loss in bandwidth efficiency since each symbol now occupies a slice of the spectrum. For coherent detection, the bandwidth slice is $R_s / 2$ and R_s for non-coherent detection. QAM and phase modulations, on the other hand, transmit within a single frequency band with a bandwidth of $R_s(1+\beta)$. Therefore, the loss in bandwidth is proportional to M, as shown in Table 2.6. Because of the loss in bandwidth efficiency, the energy efficiency as defined in (2.43) behaves similarly as that of QAM, even though the energy required is M times less. Like QAM, η_e decreases by a factor of two as M doubles for large M.

Table 2.6 Performance metrics for frequency shift keying.

Metric	Expression	Derived from
η_W, DSB (bps/Hz)	$\eta_W = \dfrac{2\log_2 M}{M}$	(2.23)
η_E, Fixed BER (%)	$\geq \dfrac{f_E(2.3)\left(\log_2 M\right)^2}{M \ln\left(\dfrac{M-1}{2P_b \log_2 M}\right)}$	(2.43)
$\gamma_p / \bar{\gamma}$	1.0	(2.37)-(2.39) $p_i = 1/M$
$\bar{P}_s$, AWGN*	$\leq (M-1)Q\left(\sqrt{\gamma \log_2 M}\right)$	(2.109)
$\bar{P}_s$, Rayleigh	$\approx \left(\dfrac{M-1}{2\bar{\gamma}\log_2 M}\right)$	(2.116)

2.4.5 Comparisons

In the previous sections, a brief overview has been given on digital modulation. In this section, the performance of various modulations is compared in terms of their BER for a given E_b/N_0, bandwidth efficiency, and energy efficiency.

2.4.5.1 BER in AWGN

BER performance of coherent MPSK is shown in Figure 2.43. Binary and quaternary PSK have the best BER performance for a given E_b/N_0. 8-PSK requires approximately 4 dB higher energy than BPSK/QPSK and as constellation size increases, doubling of constellation size results in a loss that asymptotically approaches 6 dB. Performances of MSK, GMSK, and OQPSK are not shown in Figure 2.43 but they all have equal or similar performance as that of BPSK/QPSK. Coherently differential PSK such as DCPSK and $\pi/4$-DQPSK suffer a loss of about 3 dB as compared to BPSK/QPSK.

---

* The BER is well approximated by $0.5P_s$

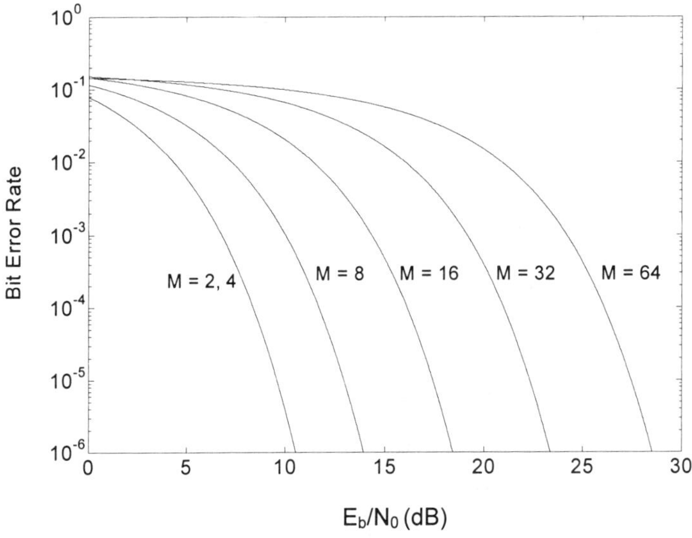

Figure 2.43 Bit error rate of coherent MPSK in AWGN.

In contrast, QAM has much better error performance for a given E_b/N_0 as shown in Figure 2.44. While the performance is equal in both cases for the binary and quaternary cases, every doubling in the constellation size results in a 2-3 dB increase in energy in contrast to the 6 dB increase in PSK. QAM, however, places a stricter constraint on the linearity of the power amplifier because of its larger peak-to-average power ratio as discussed later in this section. Also, since QAM modulation involves signal amplitude as well as signal phase, coherent detection is required.

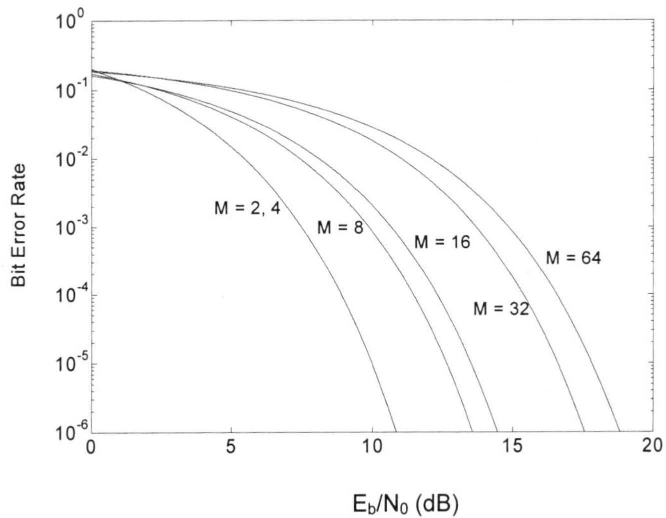

Figure 2.44 Bit error rate of coherent MQAM in AWGN.

Coherent FSK has the best BER performance for a given E_b/N_0 as shown in Figure 2.45 for constellation size greater than four. In fact, the performance of FSK approaches the Shannon bound (-1.6 dB) asymptotically as the constellation size increases. The main problem with FSK is that coherent demodulation of FSK requires a fast switching frequency reference which is costly and difficult to implement. Most FSK systems use non-coherent detection which results in a large increase in the energy requirement [14]. Furthermore, FSK has poorer bandwidth efficiency as compared to PSK and QAM.

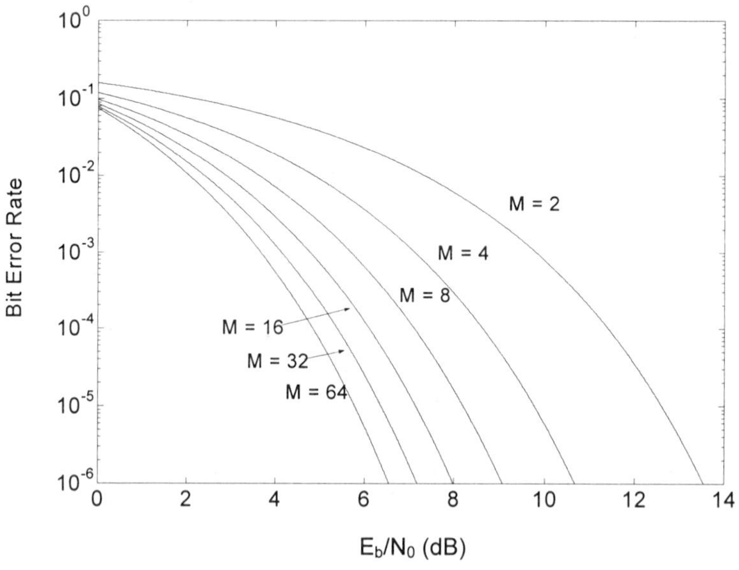

Figure 2.45 Bit error rate of coherent MFSK in AWGN.

2.4.5.2 Bandwidth Efficiency

Bandwidth efficiency of PSK, QAM, and FSK has been discussed in Sections 2.4.1-2.4.4. From Table 2.4 to Table 2.6, it is clear that PSK and QAM have the same bandwidth efficiency while coherent FSK is 2M times less efficient. However, in actual implementations, both PSK and QAM suffer from a loss in bandwidth efficiency due to spectral regrowth resulting from non-linearity in the power amplifier. Variants of PSK have been developed to mitigate spectral regrowth. These variants include OQPSK, $\pi/4$ QPSK, and CPM. Two common realizations of CPM are MSK and GMSK. Without power amplification, the spectral efficiency of OQPSK and $\pi/4$ QPSK are equivalent to QPSK, i.e. 1 bps/Hz assuming null-to-null bandwidth with an excess bandwidth of 100%. In comparison, the bandwidth efficiencies of MSK[2] and GMSK[3] are respectively 0.71 bps/Hz and 1.1 bps/Hz in a 99.5% containment bandwidth.

---

[2] Assumes modulation index of 0.5.

The advantage of MSK and GMSK becomes apparent when the modulated signals undergo non-linear amplification as shown in Figure 2.46. Both MSK and GMSK, in particular, show substantial reduction in spectral regrowth as compared to QPSK, OQPSK, and π/4 QPSK. It is for this reason that both GSM and DECT use GMSK for transmission. While not as robust as CPM to non-linear amplification, OQPSK and $\pi/4$ QPSK still offer reasonable reduction in spectral regrowth and are therefore used in many systems as well, such as IS-136 and WCDMA.

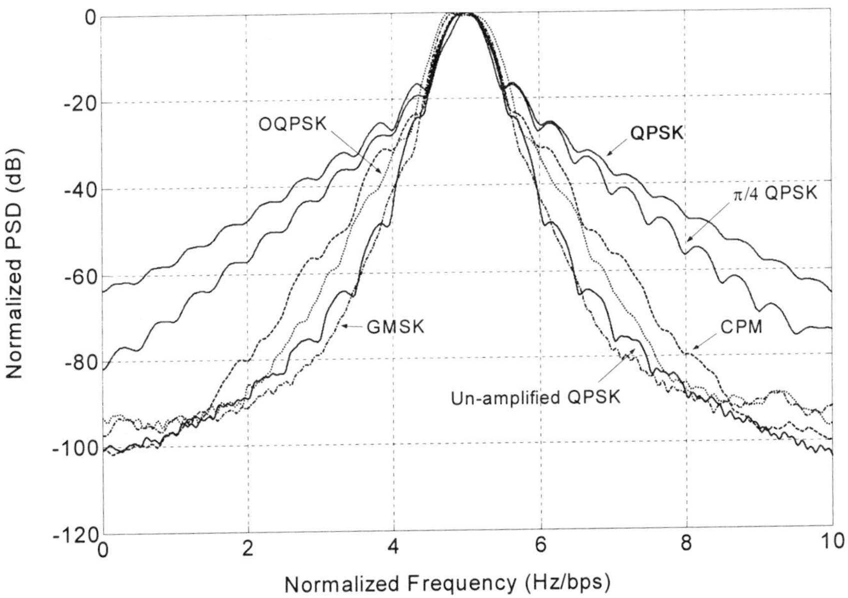

Figure 2.46 Spectral regrowth of filtered modulated waveforms due to power amplification.

2.4.5.3 Energy Efficiency

In previous sections, energy efficiency has been discussed in detail for PSK, QAM, and FSK. Figure 2.13 illustrates energy efficiency of these three types of modulation. It is clear that QPSK has the best energy efficiency. Therefore, not surprisingly, many current mobile wireless systems use some form of QPSK. Other commonly used modulations include GMSK, OQPSK, MSK, and π/4 QPSK. Their energy efficiencies measured relative to QPSK are shown in Table 2.7 below. The lower energy efficiency of MSK, GMSK, and π/4 QPSK is a compromise for lower spectral regrowth. Note the bandwidth efficiencies of OQPSK, MSK, π/4 QPSK are measured based on the null-to-null bandwidth whereas that of GMSK is based on the 99.5% containment bandwidth at –38 dBc.

---

[3] Assumes modulation index of 0.5 and BT = 0.3.

Table 2.7 Energy efficiency of common modulations.

Modulation	Energy Efficiency (Relative to QPSK)
OQPSK	1.00
π/4 QPSK	0.59
MSK	0.71
GMSK	0.95

Finally, as mentioned in Section 2.1.6, energy efficiency can also be measured in terms of the peak energy, which can be important for non-constant envelope modulations such as QAM. Figure 2.47 shows the peak-to-average power ratio for QAM which is significantly higher than PSK, CPM, and FSK. Other non-constant envelope transmissions include orthogonal frequency division multiplexing [53] and multi-code CDMA [54].

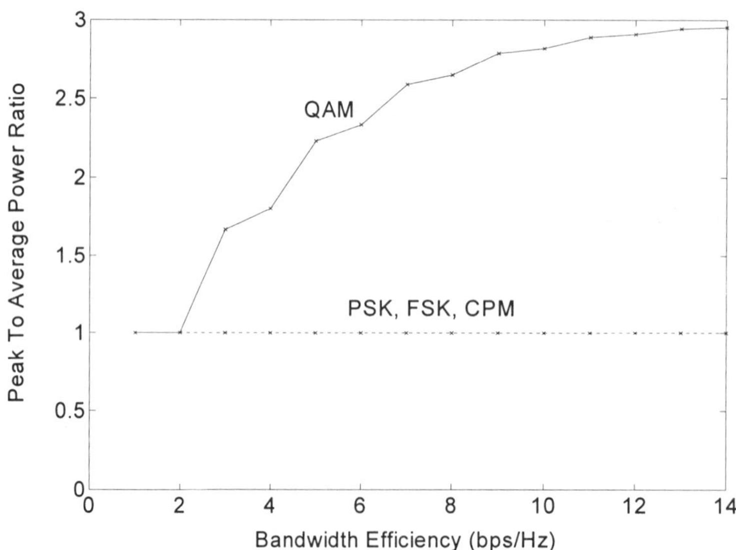

Figure 2.47 Peak-to-average power ratio versus bandwidth efficiency.

2.5 Synchronization

In recovering the baseband signal, the received signal is downconverted to baseband using a local carrier reference running at a frequency f_0. In the ideal case, f_0 should be identical to the carrier frequency in the transmitter. However, in general, both the channel and the local RF oscillator introduce an unknown phase and frequency offset. Similarly, the sampling clock of the correlator output in an ML receiver also has an unknown phase and frequency offset relative to the transmitted symbol clock due to the delay introduced by the channel as well as timing differences in the crystal oscillator used to generate the symbol clocks. Therefore, in a practice, the phase, frequency, and timing of the received signal must be synchronized at the receiver to obtain the optimum performance.

2.5.1 Effects of Imperfect Synchronization on Error Rate

If phase, frequency, and timing are not synchronized, the performance can be significantly degraded compared to the case with perfect synchronization discussed in Section 2.4. The effect of phase offset rotates the constellation points away from their ideal locations and therefore compromises the noise margin in the ML detector. The amount of phase rotation is generally characterized by the RMS phase offset. In contrast, a timing offset does not just rotate the constellation point but moves the constellation in all directions, essentially smearing the constellation point into a disk with radius D_i. A large D_i also diminishes the noise margin of the ML receiver. Finally, frequency offset inflicts the most severe degradation in error performance. Frequency offset occurs both in the timing clock as well as in the RF carrier. In an RF carrier, frequency offset also rotates the constellation points but, unlike phase offset, the amount of phase rotation grows with time and does not have a finite RMS value. In a timing clock, frequency offset tends to increase D_i, increasing the error rate. Figure 2.48 shows the three types of synchronization errors:

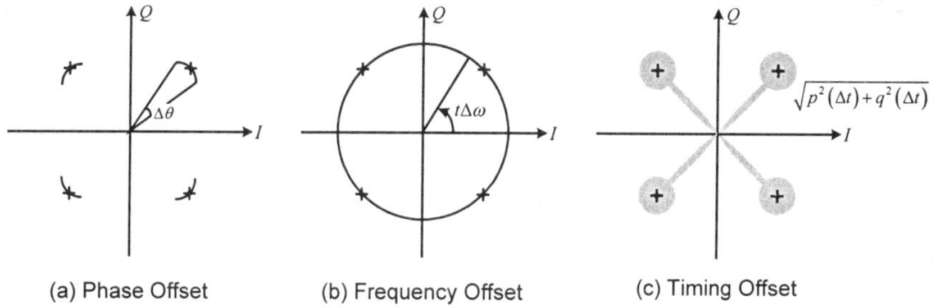

Figure 2.48 Synchronization errors.

Example 2.11 Assume BPSK modulation with a phase offset θ_{off}. Then, the received signal is represented by $\sqrt{2E_b/T_s}\cos(\theta(t)-\theta_{off})$. Since $\theta(t)$ can only be 0 or π, the received signal can be further simplified to $\sqrt{2E_b\cos^2(\theta_{off})/T_s}\cos(\theta(t))$, where it can be seen that the bit energy has been degraded by $\cos^2(\theta_{off})$. The worst case offset is at $\pi/2$, at which point the loss in E_b/N_0 is infinite. At an offset of 10 degrees the E_b/N_0 loss is only 0.13 dB. Therefore, BPSK is not sensitive to phase error. Intuitively, its insensitivity to phase error results from the wide separation between the two constellation points. However, for higher order modulations, the effect of phase error becomes more pronounced due to the tighter spacing between constellation points. □

Example 2.12 Assume again BPSK but now with an RF carrier frequency offset Δf The result from Example 2.11can be extrapolated to this case by observing the fact that Δf induces an equivalent phase offset of $2\pi\Delta ft$ The amount of offset depends on the time t and the magnitude of the frequency offset. If $\Delta f = 1kHz$, corresponding to a 1 ppm offset given a carrier frequency of 1 GHz, then appreciable error does not occur until t is large enough such that the effective phase offset approaches multiples of 90 degrees. Thus, if the error rate is observed as a function of time, in every 250 μs there will be an increase in error rate. Over an extended period of time, the average error rate can be obtained by observing that the average loss in energy is

$$\frac{1}{250\mu s}\int_0^{250\mu s}\cos^2\left(2\pi\Delta ft\right)dt = 0.5\,.$$

Thus, the average error rate becomes $Q(\sqrt{E_b / N_0})$, corresponding to a 3-dB loss in E_b/N_0. □

Example 2.12 shows that a frequency offset causes more degradation in system performance when the bitrate is low such that a packet or frame is long enough to experience the drop in signal energy. In particular, any packet larger than 250 μs will experience at least one burst of error due to 1 kHz of frequency offset. In GSM, the bitrate is 270 kbps and each time slot is 577 μs in duration. To keep the degradation of frequency offset to an adequate level, the residual offset should be kept to 10's of Hz.

In addition to phase and frequency offsets, timing offset can also cause serious degradations in the BER. Assume that the pulse shape transmit filter is a Nyquist filter with a bandwidth of $0.5R_b$, i.e. $\tilde{h}(t) = \text{sinc}\,(t/T_b)$. Then, any timing error Δt other than the ideal sampling points at $t = kT_b$ will induce ISI. Since the amount of ISI depends on the transmitted data sequence, it is not deterministic and must be treated as a random variable.

Example 2.13 For the purpose of illustration, the worst-case scenario is examined where the adjacent symbol is assumed to have the opposite sign as the symbol being sampled. Also, the effect of other symbols is neglected for simplicity. The energy loss due to Δt can then be expressed as

$$\text{Energy Loss} = -20\log\left[\text{sinc}\,(\Delta t/T_s) - \text{sinc}\,(\Delta t/T_s - 1)\right] \qquad (2.141)$$

According to (2.141), when Δt is a small fraction of T_s (e.g. less than 5%), the energy loss is a about 0.5 dB. However at larger Δt, the loss becomes significant. For instance, at $\Delta t/T_s = 30\%$, the loss is 6 dB. The worst case is at $\Delta t/T_s = 50\%$ at which the loss approaches infinity. The above analysis can be extended to other modulations and in general, it can be shown that higher order modulations are more prone to timing errors. Also, by applying a raised cosine transmit filter with high excess bandwidth, sensitivity to timing error can be reduced as discussed in Section 2.1.2. □

2.5.2 Phase Recovery

The received data can be detected non-coherently or coherently. Non-coherent detection does not require the receiver to estimate the phase of the received signal prior to demodulation whereas coherent detection requires phase estimation in conjunction with demodulation. Several techniques exist for estimating the phase of the received signal, also widely known as phase recovery. These techniques include the Costas loop [55], phase derotator, and squaring loop [56]. Since the squaring loop has essentially the same performance as a Costas loop though more complex, it will not be discussed here.

2.5.2.1 Costas Loop

The Costas loop operates on a passband signal; that is the received signal has a carrier frequency, which can be RF or IF. The Costas loop for a BPSK signal is shown in Figure 2.49. The BPSK signal is expressed as $d(t)\cos(2\pi f_0 t + \theta_i(t))$, where $d(t)$ is the received baseband signal and $\theta_i(t)$ is the input phase of the carrier. $d(t)$ is mixed down to baseband in the I and Q channels by locally generated carriers, $\cos(2\pi f_0 t + \theta_o(t))$ and $-\sin(2\pi f_0 t + \theta_0(t))$, respectively, where $\theta_0(t)$ is the phase of the local oscillator. Immediately after mixing down to baseband, the I channel holds $0.5d(t)\cos(\theta_e(t))$ plus a second order harmonic of f_0 and the Q channel holds $0.5d(t)\sin(\theta_e(t))$ plus a second order harmonic of f_0. $\theta_e(t)$ is the phase error measured as the phase difference, $\theta_i(t) - \theta_0(t)$.

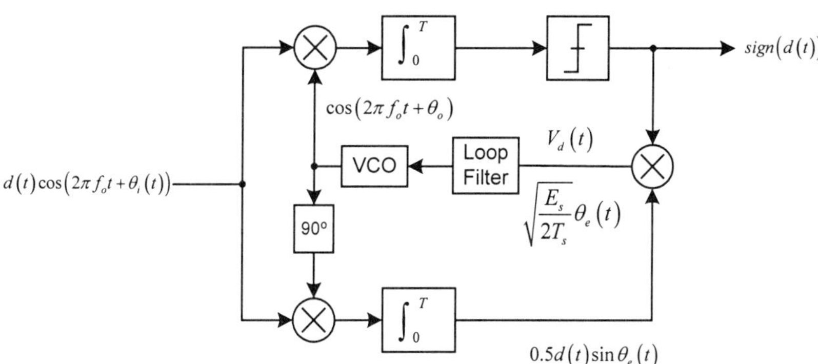

Figure 2.49 Costas loop for BPSK signals.

The goal is to make $\theta_e(t)$ equal to zero so that the received local oscillator is synchronized in phase with that of the received carrier. This synchronization can be achieved by applying $\theta_e(t)$ in a negative feedback loop whereby $\theta_e(t)$ is filtered through a loop filter and used as a correction signal for the voltage controlled oscillator (VCO) that generates the local carriers. To obtain $\theta_e(t)$, the phase detector takes the Q-channel output and multiplies it by the hard limited I-channel output sampled at T_b. This removes the data polarity from the Q-channel output and produces a signal $v_d(t)$ at the phase detector output, which is a function of only $\theta_e(t)$, i.e.

$$v_d(t) = \pm 0.5\sqrt{2E_b/T_b}\,\sin\theta_e(t)\cdot sign\left(\pm\sqrt{2E_b/T_b}\,\cos\theta_e(t)\right) \approx \sqrt{0.5E_b/T_b}\,\theta_e(t)$$

for small $\theta_e(t)$. Note that when $\theta_e(t) = 0$, the hard limiter output also decodes the received symbol according to $sign\left(\pm\sqrt{2E_b/T_b}\right)$.

When $\theta_e(t)$ is large, the Costas loop is operating in the non-linear region. The phase detector output $v_d(t)$ does not behave according to the small signal approximation, $\sqrt{0.5E_b/T_b}\,\theta_e(t)$. The large signal behavior at the phase-detector output is shown in Figure 2.50. The curves in Figure 2.50 is also referred to as the S-curves which indicates that the Costas loop can lock at $0, \pm\pi, \pm 2\pi, \cdots$. The multiple lock points can result in phase ambiguity which is resolved with differential encoding.

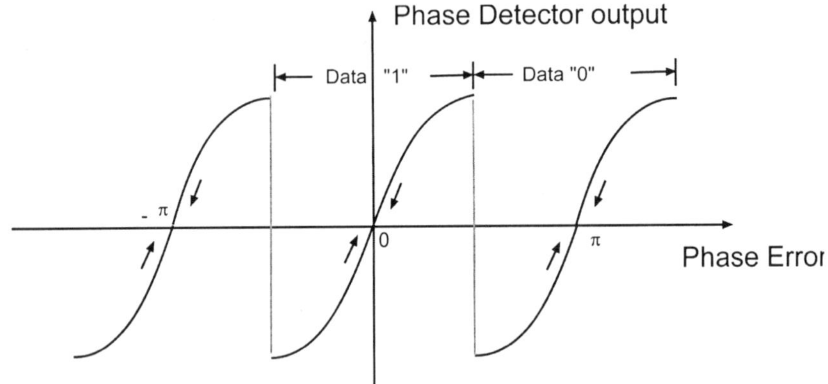

Figure 2.50 Phase detector output for BPSK signals: S-curve.

When the Costas loop is operating in the linear region, linear analysis can be applied. A linear model [56] is shown in Figure 2.51. The I-channel is ignored, assuming that most of the time a correct decision is made at the hard limiter. Such a model is often used for phase-locked loops (PLL). The following transfer function can be obtained for the linear model.

$$H(s) = \frac{\theta_0(s)}{\theta_i(s)} = \frac{K_0 K_d F(s)}{s + K_0 K_d F(s)} \tag{2.142}$$

$$E(s) = \frac{\theta_e(s)}{\theta_i(s)} = 1 - H(s) \tag{2.143}$$

where $H(s)$ is the transfer function from the input to the VCO output, $E(s)$ is the transfer function from the input to the phase detector output, and $F(s)$ is the transfer function of the loop filter. K_0 is the gain of the VCO in rad/sec/V and K_d is the gain of the phase detector in V/rad. For the Costas loop shown in Figure 2.49, $K_d = \sqrt{0.5E_s / T_s}$. The transfer function of the VCO is derived from the fact that $d\theta_o(t)/dt = K_0 v_c(t)$, where $v_c(t)$ is the control voltage at the VCO input.

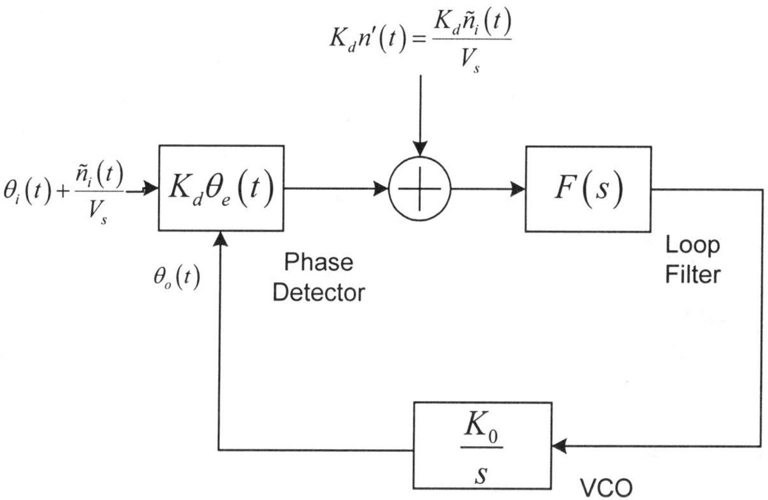

Figure 2.51 A linear model for the Costas loop.

The Costas loop can be characterized by its steady state, transient, and frequency responses. The steady-state response measures the error when the loop has reached steady state (i.e. $t \to \infty$). It can be determined from the following general expression:

$$\lim_{t \to \infty} \Delta\theta(t) = \lim_{s \to 0} \frac{s^2 \theta_i(s)}{s + K_0 K_d F(s)}. \tag{2.144}$$

If the input phase is a step function, corresponding to s^{-1}, then according to (2.144), the error signal always reaches zero independent of K_0, K_d, and $F(s)$. However, if the input phase is a ramp or a frequency step, corresponding to s^{-2}, then the steady-state error depends on K_{DC}^{-1}, where

$$K_{DC} = K_0 K_d F(0) \tag{2.145}$$

is defined as the DC loop gain. In this case, either $K_0 K_d$ should be very large or $F(s)$ must have a pole at $s = 0$ to zero out the steady-state error. In general, for an input phase response of s^{-n}, $F(s)$ must have a pole of multiplicity greater than

$n-2$ at $s=0$ to make the steady-state error zero. By definition, $F(s)$ with a pole of multiplicity m at $s=0$ results in an $(m+1)^{th}$ order loop. For example, if $m = 0$, then the loop is first order.

The transient response measures the behavior of the VCO output as it tracks the input phase. It can be found by taking the inverse Laplace transform of $H(s)\theta_i(s)$. The transient response is a strong function of $F(s)$. In general, higher order loops can track more dynamically changing input phases but tend to have stability problems as the order of the loop exceeds two.

Finally, the frequency response of the loop is also determined by $H(s)$. Ideally, $H(s)$ should be a low pass filter that filters out unwanted harmonics and input noise outside of the signal bandwidth centered on the carrier frequency. The bandwidth B_L of $H(s)$ determines the noise bandwidth and its tracking performance. B_L can be defined similar to (2.24), where if the equivalent response is unity gain, $B_L = \int_0^\infty \left|H(f)\right|^2 df$. In general, the larger the B_L, the larger the phase noise due to the input noise source $\tilde{n}_i(t)$. In the linear model, $\tilde{n}_i(t)$ is normalized by V_s and is converted to an equivalent noise source $n'(t)$ at the phase detector input, where $n'(t) = \tilde{n}(t)/V_{s.}$. If the input noise has a one-sided noise density of $2N_0$, then the equivalent noise variance is $2N_0B_L/V_s^2$ and the resulting output phase noise variance in rad^2 is simply

$$\overline{\theta_0^2} = \frac{B_L}{\gamma R_b}. \tag{2.146}$$

Unfortunately, except for a first-order loop, there is no simple relationship between B_L and the settling time of the loop τ_s, i.e. the time taken for the loop to settle to its steady-state value. For the first-order loop, $\tau_s \propto B_L^{-1}$. For higher order loops, τ_s is largely a function of the position of zeros in *F(s)* and the AC loop gain K_∞, defined as

$$K_\infty = K_0 K_d F(\infty). \tag{2.147}$$

Example 2.14 For a system, the desired E_b/N_0 is 10 dB and the RMS phase error is 5 degrees. Then, based on (2.146), the loop bandwidth B_L should be about 13 times less than the bitrate as shown below

$$\frac{R_b}{B_L} = \frac{180^2}{\gamma\overline{\theta_0^2}\pi^2} = \frac{180^2}{10\cdot 5^2\cdot\pi^2} \approx 13. \ \square$$

First-Order Loop: If $F(s) = K$, then a first-order loop results. According to (2.142), the loop transfer function is

$$H(s) = \frac{K_{DC}}{s + K_{DC}},$$

where $K_{DC} = K_d K_0 F(0)$ is the DC loop gain. Applying (2.144), the steady-state error is found to be zero for a phase step and $\Delta\omega/K_{DC}$ for a frequency step of magnitude $\Delta\omega$. Therefore, by making K_{DC} much larger than $\Delta\omega$, the steady-state error can be made arbitrarily small. However, by making K_{DC} too large increases the phase variance at the VCO output since the noise bandwidth is directly proportional to K_{DC}. Specifically,

$$B_L = \int_0^{\infty} \left| \frac{K_{DC}}{K_{DC} + j2\pi f} \right| df = \frac{1}{4} K_{DC}.$$

The transient error response due to a phase step $\Delta\theta / s$ and to a frequency step $\Delta\omega / s^2$ can be determined by taking the inverse Laplace transform of $E(s)\theta_i(s)$. The error responses are respectively $\Delta\theta e^{-K_{DC}t}$ and $K_{DC}^{-1}\Delta\omega(1 - e^{-K_{DC}t})$. In both cases, the settling time τ_s is inversely proportional to K_{DC}. However, the steady-state error response does not reach zero when there is a constant frequency offset.

Second-Order Loop: If $F(s) = (\tau_2 s + 1)/(\tau_1 s)$, then a second-order loop results. According to (2.142), the loop transfer function is

$$H(s) = \frac{2\varsigma\omega_n s + \omega_n^2}{s^2 + 2\varsigma\omega_n s + \omega_n^2}, \tag{2.148}$$

where $\omega_n = \sqrt{K_\infty / \tau_2}$ is the resonant frequency, $\varsigma = 0.5\tau_2\omega_n$ is the damping factor, and $K_\infty = K_d K_0 F(\infty)$ is the AC loop gain at $f = \infty$. Applying (2.144), the steady-state error is determined to be zero for both phase and frequency steps. However, if the phase input is $\Delta\dot{\omega} / s^3$ corresponding to a step in rate of frequency change, then the steady-state phase error has a non-zero value of $\Delta\dot{\omega}/\omega_n^2$. In most applications, $\Delta\dot{\omega}$ is generally very small and can therefore be neglected. Otherwise, ω_n can be made large to compensate for a large $\Delta\dot{\omega}$. Unfortunately, the output phase variance also increases due to an increase in the noise bandwidth,

$$B_L = \frac{\omega_n}{2}\left(\varsigma + \frac{1}{4\varsigma}\right).$$

Unlike the first-order loop, where there is only one design parameter K_{DC}, a second-order loop has two design parameters ω_n and ς. In general, the noise and steady-state performance is a function of ω_n and the transient performance (i.e. τ_s) is largely a function of ς, which depends on τ_2 of the loop filter. If ς is much less than one then the system becomes underdamped and the transient response becomes oscillatory. On the other hand, if ς is much greater than one, then the system becomes overdamped and the transient response shows a very slow rise time in converging to a steady state value. In both overdamped and underdamped systems, τ_s is excessively long. For best transient behavior, the damping factor should be chosen close to one. When $\varsigma = 1$ the system is defined as being critically damped. In general, ς is chosen to be $1/\sqrt{2}$ for the best compromise between τ_s and stability. Figure 2.52 shows the response $h(t)$ of the second-order loop due to a phase step of $\Delta\theta$. At damping factors below 0.5 substantial ringing occurs, causing the signal envelope to settle slowly. More specifically, the signal envelope shows an exponential decay time that is inversely proportional to $\varsigma\omega_n$.

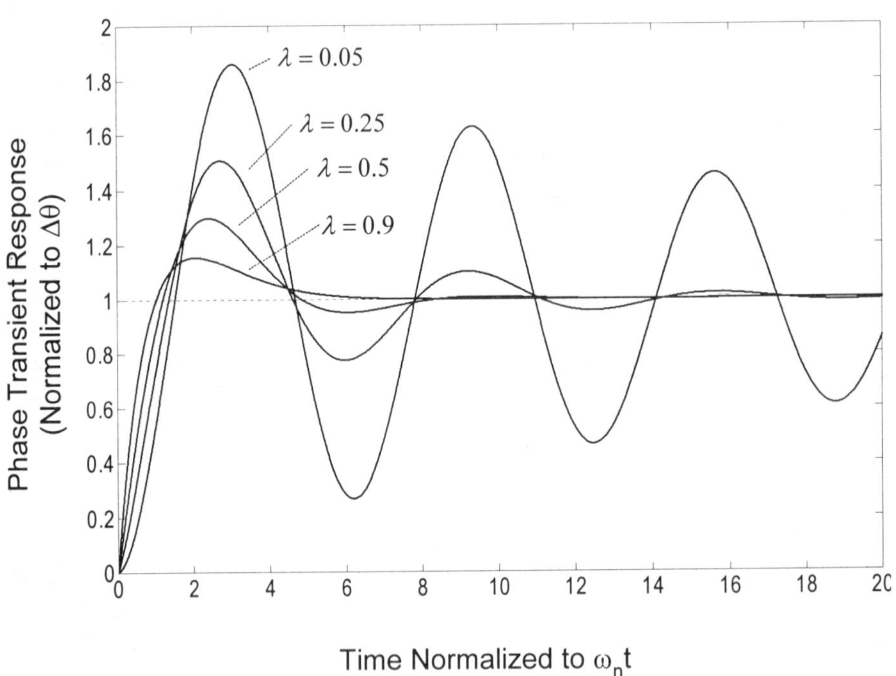

Figure 2.52 Response of a second-order loop with a phase step of $\Delta\theta$.

For higher order modulations, the phase detector of Figure 2.50 needs modification to account for the additional constellation points. A more general phase detector can be obtained by noting that the phase error can be measured with respect to the slicer output. A slicer refers to the decision block in the maximum likelihood detector.

Let the received baseband signal be $\tilde{r}(t)$ and the demodulated output be $\hat{\tilde{r}}(t)$, then the phase error can be estimated by $\arg\left\{\tilde{r}(t)\hat{\tilde{r}}(t)^*\right\}$, which can be approximated by $\text{Im}\left\{\tilde{r}(t)\hat{\tilde{r}}(t)^*\right\}$ when $\theta_e(t)$ is small. Figure 2.53a shows this type of decision-aided phase detector. Note that if $\theta_e(t)$ is greater than a critical angle θ_c shown in Figure 2.53b, then the received signal has moved into another decision region and therefore a phase ambiguity results; i.e. the phase detector cannot distinguish phase errors greater than θ_c. For MPSK, $\theta_c = 2\pi / M$. For example, in Figure 2.50, the phase detector output shows a phase ambiguity every π interval for a BPSK signal with $M = 2$.

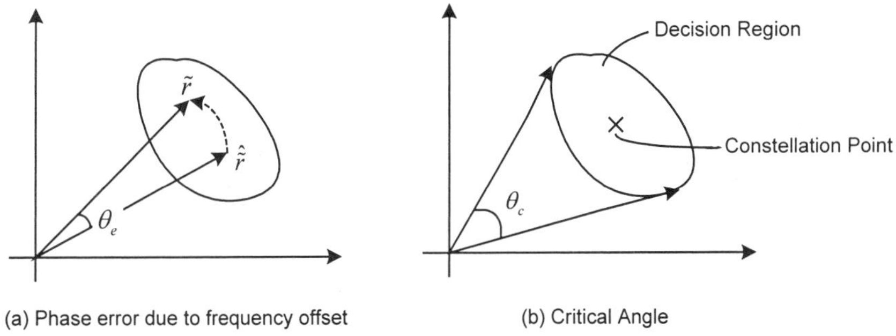

Figure 2.53 Generalized phase detector.

For MPSK, the phase ambiguity can be resolved using differential encoding/decoding. However, for QAM, since the constellation points are differentially encoded as a group by quadrant rather than individually, a slightly different approach must be taken to eliminate phase ambiguity. The modification is to use $\text{sgn}[\hat{\tilde{r}}(t)]$ as the phase reference. This is possible since QAM has a symmetric signal constellation such that the phase of the modulation itself averages to zero leaving only $\theta_e(t)$. Also since θ_c is $\pi / 2$, differential encoding can resolve the phase ambiguity.

The Costas loop for MPSK and QAM is shown in Figure 2.54. The cross-coupled phase detector follows directly from the equation $\text{Im}\left\{\tilde{r}(t)\hat{\tilde{r}}(t)^*\right\}$, where $\tilde{r}(t) = r_I(t) + jr_Q(t)$ and $\hat{\tilde{r}}(t) = \hat{r}_I(t) + j\hat{r}_Q(t)$. For QAM modulation, the slicer in Figure 2.54 should be replaced by a hard limiter to remove the effect of amplitude variations in the received signal. The linear analysis described earlier can also be applied to the generalized Costas loop shown in Figure 2.54.

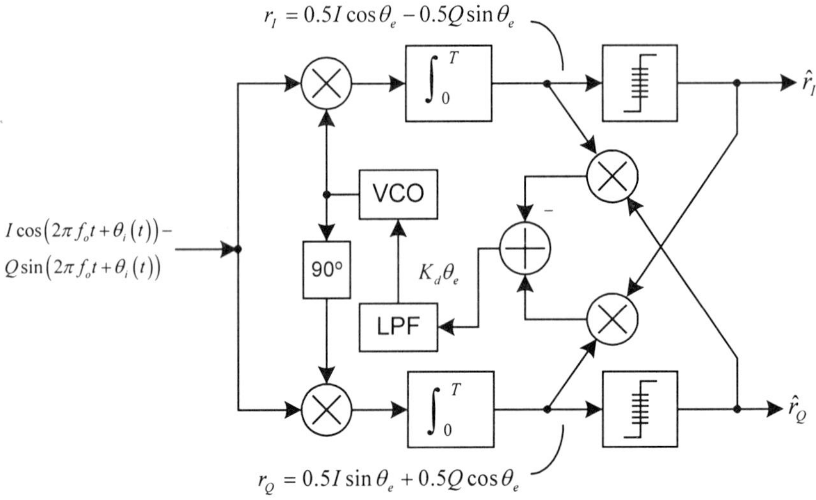

Figure 2.54 Costas loop for MPSK and QAM signals.

2.5.2.2 Phase Derotator

An alternative to the Costas loop of Figure 2.54 is the phase derotator shown in Figure 2.55. The phase derotator performs the phase correction at baseband rather than at RF or IF as in the Costas loop. The basic principle is to generate a correction signal $e^{-j\theta_e(t)}$ such that the phase error $\theta_e(t)$ in $\tilde{r}(t)$ is compensated by $\tilde{r}(t)e^{-j\theta_e(t)}$. The phase derotator shown in Figure 2.55 is essentially a complex multiplier that performs the product $\tilde{r}(t)e^{-j\theta_e(t)}$.

A direct-digital frequency synthesizer (DDFS) is used to generate digitally the correction signal $e^{-j\theta_e(t)}$. Details on the design of a DDFS are described later in Chapter 8. For now, it can be viewed as a digital VCO that takes a digital frequency control word and generates the cosine and sine waveforms at the specified frequency. For the phase derotator the digital control word is generated through a decision-aided phase detector described in Section 2.5.2. The small signal performance of the phase derotator is similar to that of the Costas loop and can be analyzed using the linear model described in the previous section. The main difference between the phase derotator and the Costas loop is that the former allows digital phase correction at a lower sampling rate since the digitization can occur at baseband. In contrast, the Costas loop requires digitization at either RF or IF, which requires a much higher sampling rate.

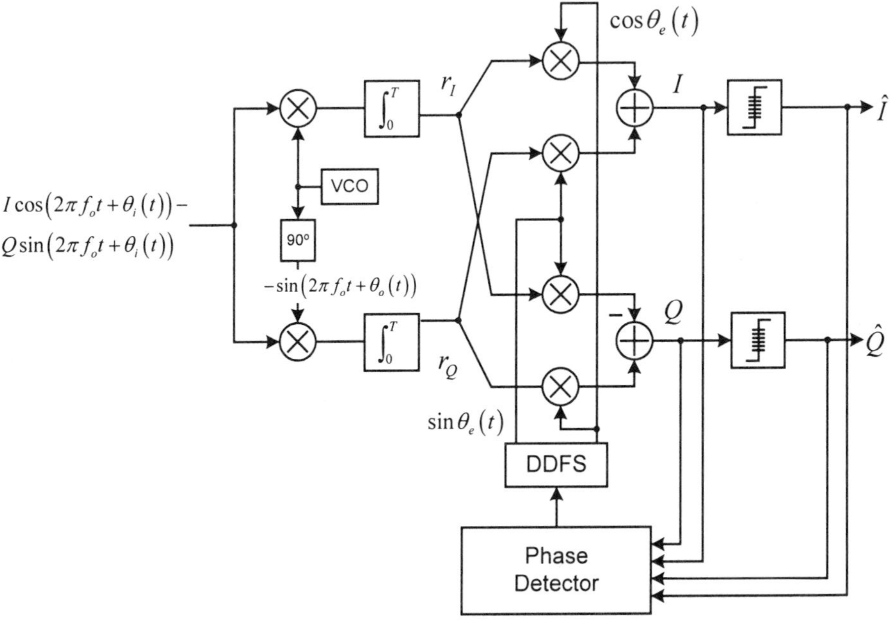

Figure 2.55 Phase derotator.

2.5.3 Frequency Acquisition

Under certain conditions, the Costas loop can also recover the frequency difference between the transmitter and receiver. If the frequency offset is less than the lock-in range $\Delta\omega_L$ of the loop, then the loop can quickly acquire the frequency offset, i.e. achieve frequency lock. Once locked, the frequency can still drift but the loop can in general hold lock to frequency drift less than the hold-in range $\Delta\omega_H$. If the frequency is larger than $\Delta\omega_L$ but less than $\Delta\omega_p$, the pull-in range, the loop can still lock onto the carrier but after an extended amount of time, know as the pull-in time T_p.

For a first-order loop, the following condition holds for frequency acquisition [56]:

$$\Delta\omega_L \approx \Delta\omega_H \approx \Delta\omega_p \approx \pm K_\infty \tag{2.149}$$

For a second-order loop, the following conditions hold for frequency acquisition [56]:

$$\Delta\omega_H \le \Delta\omega_L \approx \pm K_\infty \tag{2.150}$$

$$\Delta\omega_p \approx \pm\sqrt{2K_{DC}K_\infty} \tag{2.151}$$

$$T_p \approx \frac{(\Delta\omega)^2}{2\varsigma\omega_n^3} \tag{2.152}$$

It follows from (2.149), that for a first-order loop to acquire the frequency, the loop gain K_∞ must be large which implies that the noise bandwidth must also be large, compromising the amount of phase jitter in the recovered signal. The second-order loop can achieve a large pull-in range but with a small noise bandwidth if the DC loop gain is made large. However, having a small noise bandwidth increases the pull-in time, which may not be suitable for delay sensitive applications. In general, if the initial frequency offset is small, then a first-order loop should be used since it can lock in quickly. Since the offset is small the noise performance is not greatly degraded. If the offset is large or because of the need for improved tracking performance, a second order loop should be used but with a potential increase in the time needed to pull-in.

2.5.4 Preamble Frequency Estimation

Because the pull-in time can in some cases be excessively long, the frequency may need to be first estimated with a separate block as shown in Figure 2.56. The estimated frequency offset can be used to put the carrier recovery loop in the lock-in range such that the loop can quickly synchronize the residual frequency and phase errors. The estimation process is designed to be much faster than if the loop had to self-acquire the frequency offset (i.e. long pull-in time).

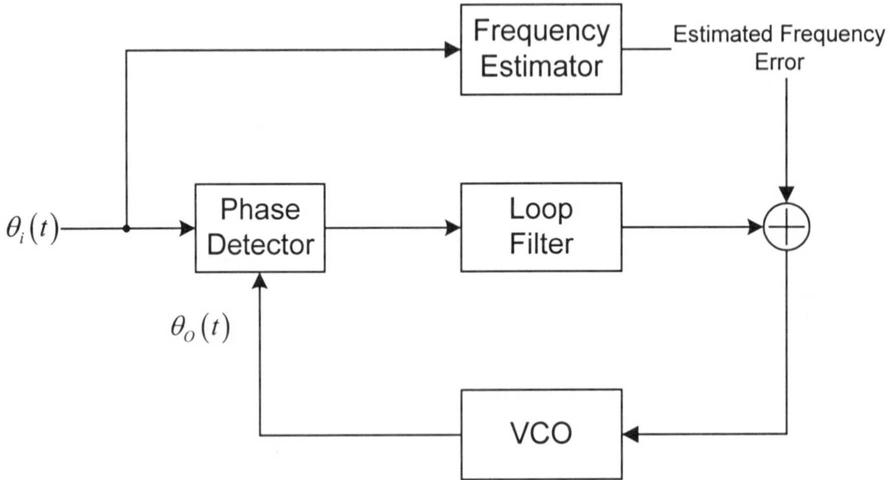

Figure 2.56 Frequency estimation.

Many estimation algorithms exist [57][58][59]. These algorithms are based on estimating the frequency transmitted on a known pilot signal, which could be a tone or a predetermined modulated bit pattern. For instance, GSM uses a pilot tone in the control channel to allow the mobile terminals to acquire the frequency of the basestation. A tone-based estimation scheme, however, does not perform well in a frequency selective channel since the tone may be greatly attenuated in a null of the channel response. This limitation does not significantly impact the performance of GSM because a control channel is available to periodically estimate the frequency error and the estimation time can be extended since the system is circuit switched.

A more robust estimation scheme is proposed in [60] where the transmitted pilot signal is based on an M-sequence designed to minimize the degradation due to multipath fading. An M-sequence is a pseudo-noise (PN) code that statistically resembles a random sequence and is described in 0. The M-sequence-based estimation scheme improves the performance over the tone-based scheme in both the acquisition time and the estimation SNR. Figure 2.57a shows the BER performance of a GMSK system operating at 2 GHz with a 10 ppm frequency offset when a tone, a square periodic pulse train, and an M-sequence is used for frequency estimation. The GMSK modulation has a BT of 0.3 and a modulation index of 0.5. The channel bitrate is 2 Mbps and the channel model consists of five equally powered paths over a delay spread of 5 μs, in which each path has independent Rayleigh fading with a Doppler frequency of 200 Hz. As the estimation time increases, the performance of all three becomes similar. However, at short estimation time (e.g. 100 symbols) the M-sequence shows appreciable improvement over the other types of pilot signal. Also, according to Figure 2.57b for the same system, an estimation SNR of 10 dB using an M-sequence is achieved 98% of the time while only 65% is achieved using a tone.

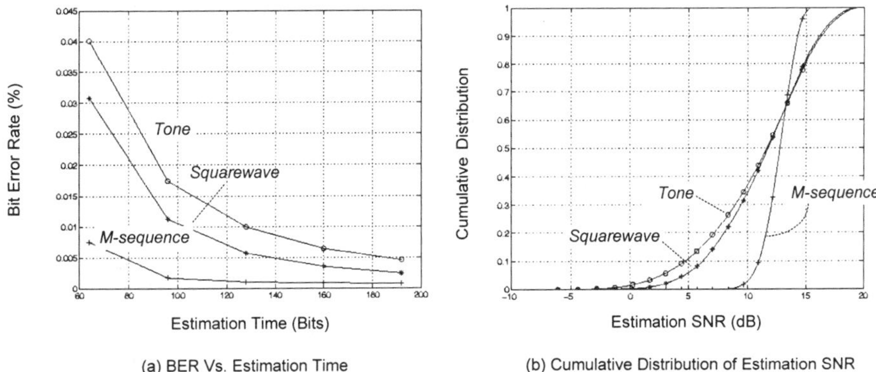

Figure 2.57 Performance comparison of frequency estimation algorithms.

The M-sequence pilot signal can be constructed by concatenating N segments of an M-sequence of period W. Because of the periodicity of M-sequences, the modulation phase is also periodic with period W. Thus, the frequency can be estimated by

$$\Delta\hat{\omega} = \frac{1}{WT_s} \arg\left\{ \frac{1}{NW} \sum_{k=0}^{NW} \tilde{r}(k)\tilde{r}^*(k-W) \right\} \tag{2.153}$$

where $\tilde{r}(t)$ is the baseband signal. Since the modulator phase has period W, multiplying $\tilde{r}(t)$ by its complex conjugate delayed by WT_s results in a phase rotation solely due to frequency error. The summation operation provides averaging to improve the estimation SNR. The robustness of the algorithm to frequency-selective fading arises from the auto-correlation property of M-sequences, which allows the estimator to separate the multipaths and optimally combine for improved estimation SNR.

2.5.5 Timing Recovery

Even if the phase and frequency are correctly tracked, without recovery of timing, the demodulated symbols may be decoded with a substantial amount of ISI or with sub-optimal SNR due to roll off at the symbol transition. Timing recovery enables the system to sample at the optimum point by synchronizing onto the received signal. In general, there are two basic types of timing recovery: coherent and non-coherent. Coherent timing recovery refers to recovering timing when the carrier phase has already been acquired whereas non-coherent refers to recovering timing when the carrier phase has not yet been acquired. Many timing recovery algorithms exist. These include the tau-dither loop [61] and the delay-locked loop [62], also known as the early-late gate correlator. This section covers the early-late gate correlator, which is most widely used.

2.5.5.1 Coherent Early Late Gate Correlator

The negative feedback principle used in phase recovery can be similarly applied to timing recovery by noting that instead of generating a sinusoidal signal at the VCO output a clock signal (clipped sinusoid) can be generated for timing recovery. A voltage controlled clock source is also known as a voltage controlled clock oscillator (VCCO) or a numerically controlled oscillator (NCO) if implemented digitally.

The phase detector in timing recovery detects the timing error defined as $\tau_e = T_i - T_o$, where T_i is the input timing and T_o is the timing of the VCCO output. If the received signal symbol timing lags the VCCO output then τ_e is positive; otherwise, τ_e is negative. A phase detector for timing recovery is shown in Figure 2.58. It consists of an early and a late correlator whose correlation intervals are $[\tau_e + kT_s, \tau_e + (k+1/2)T_s]$ and $[\tau_e + (k+1/2)T_s, \tau_e + (k+1)T_s]$, respectively. The early and late correlator outputs are clocked by an early and a late clock, where the early clock is advanced in time relative to the late clock by $T_s/2$. The period of both the early and late clocks is $T_s/2$. The rectifier (Magnitude block) removes the effect of data polarity on the phase detector output. A binary data symbol is assumed. The rectified output from the late correlator is subtracted from that of the early correlator to form the phase detector output, v_d.

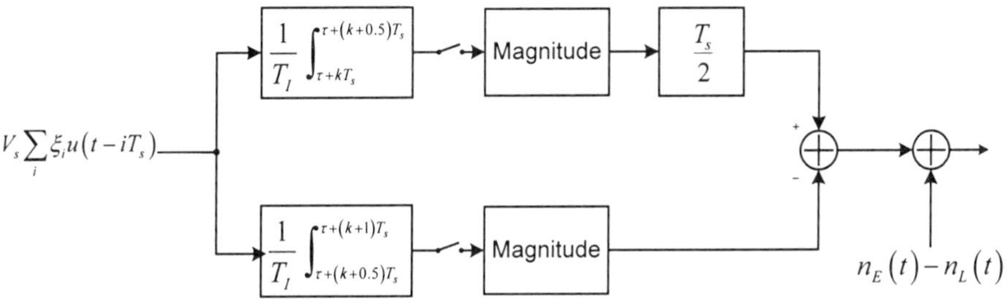

Figure 2.58 Phase detector for the early-late gate correlator.

Ideally, v_d should be of the form $K_d\tau_e$ to establish negative feedback. However, in practice, the ideal waveform can only be approximated by an S-curve, an example of which is shown in Figure 2.50 for the Costas loop. The phase detector in Figure 2.58 generates a similar S-curve. To see this, refer to Figure 2.59, where the timing of the early/late clocks are shown next to the received symbol transitions.

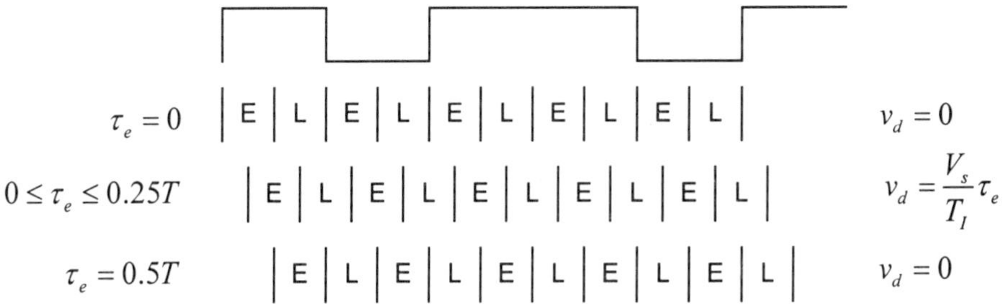

$\tau_e = 0$ | E | L | E | L | E | L | E | L | E | L | $v_d = 0$

$0 \le \tau_e \le 0.25T$ | E | L | E | L | E | L | E | L | E | L | $v_d = \frac{V_s}{T_I}\tau_e$

$\tau_e = 0.5T$ | E | L | E | L | E | L | E | L | E | L | $v_d = 0$

Figure 2.59 Phase detector (half-symbol early-late) output as a function of timing error.

When $\tau_e = 0$, the early correlator output equals that of the late correlator; thus, $v_d = 0$. As τ_e increases in the positive direction, the late correlator output will capture a symbol transition, if one exists. When this occurs, the rectified late correlator output will be less than that of the early correlator because of the polarity reversal at each symbol transition. Therefore, $v_d > 0$. Note v_d reaches a maximum when $\tau_e = T_s/4$ since beyond $T_s/4$ the rectified output of the late correlator begins to increase as more energy gets integrated in the late correlator. Therefore, v_d decreases linearly as τ_e becomes greater than $T_s/4$ and reaches zero when $\tau_e = T_s/2$. The upper half of the S-curve is then formed for $0 \le \tau_e \le T_s/2$. By using similar analysis, the lower half of the S-curve can be derived for $-T_s/2 \le \tau_e \le 0$. The complete S-curve is shown in Figure 2.60 and is periodic with period T_s.

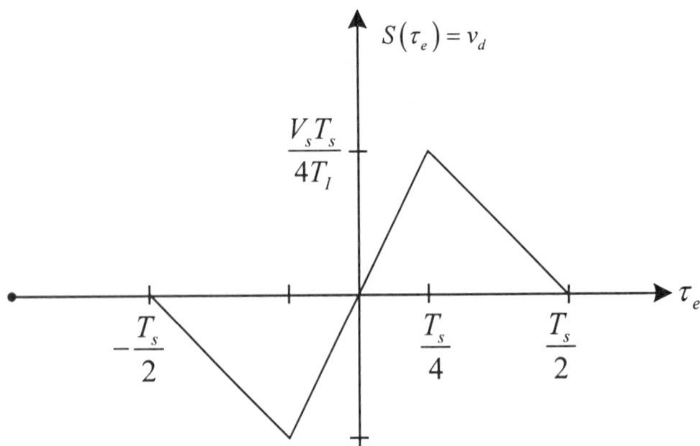

Figure 2.60 S-curve for the early-late correlator.

At the maximum, $v_d = 0.25V_s T_s / T_I$, where T_I is the correlation scale factor. Therefore, the slope at $\tau_e = 0$ is V_s / T_I, which extends from $-T_s / 4 \le \tau_e \le T_s / 4$. In this linear region $v_d = V_s \tau_e / T_I$. However, if there are no symbol transitions over an extended period of time, the phase detector output is zero. Therefore, sufficient symbol transitions must be maintained to obtain the S-curve characteristics shown in Figure 2.60. To see the effects of symbol transitions, let the transition density be denoted by d which varies between zero and one. Then, the phase detector output can be expressed by $v_d = V_s d \tau_e / T_I$, $|\tau_e| \le T_s / 4$, where the phase detector gain is $V_s d / T_I$. When d is zero, the phase detector gain is also zero.

The performance can be determined using the same linear model shown in Figure 2.51 with the exception that $K_d = V_s d / T_I$ and $n'(t) = [n_E(t) - n_L(t)] K_d^{-1}$ which has a power density of $0.5N_0 T_s^2 / V_s^2 d^2$ [56]. The resulting timing jitter due to noise can be expressed by the following expression:

$$E\left\{\left(\frac{\tau_e}{T_s}\right)^2\right\} = \frac{B_L}{2\bar{\gamma} R_b d^2}. \tag{2.154}$$

Example 2.15 Consider a digital system with transition density of 50%, a required E_b / N_0 of 10 dB, and a RMS jitter of 5%. According to (2.154), the noise bandwidth must be at least 80 times lower than the bitrate. □

2.5.5.2 Non-Coherent Early-Late Gate Correlator

Section 2.5.5.1 assumes that the phase has already been recovered. When this is not the case, the phase detector must generate the S-curve in the presence of carrier phase error at the input. A non-coherent phase detector is shown in Figure 2.61 which takes both the I and Q baseband signals to perform timing recovery. Each early and late correlator takes both I and Q as inputs, rectifies, and sums the two rectified outputs to remove the carrier phase. Note that the phase removal is only an approximate one since to remove the phase the correlator outputs should be squared. The rectifier approximates the squaring operation.

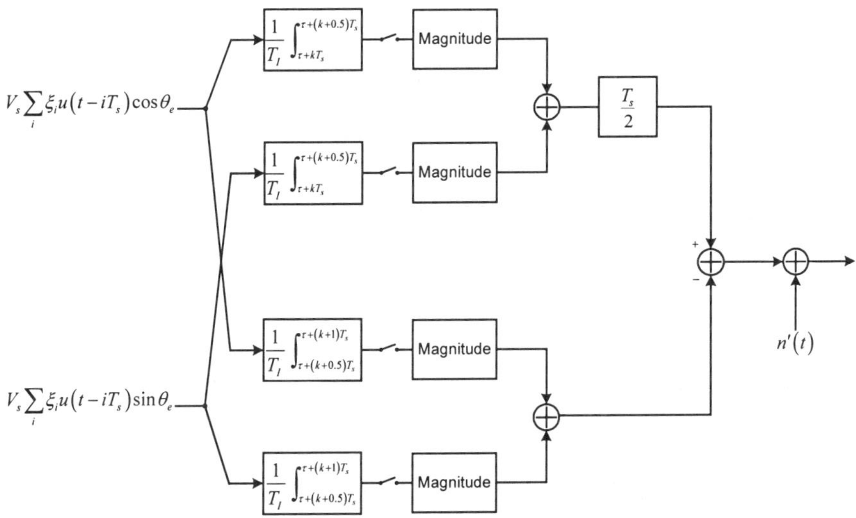

Figure 2.61 Non-coherent phase detector for the early-late gate correlator.

The S-curve has the same form as that of the coherent case but since an additional noise source is introduced via the Q channel, the SNR degrades by approximately 3 dB. Therefore, the phase jitter for a non-coherent timing recovery loop is approximately twice that of (2.154).

2.6 Forward Error Correction

As the channel condition degrades due to increased noise, interference, or signal fading, forward error correction (FEC) can be used to compensate for the associated performance degradation [63]. In particular, FEC provides coding gain G_c (described in Section 2.1.7) which helps to reduce the required E_b/N_0 for a given BER. In general, FEC is effective against uncorrelated disturbances such as white noise. If the disturbance is correlated, such as signal fading or bursty interference, then some type of interference averaging must be performed to obtain the performance gain provided by the FEC. Interference-averaging techniques include frequency hopping (0) and interleaving (Section 2.7.4.1). This section gives an overview of FEC and some of its basic principles.

2.6.1 Finite Fields

Much of the theoretical basis for FEC rests on formulations based on finite fields also known as Galois Fields (GF). A field is defined as a set in which any two elements are closed under addition, subtraction, and multiplication. The term *closed* refers to the fact that operations can be defined for elements in a set and that these operations generate elements in the same set. The order of a field is defined as the number of non-vectored elements in the field. For example, the GF field of order two, denoted by GF(2), has elements 0 and 1. An extension field is formed by groups of elements from another field. For example, the field of complex numbers is an extension field of real numbers whereby each complex element consists of a real and an imaginary part. In coding, GF(2) and its extension fields $GF(2^m)$ where $m = 2, 3, \cdots$ are often used to provide an algorithmic description of a coding and decoding process. A brief description of finite fields is provided in this section. For a more detailed treatment, the reader is referred to [64][65].

2.6.1.1 Fields

A set of elements forms a field when it satisfies the following conditions:

1) Elements are closed under addition and multiplication.
2) Elements satisfy the commutative and associative properties for both addition and multiplication.
3) Elements satisfy the distributive property.
4) There exist an additive identity and an additive inverse for each element in the field.
5) There exists a multiplicative identity for each element in the field.
6) With the exception of the additive identity, each element in the field has a multiplicative inverse

Example 2.16 A binary field or GF(2) contains two elements 0 and 1. The addition and multiplication operations can be defined to be a logical XOR and a logical AND, respectively. Since the outputs of an XOR and an AND are binary, the field elements are closed under addition and multiplication. If x and y are elements of GF(2) then multiplication of x by y is denoted by xy and the addition of x and y is denoted by $x \oplus y$. It can be easily shown using Boolean algebra that elements in GF(2) satisfy the commutative, associative property, and distributive properties. Also, the additive inverse is simply the element itself since $x \oplus x = 0$ and the additive identity is the '0' element. Finally, the multiplicative inverse and identity are both the '1' element. □

2.6.1.2 Vector and Matrix Operation in GF(2)

Following the discussion in Example 2.16, vector and matrix operations can be defined in GF(2). Let $\mathbf{x}$ be a vector in GF(2) defined as $[x_0\ x_1 \cdots x_{k-1}]$, where $x_0, x_1, \cdots, x_{k-1}$ belong to GF(2). The vector sum and dot product of two vectors $\mathbf{x}$ and $\mathbf{y}$ can then be defined as $\mathbf{x} + \mathbf{y} = [x_0 \oplus y_0\ \ x_1 \oplus y_1\ \ \cdots\ \ x_{k-1} \oplus y_{k-1}]$ and $\mathbf{x} \cdot \mathbf{y}^T = \sum_{i=0}^{k-1} x_i y_i$. An N x M matrix $\mathbf{G}$ in GF(2) is defined as $[\mathbf{g}_0^T\ \mathbf{g}_1^T \cdots \mathbf{g}_{M-1}^T]$, where $\mathbf{g}_0^T\ \mathbf{g}_1^T \cdots \mathbf{g}_{M-1}^T$ are column vectors of $\mathbf{G}$ with N elements belonging to GF(2), i.e.

$\mathbf{g}_i = [g_{i0}\ g_{i1} \cdots g_{i(N-1)}]$. Given a vector $\mathbf{u} = [u_0\ u_1 \cdots u_{N-1}]$, $\mathbf{uG}$ can then be defined as $\mathbf{uG} = [\mathbf{ug}_0\ \mathbf{ug}_1 \cdots \mathbf{ug}_{M-1}]$.

2.6.1.3 Polynomials in Finite Fields

A polynomial in a finite field is defined to have coefficients in GF(2) but roots in any of its extension fields $GF(2^m)$. A polynomial is often used to describe a vector in GF(2) where each element is a coefficient of the polynomial and the position of each element is associated with the degree of the independent variable.

Example 2.17 Let $\mathbf{c} = [c_n, c_{n-1}, \cdots, c_0]$, where $c_n, c_{n-1}, \cdots, c_0$ belong to GF(2). The corresponding polynomial representation is

$$c_n x^n + c_{n-1} x^{n-1} + \cdots + c_0.$$

□

Given two finite-field polynomials $p(x)$ and $q(x)$, the product of $p(x)$ and $q(x)$ is determined by collecting terms with equal powers and adding the corresponding coefficients in GF(2). See the example below.

Example 2.18 Let $p(x) = x^2 + x + 1$ and $q(x) = x + 1$. Then,

$$\begin{aligned} p(x)q(x) &= x(x^2 + x + 1) + 1 \cdot (x^2 + x + 1) \\ &= x^3 + x^2 + x + x^2 + x + 1 \\ &= x^3 + (1 \oplus 1)x^2 + (1 \oplus 1)x + 1 \\ &= x^3 + 1 \end{aligned}$$

□

Similarly, long division can be performed with the exception that all arithmetic operations on the coefficients are in GF(2).

Example 2.19 Let $p(x) = x^3 + x + 1$ and $q(x) = x + 1$. The long division of $p(x)/q(x)$ is shown below

$$\begin{array}{r} x^2 + x \\ x+1 \overline{) x^3 + x + 1} \\ \underline{\oplus x^3 + x^2 } \\ x^2 + x + 1 \\ \underline{\oplus x^2 + x } \\ 1 \end{array}$$

The division results in a quotient of $x^2 + x$ and a remainder of 1. □

A polynomial is defined to be irreducible over $GF(2^m)$ if it cannot be factored into polynomials of lower degree. An irreducible polynomial in the binary field with degree *m* divides $x^{2^m - 1} + 1$. A primitive polynomial is an irreducible polynomial with degree m over GF(2) that divides the polynomial $x^n + 1$ for $n = 2^m - 1$ but not for $0 < n < 2^m - 1$. Primitive polynomials are used in coding to generate extension fields $GF(2^m)$ for symbol-based codes, such as Reed Solomon codes. Primitive polynomials are also useful in generating pseudo-noise (PN) codes used extensively in spread-spectrum communications.

2.6.2 Block codes

A binary block code takes a set of K information bits and maps it into an N-bit codeword. The additional N-K bits in the codeword are known as the parity check bits, which are used by the decoder to correct errors. The code rate is defined as K/N and measures the amount of redundancy introduced into the encoded bit stream. A code is linear if given any two codewords $\mathbf{c}_1$ and $\mathbf{c}_2$, $\mathbf{c}_1 + \mathbf{c}_2$ is also a codeword. A linear code is also cyclic if given that $\mathbf{c} = [c_0, c_1, \cdots, c_{N-1}]$ is a codeword, then $[c_{N-j+1}, \cdots, c_0, c_1, \cdots, c_{N-j}]$ is also a codeword for any j. A code is systematic if the codeword has the information bits as its first K encoded bits, followed by N-K parity bits.

The performance of a block code can be measured by its minimum distance, analogous to $d_{\min}$ defined for digital modulations. The effectiveness of FEC against noise results from the larger distance between codewords as compared to the uncoded bits. Intuitively, this is because the uncoded bits taken K bits at a time form a K-dimensional space while the encoded bits taken N bits at a time form an N dimensional space with 2^{N-K} more degrees of freedom. The additional degrees of freedom allow the codewords to have greater distance between one another for increased protection against noise. To measure the distance between two codewords, we define the Hamming weight w_H of a codeword $\mathbf{x}$ and the Hamming distance $d_H(\mathbf{x}, \mathbf{y})$ between two codewords $\mathbf{x}$ and $\mathbf{y}$ as

$$w_H(\mathbf{x}) = \sum_{i=0}^{n} x_i \tag{2.155}$$

$$d_H(\mathbf{x}, \mathbf{y}) = \sum_{i=0}^{n} x_i \oplus y_i . \tag{2.156}$$

The Hamming distance measures the difference in bit patterns between two codewords. Since the code is linear, the addition of any two codewords is also a codeword. Thus, the minimum distance of a code can be found by determining the codeword with the least Hamming weight according to (2.156).

The minimum distance of a code determines coding gain which measures the effectiveness of an FEC in reducing the E_b / N_0 requirement. Coding gain has been defined in Section 2.1.7. It should be mentioned that if a decoding process uses soft decisions then the coding gain improves by 3 dB [18] over that of hard-decision decoding. Soft decisions are outputs of the maximum likelihood demodulator before decisions are made on the received symbols. Hard decisions, on the other hand, are the final symbol decisions made by the demodulator.

Other performance measures for FEC include the number of erasures ε, the number of correctable errors t, and the number of detectable errors λ. These are also functions of the minimum distance as shown below:

$$\varepsilon = \lambda = d_{\min} - 1 \tag{2.157}$$

$$t = \left\lfloor \frac{d_{\min} - 1}{2} \right\rfloor . \tag{2.158}$$

More generally, $d_{\min} \geq t + \lambda + 1$.

2.6.2.1 Encoding and Decoding of Linear Codes

A linear block code can be generated via a generator matrix $\mathbf{G}$ which consists of k row vectors, $\mathbf{g}_0$, $\mathbf{g}_1, \cdots$, $\mathbf{g}_{K-1}$. Given a block of K information bits represented as a vector $\mathbf{u}$, the codeword $\mathbf{c}$ of N encoded bits is generated using

$$\mathbf{c} = \mathbf{uG} \tag{2.159}$$

where $\mathbf{uG}$ is the matrix product of vector $\mathbf{u}$ and matrix $\mathbf{G}$ in GF(2). For a systematic code, the generator matrix has the form $[\mathbf{I}_K;\mathbf{P}]$, where $\mathbf{I}_K$ is a K x K identity matrix and $\mathbf{P}$ is a *K x (N-K)* parity matrix. The codeword then consists of the original K information bits followed by the parity-check bits formed by $\mathbf{uP}$.

For a binary linear code, a parity-check matrix $\mathbf{H} = [\mathbf{P}^T;\mathbf{I}_{n-k}]$ can be defined with dimension $(N-K) \times N$. Then, for any codeword c, the following hold

$$\mathbf{cH}^T = \mathbf{0}. \tag{2.160}$$

The condition in (2.160) results from the fact that $\mathbf{c} = [\mathbf{u};\mathbf{uP}]$ and $\mathbf{cH}^T = \mathbf{uP} + \mathbf{uP} = \mathbf{0}$. Expression (2.160) can be interpreted as a mechanism for the decoder to first compute the parity-check bits of the received codeword using its first K bits and then compare the result to the last *N-K* bits of the codeword, which correspond to the parity-check bits computed at the coder. If the two sets of parity-check bits match then no error has occurred. Otherwise, an error has occurred.

Let $\mathbf{y}$ be the received signal. Then $\mathbf{yH}^{\mathrm{T}} = \mathbf{0}$ only if $\mathbf{y}$ is a valid codeword. If detectable error has occurred then $\mathbf{yH}^T = \mathbf{S}$, where $\mathbf{S}$ is known as the syndrome of the received code vector $\mathbf{y}$. If $\mathbf{e}$ is the error vector then $\mathbf{S} = \mathbf{eH}^T$. One might then use the syndrome to identify the error pattern and correct for the error at the deccoder. However, there is not a unique map between a particular syndrome and an error pattern. To see this, let $\mathbf{y} = \mathbf{c}_1 + \mathbf{e}$ and $\mathbf{yH}^T = \mathbf{eH}^T = \mathbf{S}$. Then, a different error pattern $\mathbf{e}' = \mathbf{c}_2 + \mathbf{e}$ also generates the same syndrome since $\mathbf{y} = \mathbf{c}_1 + \mathbf{c}_2 + \mathbf{e} = \mathbf{c}_3 + \mathbf{e}$, where $\mathbf{c}_3$ is another valid codeword. Therefore, there are 2^K possible error patterns for a given syndrome value. To decode a linear code, the decoder computes the syndrome and selects the error pattern that has the least weight to decode the received code vector. The error pattern used for each syndrome can be determined a priori via algorithms such as the standard array [18].

Example 2.20 A Hamming code has the property that $N = 2^{N-K} - 1$. For N =7 and K = 4, the resulting hamming code is denoted as the (7, 4) Hamming code. This code has a generator matrix

$$G = \begin{bmatrix} 1 & 0 & 0 & 0 & 1 & 1 & 0 \\ 0 & 1 & 0 & 0 & 0 & 1 & 1 \\ 0 & 0 & 1 & 0 & 1 & 1 & 1 \\ 0 & 0 & 0 & 1 & 1 & 0 & 1 \end{bmatrix}$$

The corresponding parity-check matrix is

$$H = \begin{bmatrix} 1 & 0 & 1 & 1 & 1 & 0 & 0 \\ 1 & 1 & 1 & 0 & 0 & 1 & 0 \\ 0 & 1 & 1 & 1 & 0 & 0 & 1 \end{bmatrix}$$

Let $\mathbf{u}$ be [1101]. Then the corresponding codeword is $\mathbf{c} = [1101000]$. It can be verified that $\mathbf{cH}^T = \mathbf{0}$. The syndrome table is shown below:

Syndrome	*Error Vector*
000	0000000
001	0000001
010	0000010
011	0100000
100	0000100
101	0001000
110	1000000
111	0010000

Unfortunately, the syndromes do not uniquely determine a single error pattern. For instance, it can be shown that an error pattern of [0000110] also produces $\mathbf{S} = [110]$. For optimum decoding, the error pattern with the least possible weight is used for decoding in the syndrome table.

If the channel introduces an error $\mathbf{e} = [0000010]$ in the received code vector $\mathbf{y}$ then $\mathbf{y} = [1101010]$. The corresponding syndrome is $\mathbf{yH}^T = [010]$. Looking up the syndrome table for $\mathbf{S} = [010]$ the received code vector can be corrected by simply adding the corresponding error pattern in the syndrome table to $\mathbf{y}$, i.e. [1101010] + [000010] = [1101000].

If on the other hand, the channel introduces a double error $\mathbf{e} = [0000011]$ then the syndrome is $\mathbf{S} = [011]$ and the corresponding error pattern in the syndrome table is [0100000]. Effectively, the error pattern has now become [0100011], which shows a triple error. Therefore, a (7, 4) Hamming code cannot correct a double error. In fact, a double error may cause a triple error at the decoder. The above observation results from the fact that a (7, 4) Hamming code has $d_{\min} = 3$ and according to (2.158) the number of correctable errors is $t = 1$. □

2.6.2.2 Encoding and Decoding of Cyclic Codes

The linear block code described thus far requires a large memory if *N-K* is large and it becomes less efficient to implement the decoder. Therefore, most practical codes that are implemented are not only linear but also cyclic. Some examples of linear cyclic codes are the Hamming codes, Fire codes, Golay codes, BCH codes, and Reed Solomon codes. The latter three codes require computation in $GF(2^m)$ and will not be discussed here. Interested readers are referred to the following references [66][67][68].

Linear cyclic codes have the same linearity property as linear block codes. If $\mathbf{c}_1$ and $\mathbf{c}_2$ are codewords, then $\mathbf{c}_1 + \mathbf{c}_2$ is also a codeword. Additionally, linear cyclic codes have the property that any cyclic shift of a codeword is another codeword. A linear cyclic code is best represented in polynomials

$$c(D) = c_{N-1}D^{N-1} + c_{N-2}D^{N-2} + \cdots + c_0,$$

where $\mathbf{c} = [c_{N-1}, c_{N-2}, \cdots, c_0]$ is the codeword. It can be shown that for a linear cyclic code, each row of the generator matrix $\mathbf{G}$ can be represented as a cyclically shifted version of a specific vector, $\mathbf{g}$. The generator polynomial $g(D)$ is defined as the polynomial corresponding to $\mathbf{g}$. It can be shown directly from (2.159) that any codeword $c(D)$ can be generated from the product of the input word polynomial $u(D)$ and $g(D)$, i.e. $c(D) = u(D)g(D)$. In general, the code generator polynomial $g(D)$ has the following properties:

1. $g(D)$ is unique.
2. $g(D)$ has the form $D^{N-K} + g_{N-K-1}D^{N-K-1} + \cdots + g_1D^1 + g_0$ with degree *N-K*.
3. $g(D)$ is a factor of $D^N + 1$; i.e. $D^N + 1 \bmod g(D) = 0$ or $g(D)h(D) = D^N + 1$, where $h(D)$ is known as the parity-check polynomial and has degree *K*.
4. Any linear cyclic codeword $c(D)$ can be generated from $g(D)$ and there are 2^K such codeword polynomials.

The third property results in the cyclic shift property of the code, that a codeword cyclically shifted is also a codeword. A codeword cyclically shifted to the left by i positions to the left can be represented by $D^i c(D) \bmod D^N + 1$. The fourth property allows simpler implementation of linear cyclic codes. The primary method of generating a linear cyclic codeword $c(D)$ is shown below:

$$c(D) = D^{N-K}u(D) + D^{N-K}u(D) \bmod g(D), \tag{2.161}$$

where the left shift by N-K places the information bits in the first N-K position in the codeword while the second term generates the parity-check bits as the rest of the codeword. To see that (2.161) generates a valid codeword, observe that $D^{N-K}u(D) \bmod g(D)$ is the remainder of $D^{N-K}u(D)$ divided by $g(D)$. Then, there exists a quotient polynomial $q(D)$ such that

$$D^{N-K}u(D) = q(D)g(D) + D^{N-K}u(D) \bmod g(D).$$

Rearranging the terms in GF(2) results in (2.161).

Similar to linear block codes, decoding of cyclic codes is based on syndromes. Let the received code polynomial be $y(D)$ and the error polynomial be $e(D)$, which corresponds to the error vector. $y(D)$ can be represented by $u(D)g(D) + e(D)$. Dividing $y(D)$ by $g(D)$ and finding the remainder results in the syndrome polynomial:

$$s(D) = y(D) \bmod g(D) = e(D) \bmod g(D). \tag{2.162}$$

If there are errors in the channel, then $s(D)$ is non-zero and is indicative of the type of error. For an (N, K) cyclic code, there exist 2^K error polynomials corresponding to a particular syndrome. In general, the error polynomial with the minimum terms is selected for a particular syndrome. The straightforward decoding algorithm first generates the syndrome by using (2.162) and decodes by adding the error polynomial to the received polynomial. This method also requires storage similar to a linear block code. To eliminate the storage requirement, more efficient methods exist. One such method is the Meggitt decoder [69] based on a reference syndrome $s'(D)$ and the associated error polynomial $e'(D)$. Because of the cyclic property of the code, it is always possible to shift y by i positions to the left and arrive at the reference syndrome $s'(D)$. Decoding can therefore be accomplished based on the following two steps:

Step 1: Find i such that $D^i y(D) \bmod g(D) = s'(D)$ (2.163)

Step 2: $\hat{y}(D) = y(D) + D^{-i}e'(D).$ (2.164)

Example 2.21 Consider the (7, 4) Hamming code of Example 2.20. Let **G** in Example 2.20 be defined as $\mathbf{G} = [\mathbf{g}_0^T\ \mathbf{g}_1^T\ \mathbf{g}_2^T\ \mathbf{g}_3^T]^T$, where $\mathbf{g}_0 = [1000110]$, $\mathbf{g}_1 =$ [0100011], $\mathbf{g}_2 =$ [0010111], $\mathbf{g}_3 =$ [0001101]. Let $z(D) = D^3 + D^2 + 1$. Then, it can be shown that

$$g_0(D) = (D^3 + D^2 + D^1)z(D),$$
$$g_1(D) = (D^2 + D + 1)z(D),$$
$$g_2(D) = (D + 1)z(D),$$
$$g_3(D) = z(D).$$

Therefore $z(D)$ is the generator polynomial for the (7, 4) Hamming code.

Let $\mathbf{u} = [1101]$ with the corresponding input word polynomial of $u(D) = D^3 + D^2 + 1$. The encoded code polynomial is then $c(D) = u(D)g(D) = u(D)z(D) = D^6 + D^4 + 1$. This codeword, however, is not systematic. To generate a systematic code, (2.161) is used as shown below:

$$\begin{aligned} c(D) &= D^3u(D) + D^3u(D) \bmod g(D) \\ &= D^6 + D^5 + D^3 + (D^6 + D^5 + D^3) \bmod (D^3 + D^2 + 1) \\ &= D^6 + D^5 + D^3. \end{aligned}$$

The corresponding systematic codeword is $\mathbf{c} = [1101000]$.

If the channel introduces an error $\mathbf{e} = [0100000]$ in the received code vector $\mathbf{y}$ then $\mathbf{y} = [1001000]$. The corresponding error polynomial and received codeword polynomials are respectively

$$e(D) = D^5$$
$$y(D) = D^6 + D^3.$$

Applying (2.162), the corresponding syndrome polynomial is $s(D) = D + 1$. The syndrome polynomial corresponding to $e'(D) = D^6$ is $s'(D) = D^2 + D$. By shifting y one position to the left and applying (2.163), the resulting syndrome matches $s'(D)$. The received vector can then be decoded based on (2.164) by adding $D^{-1}e'(D)$ to *y(D)*, i.e.

$$D^{-1}D^6 + D^6 + D^3 = D^6 + D^5 + D^3,$$

which corresponds to the originally transmitted codeword $\mathbf{c} = [1101000]$. □

2.6.2.3 Performance

Historically, block codes were first applied to computer systems. A notable application of Hamming codes is for computer systems [70]. Such codes have fairly poor error correcting capability, i.e. a few correctable errors. As shown in Table 2.8, the (7,4) Hamming code has $t = 1$ and a hard-decision coding gain of only 0.6 dB. These codes are however relatively simple to implement.

Block codes have also been used extensively in recording media, such as, CD recording to combat the large bursts of error that may occur due to damages on the CD surface. One such code being widely used is the Reed Solomon code which codes a block of symbols rather than bits. Each symbol is represented by m bits. Reed-Solomon codes form a subset of a larger class of codes known as the Bose-Chaudhuri-Hocquenghem (BCH) codes, a few of which are listed in Table 2.8. The block length is generally $2^m - 1$ and the coding gain ranges between 5-7 dB for reasonably high code rates.

BCH codes have good error correcting capabilities against bursty errors because they can correct mt bit errors at a time. BCH codes are therefore also referred to as non-binary codes. In wireless communications, BCH codes have been used for Cellular Data Packet Data (CDPD) system and they are also being considered for the next-generation cellular systems. Unlike BCH codes, Fire codes are binary codes that also offer good error correcting performance against burst errors and are used for current cellular systems. One such code used in GSM is listed in Table 2.8. In general, the coding gain of block codes increases as the block size increases.

Table 2.8 Performance of selected block codes [71][72].

Codes	g(x) (Octal)	(N, K)	Code Rate	t	G_{hc} (dB)
Hamming	15	(7,4)	0.571	1	0.58
BCH	3447023271	(99,127)	0.780	4	4.94
	624730022327	(92,127)	0.724	5	6.38
	75626641375	(223,255)	0.875	4	6.41
	23157564726421	(215,255)	0.843	5	7.04
Fire	20000440400011	(224,184)	0.821	12	10.3

2.6.3 Convolutional Codes

Unlike the block code, a convolutional code does not have a fixed block length. Rather, its length can be made to some extent arbitrary without changing the coder or decoder. A convolutional coder [63] introduces memory in the transmitted information bits such that each encoded bit depends on the values of previously transmitted bits. CPM, described in Section 2.4.2, has a similar encoding scheme in which the modulator introduces memory in the transmitted phase such that the current phase is a function of previously transmitted symbols. In general, due to the memory in the encoding process, the decoder must take a sequence of received bits for decoding. By introducing the memory in a controlled fashion, the transmitted encoded data exhibit a trellis-like structure, similar to that of CPM. Thus, the distance property discussed in Section 2.4.2.2 for CPM also applies to convolutional code. As in CPM, the minimum distance between two paths in the trellis that diverge and then remerge increases as the length of the sequence increases. In convolutional coding, the minimum distance for an infinite length sequence is defined as the free distance or d_{free}. The coding gain of convolutional codes can be approximated using (2.46) with $d_{\min}$ replaced by d_{free}.

2.6.3.1 Encoder

Figure 2.62 shows a convolutional coder which takes a serial bit stream $\mathbf{u} = [u_0 u_1 \cdots]$ and splits it into K subsequences $\mathbf{u}_1, \mathbf{u}_2, \cdots, \mathbf{u}_K$ that are input to a bank of KN code generator functions $\mathbf{g}_{ij}$ to form N coded subsequences $\mathbf{x}_1, \mathbf{x}_2, \cdots, \mathbf{x}_N$. The convolutional code is formed by time multiplexing the N sub-sequences into a single coded bit stream $\mathbf{x} = [x_0 x_1 \cdots]$. In other words, the coder generates N coded bits from K information bits at a rate of $(KT_b)^{-1}$. The ideal code rate is therefore K/N. In Section 2.6.3.2, it will be shown that the code rate is actually less due to overhead in the decoding process.

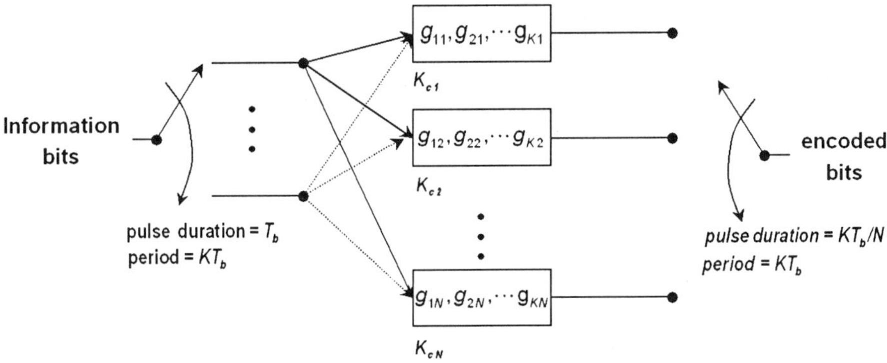

Figure 2.62 Convolutional encoder with rate *K/N*.

Each of the coded subsequences can be derived through convolution in *GF(2)*:

$$\mathbf{x}_j = \sum_{i=1}^{K} \mathbf{u}_i * \mathbf{g}_{ij}, \tag{2.165}$$

where $j = 1, 2, \ldots N$ and the '*' operator denotes linear convolution. In GF(2), linear convolution between two sequences $\mathbf{u}$ and $\mathbf{g}$ can be defined as follows:

$$\mathbf{x} = \sum_{k=0}^{\infty} u_k \beta_{n-k} \tag{2.166}$$

where the multiplication and addition are in *GF(2)*. The generator functions $\mathbf{g}_{ij}$ can be viewed as digital filters in *GF(2)* and $\mathbf{x}_j$ is the filter response due to the K input sub-sequences $\mathbf{u}_1, \mathbf{u}_2, \cdots, \mathbf{u}_K$. Each $\mathbf{g}_{ij} = [\beta_0 \ \beta_1 \cdots \beta_{K_{ij}-1}]$ is fully described by its tap values β_m, where m is an integer ranging from 0 to $K_{ij} - 1$.

Treated as finite-impulse response (FIR) filters, each generator function consists of K_{ij} delay elements or $K_{ij} + 1$ taps. Each tap in the FIR filter introduces a unit of memory into the encoded sequence. Based on this observation and from (2.166), the memory introduced by the ith input sub-sequence is the maximum of all K_{ij} for $j = 1$ *to N*. The constraint length K_c of the code is related to the largest memory introduced into the encoded bits and is defined as:

$$K_c = 1 + \max_{i,j} K_{ij}. \tag{2.167}$$

A state is defined as one of the possible values stored in the delay elements. For a delay element that stores binary values (i.e. a register) there exist $2^{K_{ij}}$ distinct states associated with $\mathbf{g}_{ij}$. The tap values of $\mathbf{g}_{ij}$ determine the state transitions and the associated output for a given input value. Note the similarity of the above observations to the phase states and phase transitions of CPM. A similar trellis can therefore be constructed for convolutional codes. The number of states in the trellis is equal to $2^{N_{mem}}$, where N_{mem} is the total amount of memory introduced into the encoded sequence and can be expressed by

$$N_{mem} = \sum_{i=1}^{K} \max_{j} K_{ij}. \tag{2.168}$$

The number of transitions leaving each state is 2^K, which also holds in general for the number of transitions entering each state. Each transition is marked by the K input bits that led to the transition and the N coded bits output by the encoder. The computational complexity is on the order of $O\left(2^{N_{mem}}\right)$ with $2^{N_{mem}}$ ACS having 2^K inputs at each stage of the trellis.

The minimum distance of the convolutional code can be found by computing the Hamming distance between paths which diverge and then remerge a common state. Based on the discussion in Section 2.4.2.2, free distance for a convolutional code can be approximated by

$$d_{free} \le d_N = \min_{ij} \left\{ d_H(\mathbf{\Gamma}_i, \mathbf{\Gamma}_j) \right\} \tag{2.169}$$

where $\mathbf{\Gamma}_i$ and $\mathbf{\Gamma}_j$ represent the ith and j^{th} path, respectively. Given that most practical convolutional codes are also linear, (2.169) can be simplified further by noting that $d_H(\mathbf{\Gamma}_i, \mathbf{\Gamma}_j) = w_H(\mathbf{\Gamma}_i + \mathbf{\Gamma}_j)$ and $\mathbf{\Gamma}_k = \mathbf{\Gamma}_i + \mathbf{\Gamma}_j$. Therefore, finding the minimum distance between any two paths is equivalent to finding the path with the least Hamming weight in the coded sequence:

$$d_{free} \le d_N = \min_{k} w_H(\Gamma_k). \tag{2.170}$$

(2.170) implies that to find the minimum distance one only needs to consider paths that diverge and remerge with the all-zero's path.

Rate 1/3 Convolutional Encoder: A rate-1/3 convolutional encoder is shown in Figure 2.63a. Since $K = 1$ and $N = 3$, the input sequence is not split into multiple subsequences but there are three coded subsequences $\mathbf{x}_1$, $\mathbf{x}_2$, and $\mathbf{x}_3$ which are time multiplexed into a single coded sequence $\mathbf{x}$. The generator functions are $\mathbf{g}_{11}$ = [100], $\mathbf{g}_{12}$ = [110], and $\mathbf{g}_{13}$ = [111], each with memory $K_{11} = 0$, $K_{12} = 1$, and $K_{13} = 2$. The constraint length according to (2.167) is thus $K_c = 3$.

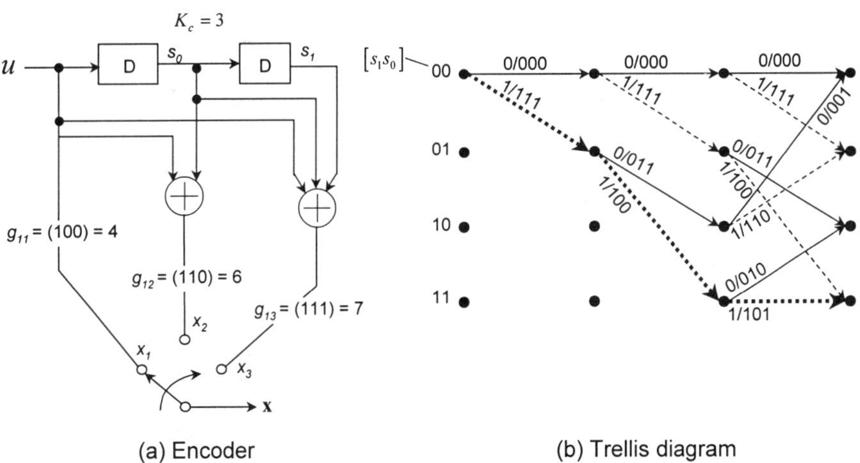

Figure 2.63 A rate 1/3 convolutional encoder. Dashed and solid lines represent an input of '1' and '0', respectively.

The trellis diagram is shown in Figure 2.63b. The number of states is $2^{K_c-1} = 4$ and the number of transitions per state is $2^1 = 2$. Each transition is marked by the associated encoder input and output at each stage of the trellis. For instance, assume that at the first stage of the trellis (i.e. $n = 1$), the initial state of the encoder is [00] and that the input sequence is $\mathbf{u} = [111]$. Then, by applying (2.165), $\mathbf{x}_1 = [111]$, $\mathbf{x}_2 = [100]$, and $\mathbf{x}_3 = [101]$. The time multiplexed coded sequence is $\mathbf{x} = [111100101]$. Given the initial state of the encoder, the trellis starts from state [00] and transitions into state [01]. The transition is marked by 1/111 indicating that an input value of "1" caused the transition and that "111" is output from the coder. In the second stage of the trellis, state [01] transitions to state [11] and 1/100 marks the transition. At the third stage of the trellis, state [11] transitions back to itself and 1/101 marks the transition. The path traced by the above set of transitions is highlighted in Figure 2.63b.

The free distance of the code can be approximated as in CPM by inspection of the trellis over a finite length of the code sequence. For instance, starting from state [00], two paths diverge and then remerge at the third stage of the trellis. Using the transitions to mark these two paths, they can be denoted by {0/000,0/000,0/000} and {1/111, 0/011, 0/001}. According to (2.169), the distance between these two paths is 6. The free distance is then at least 6 and the corresponding coding gain is 3 dB, assuming soft-decision decoding.

Rate 2/3 Convolutional Encoder: A Rate 2/3 convolutional encoder is shown in Figure 2.64. In this case, $K = 2$ and $N = 3$. The generator functions are $\mathbf{g}_{11} = [01]$, $\mathbf{g}_{12} = [00]$, $\mathbf{g}_{13} = [11]$, $\mathbf{g}_{21} = [11]$, $\mathbf{g}_{22} = [11]$, and $\mathbf{g}_{23} = [10]$. The memory of the generator functions are $K_{11} = 1$, $K_{12} = 0$, $K_{13} = 1$, $K_{21} = 1$, $K_{22} = 1$, $K_{23} = 0$. According to (2.167), the constraint length of the code is 2 and according to (2.168) the number of states is 4. The number of transitions per state is $2^2 = 4$. Therefore, the ACS at each state requires four compares rather than two compares as for the rate 1/3 code. Though more comparisons make a rate 2/3 convolutional decoder more complex, the advantage is that the bandwidth efficiency improves by a factor of two since the code rate is 2/3 rather than 1/3.

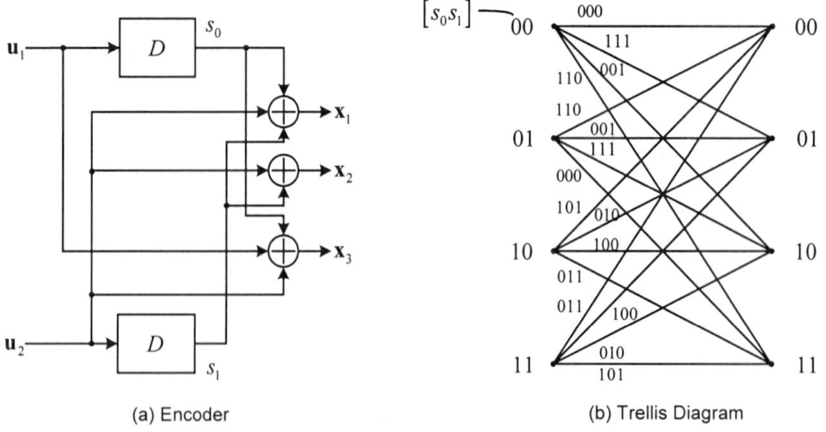

(a) Encoder (b) Trellis Diagram

Figure 2.64 A rate 2/3 convolutional coder.

In the above example, it has been determined that when $K \neq 1$, the convolutional decoder requires more comparisons per ACS. By puncturing a *1/N* convolutional code, a rate *N/(N+1)* code can be generated without the need for more comparisons. For instance, Figure 2.65 shows that by puncturing a rate 1/2 code in the second bit of every other output, the same trellis transition is formed as that of a rate 2/3 code.

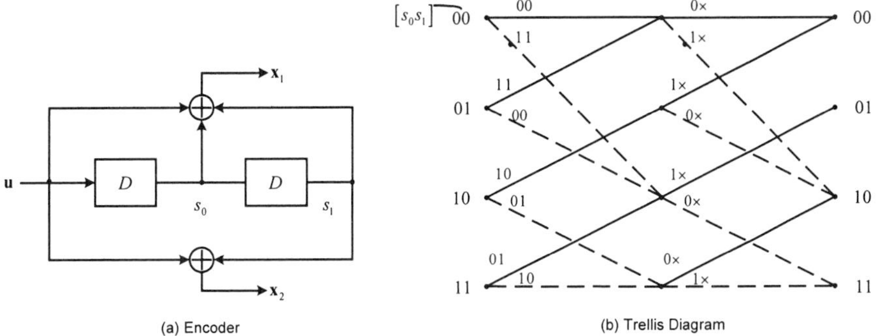

Figure 2.65 A rate 2/3 convolutional coder obtained by puncturing a rate 1/2 coder. Dashed and solid lines represent an input of '1' and '0', respectively.

2.6.3.2 Viterbi Decoder

Given the trellis structure of the convolutional code, the Viterbi algorithm [50][73] described in Section 2.4.2.1 can be used for decoding. Specifically, (2.130) is rewritten below for convolutional codes

$$\hat{\mathbf{u}}_{M-D} = \max_{\sigma_l,\sigma_{l+1}}{}^{-1}\left\{\sum_{i=0}^{M}\lambda_i(\sigma_l,\sigma_{l+1})\right\}. \tag{2.171}$$

where $\hat{\mathbf{u}}_{M-D}$ denotes the decoded input sequence, M is the total number of iterations used, and σ_l denotes the state of the coder at time index l, i.e. the lth stage of the trellis. Note that the state of the coder is based on the binary values stored after each delay element in the generator function. The transition metrics $\lambda_i(\sigma_l,\sigma_{l+1})$ can either be based on soft-decisions or hard-decisions as shown in (2.172) and (2.173), respectively:

Hard-decision based: $$\lambda_i(\sigma_l,\sigma_{l+1}) = -d_H(\mathbf{y}_i,\mathbf{x}_i) \tag{2.172}$$

Soft-decision based: $$\lambda_i(\sigma_l,\sigma_{l+1}) = \mathbf{y}_l(\mathbf{1}-2\mathbf{x}_l)^T \tag{2.173}$$

where $\mathbf{y}_l$ and $\mathbf{x}_l$ are respectively the received code vector and the code vector corresponding to transition σ_l, σ_{l+1}, at time index l. $\mathbf{y}_l$ and $\mathbf{x}_l$ are both of length N. The constant vector $\mathbf{1}$ denotes a row vector of length N with all '1' elements and

$\mathbf{1}-2\mathbf{x}_l$ maps the elements of $\mathbf{x}_l$ from {0, 1} into ±1. Therefore, (2.173) applies to binary modulation but can be extended with modification to higher order modulations.

Using the hybrid method described in Section 2.4.2.1, to decode a block of B input vectors of K bits, the input sequence is generally terminated with $K_c - 1$ zeros to bring the received sequence to the all-zero state. The decoding delay is $N_d K$ and $(B + K_c - 1)K$ input bits are needed to decode a block of BK information bits. The $K_c - 1$ bits needed to force the decoder into a known state then limit the code rate that can be achieved in practice as shown below

$$\text{Practical code rate:} \qquad r = \frac{BK}{(B + K_c - 1)N}. \qquad (2.174)$$

Rate 1/3 Convolutional Decoder: Consider the rate 1/3 convolutional code discussed earlier. If the decoder must decode a block of 200 bits then B = 200. Since the constraint length is determined to be 3 then according to (2.174) the effective code rate is 1/3.03 and shows negligible loss as compared to the ideal code rate of 1/3. The decoding depth is usually chosen to be fives times the constraint length or $5K_c = 15$ bits in this case. Therefore, the decoding delay is 15 bits.

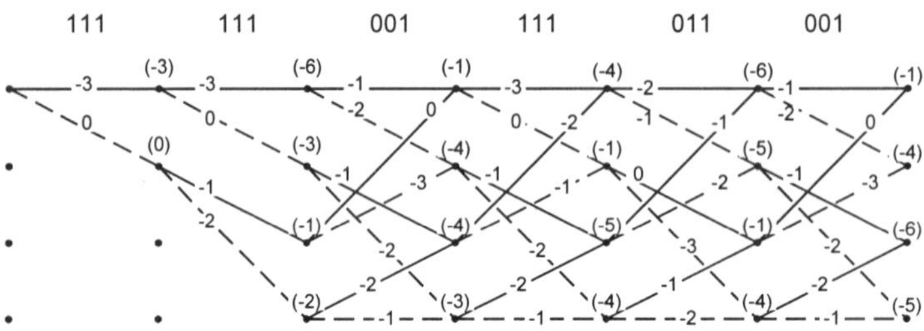

(a) Computation of branch metrics and cumlative metrics, enclosed in parenthesis.

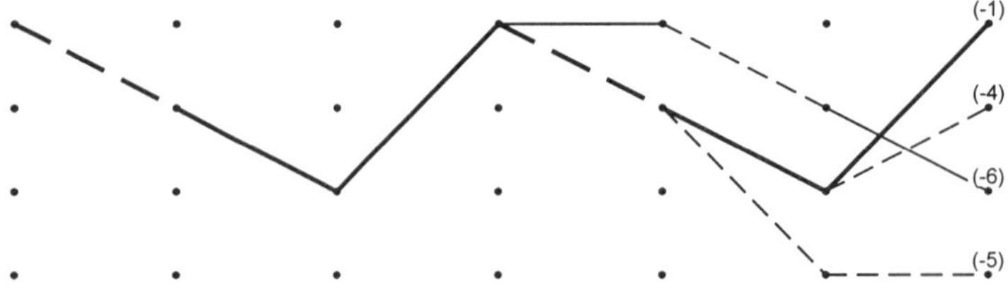

(b) Paths that survived at the end of the sixth iteration. Bold line is the path determined by traceback from the state with the maximum cumlative metric.

Figure 2.66 Illustration of the Viterbi algorithm on the rate 1/3 convolutional code of Figure 2.63. Dashed and solid transitions correspond to an encoded input bit of 1 and 0, respectively.

Assume that the transmitted bits are $\mathbf{u}$ = [100100] where the last two bits are used to drive the decoder into the all-zero state. Referring to the trellis diagram in Figure 2.63, the encoded sequence is then [111011001111011 001]. Assume that an error has occurred during the reception with an error vector [000100000000000000]. Then, the received signal at the decoder input is [111111001111011001]. Decoding of this input sequence using the Viterbi algorithm is illustrated in Figure 2.66 where the transition metrics based on (2.172) are annotated on each branch of the trellis and the accumulated metric is annotated next to each state. The paths, which have not been selected subject to condition (2.171), are in dashed lines and the surviving paths are in solid lines. At the end of the decoding cycle, the survivor path that led to the all-zero state is selected and the input that led to this path is the decoder output which is equal to the transmitted bit sequence $\mathbf{u}$. Thus, a single error event has been corrected.

2.6.3.3 Performance

Convolutional codes are widely used by many systems, including GSM and IS-95. The coding gain ranges from 1 to 5 dB for codes with a reasonable number of states as shown in Table 2.9. Most systems use the rate 1/2 and 1/3 codes. For instance, referring to Table 2.9, IS-95 uses the rate 1/2 constraint length 9 code for the forward link and the rate 1/3 constraint length 9 code for the reverse link. GSM uses a rate 1/2 code with shorter constraint length of 5 for the speech traffic channel and a constraint length 5 rate 1/3 code for the 4.8 kbps data channel. Both of these codes are listed in Table 2.9. In general, the coding gain increases with larger constraint length and lower code rate. The trade-off is that lower code rate reduces the bandwidth efficiency of the system and a larger constraint length increases the number of states and thus results in more complex designs. The coding gains listed in Table 2.9 reflect values obtained through hard-decision. By applying soft-decision, the gain can be increased by another 3 dB.

Table 2.9 Performance of selected convolutional codes. [74][75][76]

Code Rate	Generator Polynomial (Octal)	Constraint Length	d_{free}	G_{hc} (dB)
1/4	13,15,15,17	4	13	2.11
	25,27,33,37	5	16	3.01
	463, 535,733,745	9	24	4.77
1/2	15,17	4	6	1.76
	23,33	5	8	3.01
	753, 561	9	12	4.77
1/3	13,15,17	4	10	2.22
	25,33,37	5	12	3.01
	557, 663, 711	9	18	4.77
2/3	236,155,337	4	7	3.68
3/4	13,25,61,47	2	4	1.76

2.6.4 Other Codes

Thus far, commonly used coding techniques have been covered. These techniques include linear block codes, cyclic codes, and convolutional codes, all in *GF(2)*. There exist other advanced coding techniques which are briefly described below. Readers interested in more detailed analysis can refer to the references indicated below.

- **BCH and Reed-Solomon codes:** BCH codes [66] and Reed-Solomon (RS) codes [67][68] are block codes that require operation in $GF(q^m)$, where q is a prime number. These codes take a block of K symbols and generate a block of N code symbols, where each symbol consists of $m\log_2 q$ bits and represents an element of the field $GF(q^m)$. Since each symbol corresponds to $m\log_2 q$ bits, a t-symbol correcting code can effectively correct for $mt\log_2 q$ bits. Therefore, these codes can be designed to correct for burst errors with duration less than $mt\log_2 q$ bits. Convolutional codes and block codes described in the previous sections, on the other hand, can only handle random bit errors. Decoding algorithms [77][78][79] for BCH and RS codes are based on similar techniques described for binary cyclic codes but require operations in $GF(q^m)$.
- **Concatenated codes:** A high coding gain requires an increase in the minimum distance, which in general requires a longer block size and lower code rate. A single low rate code with longer block size can therefore be used to meet a high coding gain requirement. However, this requires a large increase in computation complexity. To reduce the complexity, two shorter codes are concatenated to generate a code that has an effective $d_{\min} = d_{\min 1} d_{\min 2}$ and $r = r_1 r_2$, where $d_{\min 1}$, $d_{\min 2}$, r_1, and r_2 are the minimum distances and code rates of the two codes. Such codes are also known as a concatenated code, first developed by Forney [80]. The two constituent codes are referred to as the outer code and the inner code. The latter is closest to the channel. In wireless systems, concatenated codes using the RS-code as the outer code and convolutional code as the inner code show improved performance.
- **Trellis-coded modulations (TCM):** TCM [81][82] addresses the problem of bandwidth efficiency. Block codes and convolutional codes achieve coding gain but at the price of reduced bandwidth efficiency by an amount equal to the code rate. TCM, on the other hand, does not compromise bandwidth and still achieves a relatively high coding gain. TCM maintains no loss in bandwidth efficiency by using higher order modulation to compensate for the loss due to adding redundancy in the coded bits. However, the constellation points must be suitably partitioned [83] to avoid loss in E_b/N_0 due to increasing constellation size. For instance, if the system requires a bandwidth efficiency of 4 bps/Hz, then a 32-point QAM constellation can be used by partitioning it into eight 4-point sub-constellations. The eight sub-constellations can be mapped using two information bits via a rate 2/3 convolutional code, where the three coded bits can be used to specify one of the eight sub-constellations. Two more information bits are used to modulate the 4-point constellation, for a total bandwidth efficiency of 4 bps/Hz.

The coding gain arises from the increased $d_{\min}$ within the smaller 4-point sub-constellations and the distance separation between the sub-constellations due to the trellis structure of the rate 2/3 convolutional code. For instance, if a 128-state rate 2/3 convolutional code is used to label the sub-constellations, a coding gain of 3 to 6 dB is achieved with respect to uncoded 16 QAM. Both the TCM and the uncoded 16 QAM have a bandwidth efficiency of 4 bps/Hz.

2.7 Diversity Techniques

FEC is most effective against random error statistics due to uncorrelated noise present, for instance, in an AWGN channel. When the noise source shows correlation, then diversity techniques [84] should be employed to compensate for the loss in E_b/N_0. Without diversity, the system can show substantial loss in energy-efficiency, even with coding, due to an increase in the required E_b/N_0. The gain in E_b/N_0 that can be achieved for a system with diversity over one without diversity is known as the diversity gain, defined in Section 2.1.8. Multipath fading discussed in Section 2.2.5 causes correlated errors. Specifically, fading scatters the signal energy in time, frequency, and/or space. Diversity reception effectively combines energy from the scattered components to improve the received E_b/N_0.

Depending on the inherent structure in the channel (see Section 2.2.6), different types of diversity schemes are applied and sometimes as a combination of two or more types. The different types of diversity schemes include 1) path diversity, 2) time diversity, 3) frequency diversity, and 4) space diversity. Path and frequency diversity are most effective against frequency-selective fading. Time diversity combats the burst errors introduced by time-selective fading and space diversity mitigates the energy loss due to space-selective fading. Since the channel may display selectivity in frequency, time, and space, diversity schemes may combine signals in more than one domain. Such combined schemes are also known as space-time processing [35]. More recently, interference diversity [35] has been developed to combat interference resulting from other users in a multiple-access system. The following sub-sections discuss in more detail the various diversity techniques.

2.7.1 Basic Diversity Combining

Diversity order is defined as the number of independent components in the scattered signal. The block, which processes the scattered component, is known as a diversity branch. A typical diversity receiver (Figure 2.67) consists of K_D diversity branches followed by a combiner. Ideally, the fading statistics on each of the diversity branches should be independent so that best performance can be achieved when their outputs are combined. Independence assures that if one branch is in a fade the other branches are most likely not in a fade. On average, the combiner produces a higher SNR compared to a system without diversity. However, the combiner SNR does not increase without bound as K_D increases. Rather it is limited by the diversity order of the channel.

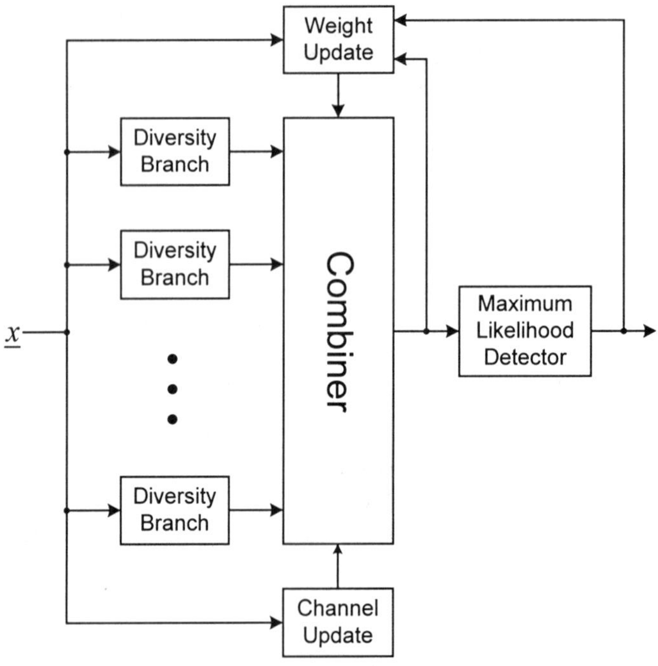

Figure 2.67 A general diversity receiver.

The design of diversity reception depends on the choice of the diversity branch and combiner for a given channel condition. In general, there are three basic types of combiners: 1) selective combining (SC) [85], 2) equal-gain combining (EGC) [86], and 3) maximum ratio combining (MRC) [20]. Their computations are shown below:

SC: $$z(n) = y_k(n) \text{ where } y_k(n) \text{ satisfies } \max_j |y_j(n)|^2 / \sigma_N^2 \quad (2.175)$$

EGC: $$z(n) = \sum_j y_j(n) \quad (2.176)$$

MRC: $$z(n) = \sum_j c_j^* y_j(n) \quad (2.177)$$

where y_j denotes the output of the sampler for the jth diversity branch and $z(n)$ is the combiner output for the nth received symbol.

The SC scheme selects the branch output with the largest SNR. The selected branch is denoted as the kth branch in (2.175). In contrast, EGC simply adds the outputs of all the diversity branches to improve the signal strength. MRC weights each branch output with the complex conjugate of the estimated channel parameter c_j. If the estimates are perfect and if there is no ISI, then MRC results in optimal combining since it combines energy from all the diversity branches, accounting for amplitude fading. Because of the zero ISI assumption, however, MRC is only optimal when the channel fading is frequency-flat or if the effect of the ISI can be eliminated. For channels that are not frequency-flat, other more effective combiners can be used and are described later in subsequent sections. Also, for diversity branches with equal received signal strength, EGC is equivalent to MRC.

The diversity gain that can be achieved for a given diversity scheme depends on the distribution of the signal at the output of the combiner. For instance, assume that the input of each diversity branch has independent Rayleigh amplitude distribution, i.e. $|c_j|$ is independent Rayleigh distributed. This occurs when the diversity branches have antenna elements separated with spacing greater than the coherence distance of the channel. The output of the MRC then has a Chi Squared distribution as a result of taking the sum of the squares of independent Rayleigh distributed signals from the output of the diversity branches. Given the Chi Squared distribution, ML detection described in Section 2.3.1 results in a BER performance and diversity gain described in (2.47) and (2.48), respectively, for binary modulation. Applying the same principle, the performance of SC and EGC can also be obtained.

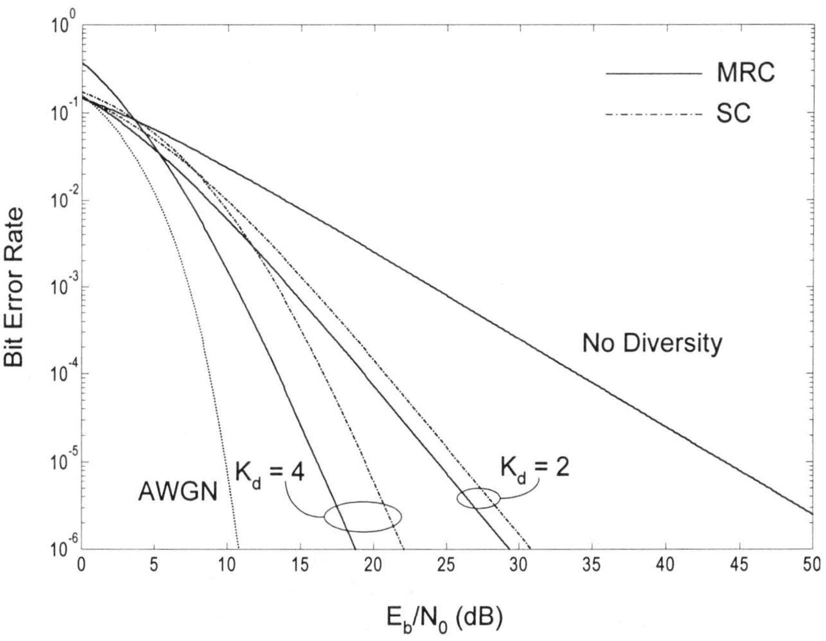

Figure 2.68 BER curves for MRC and SC of BPSK signals.

The diversity performance of BPSK in Rayleigh frequency-flat fading is shown in Figure 2.68 for different number of diversity branches K_d and combining methods. From the BER curves, MRC shows performance superior to that of SC. Results in Figure 2.68 assume independent Rayleigh fading components of equal strength in each of the diversity branches. Therefore, in this case, the performance of MRC also corresponds to that of EGC. Generally, however, EGC has slightly poorer performance compared to that of MRC when the fading components are correlated and are of different strengths. Since MRC has the best performance, the performance of MRC will be used throughout the text as a lower bound on the performance achievable with diversity, given an independent Rayleigh distribution at the input of each diversity branch. It will be seen later that this assumption in general holds for most types of diversity schemes.

2.7.2 Path Diversity

A frequency-selective channel causes time dispersion, characterized by its channel impulse response (CIR) (2.76). The channel introduces correlation on $L_c = \tau_{\max} / T_s + 1$ received symbols, resulting in ISI on the received signal. Assuming that the receiver can estimate the channel impulse response (CIR), then it can use the estimated CIR (i.e. c_j) for diversity combining. However, the combining methods described in the previous section cannot be directly applied due to the non-zero ISI. This section describes path diversity that exploits the time dispersion of the channel by optimally combining signals from different paths to achieve diversity gain. In a frequency-selective channel, the order of diversity is equal to the number of paths, each with an independent fading distribution. In general, path diversity can be achieved using four main methods 1) Rake reception [87], 2) feed-forward equalization [88], 3) decision-feedback equalization [89], and 4) maximum likelihood sequence equalization [90].

2.7.2.1 Rake Receiver

The Rake receiver [87] makes the most direct use of MRC in the presence of ISI. Assuming that there are L_c paths, a Rake receiver can then have up to L_c diversity branches. A diversity branch in this case is also called a finger. Each finger receives only one of the paths and the outputs of the K_D fingers are combined using MRC to provide an improved E_b / N_0.

In general, the paths are selected based on the total amount of desired signal power at the output of the combiner. One method would be to take K_D paths such that the sum of their power meets the combiner output power requirement. The performance of a Rake receiver can be approximated using (2.47) for BPSK and is shown in Figure 2.69. Results in Figure 2.69 assume that the CIR has equal power in all of the impulses. In practice, this assumption does not hold but it provides insight on the performance achievable through a Rake receiver.

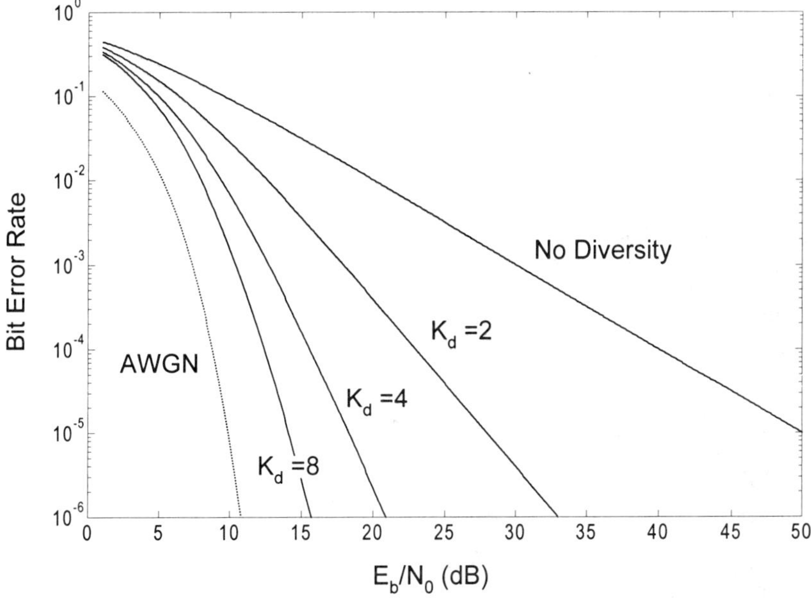

Figure 2.69 BER performance of Rake reception of BPSK signals with varying K_d.

For the Rake receiver to work optimally, each finger must be able to receive an independent fading path with zero ISI. In general, reception with zero ISI is not possible. However, it will be shown in 0 that Rake reception with negligible ISI can be achieved if the transmitted waveform is based on direct-sequence spread-spectrum. Alternative techniques also exist that minimize the effect of the ISI during diversity combining. These techniques are described in the following subsections.

2.7.2.2 Feed forward Equalizer (Zero Forcing)

The digital system shown in Figure 2.7 employs a matched filter such that the sampled output of the matched filter has zero ISI. However, due to the L_c multipaths arriving at the receiver, having a matched filter alone is no longer adequate to eliminate the ISI introduced by the channel. Additional filtering is needed to eliminate the time dispersion introduced by the channel. This is accomplished via an equalizer, the simplest being a feed-forward equalizer (FFE).

An FFE takes the output from the matched filter $\tilde{G}(f)$ in Figure 2.7 and minimizes the channel-induced ISI at the matched filter output by adjusting its own coefficients as shown in Figure 2.70. More specifically, the FFE is an FIR filter with programmable coefficients $\mathbf{w} = [w_{-\infty}, \cdots, w_k, \cdots, w_{\infty}]$ whose values are determined by minimizing an error criterion $f(\varepsilon_n)$ where $\varepsilon_n = \tilde{\xi}_n - \hat{\xi}_n$. Note that w_k denotes the value of the coefficient for the kth tap of the FIR filter. Ideally, with the optimized coefficients, the sampled output $\tilde{z}_n$ of the equalizer after ML detection should converge to the transmitted symbol; that is $\hat{\xi}_n \approx \tilde{\xi}_n$. Being a filter, the response of the FFE is determined by the convolution of its input $\tilde{x}_n$ with its coefficients $\tilde{w}_k$:

$$\tilde{z}_n = \sum_{k=-\infty}^{\infty} \tilde{w}_k \tilde{x}_{n-k} + \tilde{v}_n \tag{2.178}$$

where $\tilde{v}_n$ represents a sample of a complex Gaussian noise process.

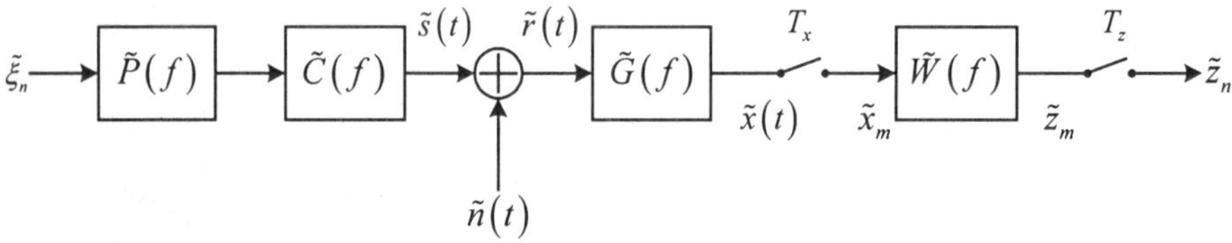

Figure 2.70 Digital system with equalization.

It is possible that the input $\tilde{x}_n$ and output $\tilde{z}_n$ of the equalizer may be sampled at different rates denoted by T_x and T_z, respectively. Since the frequency response differs for different sampling rates, the sampled response is labeled accordingly to mark the distinction. For example, the frequency response of $\tilde{x}_n$ is denoted by $\tilde{X}_{T_x}(f)$, which is the z-transform of $\tilde{x}_n$ with z evaluated at $e^{j2\pi fT_x}$. To illustrate the operation of the equalizer, the following assumptions are made:

1. The cascaded transmit and matched filter response is $\left|\tilde{P}_{T_x}(f)G_{T_x}(f)\right|^2 = 1$ for $|f| \le 0.5/T_x$.
2. The sum of the energy of each impulse in the CIR is one; i.e. $\sum_k |\tilde{c}_k|^2 = 1$. The frequency response of the sampled channel is $C_{T_s}(f)$.
3. $T_x = T_z = T_s$.
4. The equalizer has an infinite number of taps.

First, consider the zero-forcing equalizer where the error of the equalizer defined as $\varepsilon_n = \hat{\xi}_n - \tilde{\xi}_n$ satisfies the following error constraint:

$$\mathrm{E}_{ZF}(\varepsilon_n) = \varepsilon_n = 0 \text{ and } \tilde{n}(t) = 0. \tag{2.179}$$

The zero-forcing condition is equivalent to the zero ISI condition described in (2.15). To achieve zero ISI at the FFE output, the weights must be chosen such that (2.16) is met; that is $\tilde{W}_{T_s}(f)\tilde{H}_{T_s}(f) = 1$ or equivalently

$$\widetilde{W}_{T_s}(f) = \widetilde{C}_{T_s}^{-1}(f). \tag{2.180}$$

Expression (2.180) implies that a zero-forcing FFE is essentially the inverse filter of the channel response. It can be viewed as a diversity receiver with diversity branches being taps of the FFE and the combiner being the summation of all the tap outputs weighted by $\tilde{\mathbf{w}}$. To achieve zero-ISI, the number of diversity branches or taps cannot be finite. This follows from the fact that $C_{T_x}(f)$ has L_c taps and its inverse filter must be an infinite impulse response (IIR) filter with an infinite number of taps. Therefore, for a zero-forcing FFE, K_D must be infinite to achieve zero ISI. Using a finite number of taps results in residual ISI at the output of the FFE, giving suboptimal performance. In contrast, a Rake receiver requires only K_D branches to fully exploit the diversity introduced by a frequency-selective channel with L_c distinguishable multipaths.

2.7.2.3 Feed Forward Equalizer (MMSE)

A zero-forcing FFE has the additional problem of noise enhancement due to peaking in the FFE filter response. Such peaks occur when the channel exhibits nulls in its frequency response. By taking into account the effect of noise in conjunction with the channel response, noise enhancement can be mitigated. Numerous error criteria exist which take into account the effects of noise [88]. The most commonly used criterion is the minimum mean squared error (MMSE), E_{MMSE}, shown below:

$$E_{MMSE} = \min E\left\{\left|\varepsilon_n\right|^2\right\} = \min E\left\{\left|\tilde{\xi}_n - \tilde{z}_n\right|^2\right\}. \tag{2.181}$$

A MMSE FFE has coefficient values which meet the MMSE constraint stated in (2.181). After some derivation, the FFE filter response can be expressed as [88][91]

$$\widetilde{W}_{T_s}(f) = \frac{\widetilde{C}_{T_s}^{*}(f)}{\left|\widetilde{C}_{T_s}(f)\right|^2 + N_0}. \tag{2.182}$$

The FFE coefficients are a function of both the CIR and the noise power density. At significant noise levels where $N_0 >> \left|\widetilde{C}_{T_s}(f)\right|^2$, the overall response converges to $1/N_0$ while at low noise levels where $N_0 << \left|\widetilde{C}_{T_s}(f)\right|^2$, the overall response converges to that of a zero-forcing equalizer. In both cases, noise is not enhanced.

For a $2K+1$ tap equalizer, the optimum coefficients $\tilde{\mathbf{w}}_{opt}$ and the MMSE E_{MMSE} can be expressed by:

$$\tilde{\mathbf{w}}_{opt} = \mathbf{R}_x^{-1}\tilde{\mathbf{c}} \tag{2.183}$$

$$\mathrm{E}_{MMSE} = 1 - \tilde{c}^{*T} \mathbf{R}_x^{-1} \tilde{\mathbf{c}} \tag{2.184}$$

where $\tilde{\mathbf{w}}_{opt} = [\tilde{w}_{-K} \tilde{w}_{-K+1} \cdots \tilde{w}_0 \cdots \tilde{w}_{K-1} \tilde{w}_K]^T$, $\tilde{\mathbf{c}}$ is the complex conjugate of the CIR coefficients $[\tilde{c}_K^* \tilde{c}_{K+1}^* \cdots \tilde{c}_0^* \cdots \tilde{c}_{-K+1}^* \tilde{c}_{-K}^*]^T$, and $\mathbf{R}_x$ is the auto-correlation matrix of the input samples to the equalizer as shown below

$$\mathbf{R}_x = \begin{bmatrix} \tilde{x}_0 & \tilde{x}_{-1} & \cdots & \tilde{x}_{-K} & \cdots & \tilde{x}_{-2K+2} & \tilde{x}_{-2K+1} & \tilde{x}_{-2K} \\ \tilde{x}_1 & \tilde{x}_0 & \tilde{x}_{-1} & \cdots & \tilde{x}_{-K} & \cdots & \tilde{x}_{-2K+2} & \tilde{x}_{-2K+1} \\ \tilde{x}_2 & \tilde{x}_1 & \tilde{x}_0 & \tilde{x}_{-1} & \cdots & \tilde{x}_{-K} & \cdots & \tilde{x}_{-2K+2} \\ \tilde{x}_3 & \tilde{x}_2 & \tilde{x}_1 & \tilde{x}_0 & \tilde{x}_{-1} & \cdots & \cdots & \tilde{x}_{-2K+3} \\ \vdots & \vdots & \vdots & \vdots & \vdots & \vdots & \vdots & \vdots \\ \tilde{x}_{2K-2} & \tilde{x}_{2K-3} & \tilde{x}_{2K-4} & \tilde{x}_{2K-5} & \cdots & \tilde{x}_0 & \tilde{x}_{-1} & \tilde{x}_{-2} \\ \tilde{x}_{2K-1} & \tilde{x}_{2K-2} & \tilde{x}_{2K-3} & \tilde{x}_{2K-4} & \cdots & \tilde{x}_1 & \tilde{x}_0 & \tilde{x}_{-1} \\ \tilde{x}_{2K} & \tilde{x}_{2K-1} & \tilde{x}_{2K-2} & \tilde{x}_{2K-3} & \cdots & \tilde{x}_2 & \tilde{x}_1 & \tilde{x}_0 \end{bmatrix} + N_0 \mathbf{I}_{2K+1}. \tag{2.185}$$

$\mathbf{I}_{2K+1}$ denotes a $(2K+1)\times(2K+1)$ identity matrix. Both $\tilde{\mathbf{w}}_{opt}$ and $\tilde{\mathbf{c}}$ are $(2K+1)\times 1$ vectors and $\mathbf{R}_x$ is a $(2K+1)\times(2K+1)$ correlation matrix. Since the channel has a finite length of L_c+1, $x_k = 0$ if $|k| > L_c$ and $\tilde{c}_k = 0$ if $k > 0$ and $k < -L_c$. There are a total of $2K+1$ linear equations in (2.185). However, for a CIR length of L_c and $2K+1$ equalizer taps, the output sequence has a length of $2K + L_c$. Therefore, these equations cannot fully constrain the equalizer output, resulting in degraded performance due to residual ISI.

2.7.2.4 Decision Feedback Equalizer (MMSE)

To achieve the zero-ISI condition and still avoid the problem of noise enhancement, a decision-feedback equalizer (DFE) [89] can be used and has the following expression:

$$z_{n-K_2} = \sum_{j=-K_1}^{0} \tilde{w}_j \tilde{x}_{n-j} + \sum_{j=1}^{K_2} \tilde{w}_j \hat{\xi}_{n-j}. \tag{2.186}$$

The first term is a feed-forward filter that filters the current and K_1 future inputs $x_n, x_{n+1}, \cdots, x_{n+K_1}$ while the second term is a feedback filter that filters the K_2 previously decoded outputs $\hat{\xi}_{n-1} \hat{\xi}_{n-2} \cdots \hat{\xi}_{n-K_2}$. A block diagram of the FFE is shown in Figure 2.71. The lengths of the feed-forward and feedback filters are K_1+1 and K_2, respectively. To keep the system realizable the output of the equalizer is centered at the time index n. In other words, the K_1 inputs are future samples with respect to the symbol at time index n. The DFE would have processed $K_1 + K_2 + 1$ symbols before generating its output. There are a total of $K_1 + K_2 + 1$ equations in (2.186).

The output length of the equalizer due to a CIR with L_c taps is $K_1 + L_c$. To obtain zero residual ISI, K_2 should be set larger than $L_c - 1$ so that there is a sufficient number of equations to solve (2.186).

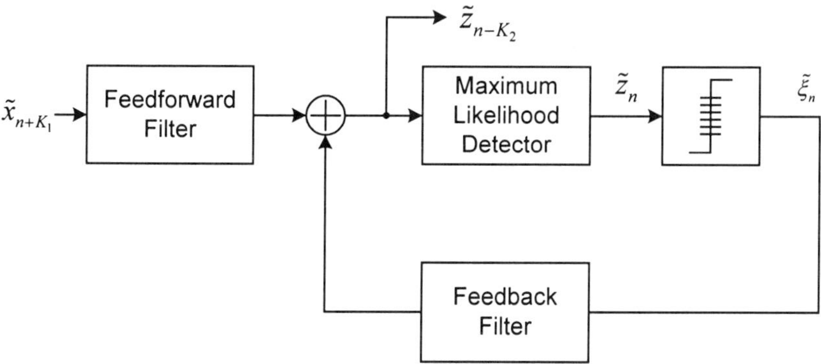

Figure 2.71 A decision-feedback equalizer.

To achieve zero residual ISI, the forward filter coefficients are constrained by (2.183) with the length of the filter set to $K_1 + 1$ while the feedback filter coefficients are determined using the following expression [91]:

$$\widetilde{w}_k = \sum_{j=-K_1}^{0} \widetilde{w}_j \widetilde{c}_{k-j}, \quad k = 1,2,\cdots,K_2 \tag{2.187}$$

Frequency response of the DFE cannot be easily derived because of the non-linear operation needed to generate the fed back decision values $\hat{\xi}_{n-1}\hat{\xi}_{n-2}\cdots\hat{\xi}_{n-K_2}$. It can be shown, however, that when K_1 is large in the feed-forward filter and K_2 is greater than $L_c - 1$, then the zero ISI condition can be achieved without noise enhancement, provided that correct decisions are fed back to the equalizer.

2.7.2.5 Maximum Likelihood Sequence Equalizer

FFE and DFE achieve diversity through a combining process, which essentially cancels out the ISI. Higher performance can be achieved if instead the multipaths are combined as in a Rake receiver. A Maximum likelihood sequence equalizer (MLSE) [90] also achieves diversity gain through combining rather than cancellation of ISI. A MLSE receiver demodulates the transmitted symbol values based on a sequence of N received symbols. In this case, diversity is achieved by taking into account the spread of the channel in the demodulation process in which the sequence of symbols affected by the spread of the channel is used to recover the transmitted data. The demodulation process can be viewed as 'combining' the information available from a sequence of N received symbols to improve the likelihood of making a correct decision, in contrast to a symbol-by-symbol demodulator.

Since the channel can be modeled as an FIR filter with coefficients equal to that of the CIR, the channel has a finite number of states equal to M^{L_c-1}, where M is the number of points in the constellation and L_c is the number of registers in the FIR. Each state is represented by the sequence of symbols stored in the register of the FIR. If the current state for the kth symbol interval is denoted by $\sigma_k = (\xi_{k-1}, \xi_{k-2}, \cdots, \xi_{k-L_c+1})$, where ξ_{k-j} denotes the symbol value at the jth tap of the FIR filter, then the next state σ_{k+1} can be represented by $(\xi_k, \xi_{k-1}, \cdots, \xi_{k-L_c+2})$. Therefore, it is clear that the next state σ_{k+1} for the kth symbol interval depends only on the current state σ_k and the current symbol value ξ_k. Since each symbol has M possible complex values, each state has M possible transitions. Furthermore, the output $\tilde{s}_k$ associated with the state transition from σ_k to σ_{k+1} can be defined as the FIR filter output based on the current state, i.e.

$$\tilde{s}_k = \sum_{j=0}^{L_c-1} \tilde{c}_j \xi_{k-j}.$$

Given the above observation, we can apply the Viterbi algorithm (2.130) to demodulate the received data. The transition metric in this case is simply

$$\lambda_k(\sigma_k, \sigma_{k+1}) = -\left|\tilde{r}_k - \tilde{s}_k\right|^2 \tag{2.188}$$

where $\tilde{r}_k$ is the received sample in the kth interval. For MLSE, the diversity branches can be viewed as the state transitions that have been processed before reaching a decision on the most likely transmitted symbol. Referring to Section 2.4.2.1, a decision is made after D steps into the Viterbi algorithm. Therefore, the number of diversity branches is on the order of DM^{L_c}. The optimum combining in MLSE can be viewed as taking all the information available in the sequence consisting of D symbols to minimize the probability of making an error.

MLSE has the optimum performance in a frequency selective channel when the noise is AWGN [90]. Its performance is limited by error events which cause an incorrect path in the trellis to have a higher path metric than the correct path. The notion of minimum distance between two paths described in Section 2.4.2.2 can be applied to MLSE. In this case, however, the designer has no control over the minimum distance since the trellis structure is set strictly by the CIR. Therefore, instead of seeing $d_{\min}$ grow with the number of states, for MLSE the minimum distance paths due to an error event may decrease with the number of states. Intuitively, the reduction in minimum distance occurs because as L_c increases, the number of possibilities for error patterns also increases, making it more likely to have paths that are close to one another. Note this problem does not occur in coding or CPM since the trellis structure can be designed appropriately to maximize $d_{\min}$. Table 2.10 lists the worst-case minimum distance for different L_c and the corresponding CIR. Note that for a two-path channel, MLSE achieves optimal performance (i.e. equal to that of an AWGN channel).

Table 2.10 Worst-case minimum distance channel. [91]

L_c	Loss (dB)	Minimum-Distance Channel
2	0	Not applicable.
3	2.3	(0.5, 0.71, 0.50)
4	4.2	(0.38, 0.60, 0.60, 0.38)
5	5.7	(0.29, 0.50, 0.58, 0.50, 0.29)
6	7.0	(0.23, 0.42, 0.52, 0.52, 0.42, 0.23)

In general, when considering path diversity, MLSE achieves the best performance and Rake reception achieves comparable performance when the fading paths do not interfere with each other. DFE suffers noticeable performance loss as compared to MLSE. For instance, the loss is on the order of 3 dB for the two-path case and degrades as the delay spread increases [91]. FFE has the worst performance and in most cases is not appropriate for wireless channels. The complexity of a diversity receiver depends on the number of diversity branches. Table 2.11 summarizes the complexity of the different schemes. Although MLSE has the highest performance, it also has the highest complexity. A Rake receiver, DFE, and FFE all have linear order of complexity. Rake stands out as having reasonable complexity yet good performance as well.

Table 2.11 Complexity of receivers with path diversity.

Receiver	Complexity (i.e. K_D)
MLSE	$O(M^{L_c})$
Rake	$O(L_c)$
DFE	$O(K_1 + K_2 + 1)$
FFE	$O(2K + 1)$

2.7.3 Frequency Diversity

In frequency-selective fading, diversity can also be achieved in the frequency domain. A channel with delay spread displays a finite coherence bandwidth in its average power spectrum. Frequency components outside of the coherence bandwidth fade independently. Therefore, frequency diversity can be implemented by transmitting a signal over multiple frequencies separated by at least the coherence bandwidth. As shown in Figure 2.72, each diversity branch is tuned to one of the transmitted frequency bands and the received signal energy from all the branches can be combined by any of the techniques described in (2.175)-(2.177).

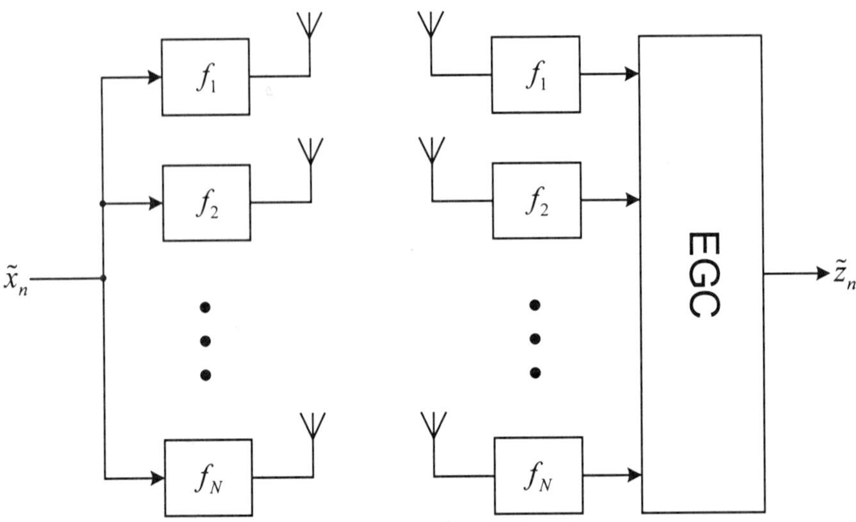

Figure 2.72 A system with frequency diversity.

2.7.4 Time Diversity

When the channel is time-selective, the signal experiences frequency-flat fading whereby the coherence bandwidth is much larger than the transmitted signal bandwidth. Essentially, there exists only one path in the CIR. The signal shows regular fading bursts with period determined by the coherence time. If multiple transmissions occur at intervals greater than the coherence time, then the received signals will experience independent fading. However, if transmissions are sufficiently long, then the receiver will experience large bursts of errors that cannot be recovered even with forward-error correction. Most FEC schemes do not perform well when the error is highly correlated.

A time diversity system exploits the time-selective nature of the channel by sending bits from the same message separated by a duration that is larger than the coherence time of the channel such that the transmitted bits experience independent fading. This can be done by shuffling bits over a sufficiently long period of time, a process often referred to as interleaving. The receiver de-interleaves or puts the transmitted bits back into the original order. The de-interleaving process randomizes the burst error. The randomized errors do not exhibit high degree of correlation and thus can be decoded reliably with an FEC decoder.

Figure 2.73 shows a block diagram of a time-diversity system that uses interleaving. The transmitter first encodes the information bits and interleaves the coded symbols before transmission on to the channel. The receiver collects the interleaved symbols and time aligns them into the original message before decoding to correct for the corrupted symbols.

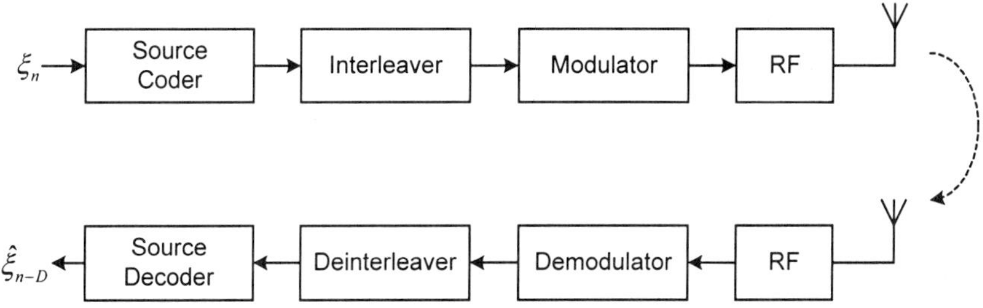

Figure 2.73 A system with time diversity.

If coding were directly applied without interleaving the message, then the duration of an error burst could easily exceed the error correcting capability of the code. For instance, if the carrier frequency is at 1 GHz and the maximum velocity of the mobile user is 100 km/hr., then according to (2.84) the maximum Doppler shift is 92.6 Hz. Given the above condition and assuming that the fading burst results in a relative signal attenuation of $\rho = 1$, then according to (2.86) the worst case fading duration is 7.3 ms. Consider a data rate of 270 kbps. Then, the fade duration spans nearly 2000 bits, large enough to make coding alone ineffective.

2.7.4.1 Interleaving/de-interleaving

There are two general types of interleavers: block interleavers and convolutional interleavers [92][93]. The former is generally used with block codes and the latter with convolutional codes. In this section, only the block interleaver is discussed. Readers interested in convolutional interleavers may refer to [92][93].

A block interleaver can be viewed as a $m \times dS$ matrix, where d is the number of bits in the coded symbol, m is the depth of the interleaver, and S is the span of the interleaver defined as the time separation between interleaved symbols. S is normalized by the symbol time. During interleaving, coded symbols are written into the matrix one row at a time while the interleaved symbols are read from the matrix one column at a time. The de-interleaver performs the inverse operation. Figure 2.74 shows a matrix representation of the block interleaver and de-interleaver.

$$\begin{bmatrix} A_1 & A_2 & A_3 & A_4 & \cdots & A_{S-1} & A_S \\ A_{S+1} & A_{S+2} & A_{S+3} & A_{S+4} & \cdots & A_{2S-1} & A_{2S} \\ A_{2S+1} & A_{2S+2} & A_{2S+3} & A_{2S+3} & \cdots & A_{3S-1} & A_{3S} \\ A_{3S+1} & A_{3S+2} & A_{3S+3} & A_{3S+4} & \cdots & A_{4S-1} & A_{4S} \\ \vdots & \vdots & \vdots & \vdots & \vdots & \vdots & \vdots \\ A_{(m-1)S+1} & A_{(m-1)S+2} & A_{(m-1)S+3} & A_{(m-1)S+4} & \cdots & A_{mS-1} & A_{mS} \\ A_{mS+1} & A_{mS+2} & A_{mS+3} & A_{mS+4} & \cdots & A_{(m+1)S-1} & A_{(m+1)S} \end{bmatrix}$$

Figure 2.74 Matrix structure of a block interleaver and de-interleaver.

Each symbol A_i contains d bits and each row in the interleaver is a block code with a block size of S symbols and a code rate r. The burst error correcting capability of such a code is [94]

$$t \leq \frac{(1-r)S}{2} \tag{2.189}$$

where t is the number of correctable symbol errors. Since the interleaved symbols are obtained by reading out columns of the matrix shown in Figure 2.74, a burst error will corrupt bits in k contiguous columns. To completely correct for the k symbol errors occurred in each row or codeword after de-interleaving, t must be at least equal to k. Therefore, the number of rows in the matrix must meet the following requirement

$$mt = \frac{\tau(\rho)}{rdT_b} \tag{2.190}$$

which simply states that the total number of coded symbols in t contiguous columns must equal to the length of the error burst, normalized to the number of coded symbols. Furthermore, to ensure that only a single error burst occurs within the span of the interleaver, the following must also hold

$$mS \leq \frac{1}{rT_b dN(\rho)} \tag{2.191}$$

which states that the total number of code symbols being interleaved should be at most equal to the interval between fades, normalized to the number of coded symbols. The design of the block interleaver then involves computing the values for m, S, d, and r to satisfy (2.189)-(2.191).

2.7.4.2 Interleaving with Frequency Hopping

If the bandwidth of the transmitted signal is smaller than the coherence bandwidth of the channel, then the received signal appears to have gone through a frequency-flat channel, when in fact the channel is frequency selective with a finite coherence bandwidth. In this situation, time-diversity using FEC with interleaving can be applied. However, if mobility is low (e.g. 1 m/s for a pedestrian) then the coherence time becomes excessively long, resulting in an interleaver with a large span that is impractical to implement.

Frequency hop offers a more practical implementation for a diversity receiver in a frequency-selective channel with slow fading statistics. Frequency hop is a form of spread-spectrum technique described in 0. In frequency-hop spread-spectrum, the transmitted signal is hopped over N_f different frequencies in a pseudo-random fashion. If the hopping frequency is such that each hop is separated by at least the coherence bandwidth of the channel, then each hop experiences independent fading.

Now consider the same system as in Figure 2.73 but with the exception that the modulator and demodulator are substituted by a hopper and dehopper. Each column in Figure 2.74 is hopped over one of N_f frequencies, which implies that each hop carries *md*-coded bits. The span of the interleaver now depends only on N_f and the design constraint for the interleaver/de-interleaver simplifies to

$$t \le \frac{(1-r)N_f\, m}{2}. \tag{2.192}$$

Expression (2.192) is derived by substituting $S = N_f$ in (2.189). The span of the interleaver is now independent of the coherence time but is instead determined by the number of available hopping frequencies. The technique just described has exploited frequency diversity to achieve time diversity.

2.7.4.3 Performance Gain

The performance gain for a time-diversity system described above can be determined using (2.47) through (2.49), with minimum modification. While (2.47) through (2.49) are valid only for binary modulation, they can be easily extended to higher order modulations. The diversity branches are essentially the columns of the interleaver matrix that are transmitted onto the channel. However, the diversity order is limited to $d_{\min}$ of the FEC since the performance can be no better than the case when error bursts in the channel are perfectly randomized. The BER, G_d, and $L_{d/n}$ are shown below for soft-decision decoding:

$$\overline{P}_b \le \left(\frac{1}{1+r\overline{\gamma}}\right)^{d_{\min}}. \tag{2.193}$$

$$G_d \approx 10\log_{10} r\left(e^{\overline{\gamma}_n}-1\right) - 10\log_{10}\left(e^{\overline{\gamma}_n/d_{\min}}-1\right). \tag{2.194}$$

$$L_{d/n} \approx 10\log_{10}\left(e^{\overline{\gamma}_n/d_{\min}}-1\right) - 10\log_{10} r\overline{\gamma}_n. \tag{2.195}$$

For hard-decision decoding, effectively, $d_{\min}$ is reduced by one half.

2.7.5 Spatial Diversity

In frequency-selective and time-selective fading, the amount of selectivity is determined by the transmission bandwidth and mobility, respectively. In space-selective fading, the amount of spatial selectivity is determined by the antenna spacing as shown in (2.95). In general, if the spacing is greater than the coherence distance d_c then the signal at the output of each antenna element will experience independent fading. The coherence distance, mentioned in Section 2.2.6.3, is inversely proportional to the angular spread of the channel. For a Rayleigh fading channel (i.e. frequency-flat fading), where the angular spread is large, the coherence distance is

0.5λ. In contrast, for a frequency-selective channel, where the angular spread can be small, the coherence distance tends to increase.

Spatial diversity exploits independent fading, which occurs on the different antenna elements. A general block diagram of a system that exploits spatial diversity is shown in Figure 2.75. It consists of a transmitter with N_t antennas separated by a fixed distance d_t and N_r receivers separated by d_r. In general, an omnidirectional antenna is used for each of the antenna elements in the array. Space-selective fading results by selecting d_t and d_r such that each transmit or receive antenna generates independent fading statistics. When such fading statistics are combined, diversity gain can be achieved. Any of the combining techniques, (2.175)-(2.177), can be used. In the case of Rayleigh fading statistics, the performance can be approximated by that of MRC, (2.47), with $K_D = N_t N_r$.

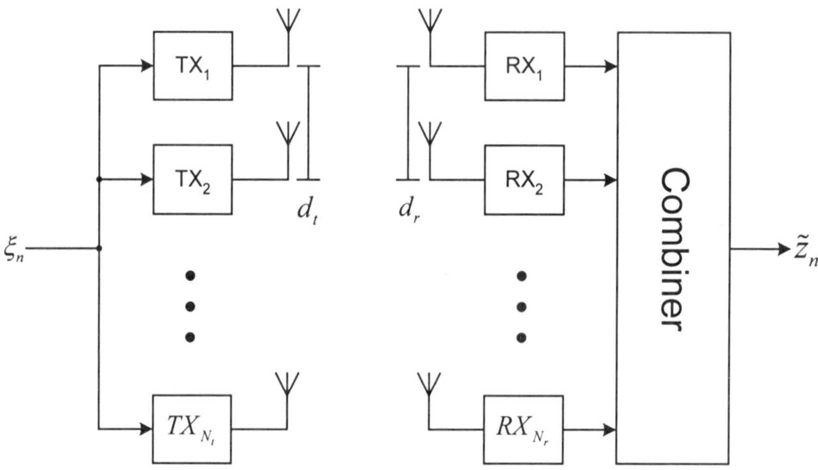

Figure 2.75 A system with spatial diversity.

In general, having multiple antennas at a mobile unit cannot be easily accommodated due to the large spacing requirement. For instance at 1 GHz, half a wavelength is 6 inches. Therefore, spatial diversity is often applied at the basestation or for fixed wireless access equipment. However, for newer systems operating at higher frequencies (e.g. 5 GHz), spatial diversity at the mobile unit becomes possible.

2.7.5.1 Switched Beamforming

Switched beamforming [34]offers selection diversity with reduced ISI or co-channel interference. As shown in Figure 2.76, the combiner forms N_t beams using N_t fixed vectored weights $\mathbf{w}_1, \mathbf{w}_2, \cdots, \mathbf{w}_{N_t}$, one for each beam. SC is used to select the beam with the maximum SNR at the receiver. In general, the beams are pointed in different directions such that the chance of isolating a strong path is maximized for a frequency-selective channel. Both ISI and CCI are minimized due to the narrow beamwidth. Switched beamforming is not very useful when the channel is frequency-flat since there a strongest path does not exist although it is still effective against CCI. Instead, in the case of frequency-flat fading, unit weight vectors should be used with antenna spacing at $\lambda/2$. While switched beamforming shown in Figure 2.76 is applied to the transmitter, it can be adapted for receiver as well.

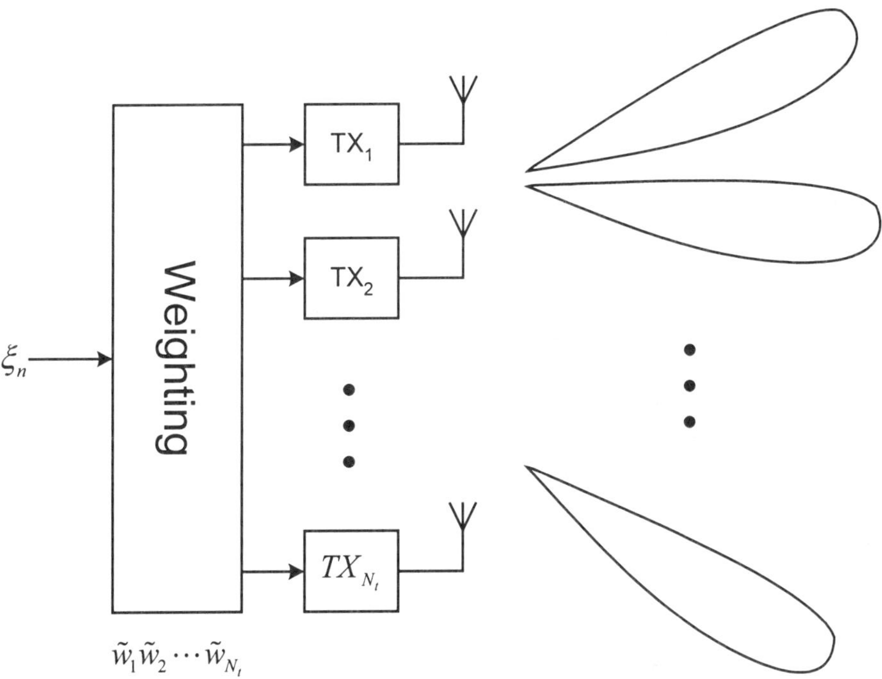

Figure 2.76 A switched beamforming system.

2.7.5.2 Adaptive Beamforming

Adaptive beamforming (Figure 2.77) [34] forces an antenna array to point to the desired signal and null out the undesirable signals, which could be ISI or CCI. By reducing the amount of ISI or CCI, improved diversity gain can be achieved via MRC. The key difference between adaptive beamforming and switched beamforming is that the former system adapts the weight vector while the latter has fixed weights. In general, an adaptive beamforming system requires N_r elements to null out N_r-1 interferers. However, it is possible that the system may lock onto a weaker signal due to errors in the channel estimation or weight-update.

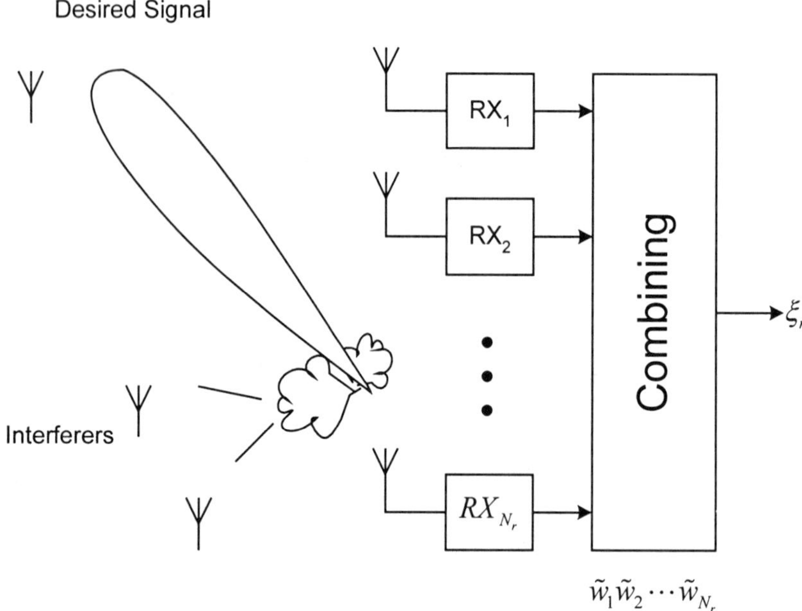

Figure 2.77 An adaptive beamforming system.

In both switched and adaptive beamforming, though the SIR has increased, the order of path diversity is reduced since fewer paths fall within the narrower antenna beams. Therefore, an increase in SIR is traded against a decrease in gain achievable through path diversity.

2.7.6 Others Types of Diversity

Several basic forms of diversity techniques have been discussed, namely: 1) path diversity, 2) frequency diversity, 3) time diversity, and 4) spatial diversity. Other types of diversity exist. These include interference diversity, transmit diversity, and space-time processing, and are discussed in the subsections below.

2.7.6.1 Interference diversity

Here a distinction is made between signal diversity and interference diversity. Signal diversity attempts to maximize the signal energy at the combiner output by exploiting independent signal components created by the channel. Interference diversity, on the other hand, attempts to minimize the interference energy at the combiner output by exploiting independent interference sources within the system. These interference sources might be co-channel transmissions within a cell or adjacent cells. Interference diversity is achieved through two main methods: 1) averaging and 2) cancellation.

Interference averaging can be accomplished via spread-spectrum discussed in the 0. In Section 2.7.4.2, FH spread spectrum has been applied to time diversity, where frequency hopping is used to generate independent fading with hopping separation greater than the coherence bandwidth of the channel. FH can be similarly applied to average out bursty interference. In particular, due to the cross-correlation property of the hopping sequence, transmission with different hopping sequences can collide only in one of the N_f hopping frequencies. Effectively, the receiver sees an average interference power that is N_f times lower than the peak interference power. Such an interference averaging system has a similar architecture as that of Figure 2.73 and is currently deployed in GSM. Direct-sequence spread-spectrum offers similar interference averaging but in the time domain.

Interference cancellation can be accomplished through techniques, such as adaptive beamforming (Section 2.7.5.2) and multi-user detection (MUD) [95]. In adaptive beamforming, weights are adapted to null out the interference using either the MSE or zero-forcing error criterion. In multi-user detection, K users are detected jointly given certain a priori knowledge on the properties of the users' signals. An example of such a signal property is the spatial response or signature a given user has at the receiver. By appropriately weighting the antennas given the specificity of the spatial signatures, each user can be detected while the interference from other users is minimized.

Experimental systems have been reported to achieve extremely high capacity (e.g. 30 bits/sec/Hz) using spatial interference diversity [96][97]. The high capacity is obtained through spatial processing that cancels out undesired user transmission based on spatial signatures. Multiple transmissions could be accomplished and combined at the receiver to achieve a high aggregate bandwidth. Spatial interference diversity can also be applied to multiple access, whereby transmission from different users are sent in the same frequency band and user separation is accomplished spatially. In essence multiple user transmission are decoded simultaneously. Such multi-user detection (MUD) can also be achieved with direct-sequence spread-spectrum whereby each user is identified with a unique PN sequence. MUD exploits the knowledge of the transmitted PN sequences to jointly detect all the users.

2.7.6.2 Transmit diversity

Transmit diversity refers to schemes in which the diversity gain is obtained in conjunction with the transmitter. An example of transmit diversity has been described in Section 2.7.4.2, where frequency hopping with interleaving and coding are employed to obtain diversity gain in a time-flat channel. By coding, interleaving, and frequency hopping, the transmitter is able to induce at the receiver time-selective fading, which results in burst error statistics that can be dealt with using a relatively small interleaver.

Similar transmit diversity is achieved for a time-flat and frequency-flat channel by intentionally delaying the transmitted signal using multiple delay elements with known delays [98]. At the receiver, the delays generate multipaths which can then be recovered via path diversity. For the same channel, transmit diversity can also be achieved in the spatial domain by interleaving coded bits onto different antenna elements at the transmitter such that space-selective fading at the receiver induces time-selective fading on the time interleaved signal [99].

In general, transmit diversity has been used to induce fading statistics at the receiver that cannot otherwise be obtained directly from the channel. However, transmit diversity can also be used to simplify the complexity of the receiver by pre-equalizing the channel. For instance, with an appropriate weight vector, a beam can be formed at the transmitter to minimize ISI or CCI at the intended receiver [100][101][102].

2.7.6.3 Space-Time Processing

In spatial processing, each antenna element in the array is multiplied by a scalar constant (Figure 2.26); i.e. spatial taps. In space-time processing [35], each antenna element is weighted by a vector instead. The elements in the weight vector correspond to time-domain taps. The simplest example is having an FFE at each antenna element output. The weight vector at each antenna element then corresponds to the taps in the FFE. If the MMSE criterion is used to generate the weight values, then the receiver is also known as a space-time (ST) MMSE receiver [103]. In a ST MMSE receiver, weights are determined to optimize the gain achieved through space as well as path diversity, whereby the FFE eliminates some of the ISI received in the individual antenna elements and the combined outputs of the FFE's provide additional diversity gain. ST MMSE is still suboptimal because outputs of the FFE still contain residual ISI's. To eliminate the ISI, the ST-MMSE can be further processed by a DFE. This principle can be extended to having a Rake or MLSE at each antenna element. The corresponding diversity receivers are known as a ST-Rake [104] and a ST-MLSE [105].

The complexity of ST processing is in general N_r times that of the time processing alone. For example, a ST-MLSE and a ST-Rake have complexities of $O(N_r M^{L_c})$ and $O(N_r L_c + N_r)$, respectively. Performing the spatial and time processing sequentially, rather than performing them jointly can reduce the complexity of ST processing. For instance, spatial combining can be first performed and then followed by an MLSE receiver, a Rake receiver, or a DFE. These solutions are suboptimal. However, the complexity is reduced by a factor of N_r.

2.7.7 Channel Estimation and Tracking

It should be clear by now that most diversity systems cannot obtain their performance gains without channel estimators and weight update blocks as shown in Figure 2.67. Without these blocks the optimal value of the weight vector $\tilde{\mathbf{w}}$ often cannot be determined. In general, information about the channel is first estimated and then tracked. The channel estimator generates the initial estimate of the channel from which initial weights can be determined. The weight-update block then updates the combiner weights to reflect changes encountered in the channel over time.

2.7.7.1 Channel Estimator

A channel estimator consists of a matched filter, matched to a known training sequence ξ_T of length N symbols. The main design consideration is the selection of a training sequence from which the channel can be accurately estimated while minimizing N since a large N compromises the throughput and delay of the system.

Since passing a signal through a filter matched to the signal ξ_T is equivalent to taking its auto-correlation, the ideal sequence is one that has an auto-correlation of the form $\delta(n-M)$, such that the matched filter output estimates the CIR, i.e. $\delta(n-M)*[\tilde{c}(n)+\tilde{n}(t)]=\tilde{c}(n-M)+\tilde{n}(n-M)$. Since the auto-correlation function is periodic with period M, the channel estimate can be averaged according to the following operation to minimize the effect of noise

$$\hat{c}_n = \frac{M}{N}\sum_{i=1}^{N/M}\tilde{y}(n-iM), \tag{2.196}$$

where $\tilde{y}(n)$ is the matched filter output and $\hat{c}_n$ is the estimate of the nth tap in the CIR.

From (2.196), it is evident that a trade-off exists on the choice of N, which cannot be made too small in consideration of noise performance yet it cannot be made too large in consideration of system throughput and delay. A good choice of ξ_T is based on the pseudo-noise (PN) sequence typically used in spread-spectrum communication. In general, M is equal to the length of the PN-sequence and the longer the sequence, the closer the auto-correlation function approximates $\delta(n-M)$. An alternate method of generating the channel estimate is to rely on the weight update block to converge to the optimal weights, given an arbitrary initial condition. However, the weights may not always converge based on this approach and if the weights do converge, the time required may be excessively long.

2.7.7.2 Error Criteria

Given the initial estimate, the combiner weights can be computed. However, these weights must be updated in response to changes in the channel, especially, for a time-selective channel. One method of tracking is based on negative feedback, where the weights are varied until a certain error criterion is driven to zero. There are four main types of error criterion: 1) mean-squared error (MSE) [106], 2) maximum likelihood (ML) [107], 3) minimum error variance (MEV) [108], and 4) maximum signal-to-noise ratio (SNR) [109]. The following discussion concentrates on the MMSE criterion, which is most commonly used in practice.

The MMSE criterion is shown in (2.181). In a diversity receiver, the estimated symbol is based on the combiner output $\tilde{\mathbf{w}}^T\tilde{\mathbf{x}}$, where $\tilde{\mathbf{w}}$ is the weight vector of the combiner and $\tilde{\mathbf{x}}$ is the input vector of the diversity branches. $\tilde{\mathbf{w}}$ and $\tilde{\mathbf{x}}$ are defined as follows

$$\tilde{\mathbf{w}}(n) = [\tilde{w}_1(n)\ \tilde{w}_2(n)\ \cdots\ \tilde{w}_k(n)]^T \tag{2.197}$$

$$\tilde{\mathbf{x}}(n) = [\tilde{x}_1(n)\ \tilde{x}_2(n)\ \cdots\ \tilde{x}_k(n)]^T \tag{2.198}$$

where $\tilde{w}_k(n)$ and $\tilde{x}_k(n)$ represent the weight and input for the kth diversity branch at time index n.

The MSE can then be expressed as

$$\begin{aligned} E\left\{\left|\varepsilon_n\right|^2\right\} &= E\left\{\left|\xi_n\right|^2\right\} - 2\,\mathrm{Re}\left\{\tilde{\mathbf{w}}(n)^H E\left\{\tilde{\mathbf{x}}^*(n)\xi_n\right\}\right\} + \tilde{\mathbf{w}}(n)^T E\left\{\tilde{\mathbf{x}}(n)\tilde{\mathbf{x}}(n)^H\right\}\tilde{\mathbf{w}}(n)^* \\ &= E\left\{\left|\xi_n\right|^2\right\} - 2\,\mathrm{Re}\left\{\tilde{\mathbf{w}}(n)^T \mathbf{r}_{x\xi}(n)\right\} + \tilde{\mathbf{w}}(n)^T \mathbf{R}_{xx}(n)\tilde{\mathbf{w}}(n)^* \end{aligned} \tag{2.199}$$

where the superscript H represents complex conjugation plus transposition of the associated matrix or vector. The MMSE is determined by taking the gradient of (2.199) and setting the result equal to zero; i.e. $\nabla_{\mathbf{w}} E\left\{\left|\varepsilon_n\right|^2\right\} = 0$. The result is shown below

$$\tilde{\mathbf{w}}^*_{opt}(n) = \mathbf{R}_{xx}^{-1}(n)\mathbf{r}_{x\xi}(n) \tag{2.200}$$

where $\tilde{\mathbf{w}}_{opt}(n)$ is the weight vector which meets the MMSE criterion. Note that the gradient of $f(\mathbf{x})$ with respect to $\mathbf{x}$ is defined as

$$\nabla_{\mathbf{x}} f(\mathbf{x}) = \left[\frac{\partial f}{\partial x_1}\ \frac{\partial f}{\partial x_2}\ \cdots\ \frac{\partial f}{\partial x_K}\right]^T, \tag{2.201}$$

where $\mathbf{x} = [x_1\ x_2\ \cdots\ x_K]^T$. By substituting (2.200) into (2.199), the MMSE can be expressed as

$$\min E\left\{\left|\varepsilon_n\right|^2\right\} = E\left\{\left|\xi_n\right|^2\right\} - \mathbf{r}_{x\xi}^H(n)\mathbf{R}_{xx}^{-1}(n)\mathbf{r}_{x\xi}(n). \tag{2.202}$$

Expression (2.200) can be used to compute the weights of a diversity combiner for a given the MMSE criterion. For example, weights for an MMSE equalizer can be derived by substituting $\mathbf{x} = \mathbf{c}^T\xi_n + \mathbf{n}$ into (2.200) and assuming that the transmitted symbols are independent with unit amplitudes.

2.7.7.3 Weight Update Algorithms

Given a time varying channel, the correlation matrices $\mathbf{R}_{xx}(n)$ and $\mathbf{r}_{x\xi}(n)$ also vary with time. Therefore, the coefficients should be updated as the channel changes. A direct method is through direct matrix inversion (DMI) [111] which inverts $\mathbf{R}_{xx}(n)$ to solve for the optimum weight vector. However, DMI is complex due to the matrix inversion. A recursive method based on the least mean squared (LMS) algorithm [107] requires much lower computation complexity.

The LMS algorithm is based on the observation that the MMSE solution occurs at the lowest point or "trough" of the error surface described by $|\varepsilon_n|^2$. Since the gradient vector defined in (2.201) is orthogonal to a given point on the error surface, a negative gradient points toward the minimum. Because of this geometric interpretation, the LMS algorithm is also referred to as the method of steepest descent.

Let $\tilde{\mathbf{w}}(n+1)$ be the updated coefficient at time index $n+1$. Then, by using the method of steepest descent, the following update rule is obtained:

$$\begin{aligned}\tilde{\mathbf{w}}(n+1) &= \tilde{\mathbf{w}}(n) - \mu\nabla_w E\left\{|\varepsilon_n|^2\right\} \\ &= \tilde{\mathbf{w}}(n) + \mu(\mathbf{r}_{x\xi} - \mathbf{R}_{xx}\tilde{\mathbf{w}}(n))^*,\end{aligned} \tag{2.203}$$

where μ is a small, positive scale factor that controls the LMS step size. Referring to (2.203), one might observe that the LMS algorithm behaves like a negative feedback control system. Whenever a component of the gradient vector is positive, the corresponding coefficient is made more negative by an amount proportional to the gradient component. The opposite occurs when the gradient component is negative. The process continues until the gradient vector settles to zero, at which point the updated coefficients result in an MMSE solution.

Expression (2.203) can be further simplified by approximating the expectation of a random variable by its sample value. For example, the auto-correlation matrix $\mathbf{R}_{xx} = E\left\{\mathbf{x}(n)\mathbf{x}^H(n)\right\}$ can be approximated by $\mathbf{x}(n)\mathbf{x}^H(n)$. By making the above approximation, (2.203) simplifies to

$$\tilde{\mathbf{w}}(n+1) = \tilde{\mathbf{w}}(n) + \mu\mathbf{x}^*(n)\varepsilon_n. \tag{2.204}$$

The error ε_n is not easily determined since in general the transmitted symbols are not known. To overcome this problem, most systems employ a training sequence, with a pre-determined bit pattern, to train the LMS algorithm until the coefficients converge to their optimum values. Thereafter, the LMS algorithm continues coefficient update through a decision-aided process, where the error is estimated by

$$\varepsilon_n = \hat{\xi}_n - z_n \tag{2.205}$$

where $\hat{\xi}_n$ is the hard-decision output of the ML-detector following the diversity receiver. Figure 2.78 shows a block diagram of the LMS coefficient update algorithm.

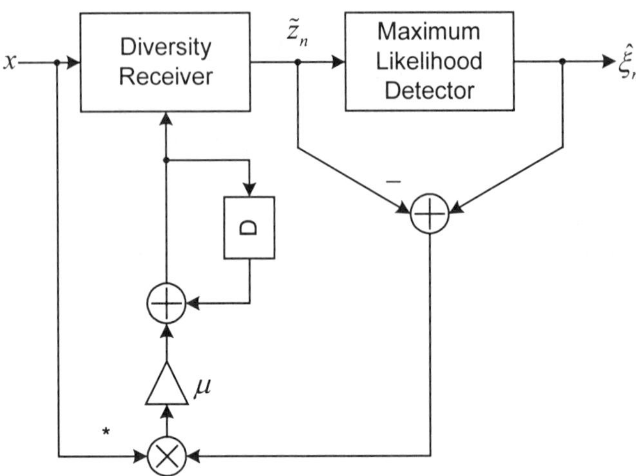

Figure 2.78 Weight-update based on the LMS algorithm.

For MLSE, (2.204) cannot be directly applied since the weights in this case are not used to combine the diversity branches but rather are used to search through the trellis for the most likely transmitted sequence. In other words, the error function for MLSE does not follow that of (2.205). Rather, the error function is defined as

MLSE:
$$\begin{aligned} \varepsilon_n &= x_n - \sum_{j=-\infty}^{\infty} \tilde{w}_j \hat{\xi}_{n-D-j} \\ &= x_n - \tilde{\mathbf{w}}^T(n)\hat{\boldsymbol{\xi}}(n-D). \end{aligned} \tag{2.206}$$

which is consistent with the transition metric (2.188) in the Viterbi algorithm.

Comparing (2.206) to (2.205), the coefficient update algorithm based on LMS is easily derived as shown below

MLSE:
$$\tilde{\mathbf{w}}(n+1) = \tilde{\mathbf{w}}(n) + \mu \hat{\boldsymbol{\xi}}^*(n-D)\varepsilon_n. \tag{2.207}$$

where the error function is based on (2.206). The symbol sequence used in the update equation (2.207) is delayed by D symbols due to the decoding delay discussed in Section 2.4.2.1.

The LMS coefficient update reduces the computation requirement from $O(N^3)$ required for DMI to $O(N)$, where N is the length of the weight vector. Because of the low complexity of the LMS update, it is widely used to update the weights of the combiner in diversity receivers. The designer should be aware, however, of several design considerations when using the LMS algorithm. First, it is recursive and therefore requires some time before the coefficients converge to their optimum values.

Recall that the LMS algorithm can be viewed as a feedback system. Therefore, its convergence time behaves like that of the synchronization loops discussed in Section 2.5, where the settling time depends largely on the loop gain. For the LMS algorithm, the step size μ (like the loop gain) determines the convergence time. By the same token, the step size also determines the noise performance of the LMS update block, whereby if the step size is made too large then the coefficients become noisy though the settling time is reduced. Also, in some cases, the loop may become unstable with too large a step size.

The stability of the loop is difficult to analyze but in general if the condition number of the auto-correlation matrix is large then the loop will have a hard time converging. The condition number of $\mathbf{R}_{xx}$ is defined as the ratio of the maximum eigenvalue of $\mathbf{R}_{xx}$ to the minimum eigenvalue of $\mathbf{R}_{xx}$. A wireless channel tends to have a large condition number. Therefore, the LMS algorithm may require an extremely long training sequence to work properly. A long training sequence is undesirable because it adds overhead in terms of both delay and throughput of the system; especially, if the system is packet switched.

Other techniques have been developed to provide systems with a faster convergence time. Many are based on the recursive least squared (RLS) algorithm [110][112]. RLS is based on a slightly different error function, which is a MSE weighted exponentially by λ^n. The constant λ, also known as the forgetting factor, provides an additional degree of freedom to control the rate of convergence. However, the RLS algorithm is generally much more complex and can also become unstable when it is implemented in fixed point.

Finally, for MLSE, the coefficient update algorithm of (2.207) does not work well in a highly dynamic channel due to the decoding delay. Channel perturbations may occur before the MLSE makes a decision on ξ_{n-D} due to the decoding delay. In this case, the update algorithm cannot track the channel perturbations and therefore incorrect symbol decisions may result. The incorrect symbol decisions may further increase the error in the LMS weight-update, creating error propagation. To overcome this problem, coefficients can be updated instead using per-survivor processing (PSP) [113]. With PSP, each survivor path keeps track of its own set of channel coefficients and therefore does not rely on the decoded outputs. Each survivor path updates its coefficient based on (2.207), with the exception that the estimated sequence and accumulated error for the corresponding path is used in place of $\hat{\xi}^*(n-D)$ and ε_n, respectively.

CHAPTER 3

Spread-Spectrum Communications

For many years, the military has used a wideband technique known as spread-spectrum to provide improved communication under hostile environments where communications can be intentionally jammed [114][115]. Besides the anti-jamming (AJ) capability, spread-spectrum provides low probability of detect (LPD) and low probability of intercept (LPI) in covert operations [116]. Historically, therefore, spread-spectrum has been used almost exclusively by the military. However, commercial applications using spread-spectrum have grown rapidly since the Federal Communications Committee (FCC) allocated three Industrial Scientific Medical (ISM) bands for license-free civilian spread-spectrum transmission. These bands comprise of 902-928 MHz, 2.4-2.4835 GHz, and 5.725-5.825 GHz. The predominant applications using these bands are based on WLAN and WLL. Other licensed bands for spread-spectrum systems are being used for the IS-95 digital cellular system [5].

The properties, which have been utilized for military communications, now find their use in commercial applications. The LPI property provides privacy in a multi-user environment where each user is assigned a code rather than a frequency or time slot as in conventional frequency division multiple access (FDMA) or time division multiple access (TDMA) systems. The LPD property implies a reduced intensity in the transmitted power spectral density which enables spread-spectrum signals to coexist with other applications used in the ISM bands. To ensure tolerable interference from unlicensed use of spread-spectrum, the FCC specifies that the transmitted power density must be less than 2 μW/Hz[4]. Finally, the anti-jamming property of spread-spectrum provides additional flexibility in bandwidth utilization to achieve a high user capacity [117][118][119] and robustness against multipath fading [120].

---

[4] For direct-sequence spread-spectrum.

These desirable properties of spread spectrum arise from modulating the information with a wideband spreading code which is uncorrelated with the data. The fundamental reasons behind the effectiveness of transmitting information in this fashion will become clear in the following sections.

3.1 Spread-Spectrum Encoding

In spread-spectrum communications, the bandwidth W_s of the digitally modulated waveform is spread to a wider bandwidth W_{ss} as shown in Figure 3.1a. The original signal before being spread is often referred to as the narrowband signal. Spreading can be accomplished by modulating the frequency, time, and/or phase of the narrowband signal with a known sequence. To attain the desired properties of spread-spectrum in AJ, LPI, and LPD, the sequence should ideally be random. The random sequence modulates frequency f_{ss}, time t_{ss}, and/or phase ϕ_{ss} as shown in Figure 3.1b. Spread-spectrum systems that modulate the frequency, time, or phase are also referred to respectively as frequency-hop (FH), time-hop (TH), or direct-sequence (DS) systems. The random sequence is referred to as the spreading code.

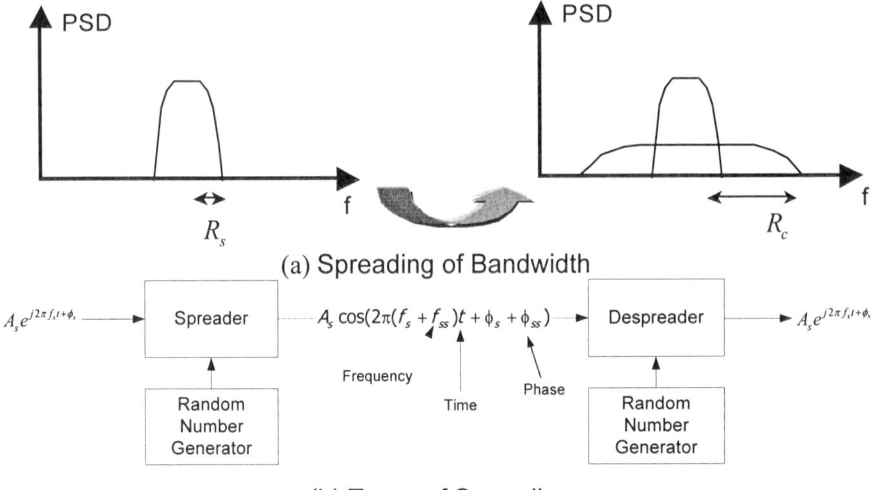

Figure 3.1 Spread-spectrum waveform encoding.

To attain the desirable features of spread-spectrum, the spreading code must satisfy the properties listed in Table 3.1. The first property indicates that the spreading code symbols should be uncorrelated. The ideal autocorrelation property of the code results in a uniform spread of the energy of jammers or interference across the spreading bandwidth. Therefore, the receiver essentially sees a flat jammer power density which results in a much lower total in-band jammer or interference power for an improved SIR performance. As mentioned in Section 2.2.2, a reduction in interference power results in an improvement of channel reuse and therefore system capacity. The second property requires that any two codes have zero cross correlation. In other words, the codes are orthogonal to one another. The orthogonality property makes spread spectrum suitable for code division multiple access (CDMA) in which each communication link is accessed via a unique spreading code. The third property provides LPI and LPD since the spread signal appears to be random and with large enough spreading bandwidth becomes buried in noise, making it difficult to detect and intercept. The commercial relevance of this last property is in providing a higher level of privacy and some degree of security. For instance, in IS-95 [5], a long PN-code provides ciphering that makes it difficult for unauthorized users to intercept and decode on-going calls.

Table 3.1 Ideal properties of the spreading code.

Property 1.	Ideal autocorrelation with zero correlation for offsets greater than a code symbol of duration T_c, i.e. $$r_{xx}(\tau)=\begin{cases}1-\frac{\lvert\tau\rvert}{T_c}, & \lvert\tau\rvert\le T_c\\ 0, & \lvert\tau\rvert> T_c\end{cases}.$$
Property 2.	Ideal cross-correlation with zero correlation between different codes, i.e. $$r_{xy}(\tau)=0\ \forall\, x\neq y.$$
Property 3.	Random over an infinite observation period.

With the exception of Property 2, the properties listed in Table 3.1 are impossible to achieve in practice because of physical limitations. The receiver must despread the spread-spectrum signal to recover the original information. To despread, the spreading code used for transmission must be known at the receiver. However, knowledge of the code would not be possible if the code were truly random. Also, for the code to have an ideal auto-correlation function implies that the code has infinite duration which would require infinite delay to synchronize the code at the receiver.

Despite these limitations, properties 1 and 3 can be approximated over finite time intervals during which the code appears random. This type of code is also known as a pseudo-noise (PN) code and is utilized in practice to generate a spread-spectrum signal with little performance loss. Several codes with simple implementations exist. Among these are the maximal-length (M) sequence [121], the Gold sequence [122], the Kasami sequence [123], and the orthogonal Walsh sequence [124]. In general, these codes have a periodicity of N_c, i.e. the code sequence repeats after every N_c code symbols. Often the period of the code is referred to as the length of the code. The cardinality of the code refers to the number of unique code sequences that exist for a given length N_c. In general, these codes exhibit conflicting cross-correlation and auto-correlation properties. For instance, the M-sequence has a near ideal autocorrelation function but poor cross-correlation properties. On the other hand, the Gold sequence has better cross-correlation properties but worse auto-correlation [122]. Similarly, Walsh codes have ideal cross-correlation but poor auto-correlation. Table 3.2 summarizes the general properties of several types of PN-codes. Note that as the length of the sequence increases for all of these codes, the code properties approach those of the ideal case listed in Table 3.1

Each code finds its particular use in a larger system. In general, for synchronization, it is desirable to have a good auto-correlation with a large peak at $\tau = 0$ and low correlation values for $\tau \neq 0$. The distinct large peak value can be utilized to detect lock quickly and with high confidence. Therefore, the M-sequence with its good autocorrelation property is ideal for synchronization purposes. However, its low cardinality, i.e. the number of codes, makes it inappropriate for applications such as CDMA where separation of users requires a larger set of available codes, one for each user. In this case, Gold and Kasami codes are more suitable given the larger number of codes available. Note also that the cross-correlation of these codes are lower than for the M-sequence, making it easier to separate out the different users in CDMA. For complete isolation between users, the Walsh code seems most suitable with its zero cross-correlation. Walsh codes are orthogonal. However, zero cross-correlation is achieved only if the codes are exactly aligned, i.e. synchronized. Otherwise, the cross-correlation is substantially worse. Exact alignment is not possible in practice even with perfect synchronization due to multipath delay spread. In a system, all of these codes might be used. For instance, in IS-95, an M-sequence is used to establish synchronization, Walsh codes are used for CDMA, and long Gold codes are overlaid on the Walsh codes to reduce the loss in orthogonality due to channel delay spreads.

Table 3.2 Properties of selected PN-codes [114].

Sequence	Period	Cardinality	Auto-Correlation	Cross-Correlation
M	N_c	$<< N_c - 1$	$r_{xx}(\tau) = \begin{cases} 1 - \frac{\lvert\tau\rvert}{T_c}, & \lvert\tau\rvert \le T_c \\ -1/N_c, & \lvert\tau\rvert > T_c \end{cases}$	$r_{xy}(\tau) \ge \frac{2\sqrt{(N_c+1)}+1}{N_c}$
Gold	N_c	$N_c + 2$	$\max r_{xx}(\tau) >> -1/N_c$	$r_{xy}(\tau) \le \frac{2\sqrt{N_c+1}+1}{N_c}$
Kasami	N_c	$\sqrt{N_c+1}$	$\max r_{xx}(\tau) >> -1/N_c$	$r_{xy}(\tau) \le \frac{\sqrt{N_c+1}+1}{N_c}$
Walsh	N_c	N_c	$\max r_{xx}(\tau) >> -1/N_c$	$r_{xy} = 0$ ✣

✣ Only if perfectly synchronized.

Figure 3.2 shows a length 31 M-sequence running at a rate of 1 MHz. The time waveform shows a period of 31 binary symbols and within each period the sequence appears to be random. The power spectral density shows a $\sin(x)/x$ shape, typical for a rectangular pulse shape; see (2.30). The peak power is 31 times or 15 dB lower as compared with a data source running at 32.25 kbps, shown as the dashed line in Figure 3.2b. The reduction in peak power directly contributes to the interference rejection capability of a spread-spectrum system.

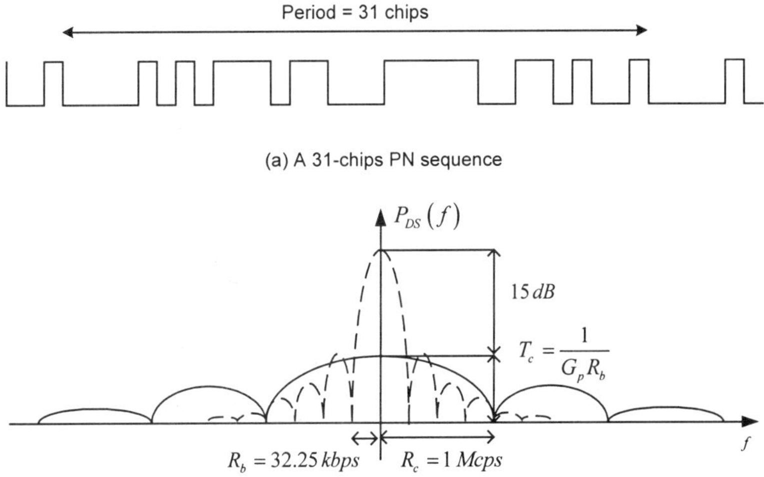

(b) Power spectral density of a 31-chips PN sequence at 1 Mcps

Figure 3.2 A length 31 M-sequence with a chip rate of 1 MSps.

3.2 Direct Sequence

In a DS spread spectrum system, to spread the information bandwidth, the transmitted information is phase modulated with a spreading code which runs at a much faster rate. The rate of the spreading code determines the resultant spread in signal bandwidth. In DS spread spectrum, each code symbol is referred to as a chip with duration T_c and its rate is referred to as the chip rate $R_c = T_c^{-1}$, measured in units of chips/sec or cps. In Figure 3.2 the chip rate is 1 Mcps.

3.2.1 Waveform generation

A DS system uses a PN code to modulate the phase of the unspread information symbols $\xi_k = I_k + jQ_k$. Let $c_I(t)$ and $c_Q(t)$ denote the PN codes for the I-channel and Q-channel, respectively. Each PN code is represented by the following expressions

$$c(t)=\sum_{k=-\infty}^{\infty} c_{\lfloor k \bmod N_c \rfloor} p_c(t-kT_c) \tag{3.1}$$

where $c_{\lfloor k \bmod N_c \rfloor}$ represents the kth chip in the PN code and $p_c(t)$ is the pulse shape for each chip. Note that the *mod* operation represents the periodicity of the PN code. Given the above definitions, the spread symbol ζ_k can be represented by

$$\zeta_k = c_I(t)I_k + jc_Q(t)Q_k. \tag{3.2}$$

In other words, the information symbols are spread in quadrature by two PN codes $c_I(t)$ and $c_Q(t)$, as shown in Figure 3.3. To see that (3.2) is a phase modulation by the PN codes, observe that the amplitude of the symbol is approximately unchanged after spreading as shown below

$$a(t)=\sqrt{\left(c_I(t)I_k\right)^2+\left(c_Q(t)Q_k\right)^2} \approx a_k. \tag{3.3}$$

The chip pulse shaping does cause fluctuations in the amplitude but the fluctuations should be negligible. The PN code effects mostly the phase of the spread symbol as shown below

$$\phi(t)=\tan^{-1}\left(\frac{c_I(t)I_k}{c_Q(t)Q_k}\right). \tag{3.4}$$

The phase now varies at the chip rate rather than the information symbol rate. The shorter transition time introduces faster phase variations, resulting in a signal with a wider bandwidth. The bandwidth spread factor is equal to R_c / R_s. With a rectangular pulse shape at the Nyquist rate, the power spectral density of direct-sequence can be found by applying (2.26) to obtain

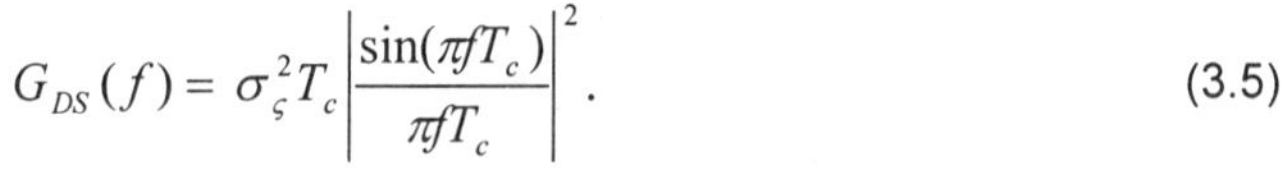

$$G_{DS}(f)=\sigma_\varsigma^2 T_c \left|\frac{\sin(\pi f T_c)}{\pi f T_c}\right|^2. \tag{3.5}$$

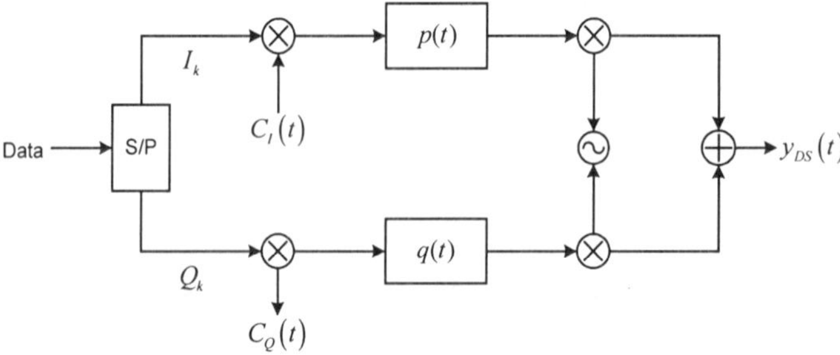

Figure 3.3 A direct-sequence spread-spectrum transmitter with quadrature spreading.

3.2.2 Processing Gain

The robustness of spread-spectrum systems can be interpreted from a geometric perspective. It can be shown that for a given observation time T and a given transmission bandwidth W_T, the total dimension of the spread-spectrum signal is $W_T T$ [17]. In a communication system, T would be the bit interval since this is the basic unit of time during which the information is transmitted. Recall the minimum distance receiver discussed in Section 2.3.2. Its performance depends on the amount of disturbance projected on an orthonormal basis with a dimension equal to the signal dimension. Given this observation, the amount of interference rejection in a system then depends on the product

$$G_p = W_T T_b = \frac{W_T}{R_b} \tag{3.6}$$

where G_p is also defined as the processing gain of the system and R_b is the bitrate.

> **Example 3.1** Take MPSK for example. Its transmission bandwidth assuming $\beta = 1$ is simply the symbol rate R_s, which is $\left(T_b \log_2 M\right)^{-1}$. The processing gain for a transmission system based on MPSK is then $\left(\log_2 M\right)^{-1}$. Clearly, such a system offers no interference suppression and in fact may enhance the effect of an interferer for constellation size larger than two. □

Given the definition in (3.6), the processing gain is derived for direct-sequence below:

$$\begin{aligned} G_p &= \frac{R_c}{R_b} = N_c \\ &= \frac{1}{\log_2 M} \underbrace{\frac{R_c}{R_s}}_{\text{Spread Factor}}. \end{aligned} \tag{3.7}$$

where according to (3.5), $W = R_c$. According to (3.7), the processing gain is directly proportional to the amount of spreading in the transmission bandwidth, R_c / R_s, and inversely proportional to the constellation size. It is desirable to have a large processing gain for an increase in interference rejection, security, and low power spectral density. Therefore, the chip rate should be made as large as possible. It is limited only by the speed and power dissipation of the underlying circuits for a given technology. The constellation size should be kept to a minimum, e.g. $M = 2$.

Substituting (3.7) into (3.5), the power spectral density can be expressed in terms of G_p:

$$G_{DS}(f) = \frac{\sigma_{\varsigma}^2}{G_p R_b} \mathrm{sinc}^2\left(\frac{f}{G_p R_b}\right) \tag{3.8}$$

where $\mathrm{sinc}(x) = \sin(\pi x)/\pi x$ and $\sigma_{\varsigma}^2 = \sigma_{\xi}^2 \sigma_c^2 = \sigma_{\xi}^2$. The variance of the PN sequence σ_c^2 is assumed to be one. As shown in Figure 3.4, direct-sequence spreading has lowered the power density of the original unspread information by $G_p \log_2 M$ and has spread the transmission bandwidth by $G_p \log_2 M$. The reduction in power spectral density allows spread-spectrum signals to operate within the constraints set forth by the FCC for the unlicensed ISM bands.

3.2.3 Interference Rejection

Perhaps the most important benefit of spread-spectrum from a commercial perspective is its ability to reject interference. Figure 3.4 shows a typical case where an interference source can seriously impair communication by lowering E_b / N_0 or γ. The effective noise density N_I introduced by the interference can be represented as

$$N_I = \frac{I_0}{W_I} \tag{3.9}$$

where W_I is the effective bandwidth occupied by the interferer with total in-ban power I_0. In this case, γ for a non-spread system, denoted by γ_{NB}, becomes

$$\gamma_{NB} = \frac{E_b}{N_I + N_0} = \frac{SW_I / R_b}{I_0 + N_0 W_I} \tag{3.10}$$

where E_b is the energy per bit and S is the signal power.

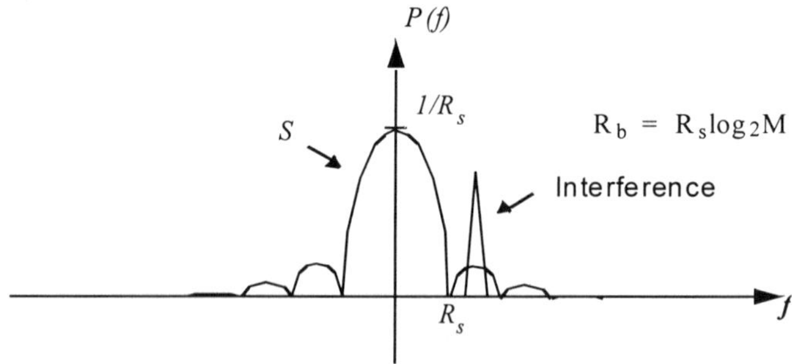

Figure 3.4 A Non-spread-spectrum system in the presence of interference.

When $W_I = R_b$, γ_{NB} becomes $S/(I_0 + N_0)$. Therefore, with a large interference power I_0, the receiver cannot provide sufficient γ for reliable communications.

A spread spectrum receiver has an improved γ in the presence of interference because the receiver performs a despread operation that recovers the original transmitted information while the interferer gets spread, resulting in a lower interference power within the information bandwidth as shown in Figure 3.5. The despread operation is performed by correlating the transmitted code with a replica code at the receiver as shown in Figure 3.6. The interference signal is uncorrelated with the PN code and contributes to only a fraction of in-band interference. To see this, the spread interference signal $i_{ss}(t)$ is represented by an orthogonal expansion using $p_c(t)$ as the basis:

$$i_{ss}(t) = i(t)c(t) = \sum_{k=-\infty}^{\infty} i_k c_{\lfloor k \bmod N_c \rfloor} p_c(t - kT_c) \tag{3.11}$$

where $i(t)$ is the interference signal and i_k is the interference value at the kth chip interval. Since i_k and c_k are uncorrelated, the power spectrum of $i_{ss}(t)$ can be expressed as

$$G_{I_{ss}}(f) = \sigma_i^2 \sigma_c^2 R_c |P_c(f)|^2 \tag{3.12}$$

where $\sigma_i^2 = I_0$ and $\sigma_c^2 = 1$ are the variance of i_k and c_k, respectively. For rectangular pulse shaping the power spectral density becomes

$$G_{I_{ss}}(f) = \frac{I_0}{R_c} \operatorname{sinc}^2\left(\frac{f}{R_c}\right) \tag{3.13}$$

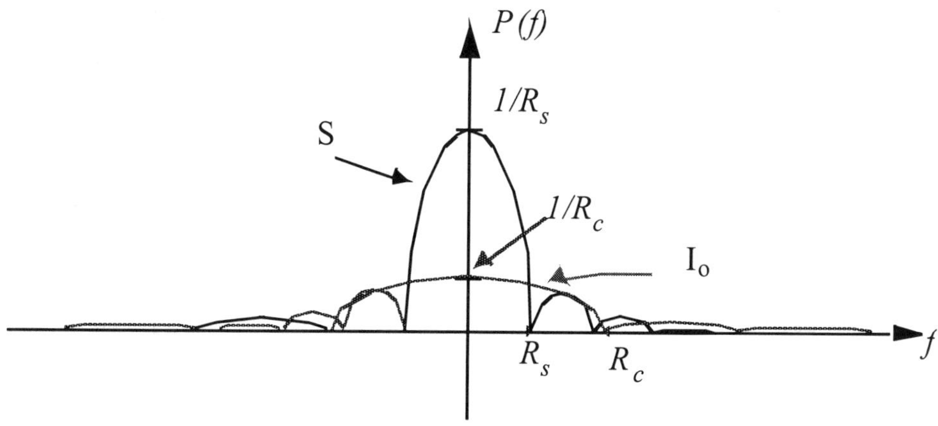

Figure 3.5 A direct-sequence system in the presence of interference.

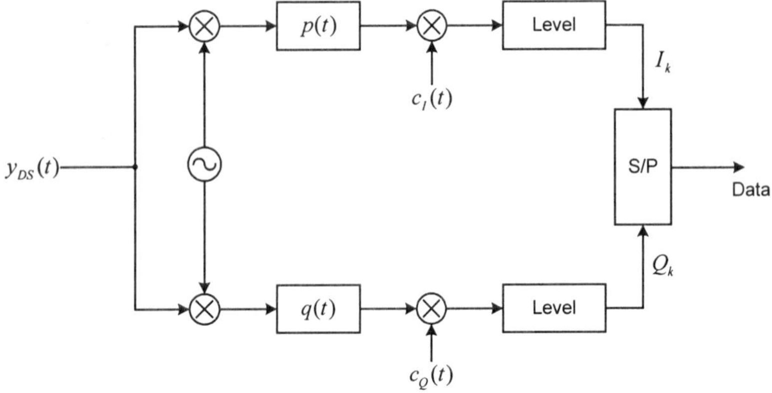

Figure 3.6 A direct-sequence spread-spectrum receiver with quadrature despreading.

From (3.13) the interference noise density after spreading is

$$N_{I_{ss}} = \frac{I_0}{R_c} \tag{3.14}$$

and the corresponding E_b / N_0 is

$$\gamma_{ss} = \frac{SR_c / R_b}{I_0 + R_c N_0} = \frac{S}{I_0 / G_p + R_b N_0}. \tag{3.15}$$

The interference power has been effectively reduced by G_p. Substituting (3.10) into (3.15) results in the following interference rejection gain compared to a narrowband system:

$$0 \le \frac{\gamma_{SS}}{\gamma_{NB}} \le \frac{R_c}{W_I} \frac{I_0 + N_0 W_I}{I_0 + N_0 R_c}. \tag{3.16}$$

When the interference power is in-band, spread-spectrum provides an interference rejection ratio corresponding to the right hand side of (3.16). However, in the case when the interference is outside of the signal transmission band but still within the spread bandwidth, a spread-spectrum system may offer a rejection ratio less than one, corresponding to the left hand side of (3.16). This is because when the interference is out-of-band, a narrowband system can completely reject it. However, a spread-spectrum system, being of wider bandwidth, does not reject it completely. Rather the interference power is averaged over the spread band and results in a lower E_b / N_0.

There is also a trade-off between noise and interference. In an interference-limited environment where noise is negligible, the maximum gain from a spread-spectrum system is R_c / W_I which implies that the chipping rate should be as large as possible.

On the other hand, in a noise-limited environment where interference is negligible, the maximum gain from a spread-spectrum system becomes unity, implying that no additional interference suppression is achieved with spreading.

To provide the highest interference rejection ratio in the former case, a spread-spectrum system should have a high chip rate which implies a high processing gain. However, to achieve a high processing gain, (3.7) indicates that for a given data rate, the chip rate must also be high. A higher chip rate places a tighter speed constraint on various components in the receiver, especially, if the IF processing is performed digitally. The implication of having a high processing gain and high chip rate on implementation will be elaborated in Chapter 5.

For most systems that employ spread-spectrum, interference tends to be the dominant source of degradation. For instance, the advantage of using spread spectrum is clear for operation over unlicensed bands since the worst-case situation may often occur due to uncontrolled interference sources. One might ask is spread-spectrum really necessary if the interference can be controlled like in most digital cellular systems where transmissions are allocated over different frequencies and time slots. Following the discussion of co-channel interference (CCI), the effectiveness of spread-spectrum stands out even for a controlled system since it can effectively improve frequency reuse. Moreover, in CDMA systems where equally powered interferers must coexist in the same frequency channel, the interference rejection capability of spread spectrum becomes essential for the operation of such a system.

3.2.4 Rake Receiver

Recall from Section 2.7 of Chapter 2, the Rake receiver [87] offers a direct application of maximum ratio combining (MRC) to obtain path diversity. In general, MRC cannot be directly applied to achieve path diversity due to the delay spread incurred by a multipath channel. The delay spread causes intersymbol interference (ISI) at the receiver output. The ISI must first be reduced or cancelled for MRC to work effectively. Ideally, ISI can be cancelled using an FFE or DFE but only partially cancellation is achieved in practice due to the finite filter length in an FFE and the error propagation in a DFE.

Spread spectrum provides interference suppression which can be exploited for path diversity via MRC. In particular, direct-sequence spread spectrum is best matched for this purpose because its wideband time waveform can resolve multipath impulses to within the accuracy of the chip duration. For instance, with a 1 Mcps system, the multipath can be resolved to within 1 μs. Furthermore, for multipaths which are more than one chip apart, the received signals from different paths will have low cross-correlation with one another due to the cross-correlation property of the PN code.

A Rake receiver with four fingers is shown in Figure 3.7. Each finger in the Rake is synchronized onto an impulse in the channel impulse response that contains four dominant paths with fading amplitudes α_1, α_2, α_3, α_4. The timing of each impulse is acquired by a search engine. The acquisition process is discussed in Section 3.6. Because of the interference rejection property of spread-spectrum, ISI from multipath impulses other than the one being received in each finger is rejected. The outputs of all the fingers are then combined, typically with MRC to obtain a diversity gain of K_D, where K_D is the number of fingers.

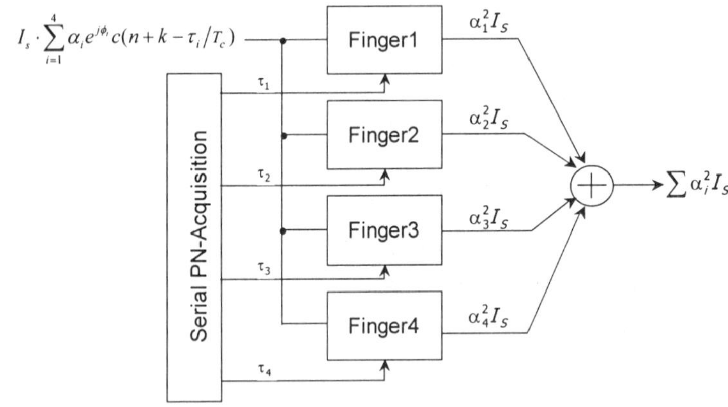

Figure 3.7 A Rake receiver.

To design a Rake receiver, one should consider three main constraints: 1) the length of the code should be long enough to capture most of the multipath dispersion, 2) the chip rate should be high enough to resolve the different paths, and 3) the processing gain should be large enough to reject the ISI in each finger. The first constraint can be expressed by

$$N_c \geq \frac{\max \tau_d}{T_c}. \tag{3.17}$$

The second constraint fixes the chip duration such that the chip rate is much greater than the coherence bandwidth

$$R_c >> B_c. \tag{3.18}$$

Usually, R_c should result in a high enough resolution (e.g. 10 times the coherence bandwidth) to resolve the multipaths. To obtain a guideline for the third constraint, we refer to (2.57) which describes the loss in E_b / N_0 in terms of the interference mean and variance. In the case of a Rake finger, the SIR is equal to the power of the multipath impulse being demodulated over the sum of the power from all the other interfering multipaths. Assume for the sake of discussion that all multipaths have equal power. Then, the SIR is simply $G_p /(K_D - 1)$. Substituting this for the value of

SIR in (2.57) and assuming that the mean of the interference is zero, the following constraint can be derived

$$G_p = N_c \geq \frac{\alpha_I \eta_W \bar{\gamma}(K_D - 1)}{1 - \alpha_I}. \tag{3.19}$$

In general, (3.19) imposes a more stringent requirement on the length of the code than (3.17).

Example 3.2 Consider a rural hilly terrain where the maximum delay spread is 20 μs with a coherence bandwidth of 100 kHz. If a multipath resolution of 1 μs is desired, the chip duration should be 1 μs. Then according to (3.17), the length of the code must be greater than 20 chips. □

Example 3.3 Let the number of fingers be $K_D = 4$. Assume BPSK is used and that the excess bandwidth $\beta = 1$. Then, the bandwidth efficiency η_W is 1 bps/Hz. If the required $\bar{\gamma}$ is 10 and the loss in $\bar{\gamma}$ due to ISI should be less than 1 dB, then $\alpha_I = 0.794$. Applying (3.19), the length of the code should be greater than 116 chips, which is longer than the constraint derived from (3.17) in Example 3.2. □

3.3 Frequency Hop

Although traditionally direct-sequence has been widely used because of its simplicity as compared to frequency hop (FH) techniques, digital technology has enabled cost-effect implementation of fast hopping spread-spectrum systems. Therefore, it is instructive to briefly describe the FH technique that has become a practical alternative to direct-sequence spreading. In fact, FH is being used in GSM and many commercial WLAN devices based on the IEEE 802.11 standard.

3.3.1 Waveform Generation

In a frequency-hop (FH) spread-spectrum system, rather than phase modulating the information with a PN code to spread the bandwidth of the baseband data as shown in Figure 3.3, the carrier frequency is varied or hopped over a set of frequencies $\{f_{H_k}\}$ with $1 \leq k \leq N_f$, where N_f is the total number of available hopping frequencies. The time between each hop is denoted by T_h and the hop rate is defined as $R_h = T_h^{-1}$. Because the bandwidth spread is achieved by hopping the carrier frequency, FH spread spectrum is more efficiently combined with M-ary frequency shift keying (MFSK). The baseband FH signal is shown below, assuming that $T_h \leq T_b$:

$$\tilde{s}(t) = \sum_{k=0}^{\infty} e^{j2\pi\left[f_{D_{\lfloor kT_h/T_b \rfloor}} + f_{H_{k \bmod N_f}}\right]t + \phi_k} \tag{3.20}$$

where the data modulation uses a set of M-ary frequencies, $\{f_{D_n}\}$. The index n denotes the current data symbol time interval. A similar expression can be derived for $T_h > T_b$. Figure 3.8 shows the block diagram for a FH transmitter as described by (3.20).

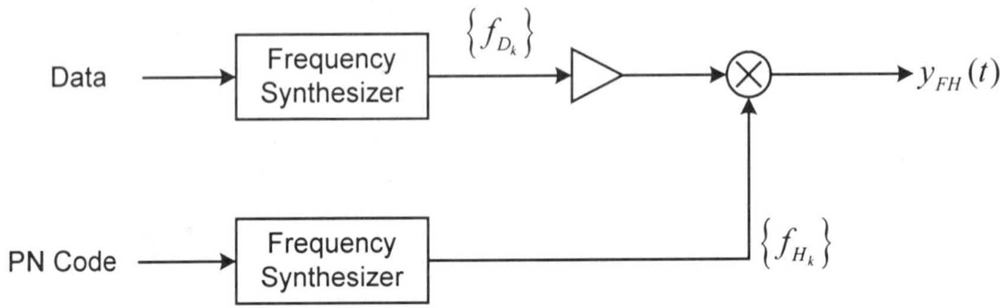

Figure 3.8 Block diagram of an MFSK frequency-hop transmitter.

Assuming a hop spacing equal to the MFSK signal bandwidth, the power spectral density for an MFSK FH system can be derived using (2.26) [114] and is shown in (3.21).

$$G_{FH}(f) = \frac{1}{RMN_f} \sum_{j=1}^{N_f} \sum_{k=1}^{M} \text{sinc}^2\left(\frac{f - f_{H_j} - f_{D_k}}{R} \right) \tag{3.21}$$

3.3.2 Processing Gain

From (3.21), W_{ss} is determined to be $MN_f R$, where

$$R = \max\{R_h, R_s\}. \tag{3.22}$$

Expression (3.22) constrains R to be at least equal to the symbol rate. Intuitively, this is because at a hop rate much lower than R_s, the FH system appears like an MFSK system with bandwidth MR_s. Such a system is also referred to as a slow frequency-hop (SFH) system. At a hop rate faster than R_s, the system is referred to as a fast frequency-hop (FFH) system and has an effective symbol rate R_h. The transmission bandwidth is then proportionally larger.

Applying (3.6), the processing gain for FH spread-spectrum can be expressed as

$$G_p = MN_f \frac{R}{R_b}$$
$$= \frac{M}{\log_2 M} \underbrace{\frac{N_f R}{R_s}}_{\text{Bandwidth Spread Factor}} \tag{3.23}$$

The expression in (3.23) is similar to that of (3.7), except that the chip rate is replaced by $MN_f R$. The PSD of an MFSK FH signal can now be expressed in terms of its processing gain:

$$G_{FH}(f) = \frac{1}{G_p R_b} \sum_{j=1}^{N_f} \sum_{k=1}^{M} \text{sinc}^2 \left(\frac{f - f_{H_j} - f_{D_k}}{G_p R_b M^{-1} N_f^{-1}} \right) \tag{3.24}$$

which decreases in magnitude and expands in bandwidth with increasing processing gain as in a direct-sequence system. See (3.7). A plot of the power density is shown in Figure 3.9.

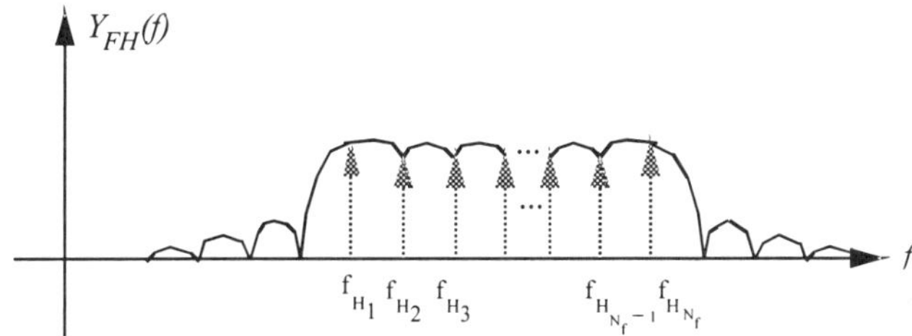

Figure 3.9 Power spectral density of frequency-hop spread-spectrum with rectangular-pulse shaping.

3.3.3 Interference Rejection

In the presence of interference, the frequency hop receiver shown in Figure 3.10 spreads the interference power across a wider bandwidth since the hopping pattern is uncorrelated with the interference signal. Derivations used in (3.9)-(3.16) produce similar results which indicate that the interference power density decreases with larger spread bandwidth

$$N_{I_{ss}} = \frac{I_0}{MN_f R} \tag{3.25}$$

and the receiver experiences proportionally less interference energy per information bit:

$$R_b N_{I_{ss}} = \frac{I_0}{MN_f R} R_b = \frac{I_0}{G_p}. \tag{3.26}$$

Using (3.10) and (3.26) the resulting interference rejection gain is

$$0 \le \frac{\gamma_{SS}}{\gamma_{NB}} \le \frac{MN_f R}{W_I} \frac{I_0 + N_0 W_I}{I_0 + N_0 MN_f R}. \tag{3.27}$$

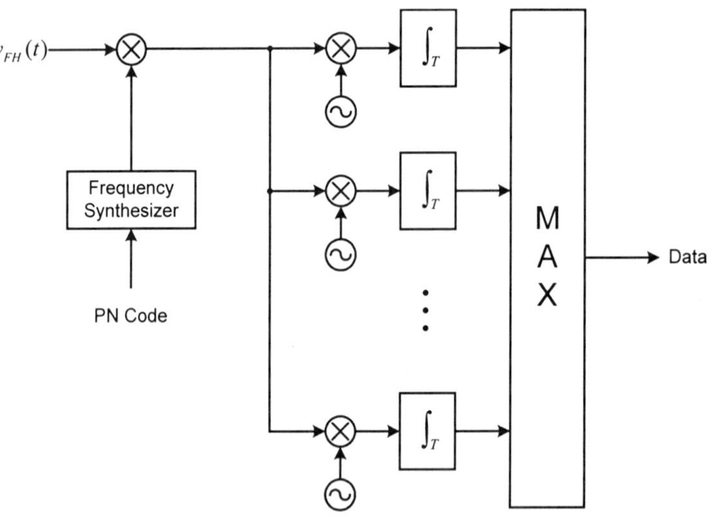

Figure 3.10 Block diagram of an MFSK frequency-hop receiver.

Expression (3.26) implies that to achieve high interference suppression, one should use a large FSK constellation size and have a large set of frequencies for hopping, provided that a large transmission bandwidth is available. Also, the hop rate should be set higher than the bit rate; i.e. use fast hopping. In contrast, interference rejection capability of a DS system degrades with increasing *M*. While it is desirable to have a large *M*, the price is paid by an *M* times increase in bandwidth. To avoid the additional increase in bandwidth, QAM or MPSK could be used with FH, but to keep the same processing gain, the hop rate must be increased *M* fold. However, due to the fast switching requirement of the synthesizer, FFH system is difficult to implement in practice.

3.3.4 Diversity

A Rake receiver for a frequency-hop system does not exist. However, as discussed in Section 2.7.4.2, frequency hop does provide frequency diversity if the hop spacing is greater than the coherence bandwidth of the channel. In this case, a FH system with

N_f results in N_f diversity branches. To obtain diversity using frequency hop proves to be effective, especially, when the channel experiences time-flat fading. By hopping with a frequency separation larger than the coherence bandwidth, a time-selective fading can be induced with a fading rate proportional to the hop rate. A slow fading channel that requires a large interleaver can then be mitigated with a shorter interleaver with a sufficiently high hop rate. See Section 2.7.4.2 for more details.

Interference diversity using frequency hop has already been discussed in Section 2.7.6.1. In a SFH system, if during a hop, the signal experiences a substantial amount of interference then most of the symbols contained in that hop could be corrupted. Being a SFH system, such a scenario introduces a significant loss in data, possible greater than 100 symbols. To safeguard against the above situation, the transmitted symbols are coded and interleaved over several hops such that the effect of a burst error corrupting one of the hops is randomized at the output of the de-interleaver. A FFH system can provide interference diversity without the need for an interleaver. However, most commercial systems employ SFH due to difficulties in implementing a fast tuned synthesizer.

Example 3.4 In GSM, a block of speech 260 bits in length is coded into 456 channel symbols. Since each time slot can take 114 bits of data payload, the 456 symbols of coded speech data could be transmitted over four time slots, each in separate but consecutive GSM frames. To improve the system capacity, GSM employs SFH whereby each frame is frequency hopped using 63 available hopping frequencies. The duration of each frame is 4.615 ms. Therefore, the potential for a strong interferer to occur during the frame time can be quite high. A strong interferer might be due to CCI or ACI. To mitigate the effect of a burst error due to a strong interference occurring inside a hop, GSM interleaves the 456 symbols of speech data over eight frames. □

3.4 Performance Comparison

It is instructive to compare the difference in performance between FH and DS spread-spectrum. For a fair comparison, the two systems are assumed to have the same bitrate and transmission bandwidth W_T. Expressions (3.7) and (3.23) imply that with proper choices of M, N_f, and R_h, frequency hop systems can theoretically achieve the same processing gain as that of direct sequence. The necessary constraint for this to occur is described below:

SFH:
$$N_f = \eta_W N_c \tag{3.28}$$

FFH:
$$N_f = \frac{N_c}{M}\frac{R_c}{R_h}. \tag{3.29}$$

Therefore, it seems always possible to find a set of parameters that make FH equivalent to a DS system. However, the equivalence does not hold in practice for a

SFH system due to its sensitivity to interferers which may inject a large amount of interference in a given hop, as discussed in the previous section. Direct-sequence does not run into this problem since every bit is protected by the spreading code. The likelihood of an interferer having sufficiently large power to disrupt the receiver within a bit duration, on the order of 1-10 μs, is less likely. For the same reason, FFH is less sensitive to the instantaneous variations in the interference power.

SFH is also less robust than DS spreading in the presence of multipath fading since optimal combining can be achieved through a Rake receiver with direct sequence whereas for SFH diversity is obtained through suboptimal combining using interleaving and coding. Also, the interleaver needs to span many hops and with a SFH system, the interleaver may become large, incurring a large delay in the overall system. The main advantage of SFH over DS is evident when the coherence bandwidth is high (e.g. 10 MHz). To resolve the multipaths, a chip rate of 100 MHz is needed according to (3.18). Such a high chip rate requires high sample rate analog-to-digital converters (ADC) and digital-to-analog converters (DAC) that dissipate excessive amounts of power leading to a shorten battery lifetime for a portable device. In contrast, a SFH system has an instantaneous bandwidth equal to the symbol rate which is many times lower than the chip rate of a DS system. Thus, only a low sample rate ADC/DAC is needed with significant power saving.

More robust system performance can be obtained with frequency hop if fast hopping is used. However, in practice FFH is not commonly used, especially in commercial systems. The reason lies in the need for an extremely agile frequency synthesizer which can be difficult to implement. With a typical bitrate of 100 kbps to 1 Mbps, the synthesizer needs to settle within 1 to 10 μs. Most analog synthesizers cannot achieve the required agility without degrading the spectral purity of the generated tone. Direct sequence, on the other hand, does not require an agile synthesizer.

Furthermore, frequency and phase synchronization become more difficult for a fast hop receiver since the synchronizing loops must settle within a small fraction of the hop time. Many FFH systems use MFSK, which allows non-coherent demodulation of the transmitted data. Direct sequence, on the other hand, is not constrained by the settling time of frequency and phase synchronization since its carrier frequency stays constant most of the time. Thus, carrier synchronization loops needs to track only slow variations in frequency drifts in the carrier. Coherent demodulation can, therefore, be used with direct-sequence to achieve an improvement in SNR over non-coherent MFSK.

Although carrier synchronization is more difficult to implement for FH, timing recovery is substantially more relaxed than that for direct sequence. A FH receiver only needs to perform symbol synchronization whereas a DS receiver must recover the PN chip timing. Chip synchronization requires circuits that must operate at 10-100 times higher speeds than that needed for symbol synchronization.

Until recently FH has not been considered as a cost effective solution for high bit rate systems requiring a high processing gain because of the difficulty in implementing a fast hopped system. As IC technology advances, more sophisticated digital signal processing techniques have been applied to frequency synthesis to implement the agile synthesizer which is the key component to achieve fast hopping [125][126][127]. For instance, [128] reports an agile frequency synthesizer that consists of a direct-digital frequency synthesizer (DDFS) and a DAC. The DDFS can change frequency instantaneously in one clock cycle while the DAC converts the digitally synthesized sinusoid into an analog waveform. As technology scales, low cost CMOS mixed signal designs such as this are becoming more prevalent and are making systems which have been traditionally difficult to implement realizable.

3.5 Code Acquisition

The processing gain of a spread spectrum system is achieved only when the desired code timing has been accurately acquired. The loss in processing gain as a function of the timing offset Δt is shown below

$$L_{G_p} = -20\log_{10}\left(1-\frac{|\Delta t|}{T_c}\right), \qquad |\Delta t| \le T_c. \tag{3.30}$$

For example, if the timing is off by half a chip, i.e. $\Delta t = T_c/2$, then the loss is $-20\log_{10}(0.5) = 6$ dB, assuming that N_c is large. The loss is significant and must be compensated by code timing recovery. The recovery of code timing is usually done in two steps. The first step is to acquire the timing of the code to within chip accuracy and the second step is to track the code timing and eliminate the residual timing error.

3.5.1 Parallel Vs. Serial

Acquisition of the PN code is accomplished by detection of the auto-correlation peak. At the receiver a PN generator produces a PN code $\hat{c}(t) = c(t+\Delta t)$ where Δt is the timing difference with the transmitted PN code $c(t)$. The correlation of the locally generated PN code with the transmitted PN code can be expressed as

$$R_{c\hat{c}}(\tau) = \left(1+\frac{|\tau+\Delta t|}{T_c}\right) = R_{cc}(\tau+\Delta t), \qquad |\tau+\Delta t| \le T_c \tag{3.31}$$

where $R_{cc}(\tau+\Delta t)$ is the auto-correlation of the PN code defined in Table 3.2. Therefore, by computing the correlation between the local and transmitted PN codes, the timing offset can be determined as the point in time at which the correlator outputs a peak. As mentioned before, an M-sequence is ideal for timing recovery because of its good auto-correlation property, where the peak is large at $\tau = 0$ and the out-of-phase correlation values are very small (i.e. N_c^{-1}). A general block diagram of a PN acquisition loop is shown in Figure 3.11, where the correlator is followed by a peak detection block which controls the timing τ of the local PN generator. The correlator can be computed based on the definition in (2.27):

$$R_{cc}(m) = \frac{1}{qN_c}\left|\sum_{k=0}^{qN_c} c_k c_{k+m}^*\right| \tag{3.32}$$

where (2.27) has been rewritten for a deterministic signal and q is the number of samples per chip.

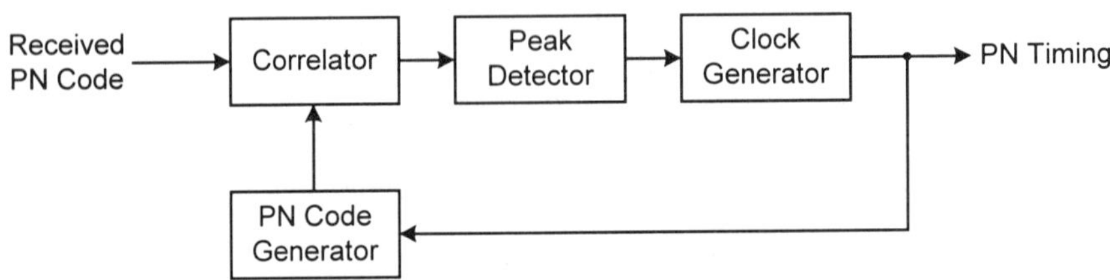

Figure 3.11 A General Block diagram of a PN-acquisition loop.

PN-acquisition algorithms are generally categorized as either serial [129] or parallel [130], depending on the structure of the correlator. In the parallel approach, the autocorrelation in (3.32) is implemented as a matched filter with coefficients c_{-k}^* as shown in Figure 3.12a. The matched filter output displays correlation peaks for each period of the PN code. The periodic peaks provide the information needed to synchronize the local PN code to the received PN code. Clearly, the complexity of a matched filter is on the order of the number of taps in the filter, which in this case is equal to the length of the PN code. The complexity, therefore, of parallel acquisition is linearly proportional to the length of the code, making it more difficult to implement for systems that require high processing gain. Currently, a matched filter is limited for a processing gain of 18 dB [131][132][133]. Though as IC technology continues to scale, integration of more taps are becoming possible.

The serial PN acquisition scheme overcomes the limitation in complexity at high processing gain by using a serial correlator. The local PN sequence is slipped at discrete increments relative to the received PN sequence until the correlator output passes a certain energy threshold. Essentially, a serial correlator is a recursive implementation of (3.32) as shown below

$$y_k = y_{k-1} + \frac{c_k c_{k+m}^*}{qN_c}, \qquad m = 0,1,\cdots,qN_c. \tag{3.33}$$

Since the correlation only occurs over the length of the code, the accumulated correlation value is dumped at the output every qN_c samples. Also, in a new correlation cycle, y_0 is reset to zero. Such an operation is also known as accumulate-

and-dump (A&D). The complexity of a serial correlator is independent of the length of the code. Figure 3.12b shows the block diagram of a serial correlator.

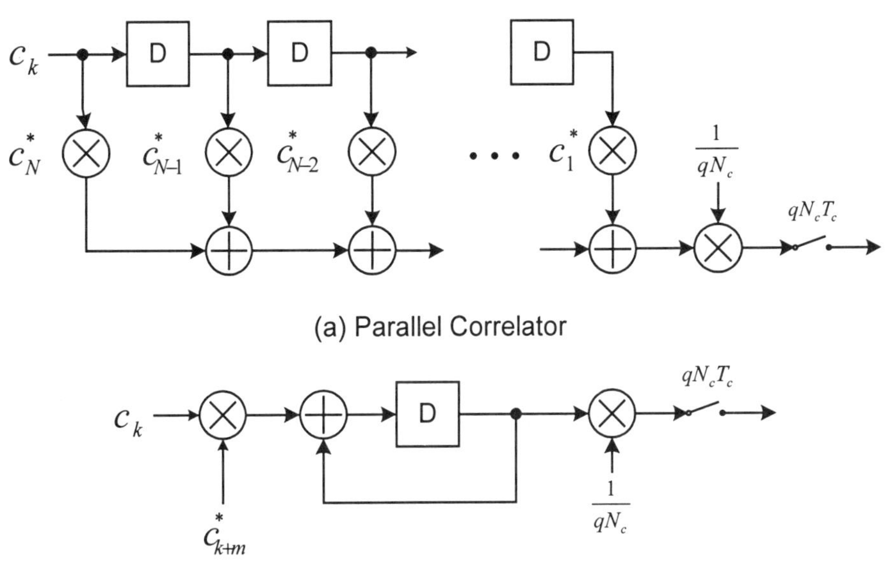

Figure 3.12 Correlator architectures.

Since the computed correlator output y_{N_c} is corrupted by noise and averaging should be performed to minimize the effect of noise. If the averaged value exceeds a given threshold then a lock is declared. Typically, a verification process follows with additional averaging over K_p correlator outputs. The average from the verification mode is then compared with a second threshold to determine if a false lock has occurred. The acquisition is referred to as single dwell if it uses a single average to determine the initial lock and a second averaging step to verify false lock. Figure 3.13 shows the flow diagram for the single-dwell PN-acquisition loop.

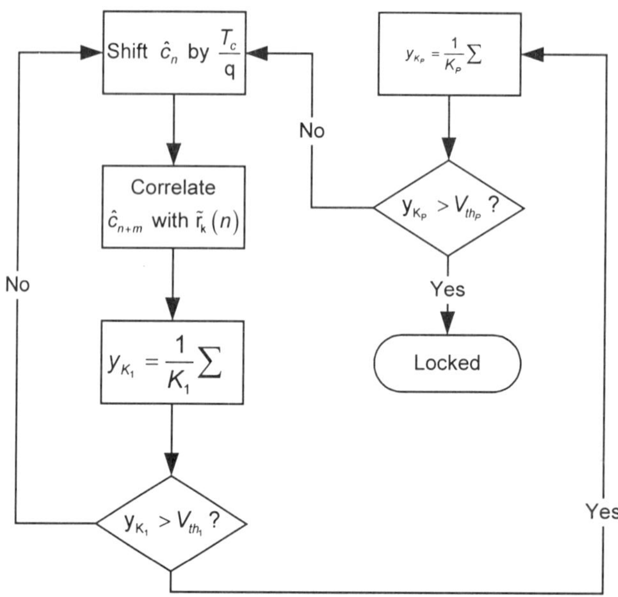

Figure 3.13 A flow diagram of the search procedure for a single-dwell acquisition loop.

While a serial acquisition scheme offers a low complexity solution even for high processing gains, it does prolong the time required to acquire the PN code. For a single dwell acquisition loop, the average time needed for acquisition is [114]

$$T_{acq}(serial) = \frac{(2-P_D)(1+K_P P_{FA})}{2P_D} qK_1 N_c \qquad (3.34)$$

where K_1 is the length in bits of the averaging used to determine initial lock, P_D is the probability that the peak has been detected correctly, P_{FA} is the probability that a false peak has been detected, and K_P is the penalty in bits associated with verifying false peaks.

According to (3.34), the acquisition time for a serial scheme is proportional to the length of the code or processing gain. Therefore, for a large processing gain, the acquisition time for the serial approach becomes prohibitively long. To alleviate the long delays caused by the long acquisition time, network architectures can be designed to ensure that acquisition time occurs only once for establishing a connection. In an IS-95 cellular network [5], for example, a control channel is dedicated to maintain code synchronization and is used to perform the call setup which establishes the initial PN timing. However, the call-setup procedure cannot be applied for a packet-switched network since the PN timing must be reacquired for each packet. The time needed to reacquire the PN timing for each packet may introduce excessive amount of overhead. To minimize the acquisition time, a matched filter can be used in the packet-switched system. With a matched filter the acquisition time is essentially independent of the processing gain and can be achieved in less then 10 symbols. Other Techniques also exist to help reduce the acquisition time. Interested readers are referred to [132][133][134][135].

Both the serial and parallel correlators are sensitive to variations in the channel and SNR. This is because lock in code timing is typically based on comparing the correlator output energy to a threshold, which is a strong function of input SNR. The probability of false locking can become especially high in the presence of strong interference. The challenge is to develop an algorithm from which a simple implementation can be derived while minimizing the false lock condition over a wide range of input SNR. Techniques to make the acquisition less sensitive to the SNR and channel include multiple dwell [136], referenced threshold [137], and slope detection [138].

- **Multiple dwell:** Since at low SNR the peak detection block tends to detect more false peaks, a single-dwell acquisition loop will have a higher acquisition time due to the large false-alarm penalty K_p, which is usually set at 10-100 bits. To improve the acquisition time due to variation in the SNR, multiple dwell schemes have been developed. The idea is to determine false lock in multiple steps, whereby each step or dwell introduces an increasing amount of averaging such that the average acquisition time is reduced for a given false alarm rate.
- **Referenced threshold:** Often acquisition is affected by not only noise but also interference from other codes, typical in a CDMA system. The interference is also referred to as multiple-access interference (MAI). When the number of users is large with equal power, the MAI can be approximated as zero-mean AWGN. However, when the number of users is small or if there is a strong interferer due to the near-far problem, then the disturbance is no longer white and zero mean. In this case, averaging does not provide a sufficient means to eliminate the effect of MAI on the peak detection process. In addition the decision threshold tends to change with the intensity of MAI. To effectively acquire the code in this situation, the interference power is estimated using a reference path, which is essentially a separate correlator that uses a PN code offset in phase from the local PN code. The estimated threshold is used in the peak detector to mitigate the variation in the threshold due to MAI.
- **Slope detection:** In reference threshold, the threshold is adapted for a given MAI level. In slope detection, the threshold is fixed but instead of using correlator energy as a metric to verify lock, the slope of the correlator output is used. In the presence of MAI, false peaks usually result in fluctuations in the slope of the correlator output y_k [138]. By measuring the slope of y_k, such false peaks can be detected with a fixed threshold. The disadvantage of slope detection is that since it uses slope as a metric, noise is enhanced and degrades the performance when the channel is noisy.

3.5.2 Non-coherent Vs. Coherent

PN-acquisition loops can be either coherent or non-coherent. A coherent loop requires that phase recovery occurs before PN acquisition whereas a non-coherent loop must acquire the PN code with no knowledge of the phase and frequency offsets.

In complex notation, the baseband signal for the nth chip of the kth symbol can be represented as

$$\tilde{r}_k(n) = r_{Ik}(n) + jr_{Qk}(n) \tag{3.35}$$

where $r_{Ik}(n)$ and $r_{Qk}(n)$ are the in-phase and quadrature components, respectively. In the non-coherent acquisition loop, the baseband signal has a residual phase offset $\Delta\phi_k$ and has the general form of $\varsigma_k e^{j\Delta\phi_k}$. Applying **(3.32)**, the correlation between the received input and $\hat{\varsigma}_k(n) = \hat{c}_I(n)\hat{I}_k + j\hat{c}_Q(n)\hat{Q}_k$ has the following expression

$$R_{\tilde{r}\hat{\varsigma}}(m) = \frac{1}{qN_c}\sum_{n=0}^{N_c}\tilde{r}_k(n)\hat{\varsigma}_k^*(n+m) \tag{3.36}$$

Expanding (3.36) and taking the square of its magnitude, the square of the autocorrelation function can be expressed in terms of the in-phase and quadrature terms as shown below:

$$\begin{aligned}|R_{c\hat{c}}(m)|^2 = &\frac{1}{q^2N_c^2a_k^4}\left[\sum_{n=0}^{qN_c} r_{Ik}(n)\hat{I}_k\hat{c}_I(n+m) + r_{Qk}(n)\hat{Q}_k\hat{c}_Q(n+m)\right]^2 + \\ &\frac{1}{q^2N_c^2a_k^4}\left[\sum_{n=0}^{qN_c} r_{Qk}(n)\hat{I}_k\hat{c}_I(n+m) - r_{Ik}\hat{Q}_k\hat{c}_Q(n+m)\right]^2.\end{aligned} \tag{3.37}$$

where a_k is the amplitude of the kth symbol. Here, $\hat{I}_k$ and $\hat{Q}_k$ denote the kth I-channel and Q-channel demodulated symbol values, respectively while $\hat{c}_I(n)$ and $\hat{c}_Q(n)$ denote the nth chip of the locally generated in-phase and quadrature codes, respectively. The demodulated data symbols in (3.37) are used to provide data-aided feedback to eliminate the effect of data symbols on the correlation result.

A serial non-coherent PN-acquisition loop based on (3.37) is shown in Figure 3.14. The input to the correlator has the form $\tilde{r}_k(n) = \zeta_k(n)e^{j\Delta\phi_n}$. Although the received baseband signal contains a residual phase offset, the phase offset can be completely eliminated because of the magnitude operation; i.e. $\frac{1}{N_c}\left|\sum_{n=0}^{qN_c} e^{j\Delta\phi_n}\right| = 1$ provided that $\Delta\phi_n$ is constant for $n = 0,1,\cdots,N_c$. If the phase offset is not constant but is varying as a result of a constant frequency offset $\Delta\phi_n = 2\pi n \Delta f$, then the correlator will have an energy loss of

$$L_{\Delta f} = -20\log_{10}\left[\frac{1}{N_c}\frac{\sin(\pi N_c T_c \Delta f)}{\sin(\pi T_c \Delta f)}\right]. \tag{3.38}$$

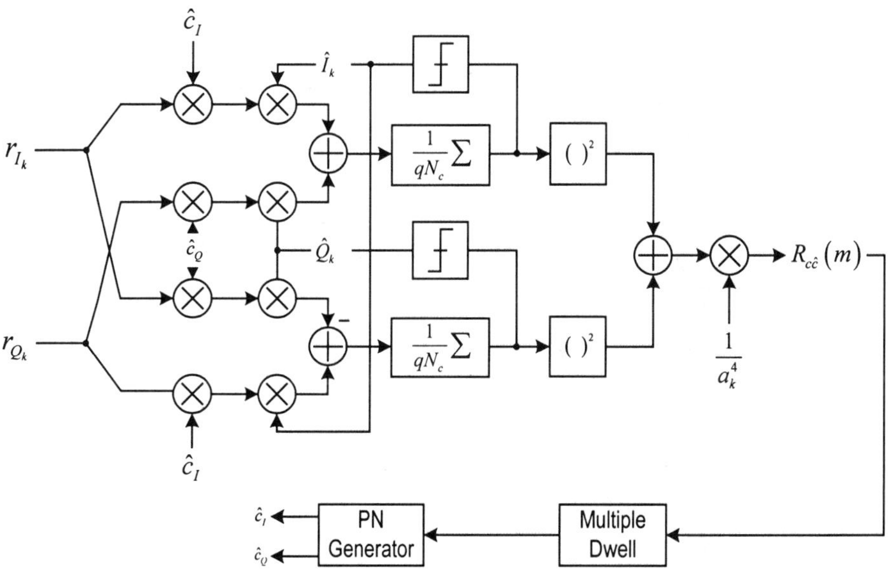

Figure 3.14 A non-coherent PN-acquisition loop.

Example 3.5 If the carrier frequency is 1 GHz and the frequency offset is 5 ppm then $\Delta f = 5kHz$. Let the chip duration T_c be 1 μs and a correlation window be of duration 128 chips. Then, by applying (3.38), the energy loss is 6.93 dB. If a more accurate crystal is used with 1-ppm frequency stability, then the loss is reduced to 0.24 dB. □

A coherent PN-acquisition loop can be derived by assuming that the residual phase offset is zero. The baseband signal then becomes $\tilde{r}_k(n) = \zeta_k(n) = I_k c_I(n) + jQ_k c_Q(n)$, where $I_k c_I(n) = r_{Ik}$ and $Q_k c_Q(n) = r_{Qk}$. Then, substituting these expressions into (3.36), the computation for the correlation simplifies to

$$R_{c\hat{c}}^2(m) = \frac{1}{qN_c a_k^2}\left[\sum_{n=0}^{qN_c} r_{Ik}(n)\hat{I}_k \hat{c}_I(n+m) + \sum_{n=0}^{qN_c} r_{Qk}\hat{Q}_k \hat{c}_Q(n+m)\right], \tag{3.39}$$

assuming that the cross-correlation terms are small for large N_c. In contrast to the non-coherent acquisition, (3.39) does not require a squaring function. Figure 3.15 shows the block diagram corresponding to (3.39) and illustrates the use of a serial correlator. While the previous architectures illustrated the application of serial correlators, it is straightforward to show that a matched filter can be applied in place of a serial correlator.

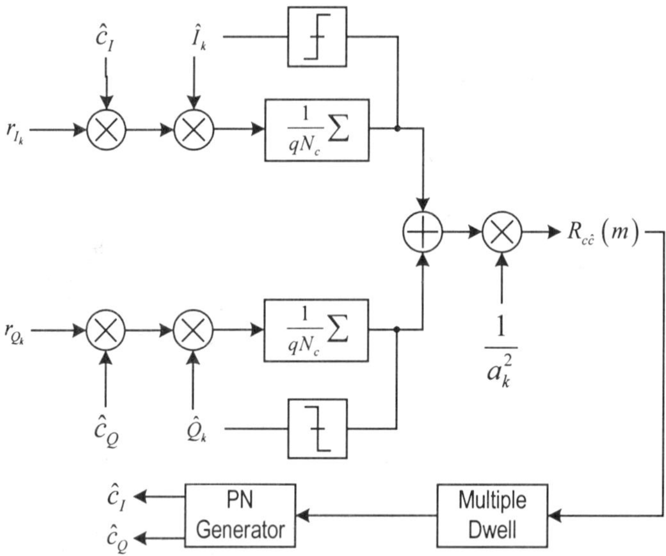

Figure 3.15 A coherent PN-acquisition loop.

3.5.3 Application to FH system

All of the above results apply to both DS and FH systems, though perhaps their applicability to DS is more direct. For FH, the above results hold given the following guidelines:

1. Set T_c to the hop time T_h.
2. Set N_c to the number of available hops N_f.
3. Set the chip waveform $c(n)$ to $\cos(2\pi f_I nT_h / q) + j\cos(2\pi f_Q nT_h / q)$.
4. Set the hop spacing between the in-phase and quadrature components to R, defined in (3.22), to maintain orthogonality between $c_I(n)$ and $c_Q(n)$.

Since the chip waveform is no longer a simple binary value (i.e. ± 1) for frequency hop, parallel acquisition is not used in practice due to the substantial hardware complexity needed to implement each hopping tone. Rather, serial acquisition is used.

3.6 PN-Code Tracking

When designing a PN-acquisition loop, the number of samples per chip or per hop, (i.e. q), is typically chosen such that the residual timing error is less than a chip or hop duration. Therefore, q should be chosen to be at least one such that the residual error is less than half of a chip or a hop. In general, the residual error is $0.5T_c / q$. A slight residual error can still cause noticeable loss in SNR as shown in (3.30). Furthermore, a crystal oscillator generally has a 1-5 ppm frequency offset, which widens the timing error and eventually push the receiver out of synchronization. PN-code tracking is therefore an essential component to keep the receiver in sync.

Example 3.6 Assume the worst case where $q = 1$. The resulting residual error is $0.5T_c$. Applying (3.30), the SNR loss is 6 dB. At $q = 2$, the loss is reduced to 2.5 dB, but still significant. □

3.6.1 Early-Late Gate Correlator

The principles developed for symbol timing recovery discussed in Section 2.5.5 can also be applied to PN-code tracking. The key difference between symbol and code timing recovery lies in the phase detector. To achieve the interference suppression provided by spread-spectrum, the phase detector in the early-late correlator of Section 2.5.5 is modified to utilize the autocorrelation property of the PN-code.

Let the baseband signal be represented by $r(t) = Ac(t) + n(t)$. For sake of illustration, the baseband signal is assumed to be real. Later, the results will be extended to complex signals. The receiver generates three local copies of the PN code, an early version $c_E(t)$, a punctual version $c_P(t)$, and a late version $c_L(t)$:

$$c_E(t) = c(t + \Delta t + 0.5mT_c) \tag{3.40}$$

$$c_P(t) = c(t + \Delta t) \tag{3.41}$$

$$c_L(t) = c(t + \Delta t - 0.5mT_c) \tag{3.42}$$

where Δt is the relative timing error referenced to the transmitter clock phase. m is typically chosen to be either 1 for half-chip correlation or 2 for full-chip correlation. The phase detector for PN-tracking utilizes the correlation between the early and late signals with the received code to generate an error signal that drives a voltage controlled clock oscillator (VCCO) in a negative feedback loop. When the feedback loop reaches steady state, the average timing error is driven to zero. The punctual code $c_P(t)$ is then synchronized with the transmitted PN code.

To derive the phase detector architecture using the early and late signals, we first take the cross-correlation between the received signal and the early and late PN-codes denoted by $R_E(\tau)$ and $R_L(\tau)$. Then, subtract $R_E(\tau)$ from $R_L(\tau)$ to form the error signal $E(\varepsilon)$, where $\varepsilon = \Delta t / T_c$. Applying (3.31) and noting that $\Delta t = 0.5mT_c$ for the early correlation and $\Delta t = -0.5mT_c$ for the late correlation, $E(\varepsilon)$ can be simplified to

$$E(\varepsilon) = A\left[R_{cc}\left(\varepsilon - 0.5m\right) - R_{cc}\left(\varepsilon + 0.5m\right)\right] + n_E(t) - n_L(t), \tag{3.43}$$

where $n_E(t) = \int c_E(t)n(t)dt$ and $n_L(t) = \int c_E(t)n(t)dt$ are the noise waveforms at the output of the early and late correlator, respectively. To see that the first term in (3.43) forms the S-curve of the phase-detector, (3.31) is substituted for the autocorrelation terms in (3.43) to obtain

$$S(\varepsilon) = A\left[R_{cc}\left(\varepsilon + 0.5m\right) - R_{cc}\left(\varepsilon - 0.5m\right)\right] = \frac{2A}{m}\left(1 + \frac{1}{N_c}\right)\varepsilon. \tag{3.44}$$

Figure 3.16 shows the S-curve graphically based on (3.44). The slope of the S-curve corresponds to the phase detector gain. With $m = 1$ or half-chip early-late correlation, the phase detector gain is a factor of two larger than full-chip correlation ($m = 2$). A larger phase detector gain helps to improve the stead-state error response of the loop. However, the trade-off is a reduction of the phase-detector tracking range.

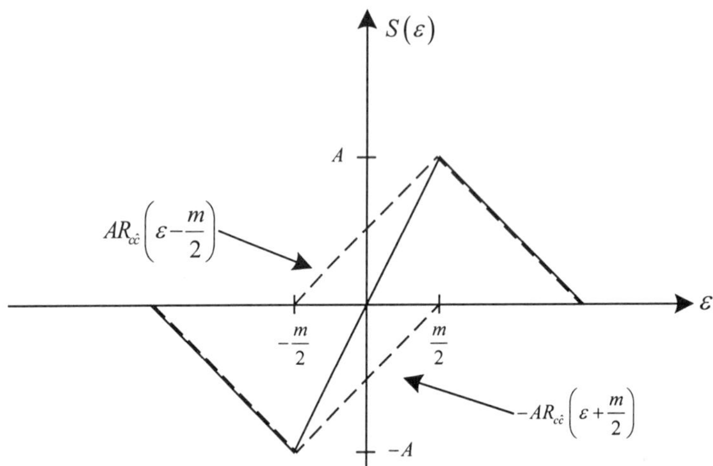

Figure 3.16 S-curve based on the early and late correlations.

A block diagram of the early-late gate correlator for PN-code tracking is shown in Figure 3.17. The noise performance of the early-late correlator can be obtained in a similar manner as in Section 2.5.5. In this case the equivalent noise source is

$n'(t) = \dfrac{n_E(t) - n_L(t)}{K_d}$. Since the early and late correlator noise outputs are independent, the variance of $n'(t)$ can be represented as

$$\sigma_{n'}^2 = \frac{2\sigma_{n_E}^2}{K_d^2} = \frac{2\sigma_{n_L}^2}{K_d^2} \tag{3.45}$$

Since the noise at the correlator output is still a white Gaussian noise, its variance is equal to $N_0 / 2T_s$. Substituting the phase detector gain from (3.44), the variance of the equivalent noise source can be expressed by

$$\sigma_{n'}^2 = \frac{m^2}{4\bar{\gamma}\left(\log_2 M\right)\left(1 + \frac{1}{N_c}\right)^2}. \tag{3.46}$$

The output timing jitter due to the noise source is determined by the noise bandwidth as shown below:

$$E\left[\varepsilon^2\right] = \frac{m^2 B_L}{4\bar{\gamma} R_b \left(1 + \frac{1}{N_c}\right)^2}. \tag{3.47}$$

Example 3.7 Let $m = 1$ for half-chip correlation. Also, assume that the PN code is 128 chips long and that the required E_b / N_0 is 10 dB. If the chip rate is 1 Mcps, then the data rate is 7.8125 kbps. Using expression (3.47), one could determine the loop bandwidth required for a given RMS timing jitter. For instance, let the target RMS jitter be 5%. Then, applying (3.47), the loop bandwidth needs to be 793.5 Hz, approximately ten times less than the data rate. □

3.6.2 Non-Coherent Vs. Coherent

So far, the tracking process assumes a real signal with no data modulation. When the signal contains data modulation, the data-aided acquisition scheme described in Section 3.5.2 can be applied for PN-code tracking. Moreover, the early and late correlators can be replaced by either a coherent or non-coherent correlator described in Section 3.5.2.

3.6.2.1 Application to FH System

The approach to tracking for frequency-hop systems depends on the hop rate. For a FSH system where the hop time is less than the symbol duration, then the guideline described in Section 3.5.3 also applies. Because the chip waveforms are no longer simple binary values but tones, the complexity can be quite high. For a SFH system where the hop rate is equal to an integer multiple of the symbol time, then each chip in the PN code is equal to or longer than the symbol duration. In this case, code tracking can be replaced by symbol timing recovery described in Section 2.5.5.

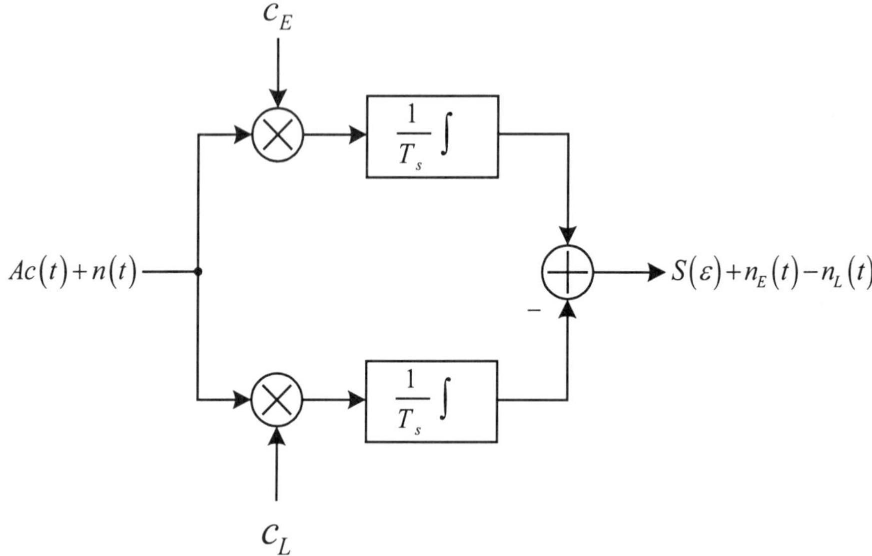

Figure 3.17 Block diagram of an early-late correlator.

CHAPTER 4

RF System Design

Chapter 2 and 0 provided an overview of methods to encode digital data onto a baseband waveform $\tilde{s}(t)$. For wireless systems, the baseband signal is typically transmitted on a carrier frequency centered at ω_0 to achieve high transmission efficiency with a reasonably sized antenna. In particular, for portable systems, the carrier frequency ω_0 is typically chosen in the 1-2 GHz range, where sufficiently small antennas (2-6 inches) can be used with a balanced trade-off between propagation loss and antenna efficiency. More recently, WLAN devices are being implemented in the 2.4 GHz and 5.8 GHz, allowing one to use even smaller antennas. The intent of this chapter is to provide an overview of the system design for radio-frequency (RF) transceivers that are responsible for up converting $\tilde{s}(t)$ to RF at the transmitter and down converting the RF signal back to baseband at the receiver.

4.1 RF Transceiver Architecture

A direct implementation of an RF transceiver based on (2.2) is shown in Figure 4.1. A transceiver implemented in this way is also known as a direct-conversion or homodyne transceiver [139][140]. In Chapter 6, other transceiver architectures will be described. At the transmitter shown in Figure 4.1a, the baseband I/Q symbols are converted to analog levels by a pair of digital-to-analog (DAC) converters. The converted analog levels modulate an RF carrier in quadrature to form two up-converted signals with 90-degree phase shift with respect to each other. These two signals are then subtracted, amplified, filtered, and transmitted via the antenna. At the receiver, the transmitted signal after passing through the channel is received by the antenna and then filtered and amplified before being mixed down in quadrature to baseband using a locally generated frequency reference centered at ω_0. Finally, the mixer output is amplified and digitized by a pair of analog-to-digital converters (ADC) to form the digital I/Q baseband outputs that are further processed by the baseband digital receiver. Note, the mixing products include a second harmonic terms in addition to the desired signal at DC. The lowpass filter filters out these unwanted high-order harmonics.

In RF designs, the various functions shown in Figure 4.1 are also referred to by other names. For example, the multiplier and adder are referred to as a mixer and combiner, respectively. The block that shifts its input by 90-degree is known as a 90-degree phase shifter. The amplifier immediately following the receive bandpass filter is referred to as a low-noise amplifier while the amplifier immediately before the transmit bandpass filter is referred to as a power amplifier.

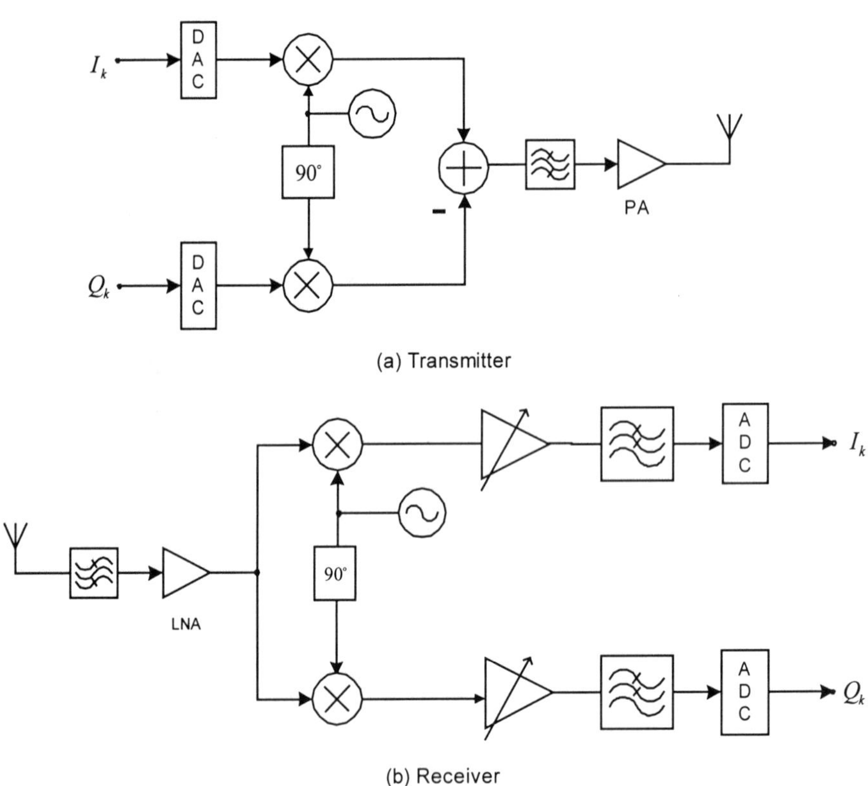

Figure 4.1 A direct-conversion RF transceiver.

An ideal RF transceiver performs up and down conversion with no noise and distortion. However, in practice, a transceiver implementation always generates some form of non-linear distortion or circuit noise. The distortion and noise introduced by a non-ideal transceiver can degrade the coverage and capacity of a radio system. The following sections discuss the RF system performance metrics useful for evaluating RF transceiver performance, in an attempt to quantify the effects of distortion and noise on the overall system performance

4.2 Noise Performance

An RF component can be modeled as a black box with a certain voltage gain G and circuit-generated noise density N_{ckt} as shown in Figure 4.2. Assuming a linear approximation for the RF component at its operating point, the output noise density N_{out} can be described by

$$N_{out} = G^2 N_0 + N_{ckt}. \tag{4.1}$$

Noise factor measures the amount of noise introduced by an RF device relative to the ambient thermal noise at its input and is defined as the ratio of the input-referred noise density to the thermal noise density at the input of the device, as shown below

$$F = \frac{N_{out}}{G^2 N_0} = 1 + \frac{N_{ckt}}{G^2 N_0}, \tag{4.2}$$

where N_{out} / G^2 is the input-referred noise density. Noise figure NF is defined as noise factor in units of dB as shown below

$$NF = 10 \log F. \tag{4.3}$$

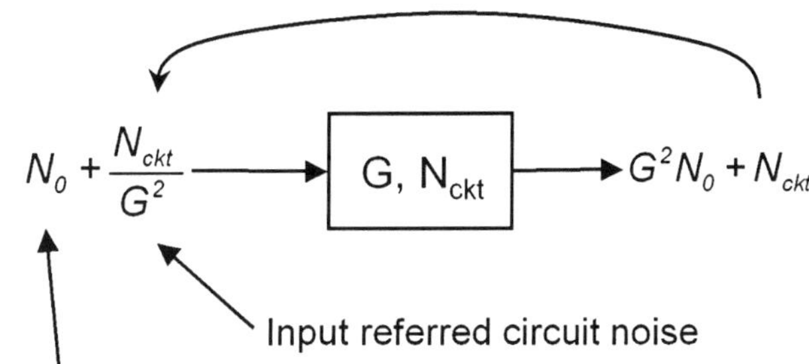

Figure 4.2 Noise figure definition.

According to (4.2) and (4.3), lower NF results in a cleaner RF signal. However, NF cannot be less than one since the total noise at the input consists of noise contribution from the circuit itself and thermal noise. The latter is always present in a system. While NF is a good figure of merit with regard to the amount of noise introduced by RF circuits, a low NF might not always result in a high SNR. To see this, (4.2) is represented as a ratio of the input SNR to the output SNR

$$F = \frac{N_{out}}{G^2 N_0}\frac{S_i G^2}{S_o} = \frac{S_i / N_0}{S_0 / N_{out}}, \tag{4.4}$$

where S_i and S_o represent the input signal power and output signal power, respectively. Therefore, the output SNR is directly proportional to the input SNR. If the input SNR is poor then even with a low NF, the output SNR can still be inadequate to meet a given BER performance.

> **Example 4.1** Let $F = 2$ and the input SNR to be 3 dB. Assume that the required output SNR is 10 dB to meet a 0.001% BER performance. Then, according to (4.4), the output SNR is only 0 dB, which is insufficient to meet the BER requirement even though the noise figure is reasonable low, i.e. 3 dB. □

Finally, designers should be careful when obtaining noise figures for systems that have different input and output bandwidths since (4.2) implicitly assumes equal bandwidth for both input and output. In spread-spectrum, for instance, the output bandwidth is smaller than the input bandwidth by a factor equal to the processing gain. In such a situation, differences in bandwidth must be taken into account.

4.2.1 Noise Figure of Cascaded Systems

While the definition in (4.2) is useful in determining the noise performance of an individual RF block, often the noise figure for the complete receiver is needed. The overall noise figure can be easily derived by referring noise from all receiver stages to the input of the first stage as shown in Figure 4.3. Then, using (4.2) to express the noise factor of each block in the cascade, the following expression can be obtained for the overall noise factor F_{total}

$$F_{total} = F_1 + \frac{F_2 - 1}{G_1^2} + \frac{F_3 - 1}{G_1^2 G_2^2} + \cdots + \frac{F_n - 1}{G_1^2 G_2^2 \cdots G_{n-1}^2}, \tag{4.5}$$

where F_i and G_i are respectively the noise factor and voltage gain for the ith block in the cascade. The total noise figure NF_{total} is obtained by applying (4.3) to (4.5).

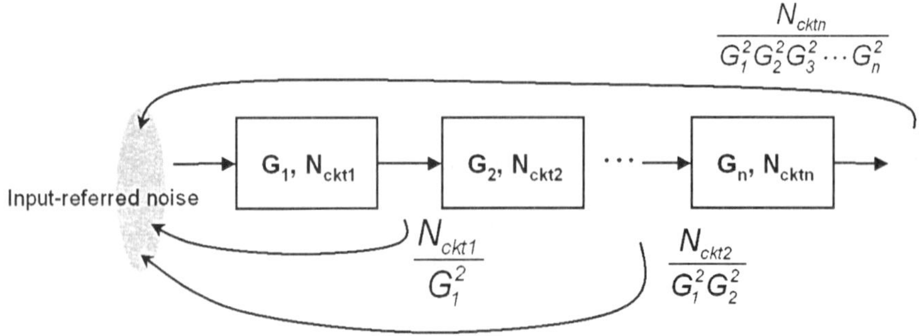

Figure 4.3 Cascaded noise figure.

Expression (4.5) reveals that to obtain a low noise figure for a cascaded system, the noise figure of the first block must be minimized while its gain should be maximized. Also, the noise figure of the last block is least important as long as the previous stages have sufficient gains to make its contribution to the input-referred noise negligible. For the receiver shown in Figure 4.1b, the above observation applies to the LNA, which should be designed with a sufficiently low noise figure and a large gain to achieve the overall NF requirement. Note that the bandpass filter in the receiver front-end violates the above design guideline because it typically introduces an insertion loss rather than gain and therefore hurts the overall noise performance.

Example 4.2 Referring to Figure 4.1b, let the noise figure and gain of the LNA, mixer, and pre-ADC amplifier be

LNA:	3-dB NF and 20-dB gain
Mixer:	10-dB NF and 10-dB gain
Amplifier:	20-dB NF and 80-dB gain

Applying (4.3) and (4.5), the total noise factor is 2.2. The corresponding noise figure is 3.4 dB. The mixer and amplifier have degraded the overall noise figure by only 0.4 dB, relative to the case with only the LNA.

Now consider a gain of 10 dB and a noise figure of 2.5 dB for the LNA. Repeating the calculation, the noise factor has worsened to 3.7, corresponding to a noise figure of 5.6 dB. □

Example 4.3 Consider the same system as in Example 4.2 but now with a bandpass filter before the LNA. The bandpass filter has an insertion loss of 2 dB and is impedance matched to the input of the LNA. To compute the overall noise figure, the noise factor of the bandpass filter must first be determined. Since the bandpass filter is matched to the LNA, the total noise power density is kT_0, where k is the Boltzmann's constant and T_0 is the ambient temperature in kelvin. In fact, kT_0 is equivalent to the thermal noise power density. See Section 4.2.3 for the derivation of kT_0. Based on this observation, the noise factor of the bandpass filter is simply 2.58. Substituting this value into (4.5) and noting that the overall gain has been reduced by the loss of the bandpass filter, the total noise factor becomes 4.47, corresponding to a noise figure of 6.5 dB. The bandpass filter has degraded the overall noise figure by 3.1 dB as compared to Example 4.2. Therefore, to achieve good noise performance, it is essential to minimize the loss before the LNA. Other such losses include cabling loss between the antenna and the bandpass filter. □

4.2.2 Antenna Noise Figure

Thus far, noise figure has been defined without regard to antenna noise. In practice, the antenna may pick up significant amount of noise from the environment. To account for noise at the antenna, the system noise figure should be defined with N_{out} in (4.1) replaced by $N_{ant}G^2 + N_{ckt}$. Substituting this expression for N_{out} in (4.2), the following expression is obtained for the system noise figure NF_{system}

$$NF_{system} = 10\log\left(F_{ant} + F_{total} - 2\right) \tag{4.6}$$

where F_{total} is expressed by (4.5) and the antenna noise factor F_{ant} is defined as

$$F_{ant} = 1 + \frac{N_{ant}}{N_0}. \tag{4.7}$$

Expression (4.6) implies that if the antenna noise factor is much greater than that of the RF circuitry, designing for a low noise figure in the front-end circuitry offers little improvement on the overall noise performance.

The antenna noise power spectral density can be colored as well as white, both of which result from random perturbations in the environment. White noise results from thermal background noise measured by kT and therefore depends on temperature. For instance, in a satellite link the antenna noise in the downlink is much higher than that of the uplink due to the large temperature differential between the uplink and downlink. In the downlink, the temperature is determined by the surface of the earth which is typically 300 degrees kelvin while the temperature in the uplink is that of space, typically only 10's of degrees kelvin.

Colored antenna noise is usually due to man-made noise found in urban areas as discussed in Section **2.2.1.2**. For analytical purposes, colored noise can be converted to an equivalent white noise density over the signal bandwidth.

Example 4.4 Consider a case in which the antenna noise is equal to the ambient thermal noise at 300 degrees kelvin, i.e. –174 dBm. Also, assume that the total noise figure of the RF receiver is 4 dB. According to (4.6), the system noise figure is then 4 dB.

Now consider another case in which the antenna noise is –150 dBm due to man-made noise at lower frequency, such as VHF. According to (4.7), the antenna noise factor is 252.2, corresponding to an antenna noise figure of 24.017 dB. Applying (4.6), the system noise factor is 252.7, corresponding to a noise figure of 24.026 dB. Clearly, in this case, the noise figure of the system is dominated by the antenna noise figure and not the receiver noise figure. Therefore, one should not put in extra effort in designing a low noise receiver in such a situation since negligible performance gain is achieved in the overall system. In fact, even if the receiver were designed to be extremely noisy, e.g. with a 15-dB NF, the system noise figure increases only by about 0.5 dB. □

4.2.3 Noise Temperature

Noise temperature measures the noise performance of an RF device based on the fact that the maximum available thermal noise power density can be expressed by

$$N_0 = kT_0 \tag{4.8}$$

where $k = 1.38 \times 10^{-23}\ J/K$ is the Boltzmann's constant and T_0 is the ambient temperature in kelvin. Expression (4.8) may be obtained by observing that the RMS noise voltage from a resistor of value R is $\sqrt{4kTR \cdot W}$, where W is the measurement bandwidth. The maximum noise power delivered to a load resistor of value R_L occurs when $R = R_L$. The maximum noise power at the load then becomes

$$\frac{4kT_0 R_L \cdot W}{R_L}\left(\frac{R_L}{R_L + R_L}\right)^2 = kT_0 W. \tag{4.9}$$

Dividing (4.9) by W results in (4.8).

Substituting (4.8) for N_0, expression for noise factor can be replaced by

$$F = 1 + \frac{N_{ckt}}{kT_0 G^2}. \tag{4.10}$$

The noise temperature T defines the temperature at which the equivalent thermal noise kT equals the input-referred circuit noise $N_{ckt}G^{-2}$. With this definition, the noise factor can then be expressed in terms of noise temperature as shown below

$$T = (F - 1)T_0. \tag{4.11}$$

4.2.4 Sensitivity

Noise figure determines the minimum signal power that can be received at a given $\overline{\gamma}$. This minimum signal level is also referred to as the sensitivity S of the RF receiver and is defined as

$$S = 10\log\left(F \cdot kT_0 \cdot R_b \cdot \overline{\gamma}\right). \tag{4.12}$$

A lower value of S implies a more sensitive receiver, i.e. high sensitivity. To achieve high sensitivity, according to (4.12), a low noise factor or noise figure is required for a given $\overline{\gamma}$. Sensitivity usually is expressed in dBm, i.e. power in dB with respect to 1 mW. Sensitivity can also be measured in terms of μV by making the following conversion

$$S\ (\text{in}\ \mu\text{V}) = \sqrt{2R_s 10^{0.1S-3}}, \tag{4.13}$$

where S is in dBm and R_s is the input impedance of the RF receiver.

Example 4.5 An RF receiver operates at 100 kbps and has a system noise figure of 6 dB. The baseband receiver requires a E_b/N_0 of 10 dB. With an ambient temperature of 290 degrees kelvin, the thermal noise term is -174 dBm/Hz. By applying (4.12), the receiver has a sensitivity of –108 dBm. Equivalently, by applying (4.13), the receiver has a sensitivity of 1.3 μV if the source resistance is 50 ohms. □

4.3 Distortion Performance

In a wireless system, a low noise figure is desirable to achieve a high sensitivity for improved transmission range, i.e. coverage. However, situations arise when the transmitter is close to the intended receiver or, due to the near-far problem, unintended transmitters are close to a receiver that has a weak desired signal level. In both cases, the receiver experiences strong signal level that generates distortion in the RF front-end because the underlying components are driven into non-linear operation region. It is therefore important to characterize the non-linear behavior of RF components to assess the effect of distortion on system performance.

An RF component may be modeled by the following expression

$$y\left[x(t)\right] = \beta_1 x(t) + \sum_{i=2}^{\infty} \beta_i x(t)^i \tag{4.14}$$

where $x(t)$ is the input waveform and β_i is the coefficient for the ith order non-linearity with $i \neq 1$. Note that β_1 is the ideal gain for the device when it is operating in the linear region.

To develop metrics that measure distortions in a receiver, a two-tone input shown below is used as excitation in (4.14):

$$x(t) = A_1 \cos \omega_1 t + A_2 \cos \omega_2 t. \tag{4.15}$$

In general, using just the square and cubic terms in the summation is sufficient to generate reasonable approximation on the circuit behavior to distortion. Using only the linear and cubic terms, and substituting (4.15) into (4.14), the following frequencies and amplitudes listed in Table 4.1 are generated at the circuit output.

Frequency components ω_1 and ω_2 carry the desired signals while all other frequency terms introduce distortion. It is interesting to note that the desired signals themselves also interfere with one another as evident in the β_3 term. Later in Section 4.3.1, it will be shown that the third-order term causes signal compression. Furthermore, it is clear that all the distortion terms increase as the signal level increases.

Table 4.1 Frequencies generated by a two-tone input.

Frequency	Amplitude
0	$0.5\beta_2\left(A_1^2 + A_2^2\right)$
ω_1	$\beta_1 A_1 + \beta_3\left(0.75A_1^3 + 1.5A_1A_2^2\right)$
ω_2	$\beta_1 A_2 + \beta_3\left(0.75A_2^3 + 1.5A_2A_1^2\right)$
$2\omega_{x=1,2}$	$0.5\beta_2 A_x^2$
$3\omega_{x=1,2}$	$0.25\beta_3 A_x^3$
$\omega_1 \pm \omega_2$	$0.5\beta_2 A_1 A_2$
$2\omega_1 \pm \omega_2$	$0.75\beta_3 A_1^2 A_2$
$2\omega_2 \pm \omega_1$	$0.75\beta_3 A_1 A_2^2$

Fortunately, most of the interfering components are distant from the desired frequency components and can therefore be adequately filtered out. The interfering frequency components include the 2nd and 3rd harmonic terms, the $\omega_1 + \omega_2$ term, the $2\omega_1 + \omega_2$ term, and the $2\omega_2 + \omega_1$ term.

The remaining terms, however, can cause substantial distortions in the RF receiver. For the direct-conversion receiver shown in Figure 4.1, the output of the mixers contains a low-frequency baseband signal that is prone to distortion resulting from the DC and $\omega_1 - \omega_2$ terms listed in Table 4.1. The distortion due to these terms is also referred to as the input-referred second-order intercept point (IIP2) discussed in Section 4.3.2. For superheterodyne receivers, distortions due to these low frequency components become less significant.

Finally, the terms $2\omega_1 - \omega_2$ and $2\omega_2 - \omega_1$ result in intermodulation distortions (IMD) which can degrade the performance of nearby channels with weak signal strength. Figure 4.4 shows an example of distortion created on adjacent channels due to IMD, where $|\omega_1 - \omega_2|$ is the channel spacing. The IMD degrades the signal quality of the adjacent channel immediately above ω_2 and below ω_1. A measure of IMD is the input referred third-order intercept point (IIP3) defined in 4.3.3.

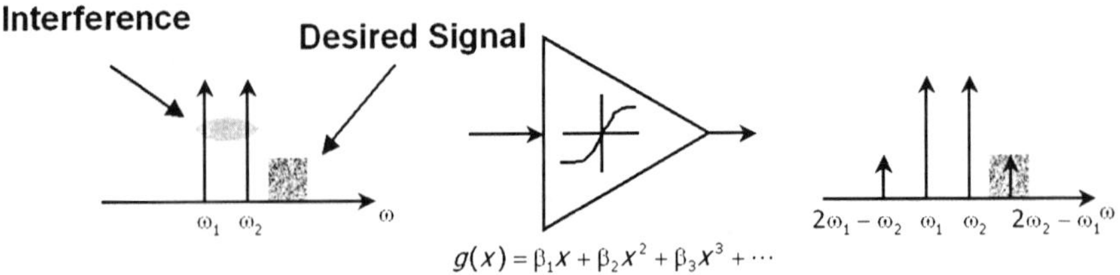

Figure 4.4 Effect of IMD on adjacent channels.

4.3.1 Input-Referred 1-dB Compression Point

Referring to Table 4.1, let's assume that the signal frequency ω_1 is the desired signal. Then, it can be seen that the β_3 term represents unwanted distortion. At first glance, it might appear that the β_3 term enhances rather than degrades the desired signal power. However, since β_3 is generally negative, the desired signal level $\beta_1 A_1$ is in fact decreased by the amount shown below

$$\rho = 20\log\left[\frac{\beta_1 A_1 + \beta_3\left(0.75A_1^3 + 1.5A_1 A_2^2\right)}{\beta_1 A_1}\right]. \tag{4.16}$$

The 1-dB input compression point (IPC) is defined as the input power of the desired signal P_{in} at which ρ becomes 1 dB as shown in Figure 4.5. By setting $\rho = 1$ and letting $P_1 = 0.5A_1^2 R_s^{-1}$ and $A_1 = A_2$, the compression point then has the following expression

$$P_{1-dB} = \frac{2}{9}\left|\frac{\beta_1}{\beta_3}\right|\frac{\left(10^{0.05} - 1\right)}{R_s}, \tag{4.17}$$

where R_s is the source resistance. Expression (4.17) depends on the ratio of the linear gain β_1 to the cubic gain β_3. A smaller β_3 results in lower distortion and therefore a higher compression point.

The IPC is a useful measure of linearity for the transmitter, in particular, for the power amplifier. In general, a high IPC is desired such that the power amplifier can be driven with a high input level before it is driven into non-linear region. By operating the PA below the IPC, spectral regrowth is minimized. An alternative measure of compression is the output compression point (OCP) shown in Figure 4.5 and is defined as $OPC = 10\log G + IPC + 10\log\left(R_s / R_L\right)$.

For the receiver, the IPC can also be applied to measure the effect of large input power level. Assuming that the desired signal is at ω_1 and that the signal at ω_2 is an adjacent channel interferer, the distortion power due to the cubic non-linearity then has the following expression

$$P_d = \frac{9}{4}|\beta_3|^2\left[P_1^3 + 4P_1^2P_2 + 4P_1P_2^2\right]R_s^3R_L^{-1} \tag{4.18}$$

where $P_1 = 0.5A_1^2R_s^{-1}$ and $P_2 = 0.5A_2^2R_s^{-1}$. The signal-to-distortion ratio (SDR) can be obtained by applying (4.18) to (4.19) as shown below

$$SDR = \frac{|\beta_1|^2 P_1R_sR_L^{-1}}{P_d} = \left[\frac{3P_{1-dB}/P_{in}}{\left(10^{0.05}-1\right)\left(1+2\alpha_2\right)}\right]^2, \tag{4.19}$$

where $P_{in} = P_1$ is the input power to the RF receiver and $\alpha_2 = P_2/P_1$ is the amount of interference power above the desired signal power P_1. Expression (4.19) indicates that the SDR degrades as the power level of the desired signal or the inference increases and that the SDR improves with a larger input compression point.

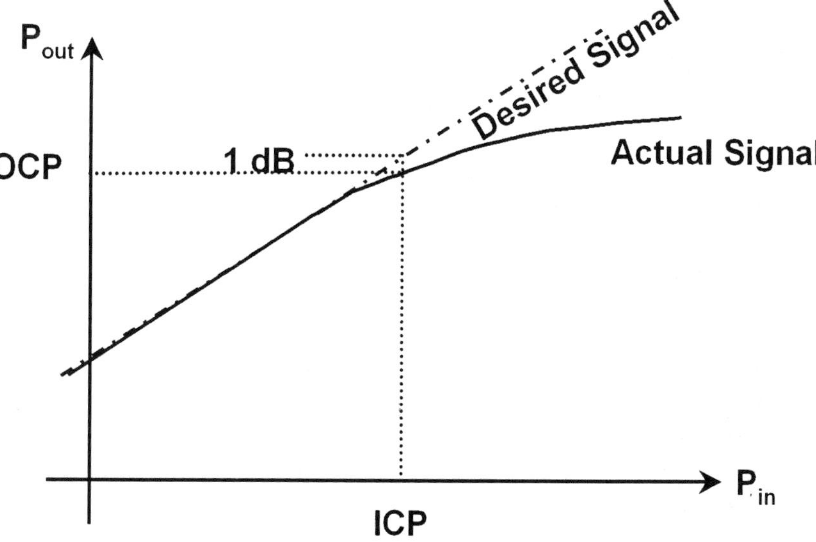

Figure 4.5 The 1-dB compression point.

Example 4.6 Consider a receiver with a 1-dB input compression of -14 dBm. Let the interferer power be 10 dB above the desired signal power of–25 dBm. Applying (4.19), the SDR is

$$10\log\left[\frac{3\cdot 10^{-1.4}/10^{-2.5}}{\left(10^{0.05}-1\right)\left(1+2\cdot 10\right)}\right]^2 = 23.4\ dB.$$

However, if the interferer power becomes 20 dB above the desired signal power, then the SDR degrades to 3.7 dB. □

4.3.2 Input-Referred Second-Order Intercept Point

The distortion due to the square term β_2 results from the DC and $|\omega_1 - \omega_2|$ frequency components. The distortion power P_d can be expressed by

$$P_d = \frac{|\beta_2|^2}{2}\left(2P_1^2 + 5P_1P_2 + 2P_2^2\right)R_s^2 R_L^{-1}. \tag{4.20}$$

Let the output power of the desired signal be $\beta_1^2 P_1 R_s R_L^{-1}$ and let $P_2 = P_1$. Then, the distortion-to-signal ratio has the following expression

$$\frac{P_d}{P_o} = \frac{9}{2}\left|\frac{\beta_2}{\beta_1}\right|^2 P_1 R_s. \tag{4.21}$$

Clearly, smaller values of β_2 with respect to the linear gain β_1 result in a lower distortion-to-signal ratio.

IIP2 is defined as the input power when (4.21) becomes unity, at which point, it has the following simple expression

$$IIP2 = \frac{2}{9}\left|\frac{\beta_1}{\beta_2}\right|^2 \frac{1}{R_s}. \tag{4.22}$$

Larger IIP2 implies lower distortion and therefore higher linearity performance. Typically, in an RF circuit the IIP2 is large since the dominant terms in the non-linear devices are the odd-powered terms whereas the even-powered terms are generally smaller.

Again assuming that the signal at ω_1 is the desired signal and that at ω_2 is an adjacent channel interferer, the SDR due to the square non-linearity then has the following expression

$$SDR = \frac{|\beta_1|^2 PR_s R_L^{-1}}{P_d} = \frac{9IIP2}{\left(2 + 5\alpha_2 + 2\alpha_2^2\right)P_{in}} \tag{4.23}$$

Expression (4.23) indicates that the SDR degrades as the interference power or the desired signal power increases. Moreover, (4.23) implies that the distortion power increases by 10 dB for every 10-dB increase in input power.

Example 4.7 Consider a receiver with an IIP2 of 25 dBm. The input consists of a desired signal and an interferer with equal power of –25 dBm. Based on (4.23), the SDR is

$$\frac{9 \cdot 10^{2.5}}{(2+5+2)10^{-2.5}} = 50\,dB.$$

However, if the interferer power becomes 30 dB above the desired signal, then SDR degrades to –3.5 dB. □

4.3.3 Input-Referred Third-Order Intercept Point

The IMD power is determined using the amplitude values of the intermodulation terms in Table 4.1 as shown below

$$P_d = \frac{1}{2R_L}\left[\frac{3}{4}\beta_3 A_1^2 A_2\right]^2 = \left(\frac{3}{2}\beta_3\right)^2 P_1^2 P_2 R_s^3 R_L^{-1}. \tag{4.24}$$

Letting $P_1 = P_2$, the distortion-to-signal power ratio becomes

$$\frac{P_d}{P_o} = \left|\frac{3}{2}\frac{\beta_3}{\beta_1} P_1 R_s\right|^2. \tag{4.25}$$

IIP3 is defined as the input power at which the IMD power equals that of the desired signal. Given the above condition and using (4.25), the following expression is obtained for IIP3

$$IIP3 = \frac{2}{3}\left|\frac{\beta_1}{\beta_3}\right|\frac{1}{R_s}. \tag{4.26}$$

Using (4.24), the SDR is obtained and has the following expression

$$SDR = \frac{|\beta_1|^2 P_{in} R_s R_L^{-1}}{P_d} = \left(\frac{IIP3}{P_{in}}\right)^2 \frac{1}{\alpha_1^2 \alpha_2}. \tag{4.27}$$

where P_{in}, in this case, is the input signal power at $2\omega_1 - \omega_2$ while $\alpha_1 = P_1 / P_{in}$ and $\alpha_2 = P_2 / P_{in}$ are the powers of the two interferers at ω_1 and ω_2, normalized to the input signal power. This definition of SDR differs slightly from that of IIP2 and P_{1dB} to account for the distortion due to the intermodulation of the two frequency components. Note, however, (4.27) is only an approximation since it has not taken into account the intermodulation products of the input frequency $2\omega_1 - \omega_2$ with ω_1 and ω_2.

Referring to Figure 4.4, if $|\omega_1 - \omega_2|$ equals the channel spacing, then α_1 and α_2 can be considered as the normalized power of the adjacent channel interferer and alternate channel interferer, respectively. In a cellular system, the adjacent and alternate channel interference powers are allowed to be much higher than that of the desired signal power. For instance, in GSM, α_1 and α_2 are 9 dB and 41 dB, respectively.

Example 4.8 Consider a receiver with an IIP3 of 0 dBm. Let the input power be –40 dBm. Also, assume that the interferer powers are similar to that of GSM. Then applying (4.27), the SDR has a value of

$$10\log\left[\left(\frac{10^{0}}{10^{-4}}\right)^{2}\frac{1}{10^{0.9\times 2}10^{4.1}}\right]=21\,dB.$$

If the IIP3 is decreased to -10 dBm, then the SDR decreases by 20 dB. □

Finally, in Figure 4.6, the IMD power is plotted against input power and the desired output power is also superimposed on the same figure. On log scale, the IMD power has a slope of three and therefore increases rapidly as the input power increases. IIP3 is determined on the graph to be the point at which the IMD power equals the ideal output power of the desired signal. In actuality, IMD crosses the desired output power much sooner due to compression. This observation generally holds true and implies that the input compression point is lower than the IIP3. If (4.17) is substituted into (4.26), it can be shown that IIP3 is about 14 dB lower than the 1-dB compression point.

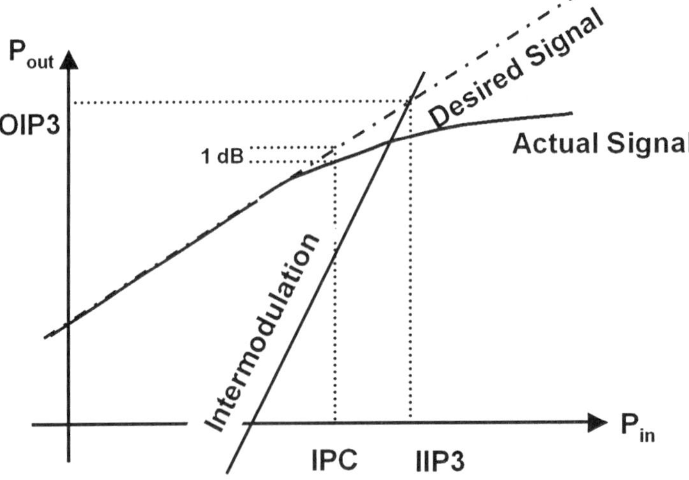

Figure 4.6 Input-referred third-order intercept point.

4.3.4 Intercept-Point of Cascaded Systems

Thus far, the distortion measures have been determined for a single stage. This section deals with intercept measures for a cascaded system of multiple RF blocks as shown in Figure 4.7, in which the output distortion power, IIP3, and voltage gain of the ith block are represented by $P_{d,i}$, $IIP3_i$, and G_i, respectively. The input power to the system is defined to be P_{in}. The total distortion at the output of the nth stage results from the distortions of the previous stages as well as the input power amplified by the previous n-1 stages

$$P_{d,n} = P_{d,n-1}G_n^2 + \frac{P_{in}^3 G_1^6 G_2^6 \cdots G_{n-1}^6 G_n^2}{IIP3_n^2}. \tag{4.28}$$

When referred to the input, the total distortion power $P_{d,in}$ at the input has the following expression

$$P_{d,in} = \frac{P_{d,n}}{G_1^2 G_2^2 \cdots G_{i-1}^2 G_n^2} \tag{4.29}$$

Let the equivalent IIP3 of the cascaded system be denoted by $IIP3_{total}$ and note that $P_{d,0}$ is zero. Also by definition, $P_{d,in}$ can be represented in terms of the total IIP3

$$P_{d,in} = \frac{P_{in}^3}{IIP3_{total}^2}. \tag{4.30}$$

Equating (4.29) to (4.30) and solving for $P_{d,n}$ iteratively based on (4.28), the total IIP3 can be expressed in terms of the voltage gains and IIP3's of the individual stages:

$$\frac{1}{IIP3_{total}^2} = \frac{1}{IIP3_1^2} + \sum_{i=2}^{n} \frac{1}{\dfrac{IIP3_i^2}{G_1^4 G_2^4 \cdots G_{i-1}^4}}. \tag{4.31}$$

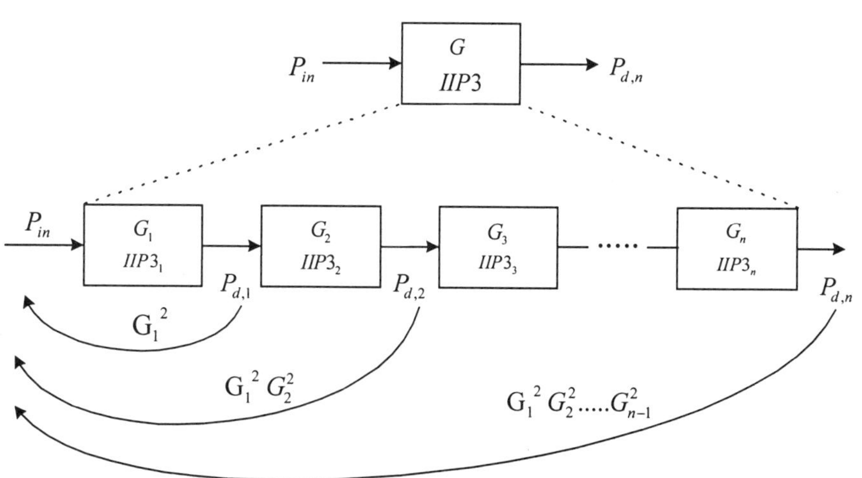

Figure 4.7 Intercept-point of a cascaded system.

It can be seen that the total IIP3 of the system depends very little on the IIP3 of the first stage. Rather, the total IIP3 depends largely on the IIP3 of the later stages and the gains before those stages. To maintain a large total IIP3, the gain should be kept low in the earlier stages while keeping the IIP3 high in the later stages. When applied to an RF system, the gain of the LNA, G_1, which appears in every term of the summation in (4.31), should be kept low while the IIP3 of the following stage such as the mixer should be kept high. Since low noise figure requires a high gain LNA, a high linearity constraint tends to compromise the noise performance of the overall system.

As a final note, the total IIP3 derived in (4.31) does not consider the effect of filtering within the cascaded stages and therefore in general presents a worst-case estimate.

4.3.5 Spurious-Free Dynamic Range

Wireless transceivers often experience a wide variation in the received signal level due to the near-far problem, discussed in Section **2.2.2.2**. A measure of the receiver's ability to handle the signal power variation is its dynamic range, defined as the ratio of the maximum power that a receiver can handle to the minimum detectable received signal power, i.e. the receiver sensitivity.

The maximum power that a receiver can tolerate before substantial distortion occurs is determined by the various distortion measures described in the previous subsections. Here, IIP3 is used to define the maximum tolerable receive power. Consider the case in which the desired signal arrives at the receiver with just enough power to be detected, i.e. its received power equals the receiver sensitivity. The maximum tolerable receive power $P_{\max}$ is defined as the amount of input power such that the associated distortion power, when referred to the receiver input, equals to the power of a minimally detectable signal. Based on this definition and using (4.25) and (4.26), $P_{\max}$ has the following expression

$$P_{\max} = \left(S \cdot IIP3^2 \right)^{1/3}. \tag{4.32}$$

The spurious-free dynamic range (SFDR) is defined as the ratio of $P_{\max}$ to S, shown below

$$SFDR = \left(\frac{IIP3}{S} \right)^{2/3}. \tag{4.33}$$

Clearly, a high IIP3 or a high sensitivity leads to a high SFDR. Generally, the SFDR places a more stringent constraint on IIP3 as compared to an alternative definition based on the ratio of IIP3 to S. In other words, given a maximum and minimum input power requirement at the receiver, the former definition of dynamic range based on the SFDR leads to a much higher IIP3 requirement as compared to that obtained by the latter definition.

Example 4.9 Consider a wireless system that can achieve a maximum range of 10 km. Suppose that there are three transceivers, A, B, C, positioned such that B is

10 km away from A and C is 1 km away from A. The dynamic range required by receiver A to accommodate transmissions from both B and C depends on the propagation condition between B and A, and C and A. Let the path loss exponents be 4 and 3 for the link between B and A and for C and A, respectively. Then, at equal transmission power, the required dynamic range is 70 dB.

The receiver sensitivity is designed to receive at 10 km range for a given transmit power level. With a system noise figure of 6 dB, a data rate of 100 kbps, and $\bar{\gamma} = 10$, the receiver sensitivity is –108 dBm. Then based on (4.33), the required IIP3 is –3 dBm. Note that the alternate definition of dynamic range results in 105 dB which is much higher than the 70 dB SFDR. □

4.4 Other Performance Measures

Thus far, the previous sections have described the key parameters that determine the system-level performance of the RF transceiver. These key parameters include the noise figure, input-referred intercept points, input-referred compression point, sensitivity, and SFDR. This section describes several other performance measures that influence the design of an RF transceiver.

4.4.1 Image Rejection

In the receiver architecture shown in Figure 4.1, the RF signal is down-converted directly to baseband. Other approaches rely on one or multiple intermediate frequencies (IF) before the RF is finally converted to baseband. Such receivers are also referred to as superheterodyne receivers that are described in **Chapter 6**. Superheterodyne receivers are subject to distortions due to image signals, a problem non-existent in homodyne receivers. Images result from aliasing where undesired signals get folded in-band and interfere with the desired signal.

The undesired frequency alias is also referred to as an image frequency, denoted by ω_{image}. For a given frequency $\omega' > \omega_{LO}$, its image frequency satisfies the following constraint

$$\left|\omega' - \omega_{image}\right| = 2\omega_{IF}, \tag{4.34}$$

where ω_{IF} is an intermediate frequency used in the superheterodyne receiver. Graphically, (4.34) implies that the image frequency of ω' is simply the frequency reflected with respect to $\omega = \omega' - \omega_{IF}$ as shown in Figure 4.8a. Expression (4.34) can be verified by showing that mixing $\omega' - 2\omega_{IF}$ with a LO at $\omega_{LO} = \omega' - \omega_{IF}$ results in a frequency component at $-\omega_{IF}$ which folds into ω_{IF}.

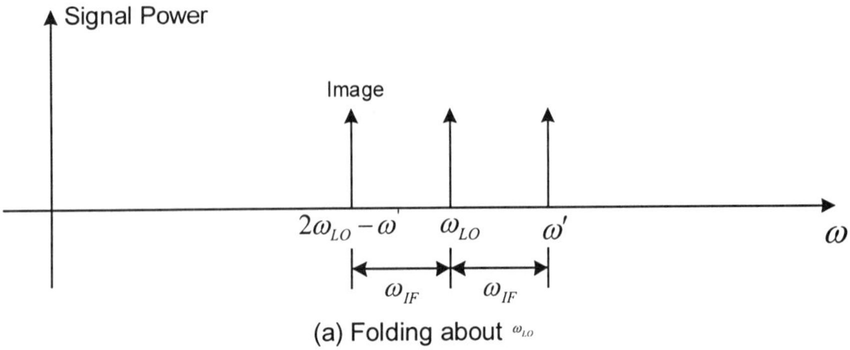

(a) Folding about ω_{LO}

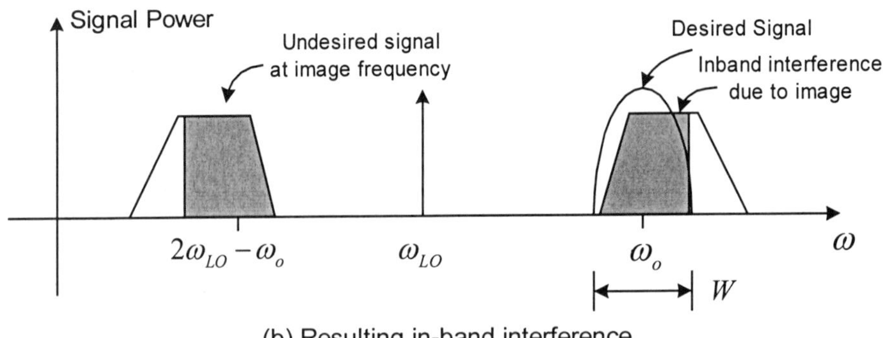

(b) Resulting in-band interference

Figure 4.8 Image frequency in the superheterodyne architecture.

Consider now the case in which the desired signal has a one-sided bandwidth W and is centered on the carrier frequency ω_0. Then, according to (4.34), any signal energy between $(\omega_0 - 0.5W) - 2\omega_{IF}$ and $(\omega_0 + 0.5W) - 2\omega_{IF}$ is folded into the down-converted signal as shown in Figure 4.8b. The signal power at the image frequency then interferes with the desired signal, degrading the system performance. Image rejection is a measure of systems ability to suppress the image power that folds into the desired signal band. Image rejection can be achieved actively or passively. Active techniques involve image cancellation during mixing, discussed in Section **9.1.4.2**, while the passive technique involves filtering of the image signal before mixing.

4.4.2 Quality Factor

RF systems typically operate on signals with a narrow passband centered on a carrier frequency much larger than the passband bandwidth (i.e. $\omega_o >> 2W$). The RF transceiver must then operate on signals with extremely small fractional bandwidth defined as

$$W_{frac} = \frac{2W_{-3dB}}{\omega_o}. \tag{4.35}$$

Typical fractional bandwidth ranges between two to five percent. To achieve the small fractional bandwidth, RF transceivers exploit tuned circuit elements that are designed to resonate at ω_o. The substantial peaking which results from resonance improves the overall transceiver performance in terms of spectral purity, noise suppression, and low power dissipation.

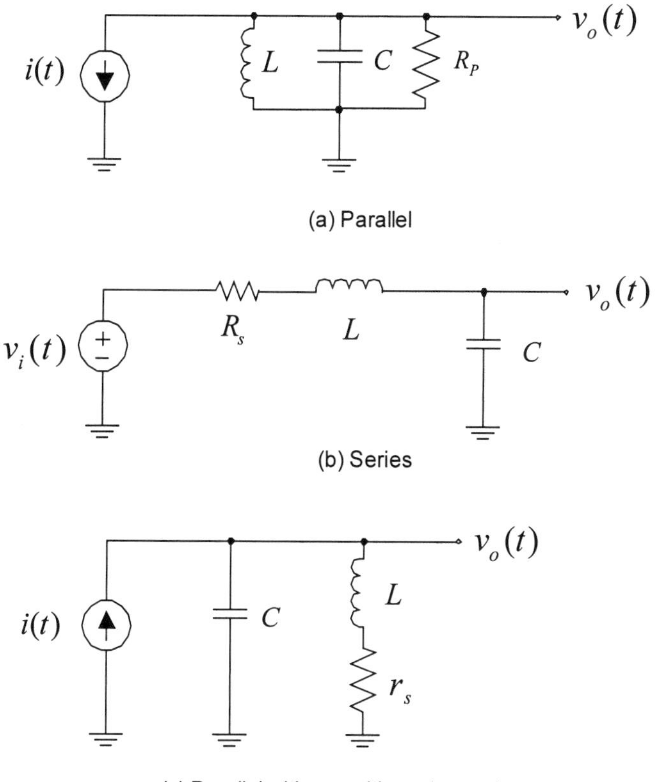

Figure 4.9 RLC tank circuits.

In theory, infinitesimally small fractional bandwidth can be achieved with perfect tuned elements such as the LC tank circuit shown in Figure 4.9 with infinitely large parallel resistance R_p and zero series resistances R_s and r_s. However, in practice, perfect resonance is not possible due to losses in the circuit, arising from parasitic elements or loading, such as R_s, R_p, and r_s shown in Figure 4.9. Quality (Q) factor measures the resonance of the tuned circuits in terms of the amount of power stored per cycle relative to the amount of power dissipated as shown in the definition below [141]:

$$Q = \omega_n \frac{\text{Energy Stored}}{\text{Average Power Dissipated}}, \tag{4.36}$$

where ω_n represents the resonant frequency or natural frequency of the system.

Example 4.10 Consider the parallel tank circuit in Figure 4.9a. The average power dissipated is simply due to the resistive element alone since the inductor and capacitor do not dissipate any power. Given the above observation, the average power dissipated can be determined by inserting an AC current source at the input of the tank. If I is the amplitude of the AC current, the average power dissipation is $0.5I^2R_p$,. The energy stored is determined by the amount of charge stored in the capacitor. The stored energy is $0.5C\left(IR_p\right)^2$. Thus, the quality factor for a parallel tank circuit is $\frac{R_p}{\sqrt{L/C}}$. In this case, a lower resistance reduces the Q factor since more loss is incurred due to more current being diverted away from the reactive elements. □

4.4.2.1 Relationship to the damping factor

The quality factor can be related to the damping factor ς defined for the second-order system discussed in Section 2.5.2.1 of Chapter 2. This is not surprising since the damping constant measures the amount of energy loss in the system whereas the quality factor measures the amount of energy retained in the system. Intuitively, then Q should be inversely proportional to ς. In fact, it can be shown that

$$Q = \frac{1}{2\varsigma}. \tag{4.37}$$

To verify (4.37), the first step is to determine the current to voltage transfer function by applying a test AC current source at the tank's input and measure the voltage at its output as shown in Figure 4.9a. Applying the Laplace transform on the current source and output voltage waveform, the following expression can be derived for the current to voltage transfer function

$$\frac{v_o(t)}{i(t)} = R_p \frac{\frac{s}{R_pC}}{s^2 + \frac{s}{R_pC} + \frac{1}{LC}}. \tag{4.38}$$

Comparing the denominator of (4.38) to that of (2.148) derived for the transfer function of the error signal in a second-order PLL, the following relationships hold

$$\omega_n = \frac{1}{\sqrt{LC}} \tag{4.39}$$

$$2\varsigma\omega_n = \frac{1}{R_p C}. \tag{4.40}$$

Substituting (4.39) into (4.40) and replace $R_p / \sqrt{LC}$ by Q, (4.37) is obtained.

Table 4.2 Resonant frequency and Q factor for the tank configurations of Figure 4.9.

	Parallel	Series	Parallel with Series Resistance
ω_n	$\frac{1}{\sqrt{LC}}$	$\frac{1}{\sqrt{LC}}$	$\frac{1}{\sqrt{LC}}$
Q	$\frac{R_p}{\sqrt{L/C}}$	$\frac{\sqrt{L/C}}{R_s}$	$\frac{\sqrt{L/C}}{r_s}$

Generalizing the above derivation for the tank configurations shown in Figure 4.9b and Figure 4.9c, we obtain Table 4.2 for their resonant frequencies and quality factors. All three tank configurations have the same resonant frequency but the three behave somewhat differently in terms of their Q. For the series tank circuit, Q decreases with increasing resistance values whereas Q increases for the parallel tank circuit. The parallel tank with parasitic series resistance in the inductor behaves like the series tank. In particular, any series resistance degrades the Q of the tank circuit. The reader may verify that if $\omega L >> r_s$, the tank configuration in Figure 4.9c can also be represented by an equivalent parallel tank circuit with $R_p = (\omega_n L)^2 / R_s$.

4.4.2.2 Relationship to Fractional Bandwidth

The transfer functions $H(s)$ for the parallel tank and series tank are $V_o(s)/I(s)$ and $V_o(s)/V_i(s)$, respectively. By using the natural frequency and Q values in Table 4.2, it can be shown that $H(s)$ has the following general expression

$$H(s) = \frac{H(j\omega_n)\frac{\omega_n}{Q}s}{s^2 + \frac{\omega_n}{Q}s + \omega_n^2}. \tag{4.41}$$

To find the frequency response, let $s = j\omega$. Expression (4.41) can be further simplified to

$$H(s) = \frac{H(j\omega_n)}{1 - \dfrac{j\left(\omega_n^2 - \omega^2\right)}{\dfrac{\omega_n \omega}{Q}}}. \tag{4.42}$$

The fractional bandwidth as defined by (4.35) can be determined from (4.42) by substituting $\omega_n + \omega_{-3dB}$ for ω and noting that at the 3-dB roll off, the amplitude of the response is $0.5H(j\omega_n)$ and $2\omega_n\omega_{-3dB} >> \omega_{-3dB}^2$. The resulting expression shows a simple inverse relationship between the Q and the fractional bandwidth

$$\omega_{frac} \approx \frac{1}{Q}. \tag{4.43}$$

For most RF systems that have a high carrier ficquency, it is necessary to have a high Q to a meet the fractional bandwidth requirement.

Example 4.11 Consider an RF transceiver that has a carrier frequency of 1 GHz and an aggregate transmission bandwidth of 25 MHz. By applying (4.35), the fractional bandwidth is 2.5% relative to the carrier frequency. Using (4.43), the required Q is 40. □

4.4.2.3 Relationship to Ringing

The transient response of a tuned system can be determined by applying a step input to (4.42) and take the inverse Laplace transform. By doing so, it can be shown that for both the parallel and series tank configurations the step response has the following form

$$v_o(t) = \frac{H(j\omega_n)}{\sqrt{4Q^2 - 1}} e^{-\frac{\omega_n}{2Q}t} \sin\left(\omega_n \sqrt{1 - \frac{1}{4Q^2}}\, t\, u(t)\right), \qquad Q > 0.5 \tag{4.44}$$

and is plotted in Figure 4.10 for different Q factors. Based on expression (4.44), the time constant τ_s of decay for the signal envelope and the ringing frequency ω_{ring} can be expressed in terms of Q and ω_n as shown below:

$$\tau_s = \frac{2Q}{\omega_n} \tag{4.45}$$

and

$$\omega_{ring} = \omega_n \sqrt{1 - \frac{1}{4Q^2}}. \tag{4.46}$$

Clearly, as Q decreases the ringing frequency increases and the signal envelope decays much faster. With a high Q, the output of the tank is simply a sinusoid with frequency ω_n and amplitude $0.5H(j\omega_n)Q^{-1}$.

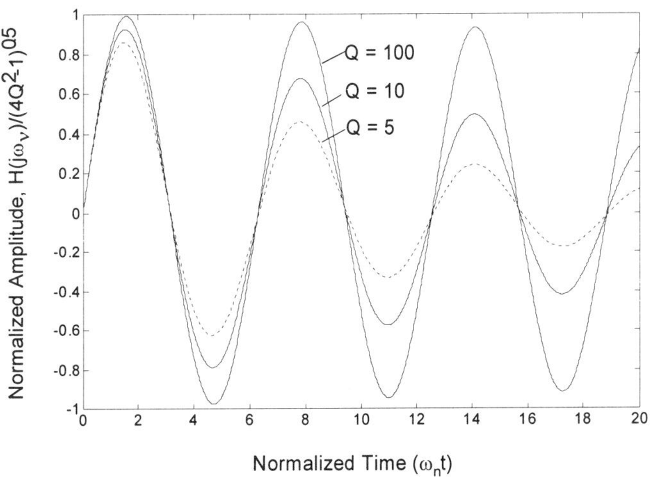

Figure 4.10 Step response of a tank circuit.

4.4.2.4 Relationship to Harmonic Suppression

In many instances, it is desirable to have a frequency-selective circuit that either passes or generates a passband signal with a small fractional bandwidth. In this case, high Q components help to suppress unwanted high-order harmonics. To see this, refer to the frequency response shown in (4.42) and note that the ratio of the fundamental component at ω_n to the nth harmonic approximately equals to

$$\left|\frac{H(j\omega_n)}{H(jn\omega_n)}\right| = \frac{Q(n^2-1)}{n} \tag{4.47}$$

for large values of Q. From (4.47), it is clear that the amount of suppression for the nth harmonic frequency is directly proportional to the Q factor.

> **Example 4.12** Consider the system in Example 4.11 where $Q = 40$. According to (4.47), the suppression for the 2nd harmonic is 35.6 dB while the suppression increases to 40.6 dB for the 3rd harmonic. □

4.4.3 Phase noise

Because of tuned elements and other non-linear effects, RF front-ends often exhibit colored noise with power density that is spectrally shaped. For instance, phase noise found in mixers and oscillators has a spectral shape that falls off inversely to frequency raised to the power n. That is If $\Phi(\Delta\omega)$ denotes the spectral density at a frequency $\Delta\omega$ offset from the carrier frequency, then the spectral density can be expressed by the following expression [142]

$$\Phi(\Delta\omega) \propto \frac{1}{(\Delta\omega)^n}. \tag{4.48}$$

Phase noise results in phase jitters and out-of-band interference that limit the performance of radio systems. The degradations caused by phase noise will be discussed a bit later in this section.

Figure 4.11 shows a typical phase noise density which generally behaves according to (4.48) with n decreasing with increased frequency offset. At high frequency above the carrier the noise power density flattens off while at small offsets from the carrier frequency the spectral density tends to decade much faster. Typical values of n range between zero and three [56].

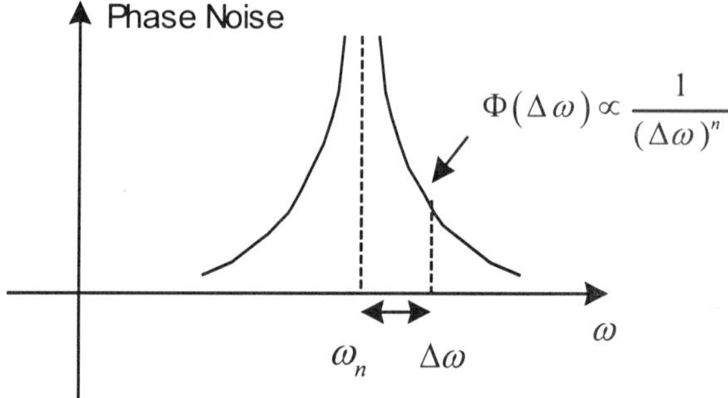

Figure 4.11 Phase noise spectral shape.

4.4.3.1 Generation Mechanism

There are two mechanisms that generate phase noise. The first results from RF device noise which has a spectral density similar to that depicted in Figure 4.11. Such device noise is often referred to as the $1/f$ noise [143][144] which is centered on DC rather than the carrier frequency. Due to non-linearity in the RF circuit, the $1/f$ noise gets modulated to the carrier frequency and results in phase noise. A mixer is an excellent example of such generation mechanism. In particular, the $1/f$ noise of the mixer circuit modulates the local oscillator (LO) and gets up-converted to the LO frequency.

The second mechanism involves the filtering action of tuned elements in RF circuits that shapes thermal noise sources to behave according to (4.48). Not surprisingly, the Q-factor of the tuned elements strongly influences the amount of noise shaping that occurs. Tuned elements are widely used to achieve the required fractional bandwidth with high gain and low noise performance. An example of such circuit is the LO shown in Figure 4.12. Since the parallel tank has a resistive loss due to R_p, a sustained oscillation can only be achieved with an external excitation that sends just enough power to overcome the resistive loss at the resonant frequency. Effectively, the external excitation in Figure 4.12 behaves like a negative resistance that cancels out R_p. The tank then has a response shown below

$$\left|H\left[j(\omega_n + \Delta\omega)\right]\right| \approx \frac{\omega_n}{2Q\Delta\omega} \tag{4.49}$$

at a frequency $\Delta\omega$ away from the resonant frequency.

If noise from the external circuit is ignored, then the only noise source in the system is R_p, which can be represented by an equivalent noise current source with a current noise density of $4kT/R_p$. The resulting voltage noise for the LO becomes

$$\overline{v_o^2(t)} \approx 4kTR_p\left(\frac{\omega_n}{2Q\Delta\omega}\right)^2 = \frac{kTR_p\omega_n^2}{Q^2} \cdot \frac{1}{\Delta\omega^2}. \tag{4.50}$$

The corresponding phase noise power density at $\omega_n + \Delta\omega$ is obtained by dividing expression (4.50) by $2R_p$.

It is customary to represent the phase noise density by referencing it to the signal power P_s measured at ω_n as shown below

$$\Phi(\Delta\omega) = 10\log\left[\frac{kT\omega_n^2}{2P_sQ^2} \cdot \frac{1}{\Delta\omega^2}\right] \tag{4.51}$$

Based on this simple noise model, the phase noise density decreases inversely with $\Delta\omega^2$.

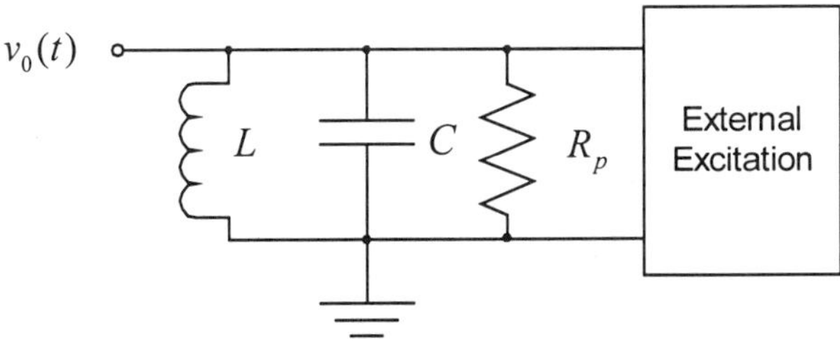

Figure 4.12 LO based on the parallel tank circuit of Figure 4.9a.

The units of phase noise density described by (4.51) is in dBc/Hz where dBc represents the power measured in dB relative to that at the center frequency. In the case of a resonant circuit, the center frequency also coincides with the resonant frequency. Note that dBc/Hz tends to cause some confusion since Hz actually refers to the linear term inside the log function. However, this is a widely accepted unit for phase noise. The phase jitter resulted from a given phase noise density can be determined by integrating the linear noise density

$$\overline{\phi^2} = \frac{180^2}{\pi^2} \int_{-\infty}^{+\infty} 10^{0.1\Phi(\Delta\omega)} d(\Delta f). \tag{4.52}$$

Taking the square root of (4.52) results in the phase jitter variance in degree^2.

Example 4.13 Consider a LO with a center frequency of 1 GHz, a signal power of 3 dBm, a Q of 10, and an operating temperature of 300 degrees kelvin. Then, based on (4.51), the resulting phase noise at 1 kHz offset from the center frequency is –80 dBc/Hz. With every decade increase in the offset frequency, the phase decreases by 20 dB. Therefore, at 10 kHz and 100 kHz offset, the phase noise becomes –100 dBc/Hz and –120 dBc/Hz, respectively.

Based on (4.51) and (4.52), the amount of phase jitter measured in RMS degree is

$$\overline{\phi^2} = \frac{180^2 kT\omega_n^2}{2\pi^3 P_s Q^2 \omega_l}, \tag{4.53}$$

where ω_l is a frequency close to the center frequency. When ω_l is zero then the phase noise introduces infinitely large RMS degree phase error. In practice, this does not occur due to parasitic resistances in the LC tank. To keep the phase noise within bound, ω_l is set to 628 rad/sec or 100 Hz. With this assumption, the phase noise generates a RMS phase jitter of 0.8 degrees. □

4.4.3.2 Spectral Leakage

Besides introducing phase jitters, phase noise also introduces out-of-band interference as shown in Figure 4.13. Ideally, an LO has a spectral density that consists of two impulses at $\pm\omega_{LO}$. However, due to phase noise, the spectral density spreads in frequency according to (4.51) and as shown in Figure 4.13. One might view this as spectral leakage where power contained in an infinitesimal bandwidth leaked out to nearby frequencies. Because of spectral leakage, power from an interfering signal spreads into the adjacent channel shown as the shaded region in Figure 4.13. The resulting ACI in the shaded region can be expressed by

$$ACI = P_{\text{int}} \int_{(n-0.5)W_c}^{(n+0.5)W_c} 10^{0.1\Phi(\Delta\omega)} d(\Delta f), \tag{4.54}$$

where P_{int} is the power of the interfering signal, W_c is the channel spacing, and n is the channel index. Since P_{int} can be much larger than the desired signal power, the ACI can substantially degrade the SIR in nearby channels if phase noise is not minimized.

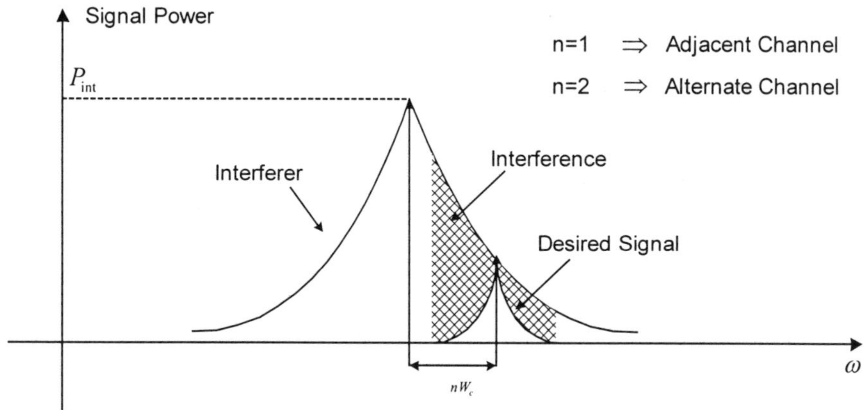

Figure 4.13 Spectral leakage due to phase noise.

Example 4.14 Assume that the phase noise is expressed by (4.51). Then, the ACI due to spectral leakage can be found by applying (4.54) to obtain

$$ACI = 10\log\left(\frac{kT\omega_n^2}{6\pi^2 Q^2 W_c}\cdot\frac{P_{\text{int}}}{P_s}\right) \tag{4.55}$$

Consider the case in which the desired signal has a signal strength of –90 dBm, the channel spacing is 200 kHz, and the power of the adjacent channel interferer is 9 dB above that of the desired signal (i.e. –81 dBm). Moreover, assume the same parameters as in Example 4.13. According to (4.55), the desired signal then experiences an ACI of –152.8 dBm as a result of spectral leakage. Therefore, amount of ACI has a much lower power than that of the desired signal.

Now consider the case with an additional interferer at the alternate channel two channels from the desired signal with a power 41 dB higher than that of the desired signal, i.e. –49 dBm. Again applying (4.54), the expression for ACI becomes

$$ACI = 10\log\left(\frac{kT\omega_n^2}{30\pi^2 Q^2 W_c}\cdot\frac{P_{\text{int}}}{P_s}\right) \tag{4.56}$$

where P_{int} is the power of the alternate channel interferer. Substituting the values for all the parameters in (4.56), the ACI resulting from the alternate channel is –127.8 dBm. □

Example 4.14 illustrates that phase noise limits the amount of interference that the receiver can tolerate over a given channel spacing. In general, a narrow channel spacing can be achieved by reducing the phase noise power density at frequencies away from the carrier. One way to achieve low phase noise is to design circuits with a high Q.

4.4.3.3 Reciprocal Mixing

While phase noise causes spectral leakage at the transmitter, it induces reciprocal mixing at the receiver. The receiver LO down-converts the RF carrier of the desired signal as well as interference to a lower frequency. Ideally, the down-conversion should not spread the interference to the desired signal frequency. However, in the mixing process phase noise in the LO modulates the interfering signal and can generate significant interference on the desired signal as shown in Figure 4.14. Results from the previous sub-section can be used to determine the amount of ACI generated by reciprocal mixing.

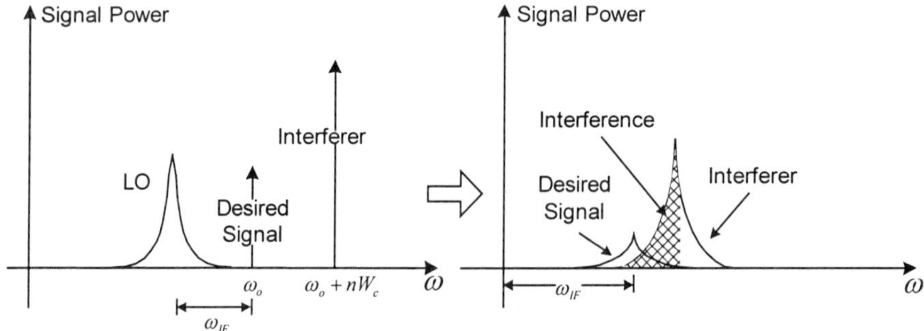

Figure 4.14 ACI due to phase noise at the receiver.

CHAPTER 5

Radio System Design

A radio system can be extremely complex and the design process cannot be easily systematized. This chapter attempts to give a systematic yet somewhat heuristic approach to the design of a radio system. The approach is not intended to provide precise results but rather simple intuitive guidelines for the radio design to meet a target specification. The purpose is to provide insights on how the high-level system requirements ripple down to constrain the underlying circuit components.

The design process is illustrated in Figure 5.1. The environment, regulatory bodies, and service providers impose constraints on all levels of the design steps. Environmental constraints have been discussed earlier in Chapter 2 and include the delay spread, Doppler spread, angular spread, noise, and interference introduced by the channel. Regulatory constraints refer to the conditions set by regulatory bodies (e.g. the FCC) on spectral usage including: the maximum transmit power density, carrier frequency, and the total available bandwidth. Service constraints refer to the QoS expected from the system. For example, if the system supports voice, then the service constraint should include the total system delay and frame-error rate.

During the system design phase, system constraints are determined to meet a given system capacity and coverage. The system design also fixes various link-level parameters such as the required E_b/N_0, fading margin, and bandwidth efficiency. Furthermore, the system design constrains various implementation parameters, such as, the length of the training sequence, the processing gain, and the margin for implementation loss. However, the system design still leaves open a few link-level parameters, such as, the amount of diversity gain required to meet a given E_b/N_0 and fading margin. These are determined during the link design through a detailed link-budget analysis. The computed link and system parameters help to form the specifications for the radio implementation. The rest of this chapter elaborates further on each of the design steps with a brief introduction to radio system issues and concepts.

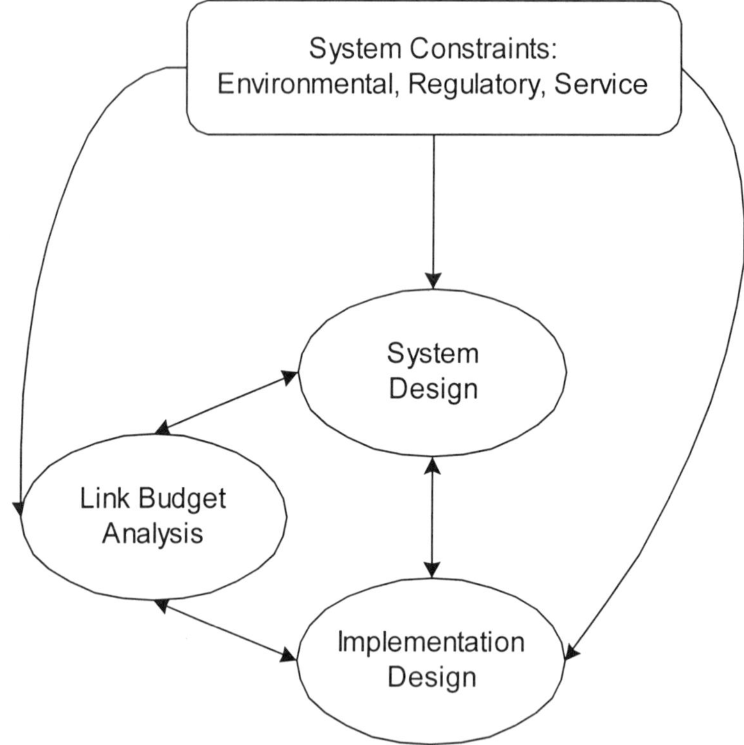

Figure 5.1 Radio design process.

5.1 Radio Systems Overview

This section gives an overview of radio systems, specifically, focusing on the multiple access schemes that support multiple concurrent transmissions and duplexing schemes that support two-way links.

5.1.1 Multiple Access

Multiple access schemes are necessary to allow multiple users to share a common spectrum. An efficient multiple access scheme is essential to obtaining a high system capacity and crucial to services that must support a large subscriber base. There are generally four main types of multiple access: 1) frequency division multiple access (FDMA), 2) time-division multiple access (TDMA), 3) code-division multiple access (CDMA), and 4) space-division multiple access (SDMA).

5.1.1.1 Frequency Division Multiple Access:

Most systems use some form of frequency-division multiple access (FDMA) as shown in Figure 5.2. Analog systems, such as AMPS, are based entirely on FDMA while digital systems are based on hybrid schemes. As mentioned earlier in Section 2.2.2, allocating a different frequency to each user provides good isolation among all the users. However, when frequencies are reused, co-channel interference (CCI) can cause substantial interference on a reused frequency channel; see (2.56). Furthermore, frequency isolation is not perfect among the different frequency channels due to spectral regrowth at the transmitter and non-ideal roll off in the receive filter. The limited isolation among adjacent channels results in adjacent-channel interference (ACI).

To minimize the effect of interference, guard bands are used to buffer the frequency channels as shown in Figure 5.2. The wasted bandwidth causes inefficiency in the channel usage. For FDMA, the channel efficiency defined as the useful bandwidth per channel divided by the total bandwidth W_c allocated for each channel is

$$\eta_c(FDMA) = \frac{W_c - W_g}{W_c} \tag{5.1}$$

where W_g is the bandwidth of the guard band. To minimize the guard band size, interference suppression techniques based on spread-spectrum or interference diversity can be used to minimize effects of ACI and CCI.

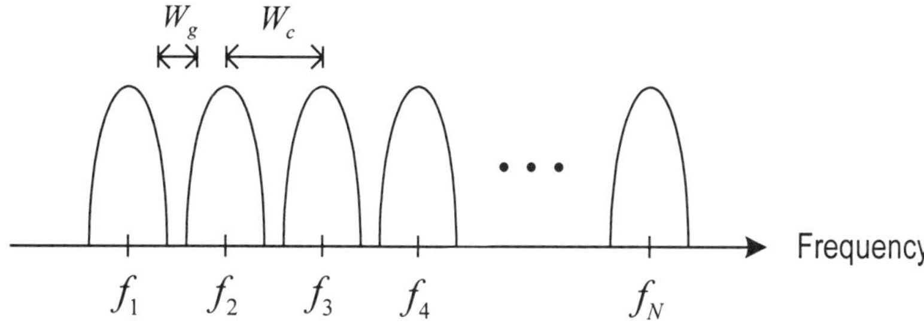

Figure 5.2 Frequency-division multiple access.

5.1.1.2 Time Division Multiple Access

Instead of allocating each channel with a frequency band, TDMA achieves user separation using time slots. Separation of concurrent transmissions is achieved by assigning each transmission a non-overlapping time slot as shown in Figure 5.3. The disadvantage of TDMA is that it requires precise synchronization to ensure that the time slots do not collide with each other. FDMA, on the other hand, does not require precise synchronization. In general, precise synchronization is difficult to achieve, especially during channel access. Therefore, a guard time, analogous to a guard band, is allocated for each slot to buffer potential collisions with adjacent time slots. The efficiency of a TDMA scheme can be described by

$$\eta_c(TDMA) = \frac{T_{slot} - T_{sync} - (L_t + L_{OH})T_b}{T_{slot}} \frac{N_{user\ slots}}{N_{slots}} \tag{5.2}$$

where T_{slot} is the duration of a time slot, T_{sync} is the time required to synchronize each slot, L_t is the number of bits used for training the equalizer or channel estimator, L_{OH} is the number of overhead bits (e.g. for tail bits in Viterbi encoding), $N_{user\ slots}$ is the number of user time slots, and N_{slots} is the total number of time slots.

For a mobile channel with a large Doppler spread, the slot duration should be reduced so that the equalizer can be re-trained more often to track channel dynamics. Since the length of training sequence depends on the delay spread and typically does not scale with mobility, TDMA channel efficiency degrades for highly mobile channels with large delay spreads. Also, if multiple time slots are interleaved to achieve time-diversity, the time slot size cannot be too large. Otherwise, the interleaving delay can be too long for realtime services.

The channel efficiency of TDMA is similar to FDMA. However, TDMA eases dynamic channel allocation, whereby a time slot can be changed on-the-fly while it is generally more difficult to do so in FDMA. Most digital systems, such as GSM and IS-136 employ both TDMA and FDMA. In such systems, the channel efficiency is equal to the product of (5.1) and (5.2).

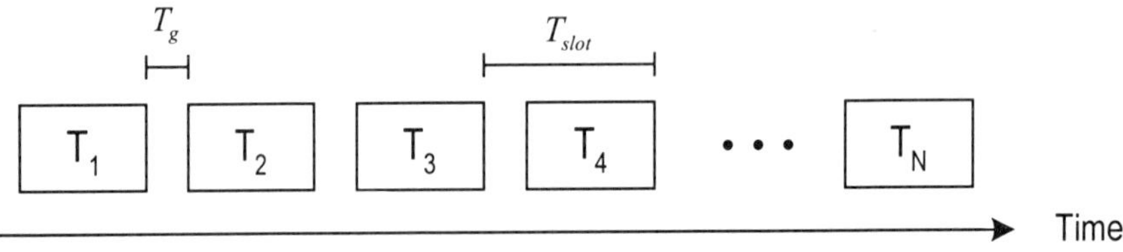

Figure 5.3 Time division multiple access.

5.1.1.3 Code Division Multiple Access

Instead of using frequency and time to isolate each channel, CDMA uses PN codes to separate concurrent transmissions by assigning a PN code to each transmission as shown in Figure 5.4. Users can share the same transmission channel and time. The disadvantage of CDMA is that it requires even more precise synchronization than TDMA to ensure that interference among users is minimized. Inadequate synchronization causes degradation in SIR as discussed in Section 3.5. Because the codes cannot maintain complete orthogonality and therefore still generate a certain level of interference in the system, guard times or guard bands are still used if CDMA is used in conjunction with TDMA or FDMA. The efficiency of a CDMA system can be described by

$$\eta_c(CDMA) = \frac{N_{user\ codes}}{N_{codes}} \tag{5.3}$$

where $N_{user\ codes}$ is the number of user codes and N_{codes} is the total number of codes.

For highly mobile applications, the channel efficiency for CDMA decreases if there is a TDMA overlay. In some systems, such as IS-95, a dedicated control channel is used instead to track the wireless channel so that training sequences are eliminated. In this case, the efficiency of a CDMA system can actually be higher than that of a pure TDMA system. Overall efficiency of a CDMA system can also be improved by reducing the guard time or guard band. The increase in interference is accommodated with an adequate amount of processing gain.

There are several other advantages that are unique to CDMA systems. First, because it is a wideband signaling scheme based on spread-spectrum, a strong error correcting code can be used without compromising the bandwidth efficiency of the system, as long as the code rate is not greater than the spread factor. In TDMA systems, higher rate codes must be used to maintain adequate bandwidth efficiency, according to (2.10). Secondly, because of CDMA's interference rejection capability, frequencies can be reused more often with a lower reuse distance. In other words, if each local area can be represented as a circle with radius R then the reuse distance can be set equal to R. High frequency reuse improves the overall system capacity. See Section 5.4.2. For the same reason, a CDMA system can better leverage SDMA without compromising frequency reuse as discussed in the next sub-section. Finally, since the capacity of a CDMA system is largely interference limited, effective interference diversity schemes can be applied to significantly improve its capacity over TDMA or FDMA systems.

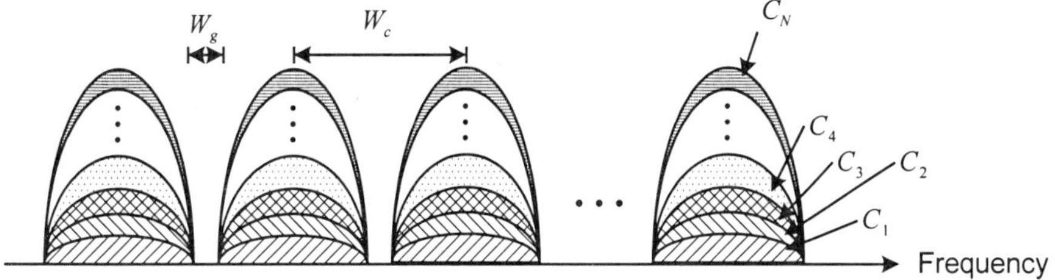

Figure 5.4 Code-division multiple access.

5.1.1.4 Space Division Multiple Access

SDMA separates users in the space domain as shown in Figure 5.5. It is generally used in conjunction with TDMA, CDMA, and/or FDMA to improve capacity and/or coverage. SDMA improves system coverage either in terms of lower transmit power or larger coverage area for the same amount of transmit power.

The most widely used SDMA system is based on antenna sectorization, which divides a local coverage area into sectors. A typical configuration with four 90° sectors is shown in Figure 5.5a. Because an antenna beam covers each sector, users within a sector can reuse its frequency. However, because users cross over into a different sector in a mobile system, the sectors must overlap. In FDMA systems or TDMA systems such overlaps cause CCI between sectors and therefore reduce the effectiveness of SDMA. CDMA can better utilize a sectorization because it can suppress or average out the interference that occurs as users move between sectors. For example, given a four sector configuration shown in Figure 5.5a, the number of users in a given area is quadrupled for CDMA systems.

Most major cellular systems, such as GSM, IS-136, and IS-95, use sector antennas. Third-generation wireless systems are considering more advanced SDMA based on adaptive beamforming as shown in Figure 5.5b to allow even higher frequency reuse and coverage. Furthermore, researchers are investigating the use of spatial multiplexing [145][146] based on spatial diversity to allow multiple users to share the same frequency channel, time slot, or code. User separation is based on spatial signatures generated by the transmit antenna array response and the channel spatial response at the receiving antenna array. The number of distinct spatial signatures that can be recovered at the receiver is proportional to the number of antenna elements. An example of such system is BLAST [96][97] which has demonstrated linear improvement in capacity with space-time processing, whereby multiple antennas are used to cancel out co-channel users. Frequencies can therefore be reused in a given area in contrast to current systems where frequencies can be reused only in areas separated by the reuse distance D. Spectral efficiency of 32bps/Hz over tens of kHz has been reported for such systems.

Because of size constraints on the antenna array, SDMA has been employed mainly for the system controller (i.e. basestation). The antenna spacing for basestations is typically on the order of 5-10 λ due to the small angular spread. Although the angular spread is higher at the mobile units to the point where the antenna spacing is on the order of 0.5λ at 1-2 GHz, it is still difficult to integrate multiple antennas into a handset that has to fit inside a shirt pocket. However, for larger mobile devices, SDMA might become applicable in the user equipment. For instance, in third-generation systems, SDMA is being considered for mobile multimedia terminals [147][148].

Despite the limited use of antenna arrays inside mobile units, SDMA and spatial diversity can still be implemented at the basestations to reduce interference experienced by the handsets. This is also referred to as transmit spatial diversity [96]

[102]. Perhaps more important to the mobile unit is the reduction of the mobile transmitter power made possible by the large antenna gains achieved with antenna arrays. The efficiency of SDMA systems is similar to that of TDMA, FDMA, or CDMA. The more advanced SDMA configurations require additional overhead in training the adaptive beamformer or interference canceller and therefore tend to have less channel efficiency.

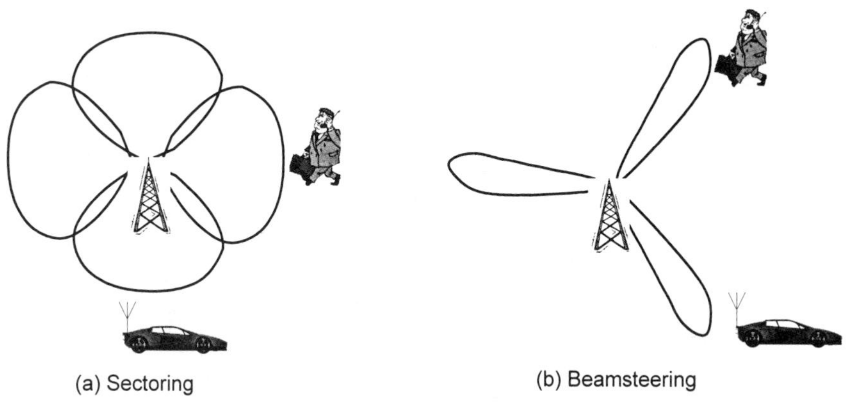

(a) Sectoring (b) Beamsteering

Figure 5.5 Implementations of SDMA.

5.1.2 Duplexing

Most radio systems are bi-directional. Frequency-division duplexing (FDD) and time-division duplexing (TDD) are two main duplexing techniques used to prevent the transmissions on either direction from colliding with each other. FDD separates the transmissions in the frequency domain while TDD provides separation in the time domain.

5.1.2.1 Time Division Duplexing

The transmission direction is defined based on the whether the radio equipment is transmitting (TX mode) or receiving (RX mode). TDD breaks up a time slot into two parts, one for the TX mode and one for the RX mode as shown in Figure 5.6a. To implement a TDD system, a relatively unsophisticated RF front-end can be time shared between the TX and RX modes. TDD also facilitates the support for asymmetric traffic, such as Internet access where the radio will be in the RX mode most of the time. In this case, a larger fraction of the time slot can be allocated to the RX mode. Finally, since both the TX and RX modes share the same frequency, frequency selectivity is similar in both modes. Channel parameters can be determined at the receiver and the transmitter can adapt accordingly without the need for two-way channel probing. For instance, a transmitter can adjust its transmit power according to the received signal strength determined locally based on measurements performed in the RX mode.

TDD does however have its disadvantages. For a given source rate, the channel rate of a TDD system must double since the RX and TX data are time multiplexed. The increased channel rate drives up the speed requirement in the radio circuits both in terms of signal processing speed and RF bandwidth. Also, equalizer complexity may increase due to the increase in channel rate that worsens the amount of ISI. Unless the channel is fairly stagnant, the equalizer must also be re-trained in RX modes. The additional training bits decrease the efficiency of the channel. Finally, synthesizer turn-around time must be shortened to minimize the time required for synchronization T_{sync}. Because of these disadvantages, systems that implement TDD have been restricted to indoor applications where the channel has a low delay spread and Doppler rate.

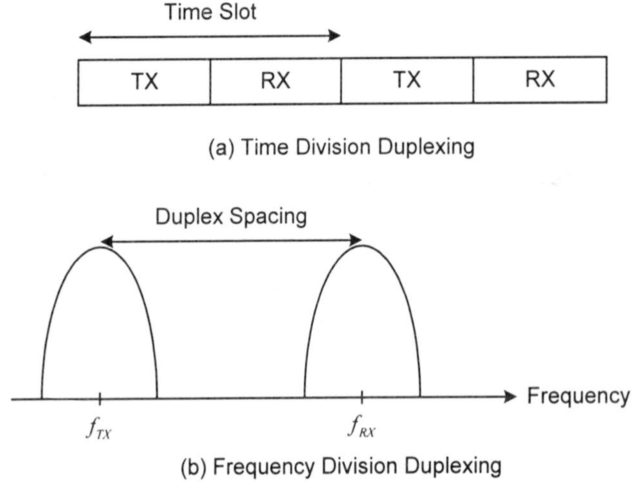

Figure 5.6 Duplexing schemes.

5.1.2.2 Frequency Division Duplexing

Rather than multiplexing the TX and RX data in time, FDD systems use two different frequency channels for separate transmissions in the TX and RX modes as shown in Figure 5.6b. FDD has the disadvantage that it requires a sophisticated RF front-end to separate out the TX and RX frequency channels because of the proximity of the transmitter to the receiver. As an example, if the transmit power is 28 dBm and the receiver sensitivity is –102 dBm then an isolation of at least 140 dB is required for an SIR of 10 dB. Designing a frequency duplexer with 140 dB of isolation for signals contiguous in frequency is impractical. Therefore, in FDD systems, the TX and RX frequency channels are separated by a large frequency gap typically 40-50 MHz. The large frequency gap makes it more difficult to perform system operations without a feedback channel since reciprocity no longer holds. That is the channel response differs substantially between the transmit direction and the receive direction.

Examples of system operations include power control [149][150] and transmitter diversity for spatial multiplexing or equalization [145][96]. Finally, FDD is not particularly suited for asymmetric channels since resizing the frequency channel is much more difficult.

Despite its disadvantages, FDD offers a higher channel efficiency since for a given source rate and slot size, only one training sequence is required and the synthesizer does not need to turn-around during operation. Moreover, the channel rate does not need to be doubled, making the equalizer simpler. FDD has therefore been employed for mobile cellular systems where the environment induces a large delay spread and Doppler spread. However, in the proposed third-generation digital cellular systems, mobile handsets are supposed to support not only FDD but also TDD so that the handset can operate indoor as well as outdoor [9].

5.2 Cellular System

Many commercial systems, such as GSM, DECT, IS-95, and IS-136, are all based on the cellular concept. Most digital cellular systems include both SDMA and FDMA as part of the multiple access. The majority of today's systems also support TDMA, including, GSM, PDC, and IS-136. In contrast, the only CDMA system that has been deployed is IS-95. However, systems based on CDMA are rapidly growing. In particular, CDMA dominates the scene in third-generation wireless.

The local coverage area in a cellular system is defined as a cell and is usually represented as a hexagon as shown in Figure 5.7. Note, however, in practice cells are not hexagonal. The coverage area is approximated by πR^2 where R is defined in Figure 5.7. Mobile units in each cell communicate via centralized radio equipment, also known as a basestation. The mobile units are notified of incoming calls by the basestation through a paging channel. When a page is received, the mobile unit answers the call through an access channel and after a verification and authentication process, the call is then established. When a mobile unit moves to a new cell, on-going calls are handed off to the new cell to maintain connectivity. Transfer of calls between cells is also known as hand-off.

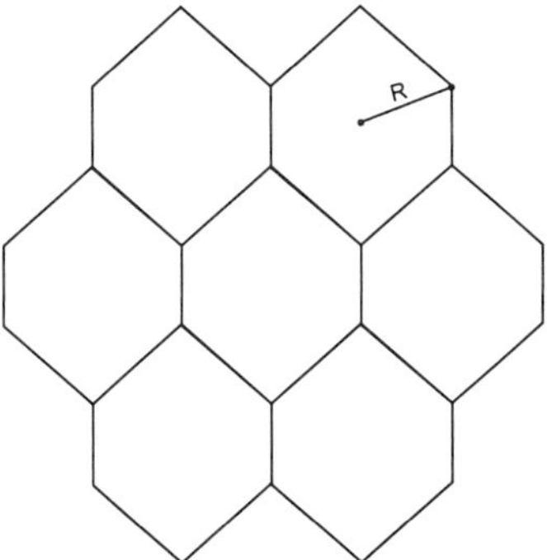

Figure 5.7 A cellular system.

5.2.1 Frequency Reuse

The system capacity of a cellular system refers to the maximum number of user channels that can be supported by the basestation or per km^2. Clearly then, by reusing as many frequencies as possible in a cell helps to improve system capacity. The frequency reuse factor K is defined as the number of cells enclosed in a cluster with an effective radius equal to the reuse distance D. Figure 5.8 illustrates several common frequency reuse plans. A larger the reuse factor results in a lower system capacity since proportionally fewer frequency channels are available for reuse in each cell. It is therefore desirable to minimize K.

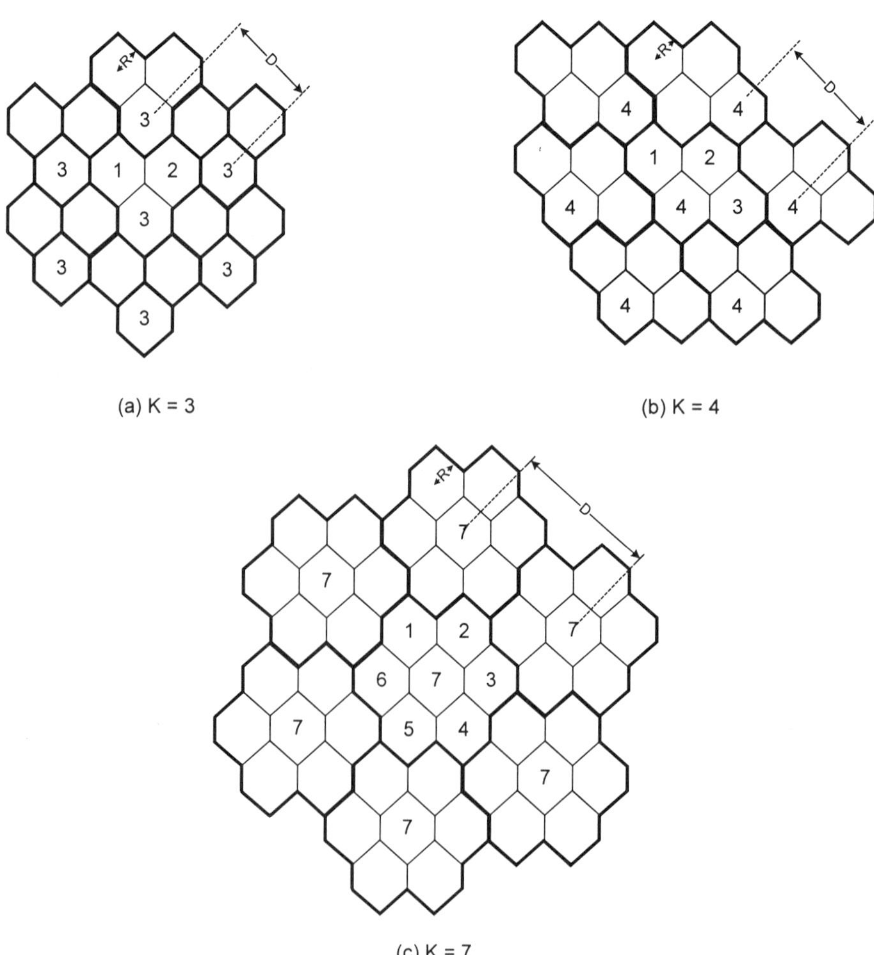

Figure 5.8 Common frequency reuse configurations.

The value of K depends largely on the co-channel interference from nearby clusters. In Section 2.2.2.1, co-channel interference has been expressed as D/R, also known as the CCI rejection ratio. To meet a given SIR requirement, D/R can be determined based on (2.56). Furthermore, it can be shown that with a hexagonal cell shape, the CCI rejection ration can be expressed in terms of the reuse factor K [38]:

$$\frac{D}{R} = \sqrt{3K}. \tag{5.4}$$

In a cellular system, it can be shown that there are always six nearby clusters for a given cluster. These are also known as tier-one clusters and are the dominant sources of CCI. In a TDMA or FDMA system, each reused frequency channel contributes CCI to six nearby tier-one clusters. At any given time then, there exist up to six co-channel interferers, i.e. $K_I = 6$. Applying (2.56) and (5.4), the reuse factor for TDMA or FDMA system can be approximated by the following simple relationship

$$K = \frac{1}{3}\left(K_I \cdot SIR_{req}\right)^{\frac{2}{n}}. \tag{5.5}$$

Example 5.1 For AMPS, the required SIR is 18 dB [38]. Assuming a path loss exponential of four, the reuse factor according to (5.5) is then approximately seven. The corresponding reuse plan is shown in Figure 5.8c. □

5.2.2 Capacity Enhancement

With the proliferation of wireless devices, capacity is becoming a major design factor. Currently four methods are used to enhance the capacity of a system: 1) smaller cell size, 2) power control, 3) discontinuous transmission (DTX), and 4) CDMA. Smaller cells can be achieved through cell splitting, whereby a larger cell is split into smaller ones. Even though capacity is not changed on a per cell basis, it is increased on a per area basis in terms of the number of users per km^2. In some systems, cells are designed to be small from the start. For an indoor environment, the cell size is generally designed to be small 10-100 m, i.e. pico-cellular. For outdoor static environments, such as outside of buildings and residential areas, cell sizes range from 100 m to 1 km, i.e. micro-cellular. Smaller cells also have the advantage of reducing the required power in the mobile unit and thus help to prolong battery life. However, smaller cells incur an excessive number of hand-offs during a call. Therefore, for mobile environments, macrocells are used with larger sizes (1-30 km) to accommodate the high mobility. GSM, IS-95, and IS-136 are all based on macrocells whereas digital cordless systems such as DECT and Personal Handy-phone System (PHS) [151] are based on picocells and microcells. In third-generation systems, cells of different sizes are supported to maximize the system capacity while meeting the mobility constraint.

According to (5.5), to minimize the reuse factor, one should decrease the required *SIR*. Power control offers a direct method in reducing the required SIR by adapting the transmitter power level such that a minimum amount of interference is generated to nearby users. In CDMA, which is interference limited, power control is particularly important to achieve high capacity. Open loop and closed loop are two types of power control generally used. In open-loop power control, reciprocity is assumed whereby the receiver measures the SIR and computes the minimum transmit power level at which to transmit. In closed-loop power control, the receiver measures the SIR and sends a power control command to the remote transmitter to adjust its transmit power level. Power control can occur both in the base-to-mobile or mobile-to-base direction. The former and latter are also referred to as the downlink and uplink, respectively. Power control is employed in many digital cellular systems, such as IS-95 and GSM.

DTX offers yet another means to minimize interference. DTX is based on the observation that about 40-50% of the time [152] the channel is idle since usually while one person is talking the other is listening. The transmitter can take advantage of this fact by lowering the transmit power when the channel has no voice activity [153]. The amount of interference is effectively reduced by the voice activity factor K_v. DTX is currently employed in most high capacity systems, such as GSM and IS-95.

Finally, CDMA provides processing gain that can reduce the effect of CCI and therefore lower the required SIR. IS-95 is a CDMA system that is currently deployed in North America and some parts of Asia. IS-95 employs CDMA together with power control and DTX. Moreover, with spread-spectrum IS-95 achieves a frequency reuse of unity, whereby all frequencies are reused in every cell. In fact, with sector antennas, IS-95 is able to achieve factional reuse. With its robustness to interference, IS-95 has touted a system capacity 4-20 times that of the current non-CDMA systems.

5.3 System Constraints

Referring to Figure 5.1, the system constraints influence all levels of the radio design. This section describes system constraints arising from environmental, regulatory, and service requirements.

5.3.1 Environmental Constraints

The transmission environment imposes constraints on the communication channel. These constraints have been thoroughly described in Section 2.2. Table 5.1 lists the key channel parameters that affect the system design, link design, and the radio implementation.

Table 5.1 Environmental constraints.

Parameter	Notation	Description
Noise Power Density	N_0	Measures the amount of Noise power per Hz.
Interference Power	I_0	Measures the amount of interference power.
Path Loss Exponent	n	Incurs $10n$ dB per decade loss in received power.
Delay Spread	τ_d	Measures time dispersion of the channel.
Doppler Spread	f_D	Measures time variation in the channel.
Duration of Fade	$\tau(\rho)$	Measures the duration of a fade at a given signal level ρ, normalized to the RMS signal level.
Angular Spread	θ_{rms}	Measures spatial dispersion of the channel.
Diversity Order	N_D	Order of diversity introduced by the channel.

5.3.2 Regulatory Constraints

Regulatory bodies, such as the Federal Communications Committee (FCC), impose conditions on what frequency to use, how much bandwidth is available, and where the system may be deployed using the assigned frequencies. Among the various regulatory constraints, the three most important restrictions are the frequency band of operation, the total available bandwidth W_A, and the maximum transmitted power. The latter is usually described in terms of the effective radiated power (ERP) which accounts for the antenna gain.

The frequency band of operation influences the design of RF front-ends as well as digital transceivers. A high carrier frequency requires high frequency circuits with high power dissipation. Moreover, A high carrier frequency induces a large Doppler spread that must be tracked in the digital transceiver. While a high transmit frequency seems to result in a more difficult design, a high frequency carrier is necessary to implement antennas with a high efficiency and small physical size.

The choice of the aggregate bandwidth impacts both the RF and digital circuit designs. A large aggregate bandwidth results in increased noise level and distortion in the RF circuits and a high sampling rate in the digital circuit, especially when a high-IF digital transceiver is implemented. The advantages in having a large aggregate bandwidth are increased capacity and user bitrate.

Lastly, a high ERP allows high system coverage but makes the design of a linear power amplifier more difficult. Further details on the impact of these parameters on the system design, link design, and implementation will be described in the following sections. Figure 5.9 illustrates the frequency allocation for selected wireless systems worldwide and Table 5.2 lists the regulatory parameters for selected systems. Note that most current digital cellular systems operate in the 1-2 GHz range with a total available bandwidth ranging from 25-50 MHz and an ERP of 10-30 dBm.

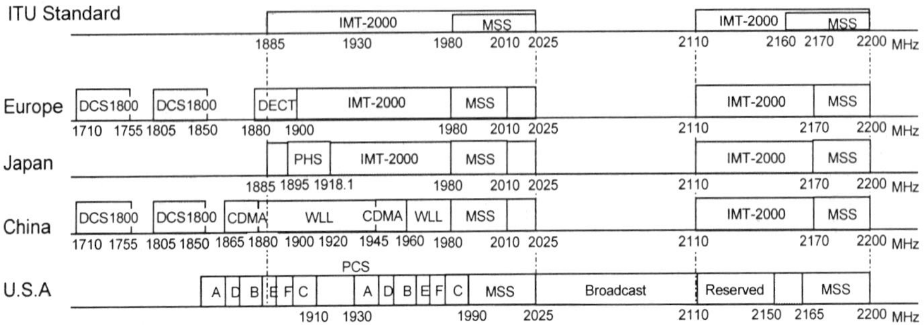

Figure 5.9 Spectral allocation for selected systems worldwide at around 2 GHz.

Table 5.2 Regulatory constraints on selected systems.

System	Available Bandwidth (Two Ways)	Maximum ERP (Uplink)
DECT	20 MHz	24 dBm
DCS1800	90 MHz	29.0 dBm (class 5)
IS-95	50 MHz	27.8 dBm
IS-136 (A-band)	30 MHz	27.8 dBm (class 4)
IMT-2000	230 MHz	10-30 dBm

Table 5.3 Service constraints.

Parameter	Notation	Description
Bitrate	R_b	Rate of information bits transmitted over the channel
End-to-end delay	T_{E2E}	Maximum end-to-end delay imposed by a given service
Frame error rate	$\overline{P}_F$	Percentage of frames in error
Packet error rate	$\overline{P}_P$	Percentage of packets in error
Bit error rate	$\overline{P}_b$	Percentage of bits in error

5.3.3 Service Constraints

The type of service provided by the wireless system constrains the radio design in terms of the required bitrate, delay, and error rate. These constraints are listed above in Table 5.3. The bitrate constrains sampling rate, channel spacing, bandwidth of synchronization loops, and transmit power. The end-to-end delay constrains processing delays in the radio, including those incurred by interleaving, source compression, FEC coding, synchronization, and filtering. An end-to-end delay refers to the delay introduced by a system from one end of the link to the other end. For example, the end-to-end delay for voice service is measured from the time a person speaks into the microphone at one end to the time the other person hears the speech coming out from the speaker at the other end. The error rate determines the required E_b / N_0 which imposes a certain set of constraints on the suitable choice for error correction and modulation.

Depending on the source, bitrate tends to vary from 4 kbps to 10 Mbps. The limits on bitrate depend largely on subjective measures. For instance, high quality voice compression requires around 13-64 kbps to achieve a mean-opinion score (MOS) between 3-5. MOS is a subjective measure that rates the quality of compressed speech on a scale of 1 to 5. For analog sources, bitrate can be defined as cW_i / S_c, where c is the conversion rate of the analog-to-digital converter (ADC) in bps/Hz, W_i is the information bandwidth, and S_c is the compression ratio. See (2.10). For instance, the GSM voice coder compresses the digitized analog speech by a factor of eight. With an audio source at 4 kHz and an ADC conversion rate of 26 bps/Hz, the bitrate for the compressed speech becomes 13 kbps.

In general, realtime (RT) services such as video and audio are sensitive to delays. Therefore, the mean end-to-end delay and RMS jitter must be kept below a certain value for RT services. On the other hand, non-realtime (NRT) services do not have a stringent requirement on delay. Rather, they require a low error rate to ensure guaranteed delivery at a reasonable throughput. Three main types of services are summarized below.

Voice: Today, the predominant service is voice, which fortunately has a fairly low bandwidth (4-8 kHz) requirement so that it can be transmitted at a reasonably low bitrate (2.4-32 kbps), depending on the compression technique. Though a high compression ratio results in a low bitrate, the price paid is quality gauged by the MOS. To achieve toll quality, the MOS should range between 4-5. Bitrate at this range varies between 13-64 kbps. The upper limit on frame error rate ranges from 20% to 30%. Subjective tests show that the maximum tolerable delay for a reasonable quality round-trip delay is about 200 ms [154]

NRT data: Data service still lags voice service in popularity though projections indicate that it is catching up fast. Current data services include file transfer, email, short-message services, FAX, and Internet access. However, the current bitrate is still limited to below 14 kbps. Systems are being upgraded to achieve bit rates as high as 384 kbps. Upgraded systems include the General Packet Radio Services (GPRS) [155] and the Enhanced Data for GSM Evolution (EDGE) [156]. Older systems include Cellular Digital Packet Data (CDPD) [157][158] and Advanced Radio Data Information Service (ARDIS) [158]. The delay for the older systems ranges from 2 to

6 seconds while the delay for the newer systems ranges from 0.1 to 1 second for short messages. The BER constraint is usually better than 0.001%. In the case of NRT data, a low BER outweighs the need for low delay.

Video/Image: Currently, video services are still confined to the wireline infrastructure. Wireless video is yet to be deployed though the ITU has already completed a standard on low bitrate video targeted for mobile wireless applications. The standard is known as H.324M and is capable of delivering video at a bitrate of 28.8 kbps. Other compression standards also exist, such as MPEG2 and MPEG4. However, these have been designed for a wireline environment with a low BER (10^{-7}) and a high bitrate (1-10 Mbps) requirement. The delay requirement for video is on the order of half a second, which is less stringent than that for voice. The deployment of wireless video has been hampered by the uncertainty of its level of adoption and cost. However, with the availability of inexpensive CMOS imagers, wireless transmission of images can now be implemented at a low cost in current wireless handsets [159]. Images can be compressed much more than video and do not have the RT requirement. For instance, using the discrete wavelet transform (DWT), compression ratio can range from 50-100 [160]. A 144x176 (i.e. QCIF) 24-bit color image can be transmitted on a 14.4 kbps link in a couple of seconds.

Table 5.4 summarizes typical delay, error rate, and bitrate constraints for voice, data, and video services.

Table 5.4 Service constraints for voice, data, and video.

Services	Delay	BER	Bitrate
Voice	< 100 ms	0.1%	8-32 kbps
Data	500 ms (typical)	0.001%	1-10 Mbps
Video	< 100 ms	0.00001%	0.1-25 Mbps

5.4 System Design

In the system design phase, system coverage and capacity are determined subject to a set of service, regulatory, and environmental constraints. Details of the system design are described below.

5.4.1 System Coverage

The large-scale behavior of the wireless channel has been discussed in Section 2.2.4. Since the transmitted signal takes multiple random paths to reach the receiver depending on the channel conditions and environment, the behavior can be at best modeled statistically. A general model is shown in (2.64) where the reflection coefficient Γ and phase ϕ are random parameters that vary with the changing environment. By taking a statistical average of (2.64) over a given terrain, the resulting signal amplitude follows a lognormal distribution shown in (2.69), with mean μ that falls off exponentially with distance raised to the power n and variance

σ^2 that depends on the particular environment. Table 2.3 lists typical values of n and σ^2 for different environments.

Given the statistical nature of path loss, the error rate cannot be guaranteed with finite transmit power over the wireless channel. Rather, the performance can only be guaranteed for a given coverage probability. Coverage is defined as the range over which the radio system guarantees a certain BER with a confidence level of P_{coverage} percent, where P_{coverage} is also known as the coverage probability.

For a given range requirement, the mean path loss $\overline{L_P}$ can be determined based on any of the analytical models shown in Section 2.2.4 or based on empirical models, e.g. $\overline{L_P} \propto d^n$. To achieve a target BER or equivalently $\overline{\gamma}$, the required transmit power P_{TX} should satisfy the following constraint

$$\overline{\gamma} = \frac{P_{TX}}{\overline{L_p} R_b N_0}, \tag{5.6}$$

provided that the variance of the path loss is zero. However, in practice there is a finite variance which necessitates allocating additional link margin to maintain the target BER. Ideally, the margin should be large enough to guarantee with 100% coverage probability. However, due to the lognormal distribution of the signal amplitude, an infinite transmit power is required to achieve 100% coverage probability. In practice, the link margin is designed to achieve a reasonably high coverage probability, such as 90%.

With a lognormal distribution, given in (2.69), the amount of link margin L_M required to achieve a given P_{coverage} depends on $m\sigma$ where m is a real constant and can be determined by the following integral equation

$$\int_{-\infty}^{\sqrt{\frac{P_{TX}}{\overline{\gamma} R_b N_0 10^{-0.1 m\sigma}}}} \frac{10}{\ln 10 \sqrt{2\pi\sigma r^2}} e^{-(20\log r-\mu)^2/2\sigma^2} dr = P_{\text{coverage}}. \tag{5.7}$$

The link margin translates to additional power required to meet a given P_{coverage} and is determined by

$$L_M = m\sigma. \tag{5.8}$$

5.4.2 System Capacity

A properly designed system must have sufficient capacity to support the projected number of users in the system. Capacity can be measured in terms of the users/cell, users/km^2, channels/cell, bps/Hz/cell, bps/Hz/km^2, offered load per cell, or offered load per km^2. Offered load A_{load} is defined as the amount of traffic offered to the system. Depending on the type of system, the definition of offered load varies. For example, in telephony systems offered load is defined in terms of the channel occupancy over a defined period of time, usually referred to as the busy hour. The offered load can be measured in units of Erlang defined as the number of calls per busy hour times the duration of the calls. Since most of the time the offered load exceeds the amount of channel resources available in the system, an incoming call could be blocked with a blocking probability P_B.

In this book, system capacity is measured in terms of spectral efficiency η_s expressed in units of bps/Hz/km^2 defined below for TDMA

$$\eta_s = \frac{\frac{W_A N_{slot}}{W_c K}\eta_{FDMA} R_b \eta_{sig}}{\pi W_A R^2} = \frac{R_b N_{slot}\eta_{FDMA}\eta_{sig}}{\pi K W_c R^2}. \tag{5.9}$$

where W_c is the channel spacing, N_{slot} is the number of time slots per frequency channel, and R_b is the information bitrate. η_{sig} is the out-of-band signaling efficiency of the system and is described by the following expression

$$\eta_{sig} = \frac{W_c}{W_A} n_{sig}. \tag{5.10}$$

where n_{sig} denotes the number of out-of-band frequency control channels.

In a CDMA system, users are separated by code channels which may be further grouped into different frequency channels and/or time slots. Its spectral efficiency is shown below

$$\eta_s = \frac{R_b N_{slot} N_{code}\eta_{CDMA}\eta_{FDMA}\eta_{sig}}{\pi K W_c R^2}. \tag{5.11}$$

where N_{code} represents the number of codes per frequency channel and η_{CDMA} is defined in (5.3). The IS-95 CDMA system is based on CDMA with FDMA. In this case, the number of time slots is one.

If expression (2.10) is applied, then a more general expression is obtained as shown below:

$$\eta_s = \begin{cases} \dfrac{r\eta_W \eta_{TDMA}\eta_{FDMA}\eta_{sig}}{\pi K R^2}, & TDMA \\ \dfrac{\eta_{TDMA}\eta_{CDMA}\eta_{FDMA}\eta_{sig}}{\pi K R^2}\dfrac{N_{code}}{G_p}, & CDMA \end{cases} \tag{5.12}$$

where

$$\eta_W = \frac{R_b N_{slot}}{r W_c \eta_{TDMA}}, \qquad TDMA \tag{5.13}$$

and

$$G_p = \frac{W_T \eta_{TDMA}}{R_b N_{slot}}, \qquad CDMA \tag{5.14}$$

The variable W_T in (5.14) represents the total transmission bandwidth and amounts to R_c and $N_{hop}W_c$ for direct-sequence and slow frequency-hop systems, respectively. For ease of notation η_{TDMA}, η_{FDMA}, and η_{CDMA} are used to denote $\eta_c(TDMA)$, $\eta_c(FDMA)$, and $\eta_c(CDMA)$, respectively. For a CDMA system, spectral efficiency depends not only on η_{CDMA} but also on η_{TDMA} and η_{FDMA}. However, the values for η_{TDMA} and η_{FDMA} could be quite different from the pure TDMA or FDMA system. For instance, if a continuous pilot channel is used for channel estimation, then no training sequence is required for each slot. A continuous code channel may be dedicated for speech traffic. In such a case, the system essentially uses no time multiplexing and therefore η_{TDMA} is close to 100%, in contrast to the 50-75% efficiency in a pure TDMA system.

Key design parameters needed to determine system capacity as defined in (5.9) and (5.12) are shown in Table 5.5.

Table 5.5 System design parameters.

Parameter	**Notation**	**Description**
Offered Load	A_{load}	Amount of usage offered to the system
Blocking Probability	P_B	Probability that the service requested is blocked
Spectral Efficiency	η_s	Bits/sec per Hertz per km^2 (information bits only)
Channel Efficiency	η_c	Percentage of user bits per packet or frame
Signaling Efficiency	η_{sig}	Out-of-band signaling efficiency
Bandwidth Efficiency	η_W	Bandwidth efficiency measured in bps/Hz
Cell Radius	R	Local area (cell) in which frequencies are not reused
Reuse Factor	K	Number of cells in a reuse cluster
Number of Sectors	K_s	Number of sectors in a local area (cell)
Activity Factor	K_v	Percentage of time link is active
Signal-to-Interference Ratio	SIR	Signal-to-interference ratio of the system

General guidelines for designing the system capacity are outlined below for both TDMA and CDMA.

5.4.2.1 TDMA System Capacity Computation

The design of TDMA system capacity may be broken down into ten general steps. First, the offered load to the system is determined. Then, the blocking probability is specified to meet a given service requirement. The number of physical channels needed to meet the offered load requirement can be estimated using the Erlang B formula. Next, the required E_b/N_0, denoted by $\bar{\gamma}_{req}$, is estimated based on the bit error rate requirement and modulation parameters. The required SIR is computed based on $\bar{\gamma}_{req}$, bandwidth efficiency, and the receiver noise figure. The number of physical channels is then bounded by a generalized constraint which helps to determine whether the offered load, blocking probability, and $\bar{\gamma}_{req}$ could lead to a feasible solution. If no feasible solution exists then the offered load, blocking probability, $\bar{\gamma}_{req}$, and other system parameters in (5.12) must be readjusted until the constraint is met. Details of the steps involved in computing the capacity of a TDMA system are elaborated below.

Step 1. Compute the offered load. In a cellular system supporting voice service, the offered load is measured in Erlangs and can be defined as the product of the call arrival rate λ during the busy hour and mean hold time of the individual call. On current wireline telephone systems, the mean hold time for a voice call is typically less than 3 min [161].

Step 2. Determine the acceptable blocking probability P_B for a given QoS requirement. For instance, typical voice services require a call block probability of 1% [161].

Step 3. Determine the total number of user channels required to support a given traffic load. Accurate estimation requires detailed traffic simulation for a given service and traffic profile. For conventional macrocellular cellular system carrying voice traffic, the relationship between P_B and the number of channels and offered load is well characterized by the Erlang B formula [161]:

$$P_B = \frac{A_{load}^{N_{ch}} / N_{ch}!}{\sum_{k=0}^{N_{ch}} \frac{A_{load}^{k}}{k!}} \tag{5.15}$$

A channel in context of (5.15) refers to a unit of channel resource that is allocated to an active call. In TDMA this unit is typically a time slot. By fixing P_B and A_{load}, the required number of channels per cell N_{ch} can be determined through a table look-up. More sophisticated models [162] are available if queuing is allowed in the system. Also, for microcellular systems where handoffs occur more often, effects of handoff on system loading must be taken into account.

Example 5.2 Consider a cellular system which receives 1000 calls per hour during its busy hour with a mean hold time of 3 min. Then, the offered load to the system is

$$1000 \times \frac{3}{60} = 50 \; Erlangs.$$

Assuming that the system requires a call-blocking probability of 1 %, by applying (5.15) the number of required channels is determined to be 64 as shown in Figure 5.10. □

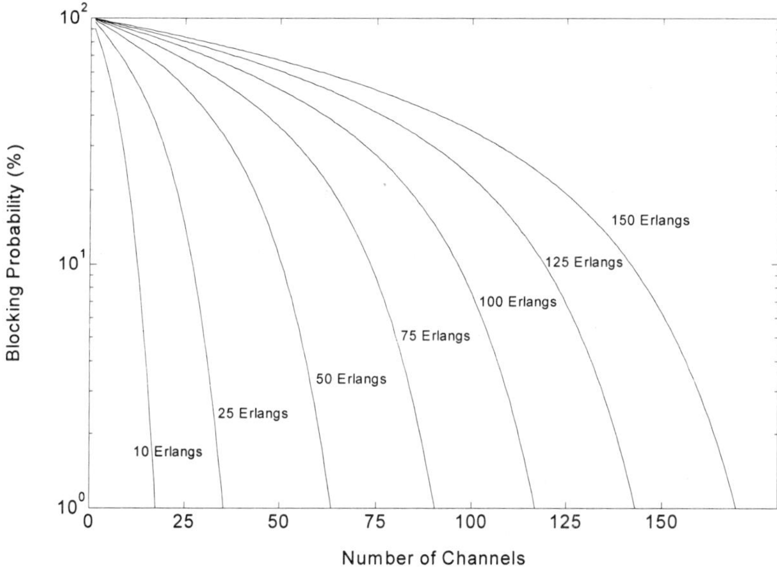

Figure 5.10 Blocking probability Vs. system loading and the number of physical channels.

Step 4. Determine $\overline{\gamma}_{req}$ to meet the target BER for a given service supported by the system. $\overline{\gamma}_{req}$ can be determined by inverting the Q function for a given modulation scheme as described in Section 2.4. $Q(x)$ can also be approximated by $0.5e^{-0.5x^2}$ at high SNR. Therefore, $\overline{\gamma}_{req}$ can be estimated using the following approximation

$$\overline{\gamma}_{req} \approx \ln\left(\frac{\kappa_1}{2P_b}\right)^{\frac{2}{\kappa_2}}, \tag{5.16}$$

where κ_1 and κ_2 are constants specific to the modulation type with $P_b \approx \kappa_1 Q\left(\sqrt{\kappa_2 x}\right)$. The approximation expressed by (5.16) always gives a worst-case estimate but approaches the actual value at low BER.

Example 5.3 To find κ_1 and κ_2 for PSK, refer to Table 2.4 and apply the approximation that $P_b \approx P_s$ with Gray encoding at high SNR. Then, $\kappa_1 = 2$ and $\kappa_2 = 2\sin^2 \frac{\pi}{M} \log_2 M$. □

Step 5. For high capacity systems, the required signal-to-interference ratio (SIR) should be designed according to the following constraint

$$SIR_{req} \le \eta_W \overline{\gamma}_{req} + \sqrt{\eta_W \overline{\gamma}_{req} \gamma_M}\,. \tag{5.17}$$

See Section 5.5 and (5.32) for the definition of SIR_{req}. The parameter γ_M denotes the amount of noise margin required to accommodate environmental and circuit noise. Typical values for γ_M lie between 3 and 10, where high performance systems require values in the low end. Details on the derivation of (5.17) are derived in Section 5.5.

Step 6. The number of required channels determined in Step 3 is analyzed to determine if this number can be supported by the given available bandwidth and if it satisfies the SIR constraint of (5.17). The general constraint is shown below

$$\left\lceil \frac{3W_A r \eta_{TDMA} \eta_{FDMA} \eta_{sig} \eta_W}{R_b \left(K_I \left(\eta_W \overline{\gamma}_{req} + \sqrt{\eta_W \overline{\gamma}_{req} \gamma_M} \right) \right)^{2/n}} \right\rceil \le N_{ch} \le \left\lfloor \frac{W_A r \eta_{TDMA} \eta_{FDMA} \eta_{sig} \eta_W}{R_b} \right\rfloor. \tag{5.18}$$

The upper limit in (5.18) results from constraining the reuse factor to be at least one and can be derived as follows. First, by using (5.9) and (5.12), the number of TDMA slots can be determined as shown below

$$N_{slot} = \frac{W_c}{R_b} r \eta_W \eta_{TDMA}\,. \tag{5.19}$$

Next, the reuse factor can be represented by the following relationship

$$K = \frac{W_A}{W_c} \frac{N_{slot}}{N_{ch}} \eta_{FDMA} \eta_{sig}\,. \tag{5.20}$$

Finally, substituting (5.19) into (5.20) and setting $K \ge 1$ result in the upper limit on N_{ch} shown in (5.18).

The lower limit in (5.18) results from constraining the SIR according to (5.17). Again substitute (5.19) into (5.20) and equate the resulting expression to that of (5.5). Then by applying constraint (5.17), the lower limit in (5.18) results.

To make use of (5.18), parameters K_I, r, η_{TDMA}, η_{FDMA}, η_{sig}, and η_W must first be determined. For cellular systems, K_I can be approximated by six since there always exist six cells close by with the same reuse frequency. Other cells further away also exist but their effects are greatly reduced. Other parameters (e.g. r and η_W) require detailed design on the baseband processor. However, as a starting point, one can apply certain heuristic of to arrive at initial values for these parameters. For example, η_W should be around 2 bps/Hz to maintain energy efficiency. See Section 2.1.6. For r and η_c, a balance should be attained. A reasonable assumption is to set the code rate at 0.5 and the TDMA efficiency at 0.75. Similarly, a value for signaling efficiency can be selected from a typical range of 0.9 to 0.99.

If the upper limit (5.18) is not satisfied then the system is capacity-limited and the call blocking probability has to be increased and/or the coverage area has to be decreased. On the other hand, if the lower limit is not met then the system is underutilized and the SIR requirement can be increased to support fewer channels with increased link quality.

Step 7. The channel spacing must satisfy the following constraint

$$\frac{R_b}{r\eta_W\eta_{TDMA}} < W_c \leq W_A, \tag{5.21}$$

where the lower limit follows from (5.19) given that the number of TDMA slots should be greater than one. Depending on the signal processing constraint, the channel spacing may be chosen based on the coherence bandwidth. To achieve path diversity at the expense of complexity, the channel spacing should be at least equal to the coherence bandwidth to sufficiently resolve the multipath components. On the other hand, if complexity introduced by path diversity cannot be afforded, the channel spacing should be lowered with respect to the coherence bandwidth but not beyond the lower bound set by (5.21).

Step 8. Determine the number of TDMA slots per frequency channel by substituting into (5.19) the value of W_c determined in Step 7. The ceiling function is applied to (5.19) such that the number of time slots has an integer value

Step 9. Determine K from (5.20).

Step 10. Determine SIR from (5.5).

5.4.2.2 CDMA System Capacity Computation:

Step 1. Repeat steps 1-4 of the procedure for TDMA capacity calculation.

Step 2. The key advantage of CDMA over TDMA is its potential for higher capacity by exploiting the interference rejection capability of spread-spectrum to allow the system to operate at low SIR, making it possible to reuse frequency in every cell. Therefore, in this step, K is assumed to be at most one.

Step 3. Analogous to (5.18), a constraint on the number of available channels can be constrained as shown below

$$N_{ch} \geq \left\lfloor \frac{W_A \eta_{CDMA} \eta_{TDMA} \eta_{FDMA} \eta_{sig} \beta}{\left(\eta_W \bar{\gamma}_{req} + \sqrt{\eta_W \bar{\gamma}_{req} \gamma_M}\right) R_b} \right\rfloor. \tag{5.22}$$

Since an assumption has been made that $K \leq 1$, N_{ch} has no upper limit in theory.

To derive (5.22), first note that according to (5.11) and (5.13), the processing gain and the number of code channels can be expressed as follows:

$$G_p = \frac{W_c \eta_{TDMA}}{R_b N_{slot}} \tag{5.23}$$

$$N_{code} = \frac{W_c N_{ch}}{W_A \eta_{sig} \eta_{FDMA} \eta_{CDMA} N_{slot}}. \tag{5.24}$$

Moreover, the SIR in a CDMA system may be approximated by

$$SIR = \frac{\beta}{N_{code}}, \tag{5.25}$$

where β is the fraction of the total amount of interference introduced by transmissions within a cell. The interference introduced within a cell is also referred to as intra-cell interference and has a typical value of 0.65 in IS-95 [4]. For simplicity, (5.25) assumes perfect power control, i.e. all N_{code} interferers arrive at the receiver with the same power as that of the desired signal. Clearly, a CDMA system is interference limited since SIR is inversely proportional to the number of code channels. Expression (5.23) is valid for direct-sequence spread-spectrum. For systems that employ frequency-hop spread-spectrum, an expression should be determined based on (3.23).

According to (5.25), the system capacity in terms of the number of users supported per cell is inversely proportional to the SIR, which implies that the system capacity can be increased if the required SIR can be lowered. Spread spectrum provides processing gain that enables CDMA to operate at low SIR. Processing gain effectively boosts the receiver SIR to the required level. Other interference mitigation techniques such as voice activity, power control, and cell sectorization can also be applied to increase CDMA capacity.

Finally to derive (5.22), constraint (5.17) is modified to take into account processing-gain as shown below

$$G_p SIR \le \eta_W \overline{\gamma}_{req} + \sqrt{\eta_W \overline{\gamma}_{req} \gamma_M} . \tag{5.26}$$

Substituting (5.23), (5.24) and (5.25) into (5.26) results in (5.22).

Step 4. The channel spacing should satisfy the following constraint

$$\frac{N_{slot} R_b}{\eta_{TDMA}} << W_c \le W_A , \tag{5.27}$$

where the lower limit follows from (5.23) given that the processing gain is usually much greater than unity. Depending on the signal processing constraint, the channel spacing may be chosen based on the coherence bandwidth. To achieve path diversity at the expense of complexity, the channel spacing should be larger than the coherence bandwidth to sufficiently resolve the multipath components. In a direct-sequence spread-spectrum system, the above constraint requires the chip rate to exceed the coherence bandwidth so that a Rake receiver can effectively combat multipath fading. See Section 3.2.4. Frequency-hop spread-spectrum has the additional requirement that the frequency separation between each hop should be at least equal to the coherence bandwidth of the channel.

Step 5. The processing gain and the number of code channels can now be determined based on (5.23) and (5.24), respectively. Numeric adjustments are needed on W_c, η_{CDMA}, η_{FDMA}, η_W, and η_{sig} to obtain integer values for the processing gain and the number of code channels. The processing gain may be further constrained by (5.26).

Step 6. Determine SIR based on (5.25).

Figure 5.11 summarizes the system capacity design outlined in Sections 5.4.2.1 and 5.4.2.2.

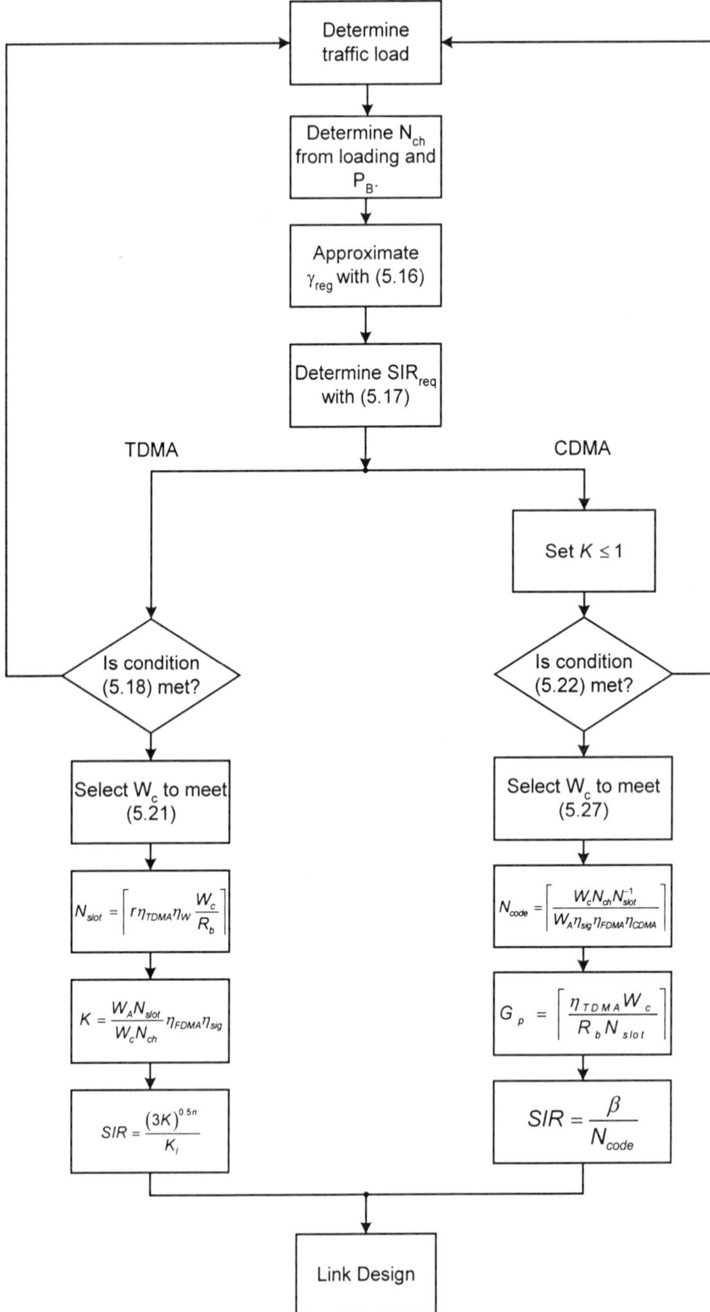

Figure 5.11 System capacity design flow diagram.

5.5 Link Design

The system design determines the parameters related to implementation. These include W_c, SIR, η_c, η_W, η_{sig}, r, G_p, N_{slot}, and N_{code}. However, link parameters must also be determined to derive a complete specification for the underlying radio implementation. This section describes a general design guideline to determine the link-related parameters listed in Table 5.6.

Table 5.6 Link design parameters.

Parameter	Notation	Description
Required E_b / N_0	$\overline{\gamma}_{req}$	The required E_b / N_0 for the complete link
E_b / N_0	$\overline{\gamma}$	The E_b / N_0 when the channel is degraded by noise, distortion, fading, and interference
Noise Margin	γ_M	Margin for noise
Antenna Gain	G_{RX}, G_{TX}	Antenna gain for the receiver (RX) and the transmitter (TX)
Signal Diversity Gain	$G_d^{(s)}$	Diversity gain obtained from signal diversity
Interference Diversity Gain	$G_d^{(I)}$	Diversity gain obtained from interference diversity
Processing Gain	G_p	Processing gain obtained from spread-spectrum
Transmit Power	P_{TX}	Power delivered by the PA to the antenna
Fading Loss	L_{fading}	Loss in E_b / N_0 due to Rayleigh fading
Path Loss	L_p	Loss in E_b / N_0 due to distance
Link Margin	L_M	Loss in E_b / N_0 due to shadowing
Signal-to-Distortion Ratio	SDR	Ratio of signal over distortion power (i.e. margin for distortion)

By applying (2.57), the required E_b / N_0 for a system to achieve a target error rate can be expressed as follows

$$\overline{\gamma}_{req} = \overline{\gamma}\left(\overline{\gamma}_M + \frac{\overline{\gamma}\eta_W}{SDR \cdot G_p G_d^{(I)}} + \frac{\overline{\gamma}\eta_W}{SIR \cdot G_p G_d^{(I)}}\right)^{-1} \tag{5.28}$$

where SDR denotes the signal-to-distortion ratio and $\overline{\gamma}$ represents the effective E_b / N_0 required by the receiver taking into account the noise margin $\overline{\gamma}_M$, SIR, and SDR. In deriving (5.28), the mean interference and distortion levels are assumed to be zero.

The noise margin in (5.28) is simply the noise factor of the receiver

$$\overline{\gamma}_M = 10^{0.1NF} \tag{5.29}$$

where *NF* in (5.29) represents the noise figure of the RF receiver. As discussed in Chapter 4, NF and *SDR* in (5.29) measure the amount of noise and distortion introduced by the RF circuits.

The error rate of a system constrains the value of $\overline{\gamma}_{req}$, which in general should be minimized for high-energy efficiency. A typical value for $\overline{\gamma}_{req}$ is between 2 to 4 for a BER of 0.01% to 0.1%. The minimum $\overline{\gamma}_{req}$ sets the limit on the maximum amount of interference and distortion that a system can handle with a given interference suppression $G_p G_d^{(I)}$, where $G_d^{(I)}$ is defined as the gain obtained from interference diversity. If a system is not interference limited, then the SIR is high and the distortion constraint on the RF circuits can be relaxed. In practice, most systems are interference limited. Therefore, RF distortion must be reduced to meet a given $\overline{\gamma}_{req}$.

The effective E_b / N_0, $\overline{\gamma}$, measures the amount of bit energy needed to overcome losses due to propagation, fading, shadowing, and thermal noise

$$\overline{\gamma} = \frac{P_{TX} G_{RX} G_{TX} G_d^{(s)} \eta_{TDMA}}{L_{fading} L_p L_M N_0 R_b N_{slot}}, \tag{5.30}$$

where $G_d^{(s)}$ denotes the diversity gain obtained from signal diversity (e.g. equalization). Moreover, if the channel exhibits no fading, then $G_d^{(s)}$ becomes the coding gain achieved with coding in an AWGN channel. Here, a distinction is made between diversity gain obtained from signal diversity and that obtained from interference diversity.

Let us now define an excess gain α_γ for the receiver E_b / N_0 as

$$\alpha_\gamma = \frac{\overline{\gamma}}{\overline{\gamma}_{req}} \tag{5.31}$$

and define the required SIR as

$$SIR_{req} = G_p G_d^{(I)} \frac{SIR \cdot SDR}{SIR + SDR}. \tag{5.32}$$

Substituting (5.31) and (5.32) into (5.28), α_γ can be expressed in a different form

$$\alpha_\gamma = \frac{\gamma_M}{1 - \frac{\eta_W \overline{\gamma}_{req}}{SIR_{req}}}. \tag{5.33}$$

Since α_γ cannot be less than zero, the following constraint holds:

$$SIR_{req} \geq \eta_W \overline{\gamma}_{req}. \tag{5.34}$$

Constraint (5.34) implies that α_γ is always greater than or equal to the noise margin:

$$\alpha_\gamma \geq \gamma_M. \tag{5.35}$$

By plotting α_γ against SIR_{req}, Figure 5.12 shows that there is a breakpoint, below which α_γ increases rapidly toward infinity and above which α_γ approaches γ_M. The breakpoint occurs when the slope of the curve is at -1 and is denoted by $SIR_{req}^{(c)}$ with the following expression

$$SIR_{req}^{(c)} = \eta_W \overline{\gamma}_{req} + \sqrt{\eta_W \overline{\gamma}_{req} \gamma_M} \tag{5.36}$$

The parameter $SIR_{req}^{(c)}$ is used to constrain the SIR during the system capacity design, for example, as shown in (5.17).

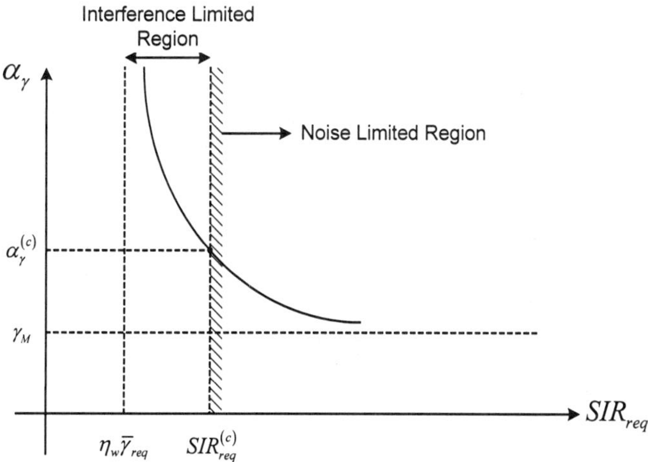

Figure 5.12 Excess gain α_γ required to compensate for noise, interference, distortion, and fading.

For $\eta_W \bar{\gamma}_{req} < SIR_{req} \le SIR_{req}^{(c)}$, the system is interference-limited and for $SIR_{req} > SIR_{req}^{(c)}$ the system is noise limited. In the former case, which occurs for high-capacity systems, a large signal gain is required to compensate for the large α_γ. According to (5.30), this gain can be accomplished by increasing any combinations of P_{TX}, G_{RX}, G_{TX}, or $G_d^{(s)}$ for a given noise power density N_0, bitrate R_b, and channel losses L_{fading}, L_p, and L_M. As discussed in Sections 2.2.3 and 2.2.4, L_p is proportional to R^{-n} where n is the path-loss exponential and L_{fading} is due to multipath fading as discussed in Section 2.2.5. L_M is the margin determined by the coverage analysis discussed in Section 5.4.1.

Given the system parameters determined from the previous sections, one can now determine the link parameters illustrated by the design guideline outlined below.

Link Design Guideline:

Step 1. Set the noise margin γ_M. Expression (5.33) implies that a larger noise margin imposes a higher α_γ, which indirectly increases the required transmit power, antenna gain, and diversity gain. For a system that requires a large coverage area, the propagation loss tends to be high. Therefore, such a system already requires substantial transmit power, antenna gain, and diversity gain to compensate for the propagation loss. Any additional margin stresses the limit of the system either in terms of the transmit power or the amount of signal processing needed to obtain higher diversity gain. For such a system, the noise margin should be set as low as possible. The minimum noise margin is determined by the noise figure of the RF front-end. With careful design, a system noise figure of 5-7 dB can be achieved for a mobile unit.

For systems with a smaller coverage requirement, a larger noise margin can be tolerated since the propagation loss tends to be lower. Examples of this type of system include digital cordless and indoor WLAN. In this case, a reasonable system noise margin ranges from 10 to 15 dB.

Step 2. Set the SDR which measures the amount of distortion introduced by RF circuits. A lower SDR degrades the capacity of the system since the distortion increases the total interference power experienced by the receiver. In general, *SDR/SIR* should be set as high as possible so that degradations due to distortion become negligible. A reasonable starting point would be to set *SDR/SIR* at ten. Several iterations may be necessary to converge on the actual value of *SDR/SIR*, such that it is not too high to implement in practice.

Step 3. Determine $\bar{\gamma}_{req}$ which represents the required E_b / N_0 in an AWGN channel. $\bar{\gamma}_{req}$ can be approximated using (5.16).

Step 4. Determine the excess gain (α_γ) necessary to accommodate for circuit noise, fading, distortion, and interference. Most wireless systems are interference limited because of the requirement for high capacity. Therefore, according to Figure 5.12, α_γ may lie within the interference-limited region and have a relatively high value. A high excess gain can result in excessive transmission power and signal processing complexity. A reasonable compromise is to choose a value for α_γ that corresponds to the breakpoint where $SIR_{req} = SIR_{req}^{(c)}$, as shown below:

$$\alpha_\gamma^{(c)} = \gamma_M \left(1 + \sqrt{\frac{\eta_W \overline{\gamma}_{reg}}{\gamma_M}} \right). \tag{5.37}$$

Step 5. Compute the required interference rejection using

$$G_d^{(I)} G_p = \frac{\eta_W \gamma_{req}}{\left(1 - \frac{\gamma_M}{\alpha_\gamma}\right) \cdot SIR} \left(\frac{1 + \frac{SDR}{SIR}}{\frac{SDR}{SIR}} \right) \tag{5.38}$$

which is derived from (5.33), (5.36), and (5.37). Constraint (5.38) implies that if the actual SIR is less than $\eta_W \gamma_{req} / \left(1 - \gamma_M / \alpha_\gamma\right)$, the amount of interference suppression must be increased to compensate for the difference. As expected, increasing the noise margin or decreasing the excess gain also increases the amount of interference suppression.

Step 6. Determine the losses due to fading and radio propagation:

a. L_M is the loss due to shadowing discussed in Section 5.4.1.

b. L_p is the path loss due to the transmit-receive distance and has a value which is inversely proportional to R^n, where R is the coverage radius and n is the path loss exponential, discussed in Sections 2.2.3 and 2.2.4.

c. L_{fading} is the loss due to small-scale fading discussed in Section 2.2.5. In general, fading loss is difficult to characterize analytically, with the exception of fast Rayleigh fading which has a loss described by (2.49). For frequency-selective fading, the loss can be significantly higher due to ISI. An analytical expression is difficult to obtain. However, the residual fading loss after adequate equalization is better characterized. For example, the worst-case loss when using MLSE is shown in Table 2.10. If a DFE is used then an additional 4-8 dB of loss may occur. Thus, for a frequency-selective channel, the residual fading loss, instead of the absolute fading loss, is used to gauge the amount of additional diversity gain that is needed on top of equalization. The additional gain may be achieved with FEC and interleaving.

Step 6. From (5.30) and (5.31), determine the transmit power P_{TX}, the antenna gains G_{RX} and G_{TX}, and the signal diversity gain $G_d^{(s)}$ based on

$$P_{TX}G_{RX}G_{TX}G_d^{(s)} = \frac{\alpha_\gamma \overline{\gamma}_{req} L_{fading} L_p L_M N_0 R_b N_{slot}}{\eta_{TDMA}}. \quad (5.39)$$

Figure 5.13 summarizes the link design guideline outlined above.

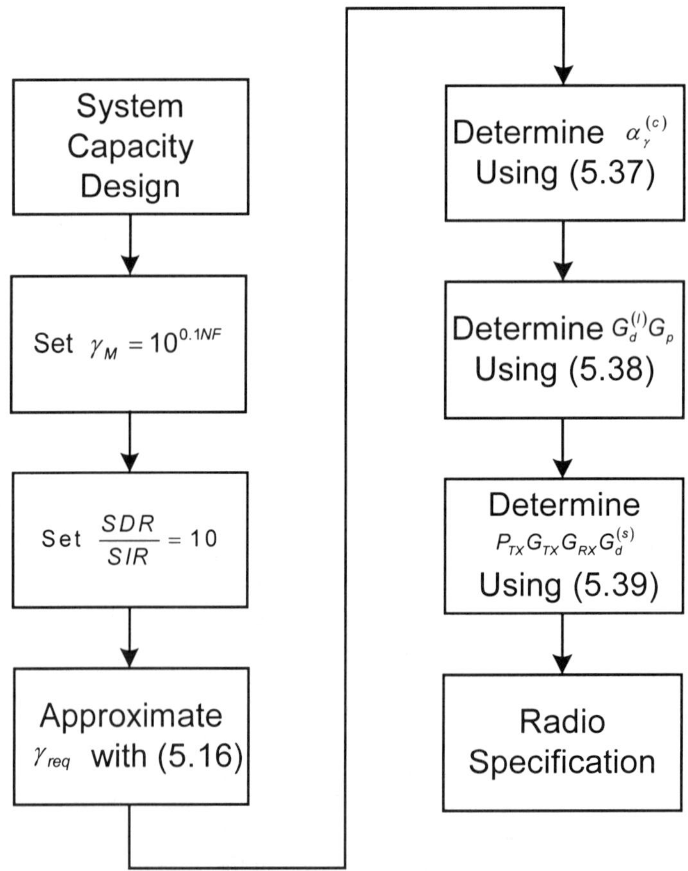

Figure 5.13 Link design flow diagram.

5.6 Implementation Specification

With the link design complete, a set of specifications can now be derived for the digital baseband processor, the RF transceiver, and the associated digital-analog interface.

5.6.1 Radio Frequency Front-end

Table 5.7 lists the main design parameters that can be used to specify the RF front-end. Most of these parameters have already been discussed in Chapter 4. This section discusses the RF parameters in the context of the constraints imposed by the link design as well as regulations.

Table 5.7 RF Design parameters.

RF Parameters	Notation	Description
Noise Figure	NF	Ratio of total input noise power to the thermal noise at the receiver input
Out-of-Band Emission	P_{OBE}	Amount of out-of-band emission at the transmitter
Input Compression Point	P_{1-dB}	The input power at which the desired output power drops by 1-dB below its ideal value
Input IP3	$IIP3$	The input power at which the 3rd order intermodulation distortion power equals the desired signal power
Receiver Selectivity	S_e	Receiver's ability to reject out-of-band signals
Phase Noise	$\overline{\phi^2}$	Noise due to phase jitter in RF circuits
Quality Factor	Q	Measures the resonance of an RF circuit
Sensitivity	S	Received signal power at which the error rate meets a given specification
Dynamic Range	$SFDR$	Measures the tolerable power fluctuation at the receiver input.
Antenna Size	A_{ant}	Effective area of the antenna

The **Noise Figure** is constrained by the noise margin according to (5.29). The noise figure is defined as the ratio of total input-referred noise power to the thermal noise power. The lower limit on noise figure is one since thermal noise always exists and cannot be eliminated. In practical circuits, the system noise figure ranges between 5-15 dB. A system noise figure less than 5-dB is extremely difficult to achieve for systems that require high dynamic range because of the trade-off between gain and distortion, which has been discussed in Chapter 4. Also, unless the antenna noise temperature is much less than the ambient temperature, the antenna noise can significantly degrade the system performance. A 10-dB noise figure is fairly easy to achieve but the price paid is loss in range. In general, a reasonable design specification includes a 6-dB noise figure for systems that require both high dynamic range and relatively large coverage area, e.g. greater than 1km.

The **Out-of-Band Emission** P_{OBE} measures the RF emission outside of the specified signal band. Out-of-band emission occurs both in the aggregate band and in the frequency subchannels as shown in Figure 5.14. In the former case, P_{OBE} must be constrained to meet regulatory requirements and in the latter case, it must be kept low enough so as not to cause distortion on the desired channel. Spectral regrowth due to non-linearity in power amplifiers and phase noise are dominant sources of out-of-band emission. A given out-of-band emission imposes constraints on power amplifier linearity, phase noise of the RF circuitry, and bandwidth efficiency of the digital modulation used.

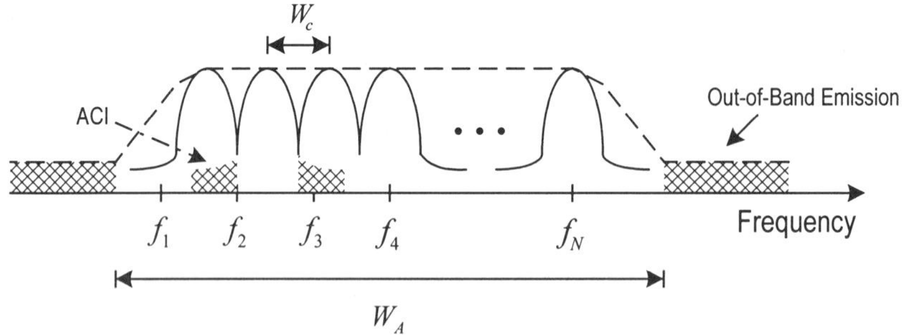

Figure 5.14 Out-of-band emission

The **input 1-dB compression point** $P_{1\text{-}dB}$ measures the input power at which the distortion power begins to reduce the desired signal level by 1 dB at the output of the receiver or transmitter. A large $P_{1\text{-}dB}$ means that the RF front-end circuitry can tolerate large signal levels before its output shows significant distortion. In general, $P_{1\text{-}dB}$ is an important measure for the transmitter, more so than the receiver. However, for CDMA systems which allow concurrent transmissions to occur in the same frequency band, a high $P_{1\text{-}dB}$ is required for the receiver front-end because of the large power fluctuations. In particular, it is quite likely that an undesired signal may have a significantly higher signal level than that of the desired signal due to the near-far problem. The undesired signal power increases the input power level, driving the receiver circuitry into compression. The SDR sets the limit on the maximum $P_{1\text{-}dB}$ that the radio can tolerate before the associated distortion begins to degrade the system performance.

The **input third-order intercept point IIP3** measures the input power at which the distortion power due to nearby channels equals the desired signal level at the output of the receiver. Like $P_{1\text{-}dB}$, IIP3 also measures the receiver linearity but in the presence of intermodulation distortion due to nearby channels. It is expected that the adjacent channels may have a large signal level compared to that of the desired signal, resulting in large adjacent channel interference (ACI). Large ACI results in high distortion or a need for high IIP3. Therefore, in a practical system, the combined effects of $P_{1\text{-}dB}$ and IIP3 should not exceed the specified SDR for satisfactory system performance.

The **Filter Selectivity** S_e is measured by the sharpness of the filter transition and the amount of stopband attenuation. The amount of distortion in the receiver can be reduced with tighter filtering to select the aggregate band as well as the frequency subchannels. The required SDR again influences the amount of selectivity needed, given the amount of ACI expected in the system.

The **Phase Noise** of RF circuits (e.g. mixers and local oscillators) introduces frequency-dependent noise, which has a high power density near the carrier frequency. The density decreases with frequency away from the carrier but can still generate distortion on nearby channels via spectral leakage and reciprocal mixing, discussed in Chapter 4. Phase noise also degrades the noise figure of a receiver due to the non-zero noise floor generated on nearby channels. The effect of phase noise together with distortion from other sources described above should be constrained such that the required NF and IIP3 are met.

The **Quality Factor** Q measures the amount of resonance needed in the RF front-end to operate over a given available bandwidth W_A and carrier frequency f_c. The quality factor may be approximated by W_A / f_c. A larger Q imposes low loss on the tuned elements in RF circuits. For most current commercial wireless systems, Q lies between 20 and 70.

The **Sensitivity** is defined as the minimum received power at which the receiver meets a given $\bar{\gamma}_{req}$ and is determined from the link design using (5.39)

$$S = \frac{\bar{\gamma}_{req} \gamma_M N_0 \cdot (uncoded\ bitrate)}{G_c}. \tag{5.40}$$

assuming no fading and interference. The losses L_{fading} and L_M become one in the absence of fading, the excess gain becomes γ_M in the absence of interference, and the signal diversity gain becomes the coding gain G_c when noise is the sole source of disturbance. The term in parenthesis in 5.40 represents the uncoded bitrate sent in the channel and refers to the rate of the actual user data, such as speech. With TDMA, the uncoded bitrate can also be represented by $R_b N_{slot} \eta_{TMDA}^{-1}$.

Generally, the receiver sensitivity ranges anywhere from –110 dBm to –75 dBm, where the larger S value indicates lower sensitivity and characterizes receivers that have a short-range requirement, such as Bluetooth with a range of less than 10 m. For other short-range systems, such as RF tags, the sensitivity may be as low as –25 dBm.

Often, the excess gain α_γ is set to unity when computing the sensitivity of the receiver since a static AWGN channel with no interferer is assumed.

The **Dynamic Range** can be measured by the SFDR, defined as the ratio of IIP3 to S; see Chapter 4. It measures the range of signal power over which the receiver maintains linearity and still achieves the desired $\bar{\gamma}_{req}$. In a mobile environment, the SFDR is an important design parameter that determines the receiver's ability to accommodate signals with large power fluctuations; i.e. the near-far problem.

The near-far problem results from the fact that in a mobile environment, locations of the transceivers are not fixed relative to the receiver. Therefore, often a strong nearby transmission may overwhelm the desired signal from a transmitter farther away. The largest power differential between the undesired and desired transmitter power marks the dynamic range requirement of the system. The dynamic range requirement can be determined by the ratio $R_{\max}^{n_1} R_{\min}^{-n_2}$, where n_1, n_2, $R_{\min}$, and $R_{\max}$ are, respectively, the path loss exponential for the far-end transmitter, the path loss exponential for the near-end transmitter, the minimum range for the near-end transmitter, and the maximum range for the far-end transmitter. Note also that because of the range differences, in general $n_2 \le n_1$. To mitigate the near-far problem, the transmission power control circuit should be designed to handle the required dynamic range and the receiver front-end should have a sufficiently large IIP3 and $P_{1\text{-}dB}$.

Antenna Size is determined by the antenna gain derived in the link design according to (5.39).

5.6.2 Digital-Analog Interface

In a similar fashion, link design parameters and regulatory constraints, such as γ_{req}, SDR, W_A and W_c also confine the design parameters associated with the digital-analog interface circuits shown in Table 5.8.

Table 5.8 Digital-analog interface design parameters.

Baseband Parameters	Notation	Description
RX sampling rate	$F_s^{(RX)}$	Digital receiver sampling rate
TX sampling rate	$F_s^{(TX)}$	Digital transmitter sampling rate
RX input resolution	N_{in}	Digital receiver input resolution
TX output resolution	N_{out}	Digital transmitter output resolution
AGC Dynamic Range	D_R	Dynamic range of the AGC

The **RX Sampling Rate** corresponds to the conversion rate of a digitized down-converted RF signal. A receiver digitizes the down-converted RF signal either at the intermediate frequency (IF) or at baseband. With low-IF sampling the sampling rate is on the order of W_A and with high-IF sampling the sampling rate is on the order of f_o. With current commercial systems, IF sampling requires 1-100 MHz while digital baseband sampling requires 1-4 MHz. In third-generation systems, the IF and baseband sampling rates are increased to 500 MHz and 4-64 MHz, respectively. The RX sampling rate determines the conversion rate of the ADC used to digitize the downconverted RF signal.

The **TX Sampling Rate** corresponds to the conversion rate of a digitally modulated signal to an RF signal. Similar constraints mentioned above apply to the transmitter sampling rate. The difference lies in that digitization occurs at the output rather than the input.

The **RX Input Resolution** measures the number of bits used to quantize the digitized RF front-end output. The RX input resolution N_{in} is equivalent to the effective number of bits in the ADC used to digitize the down-converted RF signal. Having an inadequate number of bits, the receiver will experience an unacceptable level of quantization noise that degrades the required error-rate performance. This implies that $\bar{\gamma}_{req}$ is not met for a given service. In general, the receiver input resolution lies between 1-12 bits. A lower bit resolution could be used but would require higher sampling rate to compensate for the loss in SNR.

The **TX Input Resolution** measures the number of bits used to represent the digitally modulated waveform at the transmitter. Constraints similar to the ones described for the RX input resolution also apply to the TX input resolution. The amount of resolution affects mostly the amount of distortion introduced on the modulated waveform and is therefore constrained by the SDR specification. The distortion due to quantization can be measured in terms of the mean squared error with respect to the ideal unquantized signal.

The **AGC Dynamic Range** measures the range of gain control needed in the RF receiver to adjust the RF signal amplitude to within the ADC input swing. Without automatic gain control (AGC), the full range of the ADC may be exceeded or not fully utilized. Thus, in order to accommodate the dynamic range of the front-end circuitry, the ADC input must be conditioned properly such that the swing at the output of the front-end circuitry can be adjusted to within the operation range of the ADC. The AGC circuit provides the range of gain control necessary to accommodate the wide variation in RF signal power. In general, the total dynamic range provided by the ADC and AGC should be large enough to handle the SFDR of the complete receiver.

5.6.3 Digital Transceiver

The link design parameters and regulatory constraints confine the digital transceiver design parameters shown in Table 5.9.

Table 5.9 Digital transceiver design parameters.

Baseband Parameters	Notation	Description
Constellation Size	M	Constellation size of the digital modulation
Mod. Minimum Distance	$d_{\min}^{(m)}$	Minimum distance of the digital modulation
Interleaver Size	mdS	Size of the interleaver in coded bits
Code Rate	r	Rate of forward-error correcting code
Code Minimum Distance	$d_{\min}^{(c)}$	Minimum distance of the FEC
Constraint Length	K_c	Constraint length of the convolutional code
Diversity Branches	K_D	Number of diversity branches
Chip Rate	R_c	Chip rate of DS spread-spectrum
Hop Rate	R_h	Hope rate of FH spread-spectrum
Number of Hops	N_f	Number of hops in FH spread-spectrum
OTA Message Duration	T_{OTA}	Over-the-air message duration
Sync Time	T_{sync}	Time required for receiver synchronization
Noise Bandwidth	B_L	Noise bandwidth of synchronization loops
Training Length	L_t	Length of the training sequence

The **Constellation Size** is determined by the bandwidth and energy efficiency of the system. As discussed in Section 2.1.6, for an energy-efficient system the constellation size should be either binary or quaternary. However, for a short-range wireless system, a larger constellation size may be used. The loss in energy efficiency due to the increase in $\overline{\gamma}_{req}$ is tolerated given the lower path loss associated with shorter distance. Also if a non-linear amplifier is used, then modulation efficiency should account for spectral regrowth discussed in Section 2.4.5.2.

The **Modulation Minimum Distance** is constrained by $\overline{\gamma}_{req}$ according to (2.110) for a given error rate.

The **Interleaver Size** is constrained by the required signal and interference diversity gain determined in the link design. A portion of the diversity gains can be achieved via interleaving combined with coding as discussed in Section 2.7.4. In particular, interleaving proves effective for time-selective channels. For best result, the interleaver size is chosen large enough to capture an error burst due to fading according to (2.86). However, the interleaver introduces a delay that is proportional to its size and therefore must be designed with a delay that does not exceed the required end-to-end delay. Other processing delays, such as vocoder and traceback delays should also be taken into account as part of the end-to-end delay. For time-flat but frequency-selective channels, where the coherence time is long, frequency-hop spread-spectrum can be employed to reduce the required interleaver size and still obtain effective diversity gain with a relatively low interleaver delay.

The **Code Rate** is determined based on (5.18) in the system capacity design. This value should be reconciled with the requirements on the interleaver size and the bandwidth efficiency. When the channel is AWGN, such as a microwave point-to-point link, the code rate is largely constrained by the coding gain defined in (2.46). Therefore, in an AWGN channel, the diversity gain requirement is essentially replaced by a corresponding requirement on coding gain. This occurs because an AWGN channel has zero diversity order and thus offers no diversity gain.

The **Code Minimum Distance** is constrained by the diversity gain when the channel is degraded by fading according to (2.194) or by the coding gain when the channel is degraded by AWGN. See (2.46).

The **Constraint Length** of the convolutional code is constrained by the required coding gain or diversity gain. With a longer constraint length, convolutional codes tend to have a larger minimum distance, also known as the free distance, as shown in Table 2.9. In general, the coding gain of a convolutional code is proportional to the constraint length of the code.

A **Diversity Branch** processes information that has experienced independent fading statistics. The number of diversity branches needed depends on the type of equalization. For a Rake receiver, the number of branches necessary to meet a given diversity gain can be approximated using (2.48). For MLSE and DFE, the number of branches is related to the memory of the channel or the length of the channel impulse response L_c. Maximum diversity gain is achieved only when the number of states equals M^{L_c-1} for MLSE and the number of feedback taps is greater than $L_c - 1$ for DFE.

The **Chip Rate** of a direct-sequence spread-spectrum system is constrained predominantly by the degree of multipath resolution. Given a frequency-selective channel, the coherence bandwidth determines the chip rate as discussed in Section 5.4.2.2. In general the delay spread of the channel varies between 100 ns to 20 μs. The chip rate therefore may need to be as high as 100 Mchips/sec for short delay spreads. The high chipping rate increases the speed requirement of the digital processing. Moreover, it also requires a high-speed ADC and DAC to interface the despreading and spreading digital circuits to the RF front-end. Both high-speed ADC/DAC and digital processing increase the power and complexity of the overall receiver. Frequency hop radios, on the other hand, do not require high speed processing though they do require a frequency agile oscillator if fast hopping is needed. See the discussion on hop rate below.

The **Number of Chips** in a direct-sequence spread-spectrum system is determined by the processing gain derived during the system capacity design. The number of chips is related to processing gain according to (3.7).

The **Hop Rate** of a frequency-hop system is constrained mostly by the fading rate of the channel. The hop rate is proportional to the symbol rate according to (3.22). Having a faster hop rate improves the system performance in a slowly varying channel, provided that the hop spacing is greater than the coherence bandwidth. In particular, the size of the interleaver can be reduced substantially since each hop takes shorter time. However, a fast hop rate also requires a short settling time for the frequency synthesizer, making it harder to implement.

The **Number of Hops** in a frequency-hop system is determined by the processing gain derived from the system design according to (3.23). For an FFH system, the limitation on the number of hops by the available bandwidth W_A becomes more pronounced.

The **Over-The-Air Message Size** is defined as the duration of the message that is actually sent over the air. Section 2.1.3 has described a number of message units in terms of bits, symbols, frames, and packets. Different services impose different requirements on error rates of these message units, i.e. BER, SER, FER, or PER. BER and SER are generally used to characterize the performance of the physical layer while FER and PER are used to measure the performance of a given service. In realtime services such as video and voice, the message units are typically referred to as frames while in non-realtime services, the message units are typically referred to as packets. The PER and FER depend on both the noise level as well as the size of the packets or frames as shown in (2.32) and (2.33).

In a Rayleigh fading channel, the error tends to be bursty such that (2.32) and (2.33) no longer hold. Therefore, they cannot be utilized to design the frame or packet size. However, it is still true that the PER and FER are lower for smaller packet or frame sizes. This follows intuitively from the observation that if the message size were much smaller than the coherence time of the channel, the channel would appear to be static. Most of the packets may lie outside of a fade and have good error rate performance. The performance degradation for the packets that lie within a fade can be compensated via diversity combining such as interleaving. While burst error within a large packet or frame can still be handled via interleaving, the delay introduced often becomes unacceptable. A general rule of thumb to achieve reasonable diversity gain with low delay is to set the over-the-air (OTA) message duration, denoted by T_{OTA}, to be 10% of the reciprocal of the maximum Doppler frequency [163]:

$$T_{OTA} \approx \frac{0.1}{\max f_d}. \tag{5.41}$$

Example 5.4 In a TDMA system, a set of time slots is grouped into a larger unit, referred to as a TDMA frame. Each user is assigned a time slot within a frame. A speech frame for a given user is partitioned into multiple time slots for transmission on multiple TDMA frames. To combat fading, the time slots are chosen according to (5.41) and are interleaved across multiple TDMA frames to obtain diversity gain. In this case, the over-the-air message unit is the individual time slot. In GSM, the carrier frequency is 900 MHz or 1.8 GHz. At vehicle speeds of 120 km/hr the maximum Doppler is 100 Hz for a 900 Hz carrier and 200 Hz for a 1.8 GHz carrier. According to (5.41), the OTA message size is then 1 ms and 500 μs, respectively. The computed slot times agrees with the actual slot size of 577 μs used in GSM. □

The **Sync Time** represents the time required by the receiver to synchronize onto the transmitted message. Once the slot size has been determined, the channel overhead can be computed using the channel efficiency η_c derived from the system design as shown below

$$\text{Channel Overhead} = T_{OTA}\eta_{TDMA}. \tag{5.42}$$

A portion of the channel overhead is assigned to the synchronization of symbol timing and carrier phase. However, if the system has an out-of-band signaling channel dedicated to synchronization, as in most cellular systems, then the Sync Time does not add to the channel overhead. Sync Time has a larger impact for packet radio systems in which synchronization may be required on a per packet basis.

The **Noise Bandwidth** of a synchronization loop is constrained by the Sync Time and the error-rate performance. Although clearly a shorter Sync Time helps to reduce the channel overhead, the Sync Time cannot be made arbitrarily small. Referring to Section 2.5, the time required for synchronization varies inversely with the noise bandwidth. Therefore, the Sync Time cannot be reduced beyond the point where noise begins to impair the error-rate performance.

The **Training Length** affects the TDMA efficiency and therefore the channel overhead. While a short OTA message size together with interleaving helps to mitigate the effects of Rayleigh fading, it does very little to improve the performance in the presence of frequency-selective fading. In the latter case, equalization is required to achieve the diversity gain. Recall the discussion from Section 2.7.7 that in order for the equalizer to eliminate the effect of ISI, the channel impulse response (CIR) must be estimated. In most practical methods, a training sequence is used for this purpose and should be at least as long as the CIR for an accurate estimate. With an LMS-based channel estimator (2.203) the training sequence may be 10-100 times as long as the CIR. Therefore, the LMS-based estimator is generally applied for a quasi-static channel where the OTA message size can be made larger to accommodate the long training sequence. For mobile channels, a direct estimate based on (2.196) should be used.

5.7 Summary

Figure 5.15 summarizes the radio design flow. At the start of the design, system constraints are identified and defined. System parameters are then designed to meet the given constraints. System parameters include the blocking probability, the required SIR, the reuse factor, and the number of physical channels. The next step in the design flow is to determine the link parameters that satisfy the system constraints. These link parameters include the required E_b/N_0, the transmit power, the antenna gain, and the diversity gain. Finally, the system and link parameters specify constraints on the implementation parameters such as chip rate, hop rate, noise figure, antenna gain, and diversity gain. It may be possible that the system constraints are too stringent for practical implementations in terms of area, size, and power. Therefore, a few iterations may be necessary to finally converge on a set of implementation specifications that will meet a given set of system requirements. For example, as shown in Figure 5.15, if the actual SIR is too low requiring unrealistically high interference suppression, the system could be redesigned with a smaller coverage radius to achieve the system capacity with a more relaxed interference suppression.

While this chapter has described a systematic approach to the radio design, it would be unrealistic to expect that all radio designs will follow the depicted approach. For instance, the system capacity design discussed in Section 5.4 is applicable mainly for a cellular system providing voice services. The capacity design would have involved a different set of trade-offs for data services or other types of network topology such as peer-to-peer networking. However, irrespective of the type of system, a set of principles have been described that can be applied to guide the design of a radio from system concept to final implementation.

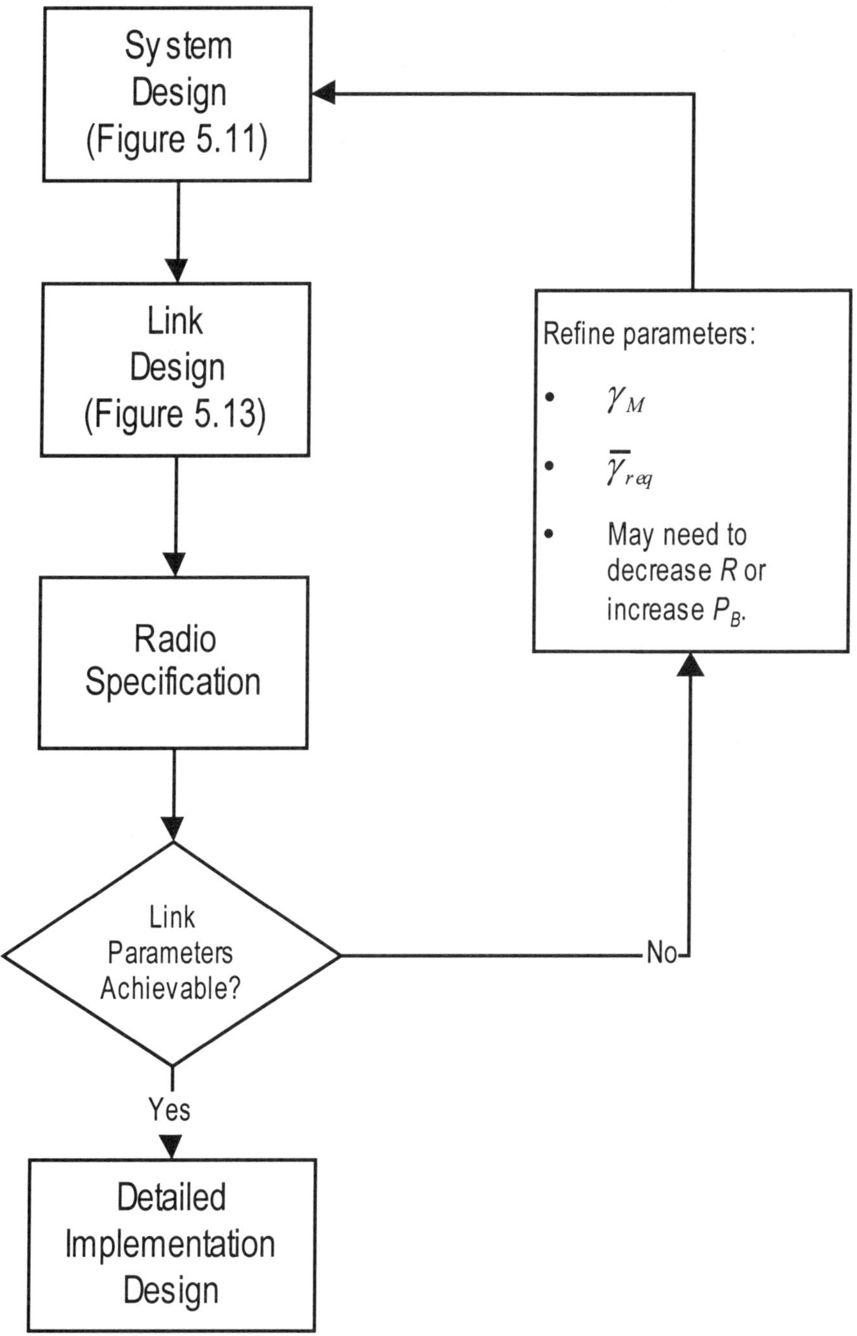

Figure 5.15 Summary of the radio design flow.

CHAPTER 6

Digital Radio Architecture

Up until the early 80's, radios were implemented completely in analog, using predominantly FM. A general FM radio transceiver is depicted in Figure 6.1. The RF signal consists of carrier centered at ω_0 and with sidebands generated by $m \cdot i(t)$

$$s(t) = A\cos\left(\omega_0 t + m\int i(t)dt\right) \tag{6.1}$$

where m is the modulation index and $i(t)$ is the transmitted information. At the transmitter, $i(t)$ can be frequency modulated using a VCO that is locked by means of a phase-locked loop (PLL) using $i(t)$ as the reference signal. Like the Costas loop, a PLL may be represented as shown in Figure 2.51. The VCO may be running at the RF carrier frequency ω_0 or at some intermediate frequency (IF). In the latter case, addition up-conversion is needed to convert the IF signal up to ω_0. The RF signal is then bandpass filtered and radiated from the antenna. At the receiver, the received signal (6.1) is first down converted to either DC or IF, corresponding to a direct-conversion receiver (homodyne) or superheterodyne receiver, respectively. Although a PLL approach can be used to demodulate the signal as well, a simpler implementation based on a discriminator is generally used. The discriminator effectively differentiates the down-converted signal such that $i(t)$ resides on the signal envelope that can then be extracted with an envelope detector.

All of the transceiver processing mentioned above for a conventional analog system is done in the analog domain. In contrast, a digital radio transceiver performs functions in both digital and analog domains as shown in Figure 1.3, where the digital transceiver performs functions such as digital modulation and the RF transceiver performs analog up/down conversion. An ADC and DAC interface ties the digital and analog processing together. A critical design choice for digital radio architecture lies in finding an appropriate partition between digital and analog processing. In other words, where should one place the ADC or the DAC.

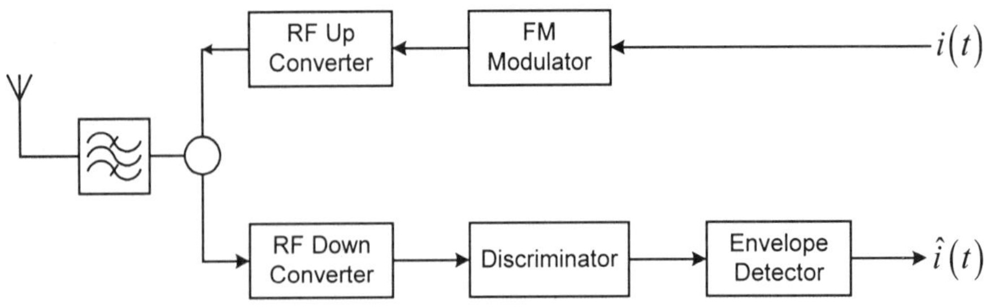

Figure 6.1 An FM radio transceiver.

This chapter addresses the issue of digital-analog partitioning and gives an overview of the architecture design trade-offs that exist in the RF front-end and the digital backend. The block that handles the digital processing is often referred to as a baseband processor since it typically handles the signal processing at baseband. However, current trend is to process digitally at IF or even RF. Therefore, a baseband processor has become a misnomer. To maintain consistency of terms, we refer to the digital backend as a digital transceiver. Most of the discussions in this chapter focus on design issues related to the partition between an RF front-end and a digital transceiver in the user terminal and not in the infrastructure equipment. However, many of the issues discussed can easily apply to the more complex infrastructure equipment, such as a basestation.

6.1 RF Front-End Architecture

The RF front-end mainly serves as a frequency translator between the antenna and the digital transceiver. An ideal RF front-end has the following characteristic

$$y(t) = \Re\left\{h(t) * x(t)e^{\pm j\omega_0 t}\right\} \tag{6.2}$$

where $x(t)$ and $y(t)$ are the input and output of the RF front-end, respectively. The transmitter uses $e^{+\omega_0 t}$ for up conversion and the receiver uses $e^{-j\omega_0 t}$ for down conversion. Filtering represented by $h(t)$ controls out-of-band emission at the transmitter and rejects out-of-band interference at the receiver. Expression (6.2) characterizes a perfect frequency translator. However, as discussed in Chapter 4, (6.2) is never achieved in practice due to non-linear distortion and noise found inherent in the RF front-end. While much of the distortion and noise result from the underlying circuits, RF transceiver architecture determines. RF transceivers may be based either on the superheterodyne architecture or the homodyne architecture.

6.1.1 Homodyne Receiver

The most direct implementation of an RF receiver is based on the homodyne architecture [139][140][164], also referred to as the direct-conversion receiver (DCR) described in Chapter 4. A DCR is also sometimes referred to as a zero-IF receiver.

The key advantage of a DCR lies in its simplicity that could result in a low cost and low power solution. As shown in Figure 4.1b, a DCR requires only ten key elements, namely: an RF filter, LNA, quadrature oscillator, phase shifter, two mixers, two lowpass filters, and two variable gain amplifiers (VGA), all of which with the exception of the RF filter could be integrated on-chip. For instance, since a DCR directly converts the RF signal down to baseband, an analog lowpass filter can be more easily implemented on-chip due to the lower Q requirement. Moreover, a DCR does not have the image problem that is inherent in a superheterodyne receiver as discussed in Chapter 4. Therefore, the bandpass filter at the receiver front-end, which is an off-chip component, could be eliminated. However, for most wireless systems, a bandpass filter is still needed to meet the high dynamic range requirement, in some cases up to 100 dB.

For high performance systems where a high dynamic range and low noise are necessary, it is difficult to meet both requirements in one stage, wherein lies one of the disadvantages in a DCR. Other disadvantages include its high sensitivity to LO re-radiation and $1/f$ noise. Perhaps, the most difficult problem associated with a DCR is LO re-radiation, which refers to the unwanted radiation of the LO signal from the synthesizer circuit. The re-radiation can be significant since the passband of the RF filter is centered on the LO frequency. Therefore, the RF filter provides essentially no isolation on the re-radiated LO signal which not only causes interference to other radio links but also generates self-interference when the re-radiated signal returns to the original receiver due to multipath fading. The worst multipath is due to near-field reflectors that have near perfect reflection coefficients, e.g. $|\Gamma| = 1$, where Γ is defined in (2.65). These reflections arrive at the receiver with minimum loss and generates substantial DC offset when they are re-mixed with the LO. Figure 6.2 illustrates the LO re-radiation problem.

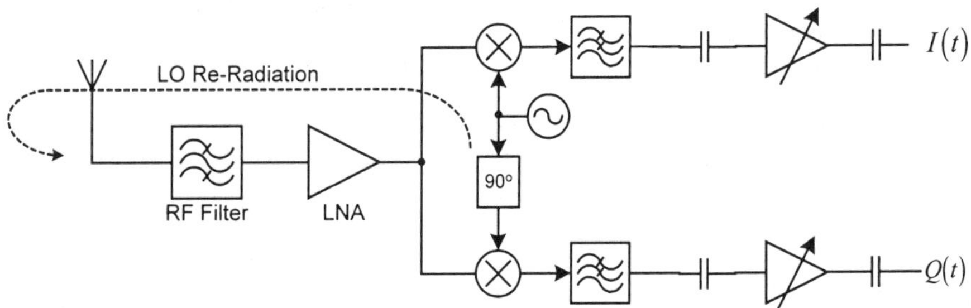

Figure 6.2 LO re-radiation in a DCR.

Example 6.1 Consider the case in which the LO signal in a DCR outputs 3 dBm power at 1 GHz. The LO may re-radiate via the LO-RF port through the LNA out the RF filter and finally to the antenna. Let the LO-RF isolation be 20 dB, the reverse isolation of the LNA (i.e. S_{12}) be 20 dB, and the antenna gain be 0 dBi. Then, the LO signal re-radiates out of the antenna at –37 dBm. Assume that there is a near-field reflector 1 meter away with $|\Gamma| = 1$. Applying (2.60), the path loss is 32.4 dB. The reflected signal arriving at the antenna has twice this path loss and generates a carrier power of –101.8 dBm.

The input-referred DC offset in μV can be computed by assuming a 50 ohms input load using the following expression

$$V_{rms} = 10^{0.05 P_{dbm} + 5.35} \qquad (\mu\text{V}). \tag{6.3}$$

Applying (6.3), with P_{dbm} of –101.8 dBm, the resulting DC offset is 1.82 μV. If the receiver sensitivity is measured to be –105 dBm (or 1.26 μV), then the re-radiation has reduced the receiver dynamic range by 3.2 dB. Note that the loss in dynamic range is directly proportional to the amount of isolation provided by the receiver circuit between the LO and any radiating element, antenna being the predominant EM radiator. In practice, the amount of isolation is largely a function of circuit topology and layout. The isolation could be much worse than the numbers mentioned earlier in this example. For instance, if the isolation were 20 dB less, then the reduction in dynamic range would be 23.2 dB. □

A second source of DC offset is from the second-order non-linearity in circuit elements that behaves like a squaring function. Given that the LO frequency equals the carrier frequency, any signal passing through a second-order non-linearity appears at the output of the receiver as a DC component plus a second harmonic. The latter can be filtered out but the former is much more difficult to eliminate because the baseband signal generally has significant spectral content near DC. Therefore, a large IIP2 is needed to minimize the effect of a second-order non-linearity.

Example 6.2 Consider the same system as in Example 6.1 with a strong received signal of –20 dBm. Then, according to (4.21), the resulting input-referred distortion power is –40 dBm – IIP2. If IIP2 is 50 dB, which is reasonable large, then the input-referred distortion power becomes –90 dBm. The SDR is 70 dB. The corresponding input distortion voltage is 7.1 μV. If the gain of the receiver were 60 dB, then 7.1 mV would result at the output of the receiver. The large offset voltage could cause a loss in bit resolution at the input of the ADC and should therefore be reduced by using offset cancellation circuitry or AC coupling capacitors. □

Currently, several methods have been tried to overcome the DC offset problem. An obvious solution is to apply a large enough AC coupling capacitor so that the DC offset is blocked by the coupling capacitor. To have negligible loss in the desired signal power, the capacitor value should be chosen large enough such that the corner frequency is at most 1 % of the signal bandwidth [139]. For PSK and QAM which have a strong DC content, line coding may be used to introduce correlation in the modulated signal so that DC energy is diverted to higher frequencies. For MFSK, direct AC coupling could be used since MFSK signals have little energy around DC.

Unfortunately, unless the symbol rate is extremely high, the size of the AC coupling capacitor tends to be large, on the order of ηF or μF. The large capacitor makes it impractical for single-chip integration. Moreover, the large capacitor causes long recovery time due to the large time constant, which can be on the order of ms. The long recovery time results in substantial over-the-air overhead, especially, for packet-switched systems.

Example 6.3 Consider a TDMA system with a symbol rate of 200 kbaud. To reduce DC offsets, AC coupling capacitors are inserted at the output of the DCR as shown in Figure 6.2. Assume that the output drives a 1 Mohms load, typical for low-frequency CMOS input impedance. Then, to achieve a corner frequency of 200 Hz, a 0.8 nF capacitor is needed:

$$C = \frac{1}{2\pi R f_c} = \frac{1}{2\pi 10^6 \cdot 200} = 0.8 \cdot 10^{-9}.$$

The time constant associated with this AC capacitor is 796 μs. Note that the above calculations assume a 0.1% corner frequency. □

Example 6.4 Consider a CDMA system with a symbol rate of 16 Mcps. Again, AC coupling capacitors are inserted at the output of the DCR as shown in Figure 6.2. Assume that the output drives a 1 Mohms load. Then, to achieve a corner frequency of 16 kHz, a 9.95 pF capacitor is needed:

$$C = \frac{1}{2\pi R f_c} = \frac{1}{2\pi \cdot 10^6 \cdot 16 \cdot 10^3} = 9.95 \cdot 10^{-12}.$$

The time constant associated with this AC capacitor is approximately 9.95 μs. Thus with a wideband system, both the size of AC coupling capacitors and their associated time constants can be made smaller. □

Another straightforward method to minimize DC offset is to use differential circuits throughout the RF receiver. Differential circuits are balanced and therefore tend to have high IIP2. Also, DC offsets due to re-radiation most likely has equal power on both signal lines of the differential pair and therefore can be cancelled out. However, the disadvantage of using differential circuitry is that twice as much hardware is needed, resulting in higher power dissipation.

Finally, the DC offset may be estimated and cancelled either digitally or with specialized analog circuits [165][166][139]. Offset cancellation relies on the availability of system idle time, during which the receiver estimates the offset value needed for offset cancellation. Digital offset cancellation may require additional ADC resolution to provide sufficient dynamic range since cancellation does not occur until after the ADC. The optimum solution would implement compensation in both the digital and the analog domain.

In summary, though DCR offers the potential for a very simple RF receiver implementation that lends well to single-chip integration, it has several problems that make it difficult to realize in practice. Specifically, it requires extremely linear LNA and mixer, a quadrature LO that must operate at the carrier frequency, and high isolation of the LO signal. Also, because filtering of individual channels does not occur until after the mixer, any strong interferer, such as the alternate channel interference can generate sufficient IMD that reduces the dynamic range of the receiver. Current commercial products that implement DCR are typically for low-end systems. Some examples are chip-sets for digital cordless telephone [167][168] and pagers [169]. Higher performance front-ends using the homodyne architecture are beginning to emerge. For example, researchers in Europe have demonstrated a DCR for wideband CDMA [170] and researchers at Bell Labs have demonstrated a DCR for a 5-GHz WLAN [171]. Table 6.1 summarizes the pros and cons associated with the DCR architecture.

Table 6.1 Trade-offs associated with DCR.

Advantages	Disadvantages
• No image reject required	• High dynamic range requirement
• Simple => low cost & power	• High IIP2 requirement
• Ease of on-chip integration	• Sensitive to LO re-radiation
• Channel select at baseband	• Sensitive to $1/f$ noise
	• High frequency synthesizer
	• Sensitive to DC offsets.

6.1.2 Super-Heterodyne Receiver

A superheterodyne receiver uses multiple IF stages to down convert the RF signal to baseband. Typical superheterodyne receivers use only one IF stage or dual conversion, as shown in Figure 6.3, to avoid the added cost associated with additional stages. A dual conversion superheterodyne receiver can be represented by

$$y(t) = \Re\left\{x(t)e^{j\omega_0 t}e^{-j(\omega_0-\omega_{IF})t}e^{-j(\omega_{IF}-\omega_c)t}\right\}, \tag{6.4}$$

where ω_{IF} and ω_c represent the IF frequency and desired channel frequency in rad/sec. The $\omega_0 - \omega_{IF}$ term is the LO frequency for the IF stage (i.e. OSC1 in Figure 6.3) and the $\omega_{IF} - \omega_c$ term represents the LO frequency required for channel selection, i.e. OSC2 in Figure 6.3. Expression (6.4) denotes down conversion using high-side injection, meaning the LO frequency is injected above the frequency being down converted. A similar expression may be obtained for low-side injection where the LO frequency is injected below the frequency being down converted.

In contrast to DCR, the dual-conversion superheterodyne is more complex and has five addition blocks: the image reject (IR) filter, OSC1, a mixer, and an IF amplifier and filter. The IR filter must provide sufficient rejection at the image frequency. The folding of image into the signal band is a major disadvantage of the superheterodyne architecture. A thorough discussion of image rejection has been described in Section 4.4.1. Also, if the IF frequency is high, then the IF filter is difficult to integrate on-chip. Typically, external SAW filters have been used for the IF filter as well as for the RF and IR filter. Therefore, the superheterodyne receiver does not lend well to single-chip integration. Recently, however, integrated receivers with on-chip IR mixers have been demonstrated [172][173] that performs IR as part of the mixing function. The trade-off is that it requires roughly twice the hardware and power as compared to a conventional design that uses separate IR filter and mixer combination. IR mixers will be discussed in Chapter 9.

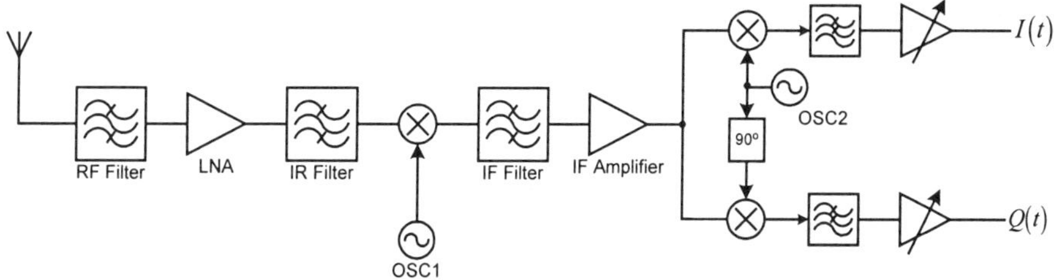

Figure 6.3 A dual-conversion superheterodyne receiver.

So why use superheterodyne receiver with the added complexity and more off-chip components? The main reason lies in the good frequency isolation provided by this architecture that minimizes LO re-radiation, which had been a major problem with a DCR. Good isolation results from the fact that none of the LO frequencies is at the carrier frequency. Therefore, any frequency that radiates out towards the antenna gets significant attenuation from the IR and RF filters, centered on the carrier. Also, both LO's from OSC1 and OSC2 are lower in frequency than the LO generated by the DCR, making them less efficient in re-radiation. DC offsets, therefore, pose fewer problems in a superheterodyne receiver. It follows from the above reasoning, that a superheterodyne receiver is less sensitive to other DC distortions due to second-order non-linearity and $1/f$ noise. The required IIP2 is therefore much lower than that required by a DCR.

A high dynamic range also drives designers to use the superheterodyne architecture. It is extremely challenging to achieve a dynamic range as high as 100 dB for a wireless channel using a one-step down conversion architecture, such as DCR, because of the high power blockers arising from adjacent and alternate channels. In a DCR, all channels within the passband are down converted in quadrature and out-of-band signals are filtered out by an analog baseband filter. In a conventional design, the lowpass filter does not perform any channel selection since that would require a tunable RF oscillator at the carrier frequency or it would require a tunable analog lowpass filter. Rather, the channel selection is done digitally as part of the digital transceiver. Therefore, all channels get translated to baseband intact with virtually no filter selection. Potentially, high-powered adjacent and alternate channel powers may be felt by the VGA that follows the lowpass filter as shown in Figure 6.2. Referring to (4.31), it is clear that the linearity of later blocks in the receiver dominates the overall linearity of the receiver. These blocks include the VGA and the mixer that precedes it. With the high-powered blockers, the VGA's and mixers must have a high IIP3 to prevent excessive distortions in the baseband signal. However, the high IIP3 makes the design of VGA's and mixers difficult given the high gain requirement.

By breaking up the down conversion into multiple stages, the superheterodyne receiver helps to relax the dynamic range requirement on the RF components. This is possible because at each IF stage, a filter suppresses high-powered blockers for each of the received channels. In other words, channel selection is performed in stages rather in one step as in a DCR. The lower LO frequency in the front-end helps to make the channel selection possible since tunable oscillators are much easier to realize at a lower frequency. Table 6.2 compares the advantages and disadvantages of superheterodyne receiver architecture.

Table 6.2 Trade-offs associated with the superheterodyne receiver.

Advantages	Disadvantages
• Lower dynamic range required	• Complex
• Lower IIP2 required	• High image rejection required
• Good LO isolation	• Harder and sometimes impractical to integrate on-chip
• Less prone to $1/f$ noise	
• Lower frequency synthesizer	

Example 6.5 Consider a wireless system that requires a dynamic range of 100 dB. Given that the sensitivity of the system is specified to be –105 dBm, the overall IIP3 must then be at least -5 dBm. Let's look at how the above constraints influence the design of a DCR and a superheterodyne receiver.

DCR: Assume that the VGA has a gain of 60 dB and that the rest of the gain (40 dB) comes from the LNA, mixer, and fixed amplification preceding the VGA. Also, for the sake of illustration, the preceding gain stages are assumed to have an effective IIP3 of –3 dBm. Since the VGA has a high gain of 60 dB, it would be difficult to achieve also a high IIP3. A reasonable estimate would be –10 dBm. Then, substituting –10 dBm into (4.31) results in the following expression

$$\frac{1}{IIP3_{total}^2} = \frac{1}{10^{-0.6}} + \frac{1}{\frac{10^{-2}}{10^8}},$$

and leads to a total IIP3 of –50 dBm. The dynamic range that is actually achieved with this design is 65 dB.

Superheterodyne: Given the same set of assumptions as for the DCR but with a 30 dB attenuation on the undesired channels, expression (4.31) can be applied again to obtain the following

$$\frac{1}{IIP3_{total}^2} = \frac{1}{10^{-0.6}} + \frac{1}{\frac{10^{-2}}{10^8 \cdot 10^{-6}}}$$

which leads to a total IIP3 of –20 dBm. The superheterodyne design achieves a SFDR of 85 dB that is much closer to the 100 dB requirement. The 30 dB improvement results from the attenuation introduced by the IF and channel selection filters, i.e. the 10^{-6} term. Although the computation illustrated above takes on somewhat of a simplistic view of how filtering effects the IIP3, it does demonstrate, on a first order, the effectiveness of selective filtering available in a superheterodyne architecture. □

Example 6.5 illustrates the trade-offs on dynamic range between the DCR and the superheterodyne receiver. In general, superheterodyne receivers permit more design flexibility as compared to a DCR. The following sub-sections describe three variants of the superheterodyne receiver architecture: namely, the high IF, low IF, and wideband IF receivers.

6.1.2.1 High IF

In a high-IF superheterodyne design, the LO frequency is chosen to obtain a high IF frequency such that the image is far apart from the desired signal. Recall from (4.34) that the image is separated from the desired signal by two time the IF frequency. The large frequency separation facilitates the filtering required to provide sufficient suppression at the image frequency as shown in Figure 6.4. The IR filter is designed to generate an attenuation of R_{ATT} at the image frequency. Depending on the application, R_{ATT} could be high, for example 80 dB. Given that the RF signal consists of multiple channels in an aggregate band, the attenuation is designed for the worst case corresponding to the frequency at the lower edge of the band, i.e. ω_L in Figure 6.4.

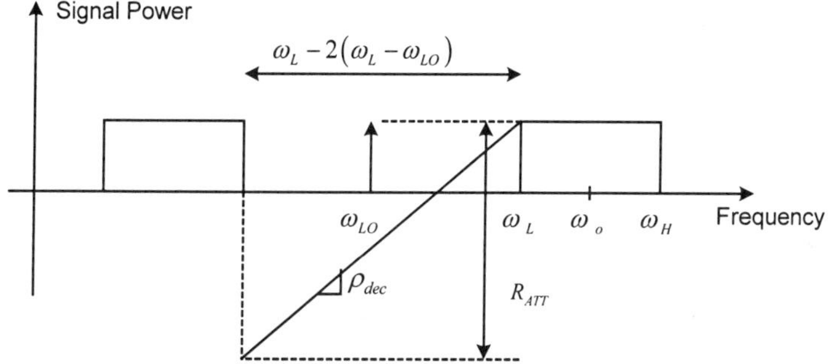

Figure 6.4 Image suppression in a high-IF receiver.

Besides achieving higher image suppression given the large image frequency separation, a high-IF design also requires a lower oscillator frequency for converting down to IF in the LO of the first stage, i.e. OSC1. Lower frequency facilitates high Q and low phase noise for the LO. The disadvantage, however, is that with a high IF, the IF filter and amplifier must operate at fairly high frequency, making it more difficult to implement the IF filter on-chip. Despite, its disadvantages in terms of higher cost and need for external off-chip components, many high-performance commercial radios still use the high-IF architecture, though more developments have demonstrated IF at 70 MHz [174][175] and even 100's kHz [266].

Let us now derive some basic guidelines for designing the IF and LO frequencies to meet a given image suppression. In the RF receive chain, the IR filter precedes the mixer to provide image suppression. The selectivity of the IR filter depends largely on the rolloff ρ_{dec} in dB/decade within its transition band. Using (4.34), the image frequency of ω_L can be related to the LO frequency ω_{LO} by

$$\omega_{image} = \omega_L - 2(\omega_L - \omega_{LO}). \tag{6.5}$$

If the required attenuation is denoted by R_{ATT}, then the following general expression can be obtained that relates the LO frequency to ω_L, ρ, and R_{ATT} as shown below

$$R_{ATT} = \rho_{dec} \log \sqrt{1 + \left(\frac{2 f_{LO} - f_L - f_o}{0.5 W_A} \right)^2}. \tag{6.6}$$

Substituting $f_0 - 0.5W_A$ for f_L, the required LO frequency is found to be

$$f_{LO} = f_o - \frac{\sqrt{10^{2R_{ATT}/\rho_{dec}} - 1} + 1}{4} W_A, \tag{6.7}$$

where W_A is the aggregate bandwidth in Hz. By applying (6.7), the IF frequency can then be expressed by

$$f_{IF} = \frac{\sqrt{10^{2R_{ATT}/\rho_{dec}} - 1} + 1}{4} W_A. \tag{6.8}$$

The following constraints must be considered when applying (6.7) and (6.8):

$$\frac{2f_o + W_A}{4} \le f_{LO} \le \frac{2f_o - W_A}{2}. \tag{6.9}$$

The lower limit on the LO frequency implies that the LO frequency should not be too low such that the image frequency begins to fold back into the region where the filter attenuation is low. The upper limit implies that the LO frequency cannot be too high as to exceed the lower band edge of the transmitted RF signal. In the limiting case, when a tone is received, the maximum IF or minimum LO frequency becomes half of the RF carrier frequency. If the limits of (6.9) are substituted for f_{LO} in (6.6), the filter attenuation is found to be constrained by

$$\rho_{dec} \log \sqrt{2} \le R_{ATT} \le \rho_{dec} \log \left(2 \frac{f_o - W_A}{W_A} \right). \tag{6.10}$$

The above derivations assume low-side injection but can easily be extended to high-side injection.

Example 6.6 Consider a system that requires an RF carrier of 1 GHz with an aggregate bandwidth of 25 MHz. Let the desired image suppression be 80 dB and assume that the IR filter has a rolloff of 50 dB/decade, i.e. ρ_{dec} = 50 dB/dec. Then, the LO frequency can be computed by applying (6.7) as shown below

$$f_{LO} = 1000 - \frac{\sqrt{10^{2\cdot 80/50} - 1} + 1}{4} \cdot 25 = 745\ MHz$$

By applying (6.8), the IF frequency is found to be 255 MHz. If an IF of 80 MHz is chosen instead, then the required rolloff becomes

$$\frac{80}{\log \sqrt{1 + \left(\frac{4 \cdot 920 - 4 \cdot 1000 + 25}{25} \right)^2}} = 74.5 \text{ dB.}$$

□

6.1.2.2 Low IF

While high-IF designs achieve high image suppression, it does have a few shortfalls, including the difficulty of integrating an IF filter on-chip. The low-IF design avoids this problem by using a low IF frequency such that the IF filter can be implemented on-chip using active filters. Although the low-IF receiver resembles a DCR, it achieves superior frequency isolation due to the small but non-zero frequency separation between the IF LO and the RF carrier. Thus, it does not exhibit severe DC offset or LO re-radiation that plague the DCR. However, with a low IF frequency the LO frequency must run at a higher frequency, making the LO design more difficult. Moreover, a low IF implies that the image is closer to the desired signal such that it cannot be rejected completely with an IR filter. In this case, complete image rejection usually requires an image reject mixer in addition to an IR filter. The image reject mixer increases the complexity and power dissipation of the receiver since both in-phase and quadrature signals are needed. A few research prototypes [172][174] have been developed based on the low-IF architecture.

Example 6.7 Consider the same system in Example 6.6 but now with a low IF using a LO frequency of 986.5 MHz, which is 1 MHz from the lower band edge. According to (6.6), the image suppression is

$$R_{ATT} = 50 \log \sqrt{1 + \left(\frac{2 \cdot 986.5 - 987.5 - 1000}{0.5 \cdot 25} \right)^2} = 9.3 \text{ dB.}$$

Clearly, the low-IF design offers very little image suppression with the IR filter, even if a filter with 80 dB/decade rolloff is used. Higher rejection is possible with an image reject mixer or a filter with higher selectivity. □

Figure 6.5 illustrates the configuration that results in minimum image suppression for a low-IF receiver. The LO frequency is set just high enough such that the image is not in-band. That is the IF frequency is slightly higher than

$$\frac{f_H - f_L}{2} = \frac{W_A}{2}. \tag{6.11}$$

It will be shown later in Chapter 9 that with an image reject mixer, the IF frequency can be made even lower than that of (6.11).

> **Example 6.8** Given the same system as in Example 6.6, the RF signal extends from 987.5 MHz to 1012.5 MHz. Substituting these values into (6.11), the IF frequency should then be larger than 12.5 MHz. Such limitation in IF frequency could be eliminated if single-sideband down conversion is used as discussed later in Chapter 9. □

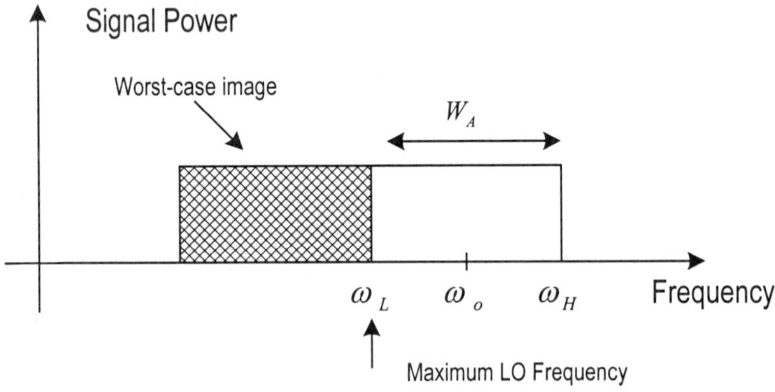

Figure 6.5 Image suppression in a low-IF receiver.

6.1.2.3 Wideband IF

The wideband IF architecture is a mix between the high-IF and the DCR architectures. Like the high-IF architecture, it uses a LO that converts the RF signal to a relatively high IF. However, it does not perform any channel selection in the IF stage. Rather a lowpass filter is used that passes the entire aggregate band at IF and only filters out frequency components above the upper band edge. Recall that the major disadvantage with a high-IF receiver is the need for an off-chip IF filter. A lowpass filter with a wide bandwidth can be more easily integrated on-chip. Note, that the RF filter still presents an obstacle to a completely integrated solution, though this is a problem common to all architectures.

The downside of the wideband-IF architecture is that, like the DCR, it has the problem of achieving a high dynamic range since if high-power blockers exist, they will generate signal distortion in all stages of the receiver chain. The upside is that LO re-radiation is not an issue in the wideband-IF receiver because of the frequency isolation provided by the IF stage. Figure 6.6 shows the wideband-IF receiver architecture. The RF signal is down-converted to a fixed IF. For an integrated solution, instead of an IR filter, an IR mixer is used to suppress the image. The IR mixer is described in Section 9.1.4.2. Channel selection is performed in the same fashion as a superheterodyne receiver with a low-frequency synthesizer. The IF to baseband down-conversion is performed with a complex mixer as shown in Figure 6.6.

Like the low-IF architecture, radios that implement the wideband-IF architecture are not yet wide spread and have been demonstrated only in research prototypes. Given its flexible architecture and relatively good performance, it is a strong contender for future multi-standard radio that requires multiband operation [173].

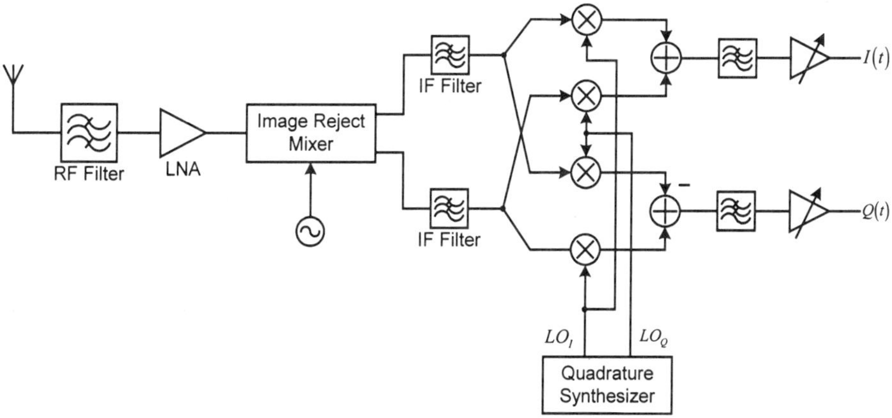

Figure 6.6 A wideband-IF receiver architecture.

6.1.3 Direct-Conversion Transmitter

Like the receiver, the most straightforward implementation for a transmitter is to use a one-stage up conversion as shown in Figure 4.1a. The advantage of this architecture lies in its simplicity and potential for an integrated solution. However, the front-end RF filter presents a barrier to a complete single-chip integration of the transmitter since integrated high Q filters at the carrier frequency is difficult to implement. Note that an RF filter is needed before the PA to filter out unwanted spurs generated by the up-conversion mixers. Any spurs in-band or out-of-band can drive the PA into compression and reduce the overall efficiency of the PA.

VCO pulling is a more serious problem that makes it very difficult to integrate a direct-conversion transmitter. VCO pulling results from lack of isolation between the PA output and the VCO output since both are centered at the same carrier frequency. The high power output from the PA couples back to the VCO and modulates it and may pull it out of lock.

To mitigate VCO pulling, one may resort to using two LO's, which are offset from one another in frequency but when mixed generate the desired carrier frequency. This method, also referred to as offset VCO [175] minimizes VCO pulling by choosing the LO frequencies such that neither LO's have the same frequency as the RF carrier.

6.1.4 Dual-Conversion Transmitter

Analogous to superheterodyne receivers, a dual-conversion transmitter employs an IF stage before up converting to RF as shown in Figure 6.7. The IF stage helps to isolate the PA output from the LO's. An IF filter is often required to eliminate spurs generated by the IF stage. To achieve adequate image suppression, a high IF frequency is often chosen such that the image is far enough away from the desired transmitted signal. The dependence of the image frequency on IF is evident in Figure 6.8 which shows the image centered on $\omega_{LO} - \omega_{IF}$ and the up-converted signal centered on $\omega_{IF} + \omega_{LO}$. The separation between the center frequencies is precisely twice the IF. To avoid negative frequency components folding in-band, the IF should satisfy the constraint below

$$\omega_{IF} \geq \frac{W_A}{2}. \qquad (6.12)$$

To efficiently reject the image signal, the IF frequency should be chosen large enough so that a filter can be implemented to provide a sufficient roll off ρ_{dec} as shown in Figure 6.8. Expression (6.6) can applied to determine the filter attenuation by substituting $f_L = f_{LO} + f_{IF} - 0.5W_A$ and $f_o = f_{LO} + f_{IF}$.

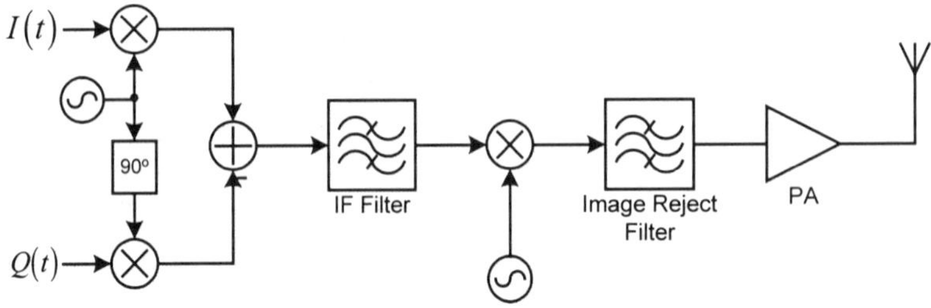

Figure 6.7 A dual-conversion transmitter architecture.

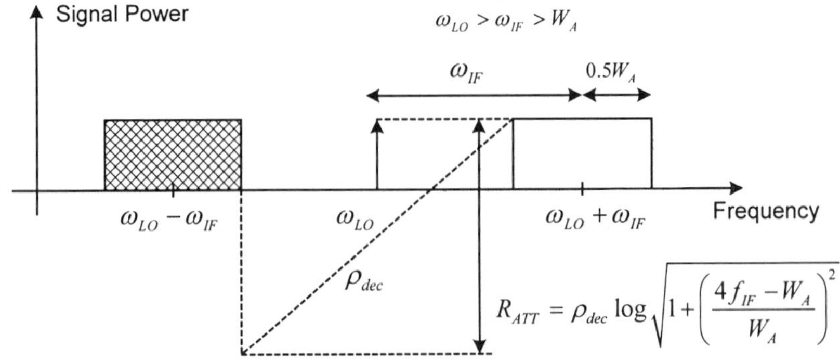

Figure 6.8 Image frequency in a dual-conversion transmitter.

6.2 Analog-Digital Partition

Just twenty years ago, very few digital components were found in radio devices. Most of the digital components were used peripherally to maintain system control and human-machine interface. The shift to digital began in the late 80's when new digital standards, such as GSM and IS95, were formed and when unlicensed spread-spectrum bands were issued by the FCC. The new standards together with the advent of CMOS technology drove the development digital radios that are essential to achieve the performance-cost ratio demanded by emerging wireless applications.

In contrast to traditional analog radios, digital radios consist of not only of an RF front-end that performs the frequency translation but also a digital transceiver that performs all of the communications processing. An ADC/DAC interface forms the boundary between digital and analog. A major design issue in digital radios involves the partition of analog and digital processing; that is, where to put the ADC/DAC interface. In a conventional digital radio architecture, all of the digital processing is performed at baseband but the current trend is to push the digital processing further up the RF chain toward the antenna. Therefore, digital processing is no longer limited to baseband but also extends to IF and even RF. This section discusses the various possible partitions for digital radios and the associated trade-offs.

6.2.1 Digital Radio Transmitters

We begin our discussion with the transmitter. For illustrative purpose, the dual-conversion transmitter of Figure 6.7 is used as an example, though the principles discussed can be easily extended to the direct-conversion architecture.

6.2.1.1 Digital Baseband Radio Transmitter

The digital baseband transmitter, shown in Figure 6.9, implements the most basic and commonly used architecture that generates the in-phase and quadrature signal components digitally. As depicted in Figure 6.9, the digital bit stream $d(t)$ gets converted to a parallel format which maps to I and Q symbols, each represented by $\log_2 M$ bits. The digital I-Q symbols are up-sampled, interpolated with zeros. The interpolated signal is pulse shaped by a pair of digital filters based on the principles discussed in Section 2.1.2. Before the filtered outputs are converted by the DAC's, they are first sent to a pre-distortion filter to compensate for the distortion due to the sample-and-hold operation in the DAC's. The outputs of the pre-distortion filters are then digitized by a pair of DAC's followed by analog filters, which attenuate images present at multiples of the sampling frequency. The filtered signals are then up-converted by an RF transmitter as shown in Figure 6.9.

This architecture presents the simplest approach to implement a digital transmitter since the required digital processing is performed at modest speed and complexity. The bulk of the computation complexity resides in the digital transmit filters which requires N_{TF} taps. The number of taps depends largely on the transition band of the filter. Sharp transitions result in a large N_{TF}. Recall from Chapter 2, a typical transmit filter used has a square-root raised cosine impulse response with a transition that varies inversely to the excess bandwidth. Therefore, for a square-root raised cosine filter, N_{TF} is largely a function of the excess bandwidth.

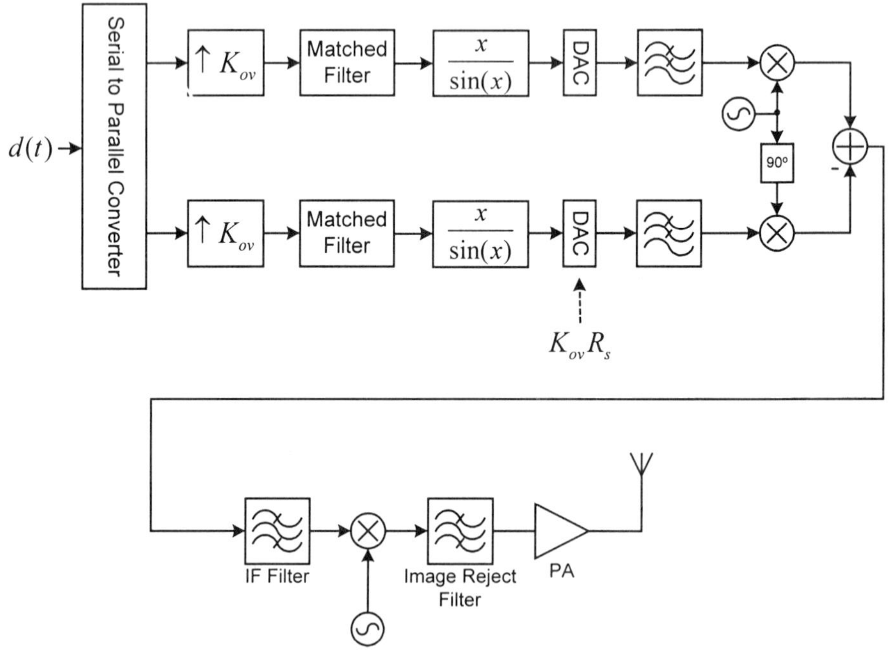

Figure 6.9 Digital baseband transmitter architecture.

Given the Nyquist criteria (2.12), the sampling rate must be at least twice the bandwidth of the transmit filter to avoid aliasing. To minimize the amount of image power which is folded into the signal band, it is necessary to over-sample the digital waveform by K_{ov}, defined as

$$K_{ov} = \frac{F_s}{R_s} \tag{6.13}$$

where F_s is the sampling rate. Higher over-sampling ratio K_{ov} implies that the images are further apart, facilitating the design of analog filters to reject the images. The amount of rejection is determined by the same set of constraints that guide the design of sufficient image rejection in the superheterodyne design. Though in the transmitter, the image power is well behaved with much smaller power fluctuations in contrast to that experienced at the receiver. Therefore, the separation of the image from the desired signal does not need to be as far apart at the transmitter. In general, the frequency separation between the image and the desired signal can be expressed by

$$\Delta_{image} = \left(K_{ov} - 1 - \alpha\right) R_s . \tag{6.14}$$

Note that when the over-sampling ratio K_{ov} is one and the excess bandwidth α is zero, there is no separation between the image and the desired signal.

> **Example 6.9** Consider a system with a baud rate of 200 kbaud and an excess bandwidth of 35%. Assume that the over-sampling ratio is two. Then, based on (6.14), the image separation is 130 kHz, making it harder to implement an analog filter with a fast enough roll off over the narrow frequency range. By increasing the over-sampling ratio to four, the image separation becomes 530 kHz. □

Over-sampling requires interpolation in the digital waveforms to convert the original symbol rate to $K_{ov}R_s$. Interpolation involves inserting $K_{ov}-1$ zeros into each symbol before filtering. The interpolation function is denoted by $\uparrow K_{ov}$ shown in Figure 6.9. Thus, the output of the filter runs at a sample rate of $K_{ov}R_s$.

The pulse-shape transmit filter is followed by a pre-distortion filter to compensate for the distortion introduced by the sample-and-hold circuitry in the DAC. A sample-and-hold has an impulse response that is a step function of duration equal to the sampling interval, i.e. F_s^{-1}. The PSD of such a step function has a $\left(\sin(x)/x\right)^2$ characteristic. Therefore, it introduces undesirable roll off in the signal band. The pre-distortion filter introduces peaking in the signal passband to compensate for this distortion. In particular, the pre-distortion filter has a frequency response with a $x/\sin\left(x\right)$ spectral shape that inverts the $\sin(x)/x$ transfer function of the DAC.

Compared with other approaches described later, the digital baseband architecture results in the lowest sampling rate and complexity. Lower sampling rate facilitates the design of low power DAC's for a given required bit resolution. The complexity results from the four linear-phase finite-impulse response filters (FIR) that implement pulse shaping and pre-distortion. Each tap of the filter consists of a multiply, an addition, and a memory element. The complexity is broken down into two components: the number of operations N_{op} and amounts of memory storage N_{mem}. An operation is defined as an addition or a multiplication. The complexity and sampling rate requirements for the digital baseband design are shown below:

Complexity:

$$\begin{aligned} N_{op} &= N_{\otimes} + N_{\oplus} = 4\left(N_{TF} + N_{DAC}\right) \\ N_{mem} &= N_{\mu} = 2\left(N_{TF} + N_{DAC}\right) \end{aligned} \tag{6.15}$$

Sampling Rate:

$$K_{ov}R_s \tag{6.16}$$

where $N_{\otimes}$, $N_{\oplus}$, N_{μ}, N_{TF}, and N_{DAC} denote the number of multiplies, the number of additions, the number of memory elements, the number of taps in the transmit filter and the number of taps in the $x/\sin(x)$ filter, respectively.

The complexity defined in (6.15) assumes a tapped delay line implementation for a digital filter. In general, the complexity and amount of memory access depend on implementation. For instance, if instead a serial scheme is used, a much lower complexity can be obtained. Moreover, a serial scheme with shared memory results in about half as much memory access when compared with a tapped-delay line implementation. More detailed discussions on implementation choices available for digital transceivers are described in Chapter 8.

Example 6.10 Consider a system that has a baud rate of 200 kbaud and that N_{TF} = 64 taps and N_{DAC} = 50 taps are required for the pulse-shaping filter and the pre-distortion filter, respectively. To obtain good image rejection an over-sampling ratio of four is used. Then, the amount of processing required can be obtained by multiplying the number of operations by the sampling rate

$$4\cdot(64+50)\cdot 4\cdot 2\times10^5 = 364.8 \text{ MOPS}.$$

Likewise, the amount of memory access can be determined by multiplying the number of memory cells by the sampling rate

$$2\cdot(64+50)\cdot 4\cdot 2\times10^5 = 182.4\times10^6 \text{ access/sec}.$$

The DAC's must operate at 800 ksamples/s. In this example, MOPS denotes Mega operations per second and access/sec denotes memory accesses per second. In many cases, a memory access is also considered an operation and therefore it can be combined to form a total MOPS metric. In this case, the total MOPS metric becomes 547.2 MOPS. □

6.2.1.2 Digital IF Radio Transmitter

A more complex architecture is the digital IF transmitter in which the DAC's are pushed further up in the RF chain, immediately before the IF mixer as shown in Figure 6.10. Digital processing of IF up conversion offers many advantages associated with digital processing, which include improved sensitivity to IC process variations and increased flexibility in adapting the IF frequency by simply setting appropriate parameters digitally. Thus, the RF LO can be implemented by a fixed RF oscillator. Furthermore, only a single DAC is required as compared to a dual DAC in the digital baseband architecture, though this DAC must run at a much higher sampling rate.

The price paid for the improved performance and flexibility is high. Because of the image rejection requirement, discussed in Section 6.1.4, the IF should be set reasonably high. The high IF incurs extremely high processing requirement on the digital circuits. In the worst case, the maximum IF is $0.5\omega_0$ according to (6.8). The corresponding sampling rate depends on the separation of image frequencies from the desired IF frequency. At the Nyquist rate, the image is close to the desired signal and is difficult to filter out with the IF filter that follows the DAC. To make it easier on the analog filter, the sampling rate should be larger than the Nyquist rate with an IF over-sampling ratio denoted by $K_{ov,IF}$ as shown below

$$K_{ov,IF} = \frac{F_{s,IF}}{\omega_{IF}}. \tag{6.17}$$

$K_{ov,IF}$ lies between two to ten. Since $\omega_0 >> R_s$, a digital IF transmitter requires substantially higher processing than a baseband design.

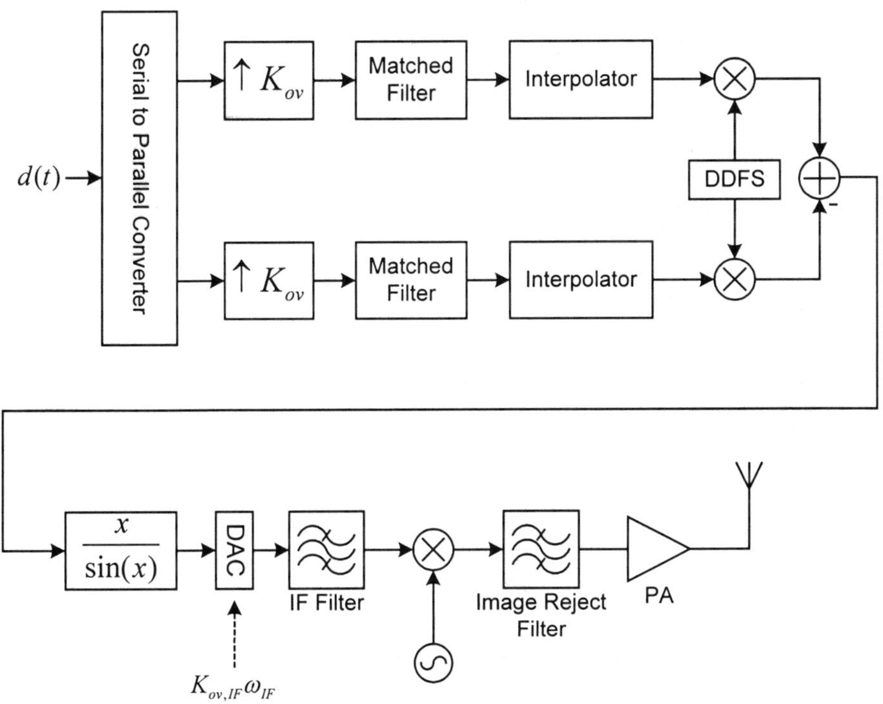

Figure 6.10 Digital IF transmitter architecture.

The IF up conversion requires two interpolation filters, a quadrature oscillator and mixers followed by a subtractor, and a pre-distortion filter. Digitally, the mixer corresponds to a multiplier and the oscillator corresponds to a direct-digital frequency synthesizer (DDFS). The interpolation filter performs the same function as the interpolator in the pulse-shaping filter but it requires an interpolation factor that is much greater. Details on the implementation of the filters and DDFS are described in Chapter 8. Computation requirements for the digital IF architecture are summarized in Table 6.3.

Table 6.3 Complexity requirement breakdown for the digital IF transmitter.

Components	$N_{\oplus}$	$N_{\otimes}$	N_{μ}
Pulse-shaping Filter	$2N_{TF}$	$2N_{TF}$	$2N_{TF}$
Mixer		2	
DDFS		3	$2+\frac{F_{s,IF}}{\Delta f}$
Interpolation Filter	$2N_I \log_{N_I} \frac{F_{s,IF}}{K_{ov}R_s}$	$2N_I \log_{N_I} \frac{F_{s,IF}}{K_{ov}R_s}$	$2N_I \log_{N_I} \frac{F_{s,IF}}{K_{ov}R_s}$
Pre-distortion Filter	N_{DAC}	N_{DAC}	N_{DAC}

The memory storage requirement for DDFS arise from the need to store digital values of the sine and cosine waveforms for a given phase resolution that depends on the IF sampling rate $F_{s,IF}$ and its frequency resolution Δf, i.e. proportional to $F_{s,IF}/\Delta f$. A direct implementation of the interpolation filter requires as many taps as the interpolation ratio, which in this case is $F_{s,IF}K_{ov}^{-1}R_s^{-1}$, a large number. However, more efficient implementations [176] are possible that performs the interpolation in multiple stages, each stage using a N_I tap filter with interpolation ratio of N_I. Then, fewer taps are needed, approximately $2N_I \log_{N_I}\left(F_{s,IF}K_{ov}^{-1}R_s^{-1}\right)$.

Consider the same system as in Example 6.10 but with a carrier frequency of 1 GHz. The processing requirement is simply the number of operations multiplied by the sampling rate. While the above rule easily applies to the pulse-shaping and pre-distortion DAC filters, it must be applied with care to the interpolation filter. Since each stage of the interpolation filter operates at a different sampling rate, the total processing rate is the sum of all the processing occurred in each stage:

$$\text{Interpolator Processing} = \sum_{k=1}^{\log_{N_I}\left(\frac{F_{s,IF}}{K_{ov}R_s}\right)} N_I^{k+1}K_{ov}R_s = \frac{N_I^2\left(F_{s,IF}-K_{ov}R_s\right)}{N_I-1} \tag{6.18}$$

Example 6.11 Let $N_I=2$ and $K_{ov}=K_{ov,IF}=4$. With a high-IF design, the IF frequency is approximately 500 MHz. Then, by applying (6.17)-(6.18) and the computation complexity listed in Table 6.3, the computation requirement can be determined as shown below

$$4\cdot 64\cdot 8\times 10^5+(2\cdot 50+5)\cdot 2\times 10^9+4\cdot 2^2\cdot 2\times 10^9=242.2\ \text{GOPS}$$

where the sampling rate of the DAC is 2 Gsample/s. The first term stems from the processing required by the two pulse-shaping filters. The second term represents the processing required by the pre-distortion filter, and the multiplications in the mixers and the DDFS. Finally, the third term represents the processing required by the two interpolation filters.

Assuming a frequency resolution of 10 kHz for the DDFS, the memory access requirement can be determined as shown below

$$2\cdot 64\cdot 8\times 10^{5}+50\cdot 2\times 10^{9}+\frac{2\times 10^{9}}{10\times 10^{3}}\cdot 2\times 10^{9}+2\cdot 2^{2}\cdot 2\times 10^{9}$$

$$\approx 4\times 10^{14}\, access/\sec.$$

The third term represents the memory access required by the DDFS which dominates the overall memory access requirement. □

Example 6.11 illustrates that the computation complexity has increased by 620 times for the IF architecture as compared to the baseband architecture in Example 6.10. The sampling rate has also increased significantly by a factor of 2500. At 2 Gsample/s, DACs can be implemented but requires more costly technology such as GaAs and requires extremely high power, up to 10's of Watts [177]. Most notably, the memory access speed has increased by a factor of a million as compared to that of the digital baseband architecture. In practice, the memory access requirement can be relaxed a bit by truncating the phase accumulator in the DDFS. More details on the implementation of DDFS can be found in Chapter 8.

6.2.1.3 Digital RF Radio Transmitter

The ultimate radio transmitter implements digitally all blocks up to the RF filter before the PA as shown in Figure 6.11. The main advantage of this architecture over the digital IF architecture lies in the all-digital implementation of the LO, which brings tremendous flexibility to the overall system. A digitally synthesized LO provides high frequency agility with virtually zero settling time and a wide linear tuning range. The purity of the LO can be controlled by appropriately setting the bit resolution in the DDFS. Most importantly, given that the LO is implemented as a DDFS, there is no longer an issue with LO pulling due to the transmitted signal feeding back to the control input of the VCO. The primary trade-off for all of these benefits is the slightly increased processing requirement as compared to a digital high-IF design. Given that the highest digital frequency is at ω_0 rather than $0.5\omega_0$, the sampling rate for a digital RF implementation doubles compared to that of digital IF implementation. Therefore, the computation processing and memory access requirement approximately doubles as well.

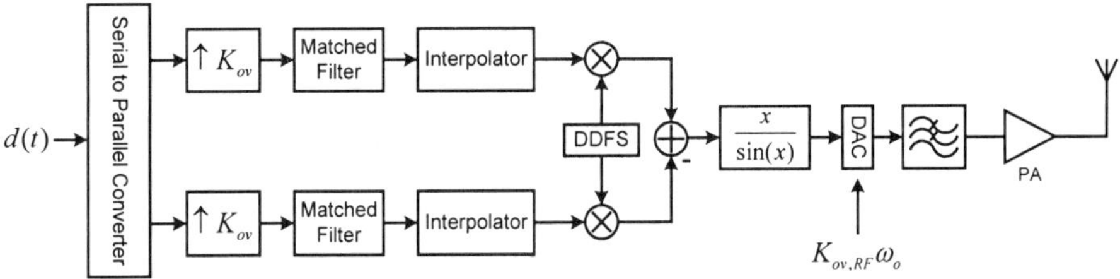

Figure 6.11 Digital RF transmitter architecture.

Example 6.12 Consider the same system as in Example 6.11 but assuming a digital RF architecture. The processing requirement may be computed as before but now substituting 4 Gsample/sec. For the RF sampling rate:

$$4 \cdot 64 \cdot 8 \times 10^5 + (2 \cdot 50 + 5) \cdot 4 \times 10^9 + 4 \cdot 2^2 \cdot 4 \times 10^9 = 484.2 \text{ GOPS}$$

which is approximately twice the processing required in Example 6.11. □

Although only the high IF architecture has been discussed, it is still possible to implement a digital low-IF transceiver with additional digital processing that performs single-side band mixing. Examples of single sideband processing are illustrated with low-IF digital receivers described in Chapter 9.

6.2.2 Digital Radio Receivers

Like digital radio transmitters, digital radio receivers have the same three main architectures: digital baseband, digital IF, and digital RF.

6.2.2.1 Digital Baseband Radio Receivers

The digital baseband receiver architecture places most of the burden on the RF front-end for the down-conversion. Digital processing is performed only at the baseband outputs for channel selection and matched filtering as shown in Figure 6.12. No pre-distortion filters are needed for the ADC but analog filtering is still needed before the ADC to provide anti-aliasing. While additional filtering requires more hardware, one could take advantage of the anti-alias filtering to relax the filtering requirement for channel selection.

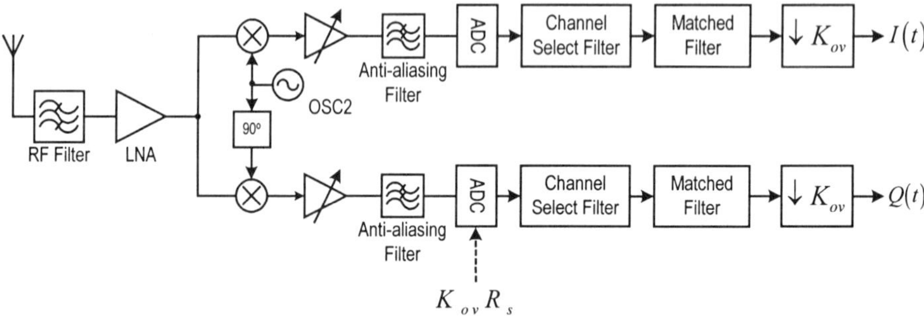

Figure 6.12 Digital baseband radio receiver architecture.

The advantage of the baseband architecture lies in its relatively relaxed constraint in the ADC and digital processing. The ADC samples at a low rate of $K_{ov} R_s$ and the digital processing consists of straight forward filtering operation needed for channel selection and matched filtering. In total, a receiver uses four digital filters, two for the I-channel and two for the Q-channel. The number of taps for the channel selection filter depends on the width of the transition band and the stopband attenuation. Denoting the number of taps for the channel select filter by N_{CS}, the total number of operations and number of memory elements can be represented by

$$N_{op} = N_{\otimes} + N_{\oplus} = 4(N_{TF} + N_{CS})$$
$$N_{mem} = N_{\mu} = 2(N_{TF} + N_{CS}). \quad (6.19)$$

Example 6.13 Consider again the same system as in Example 6.10 and assume 50 taps for channel select filter and an oversampling ratio of four. Then, the amount of processing required is

$$4(64+50)\cdot 8\times 10^5 = 364.8 \text{ MOPS}$$

and

$$2(64+50)\cdot 8\times 10^5 = 182.4\times 10^6 \text{ access/sec.}$$

The ADC's operate at 800 Msamples/s. □

The baseband architecture for the receiver requires the least complexity of all three architectures. However, this architecture takes the least advantage of digital processing for increased flexibility, programmability, and robustness to undesirable RF noise and non-linearity.

6.2.2.2 Digital IF Radio Receivers

A digital IF radio receiver balances the amount of processing performed in the digital and analog domain. Its architecture is shown in Figure 6.13. The output of the IF filter is digitized and the resulting digital IF signal is down-converted in quadrature with a DDFS. The baseband I and Q signals are sampled at a much higher rate than the original baud rate and therefore must be decimated, the exact opposite of interpolation performed at the transmitter. A decimation filter performs the decimation function by effectively deleting output samples. The ratio of the input sample rate $F_{s,IF}$ to the output sample rate $K_{ov}R_s$ is defined as the decimation ratio.

If each stage of the decimation filter has N_D taps and a decimation ratio of N_D, then the total number of taps required by the overall filter is $N_D \log_{N_D}\left(F_{s,IF}K_{ov}^{-1}R_s^{-1}\right)$, similar to the expression for an interpolation filter with the exception that N_I is replaced by N_D. The output of the decimation filter is further processed by a matched filter which has coefficients matched to the transmit filter to achieve the optimum SNR as discussed in Section 2.1.2.

The minimum IF sampling rate should be twice $0.5\omega_{IF}\pi^{-1} + 0.5W$, where W denotes the signal bandwidth being down converted. However, the IF sampling rate is generally chosen to be higher than the minimum to improve the amount of image suppression as shown below

$$F_{s,IF} = K_{ov,IF}\left(\frac{\omega_{IF}}{2\pi} + 0.5W\right) \quad (6.20)$$

where $K_{ov,IF}$ is the IF oversampling ratio. Clearly, the sampling rate of the ADC depends on the choice of the IF frequency. With a high IF, a high sampling rate is required for the ADC but with potentially high image rejection. On the other hand, a low IF results in a more relaxed sampling rate constraint, on the order of the channel baud rate. However, a low IF receiver makes image rejection more difficult. Often more complex I/Q processing such as an image-reject mixer is needed to achieve a high image reject ratio [172]. Therefore, for an efficient design, the choice of IF must take into account the above trade-offs.

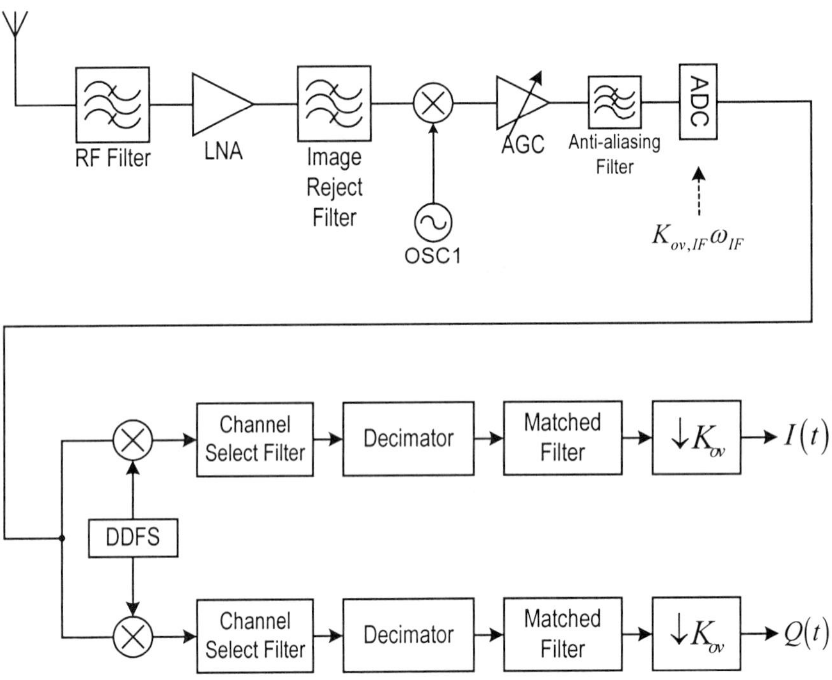

Figure 6.13 Digital IF radio receiver architecture.

The computation complexity of a decimation filter has a similar expression as that for an interpolation filter with the exception that N_I is replaced by N_D as shown below

$$\text{Decimator Processing} = \sum_{k=1}^{\log_{N_D}\left(\frac{F_{s,IF}}{K_{ov}R_s}\right)} N_D^{2-k} F_{s,IF} = \frac{N_D^2\left(F_{s,IF} - K_{ov}R_s\right)}{N_D - 1}. \tag{6.21}$$

Example 6.14 Consider the same system as in Example 6.13 and with a carrier frequency at 1 GHz. Let $N_D = 2$ and $K_{ov} = 4$. Then, applying (6.21) and the computation complexity listed in Table 6.3, the total computation requirement for the digital receiver can be determined. Both high IF and low IF are considered.

High IF: With $K_{ov,IF} = 4$ and an IF of 500 MHz, the sampling rate is approximately 2 Gsamples/s according to (6.20). The processing needed in this case is

$$4 \cdot 64 \cdot 8 \times 10^5 + (4 \cdot 50 + 5) \cdot 2 \times 10^9 + 4 \cdot 2^2 \cdot 2 \times 10^9 = 442.2 \text{ GOPS}$$

which is the same as that of the digital IF transmitter.

Low IF: With an IF frequency equal to the symbol rate, then according to (6.20), the sampling rate should be at least two times the baud rate, i.e. with $K_{ov,IF} = 2$. To relax the constraint of the analog anti-aliasing filter, $K_{ov,IF}$ is increased to eight, resulting in 3.2 MHz IF sampling rate. The processing needed in this case is

$$4 \cdot 64 \cdot 8 \times 10^5 + (4 \cdot 50 + 5) \cdot 3.2 \times 10^6 + 4 \cdot 2^2 \cdot (3.2 - 0.8) \times 10^6 \approx 1.8 \text{ GOPS}$$

where the sampling rate of the ADC is 3.2 Msamples/s. The processing requirement is reduced significantly compared to the high IF design and is now comparable to that of a digital baseband receiver. □

6.2.2.3 Digital RF Radio Receivers

The digital RF radio receiver architecture, also widely known as *software radio*, stresses the state-of-the-art in digital processing by pushing most of the burden onto the digital circuits whereby the ADC is placed further up the receiver chain immediately after the LNA. Given that the received signal power experiences a wide variation in the wireless environment, the ADC must be able to handle an extremely large dynamic range, up to 100 dB in some cases, e.g. CDMA. In more aggressive approaches, the ADC is in fact placed immediately after the RF bandpass filter, stressing even further the dynamic range requirement due to the increased out-of-band interference power.

Figure 6.14 shows a digital RF radio receiver that performs a direct conversion from the RF carrier down to baseband using a quadrature digital mixer and DDFS followed by a channel select, decimation, and matched filter. The number of operations required is approximately the same as that of a high-IF design. However, the amount of computation roughly doubles because of the doubling in sampling rate.

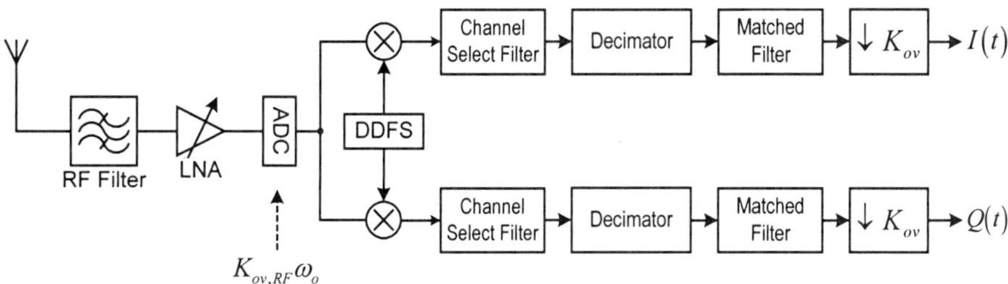

Figure 6.14 Digital RF radio receiver architecture.

The challenge in this architecture arises from the need for a high dynamic range together with a high sampling rate in excess of several GHz, depending on the RF carrier frequency. Needless to say, the high sampling rate makes it difficult not only for the design of the ADC but also digital circuits that must operate at GHz clocking speed.

Example 6.15 Consider a system with a channel baud rate of 200 kHz centered at 1 GHz RF frequency. Also, assume that the ADC is placed after the LNA which can handle up to 100 dB in dynamic range. Then, the ADC requires 17 bits sampling at 4 GHz, given an RF over-sampling ratio of four. The bit resolution is obtain by using the following rule

$$N_{bit} = \left\lceil \frac{DR}{6} \right\rceil. \tag{6.22}$$

Consider now the same system but assume that the ADC is placed immediately after the antenna. Now, in addition to handling signal variation in-band, the ADC must also handle the out-of-band interference fluctuations which can be as high as 50 dB. In this case, the ADC requires 25 bits and a sampling rate of 4 GHz. □

Example 6.15 shows the need for ADC's with both high bit resolution as well as high sampling rates, making it very challenging to implement with reasonable cost and power. Current state-of-the-art ADC's are implemented with high cost process such as GaAs and are still far from meeting the dynamic range requirement while dissipating in excess of 10 W [178].

Despite the many implementation difficulties, the digital RF radio receiver offers many advantages. For instance, a digital RF radio enables interoperability across multiple standards with simple reprogramming of the digital processing to handle different air interfaces. Though a low power version of such a radio might not be possible in the foreseeable future, a high power version might be achievable Therefore, this architecture may very well be implemented for basestations where flexibility and programmability are desired but low power is not a necessary requirement.

6.3 Digital Transceiver Architecture

A digital transceiver performs the actual communication tasks, such as error correction and modulation/demodulation. Traditionally, a digital transceiver has been referred to as a digital *modem* since most of its function is to perform modulation and demodulation. The term digital modem, however, is quickly becoming a misnomer since modulation and demodulation make up only a small part of today's digital transceivers. Therefore, throughout this book the term *digital transceiver* is used instead.

There is a wide variety of digital transceiver architecture but almost all of them can be broken down into several key components listed below.

- A *Modulator* takes a serial bit stream and maps it to digital symbols, each representing a constellation point. The digital symbols are also filtered by a matched filter.
- *Diversity transmitter* segments a message and transmits each segment such that individual segments experience independent channel fading. Often, instead of taking parts of a message, a diversity transmitter duplicates the entire message into multiple instances and induces diversity on the individual copies. Examples of diversity transmitter include spatial/time interleaving with coding, frequency-hopping, and direct-sequence spreading.
- *Demodulator* filters the received symbols with a matched filter and decodes the transmitted bit sequence from the matched filter output.
- *Diversity receiver* consists of a number of diversity branches, each responsible for receiving data from a diversity path induces either by the channel itself or by the diversity transmitter. The outputs of the diversity branches are then combined to improve the BER performance. Examples of diversity reception include Rake receiver, de-interleaving plus FEC decoding, antenna array combining, MLSE, and DFE/FFE.
- Synchronization loops recover the transmitted phase, frequency, and timing. Examples of synchronization loops include the Costas loop, PN-acquisition loop, and early-late gate correlator.

Detailed functional descriptions of the above components have been described in Chapter 2.

6.3.1 Digital IF Transceiver

Depending on the system requirement such as BER and delay, the digital transceiver can take on drastically different forms. To illustrate the interaction of the main components in a complete transceiver, three basic configurations are shown for the digital IF architecture. Other configurations for the baseband architecture will be illustrated later in Chapter 8 and Chapter 9.

6.3.1.1 Narrowband Transceiver

Figure 6.15 shows a narrowband digital IF transceiver where the transmitter has the same architecture as the digital portion of the block diagram shown in Figure 6.10. The IF down-converter portion of the digital receiver has already been discussed in Section 6.2.1.2. The output of the IF down-converter feeds into the frequency-acquisition block which estimates the frequency error in the down-converted signal. The DDFS takes the frequency error as input to control its output frequency to compensate for this error. Once the error has been compensated, the down-converted signal can then be used to establish frame synchronization and diversity reception.

Only path diversity is shown here. However, it is possible to perform spatial diversity by having multiple antennas, each feeding a separate radio transceiver. The channel estimate is derived from the down-converter output and is used in the diversity combiner. The output of the path diversity receiver then feeds into the demodulator that generates the received data. A time diversity block processes the received data to obtain the original bit sequence. The time diversity processing may consist of, for instance, a de-interleaver followed by an FEC decoder. To obtain phase and symbol timing recovery, the soft-decision output from the path diversity block is used in combination with the hard-decision output of the demodulator to form the necessary error signal that drives the DDFS and NCO, respectively, in a negative feedback loop. The feedback action drives the phase and timing errors to zero.

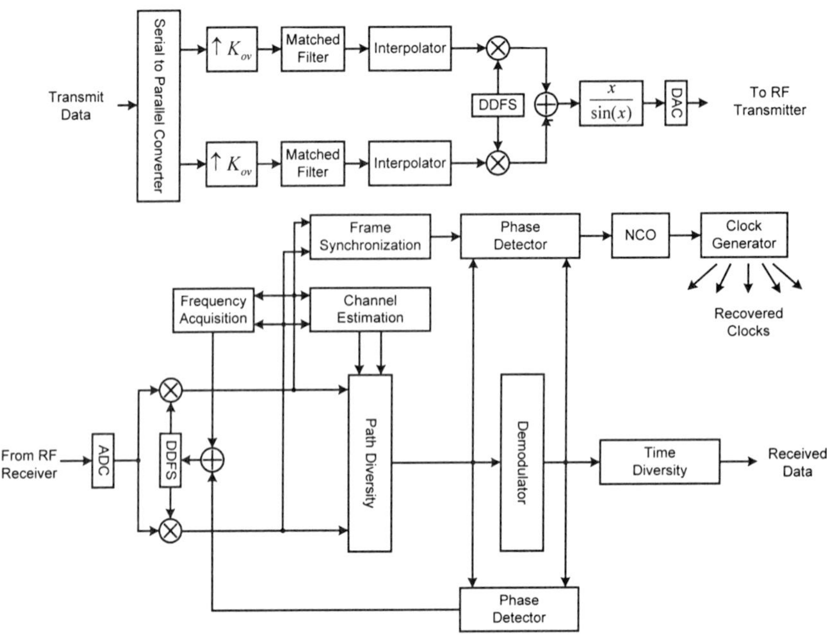

Figure 6.15 A digital IF narrowband transceiver.

6.3.1.2 Direct-Sequence Transceiver

The architecture for a direct-sequence digital IF transceiver can be derived from the narrowband architecture shown in Figure 6.15. The frame synchronization block is replaced by a PN-acquisition block, which by locking on to the transmitted PN code derives the frame as well as symbol timing. The chip timing is achieved with the timing recovery loop implemented based on an early-late gate correlator. Unlike the narrowband case, where path diversity is achieved using techniques such as MLSE and FFE/DFE, direct-sequence spread spectrum uses a Rake receiver to achieve path diversity. An example of an all digital IF BPSK direct-sequence spread-spectrum transceiver is shown in Figure 6.16. The transceiver takes digitized IF received signal and converts it to baseband via a digital Costas loop. It also implements digitally an early-late correlator to track the PN code and a serial PN-acquisition loop to acquire the initial timing of the PN code.

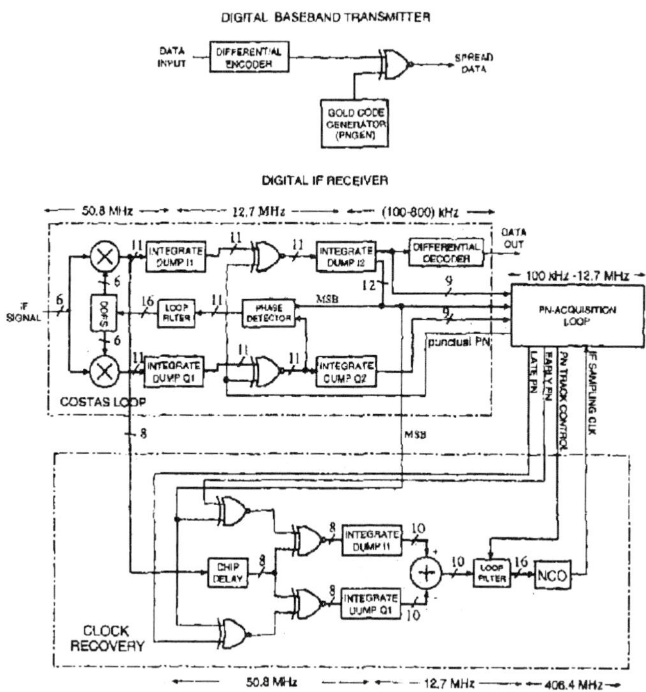

Figure 6.16 An all digital IF BPSK direct-sequence spread-spectrum transceiver; from Chien [179], © 1994 IEEE.

6.3.1.3 Frequency-Hop Transceiver

Similarly, the architecture for a frequency-hop transceiver can be derived from the narrowband architecture. In the FH architecture, frequency acquisition and frame synchronization are performed jointly. In general, a synchronization preamble precedes each packet or data stream that allows the hop-acquisition block to lock on both in frame timing as well as in frequency. If the hopping rate is slow, on the order of thousands of symbols, fairly complex path diversity combining, such as DFE, can be implemented. If the hopping rate is fast, on the order of several symbols, it is usually not feasible to implement an equalizer due to the lack of time to track the channel impulse response between hops. A fast hopper typically uses frequency diversity in conjunction with time diversity combining (i.e. interleaving plus coding) to achieve the diversity gain.

6.4 Typical Radio Handset Architecture

Most commercial handsets use the baseband radio architecture for both the transmitter and receiver as shown in Figure 6.17. The RF front-end can be either direct-conversion or super-heterodyne with analog-digital interface occurring at the baseband boundary. At the user end, there is a man-machine interface (MMI) together with ADC to convert analog media to digital data and DAC to convert received data to the corresponding analog media. The current predominant media is voice though data, image, and video are becoming available and are growing in popularity. For voice, the media converter consists of a microphone for the DAC and speaker for the ADC. The MMI provides a user interface between the handset core and the user display and keyboard.

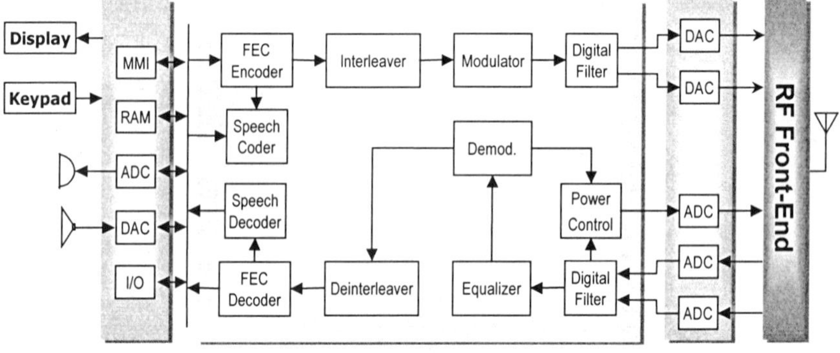

Figure 6.17 Typical commercial radio handset architecture.

The core of the handset consists of a baseband transmitter and receiver. The digital transmitter consists of a source coder that compresses the digitized media data to improve system capacity. The FEC encoder provides protection on the compressed source data against noise in the channel. An interleaver is generally included to provide diversity against a flat fading channel. The interleaved coded symbols are then digitally modulated and filtered with a pulse-shaping filter with an excess bandwidth that ranges between ten to twenty percent.

The digital receiver takes the digitized baseband data from the RF front-end and filters them with matched filter to obtain optimum noise performance. An equalizer provides diversity gain against a multipath channel. The equalized signal is then demodulated and de-interleaved. Any burst error before the de-interleaver is spread apart to enable the FEC decoder to achieve maximum coding gain. The decoded bit stream is then decompressed by the source decoder to reconstruct a digital representation of the transmitted analog media.

Currently, a commercial handset utilizes a mixture of technology in its implementation. The digital baseband transceiver is implemented in digital CMOS technology, with a feature size of 0.25 μm. The digital-analog interface in some cases are integrated with the digital baseband transceiver and depending on the performance requirement, a more expensive mixed-signal CMOS process may be used. The current trend is to integrate both digital baseband and analog interfaces on the same chip using a low-cost digital process when possible. The same holds true for the RF front-end. However, in reality, almost all commercial radio products are still using bipolar technology for most of the RF front-end and GaAs technology for the power amplifier and the TR (transmit-receive) switch. The level of integration, especially for high-performance systems, is still far from a single-chip RF solution. Some are still based on a chipset solution with a synthesizer chip, IF chip, and front-end chip, along with a handful of discrete passives and SAW filters for the RF filtering and image rejection.

Nevertheless, both the commercial sector and research community are driving hard toward a single-chip radio solution, in particular using CMOS. The driving force behind this move to a single-chip CMOS radio lies in the potentially huge reduction in cost of radio production by leveraging the scale-of-economy that exists for a CMOS process, especially a digital CMOS process. A few research works have already demonstrated a single-chip CMOS radio [167][168][180], some of these will be discussed in Chapter 9. Commercially, the first CMOS radio product for a single-chip 900 MHz CMOS radio intended for digital cordless phone has recently been reported in [168]. Other more complex and high-performance chips will most likely follow. The next three chapters concentrate on specific examples of highly integrated radios as well as the underlying building blocks, including advanced digital processing such as spatial diversity combining. Finally, the last chapter discusses some of the key issues that must be overcome in order to achieve single-chip integration of complex digital circuitry along with sensitive RF circuits on the same substrate.

CHAPTER 7

RF Circuits Design

The previous chapters gave an overview of the communications theory, RF and radio system design issues, and digital radio architectures. This chapter delves into the specifics of RF circuits design. In particular, it covers the design of key RF circuit components, including: low-noise amplifier, mixer, oscillator, phase shifter, synthesizer, and power amplifier. Other components such as antennas will not be covered. Interested readers on antennas are referred to [181] and [182] for more information.

While designers have developed many useful design metrics for RF circuits, it is not clear how one should constrain these metrics such as NF and IIP3 to satisfy a given system performance. Chapter 5 describes a methodology that maps system constraints to criteria used to determine implementation-specific parameters. This chapter focuses on the design flow and trade-offs in applying the RF specifications discussed in Section 5.6.1 to circuit implementation.

Figure 7.1 illustrates a typical RF circuits design flow that begins with the radio system design summarized in Figure 5.11 to generate the RF system specifications summarized in Table 5.7. These RF system specifications are used to determine the suitable RF circuits and the associated circuit parameters such as device size. With initial circuit parameters, the RF circuits are then simulated with a circuit simulator, such as ADS by HP. The simulator validates that the circuit performance is adequate to meet the system specification and also fine tunes the device sizes for IC layout. When digital transceiver circuits are integrated along with the RF circuits, the coupling from digital circuits must be simulated to ensure that adequate isolation has been provided. To simulate in a reasonable amount of time, digital circuits are often modeled by equivalent capacitors and transistors operating at the target clock rate. The simulated circuit is then laid out and layout parameters are extracted to re-simulate the circuit for more accurate results. Finally, the RF circuit is integrated along with the digital circuits in the final layout and submitted for IC fabrication.

Often, the fabricated radio chip will not be suitable for production the first time due to lack of a precise model for the relatively large digital circuit along with the sensitive RF circuit. The fabricated chip is tested and problems encountered are identified and fixed in the layout before re-submitting the IC for fabrication. Sometimes several chips may need to be fabricated before obtaining working silicon with sufficiently high yield. In some cases, even before the digital circuit is integrated with the RF, the RF circuits are first fabricated as a standalone test chip a few times to fine tune the layout as well as the RF device model. This is needed because there often lacks a precise RF device model to ensure that the simulated RF circuit will actually achieve the target performance.

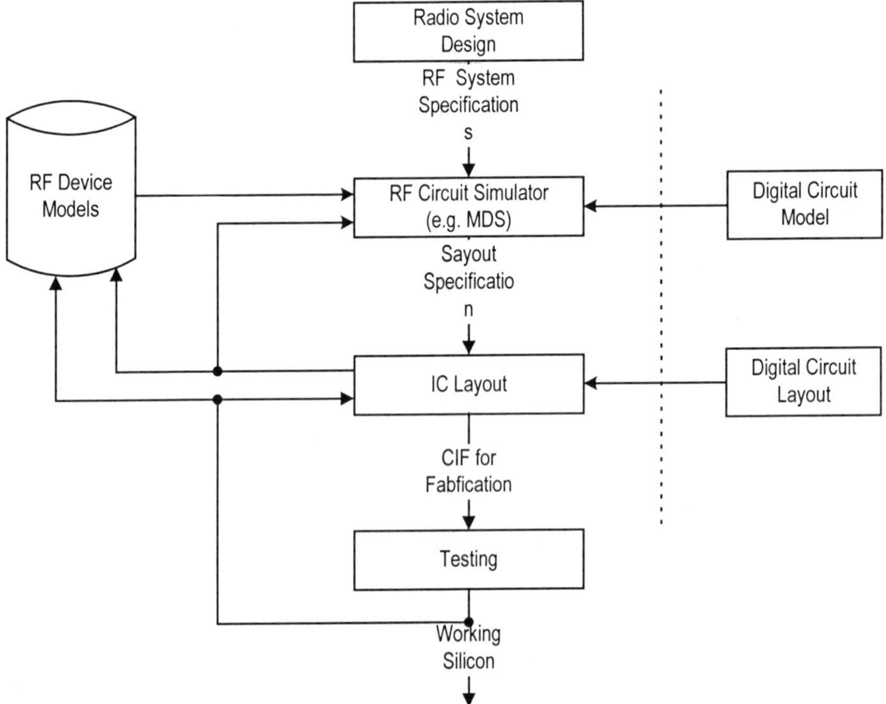

Figure 7.1 RF design flow.

Currently, BJT and GaAs technology prevail in commercial RF IC's. On the other hand, complementary MOS (CMOS) technology dominates digital and some mixed-signal designs. Because of the high-volume production and ease of scalability in CMOS designs, there is a push to integrate all functions both digital and RF on a single CMOS chip. While most RF CMOS works are still in the research prototype stage, commercial products are beginning to appear [168]. This chapter focuses primarily on CMOS technology and circuit examples, leading to case studies in Chapter 9 on integrated CMOS digital radio IC's. Relevant discussions on BJT technology and chip sets will also be included for comparison purposes.

7.1 Transistor Models

A MOS transistor (MOST) has the general model shown below for the active region [141]

$$I_D = \frac{\mu C_{ox} W}{2L} \frac{\left(V_{GS} - V_t\right)^2}{\left(1+\rho\right)}, \tag{7.1}$$

where I_D is the drain current, μ is the mobility of the carrier in $m^2/\left(Vs\right)$, C_{ox} is the oxide capacitance in F/m^2, W is the width of the gate, L is the channel length[5], ρ is the short-channel parameter, V_{GS} is the gate-source voltage, and V_t is the threshold voltage. The short-channel parameter has the following expression

$$\rho = \frac{V_{GS} - V_t}{LE_{sat}}, \tag{7.2}$$

where E_{sat} is the field strength in V/m at which the carrier has reached velocity saturation. As L decreases, ρ increases as a result of saturation in drain current. Below 0.5 μm, this short-channel effect becomes increasingly noticeable. For large enough ρ, the current varies linearly with $V_{GS} - V_t$ rather than quadratically as in a long-channel device where $\rho \to 0$.

Expression (7.3) below summarizes the transconductance of a MOS transistor.

$$g_m = \frac{\mu C_{ox} W}{L}\left(V_{GS} - V_t\right)\frac{\left(1+0.5\rho\right)}{\left(1+\rho\right)^2} = \sqrt{\frac{2\mu C_{ox} W I_D}{L}}\frac{\left(1+0.5\rho\right)}{\left(1+\rho\right)^{1.5}}. \tag{7.3}$$

For long-channel device, the transconductance varies linearly with $V_{GS} - V_t$ but becomes a constant as channel length decreases corresponding to a larger ρ. A constant transconductance is undesirable since it implies that the transconductance has reached a plateau and can no longer be controlled by increasing the drain current.

BJT has a similar I-V characteristic but follows a substantially different model as shown below

$$I_C = \frac{qA_{BE}Dn_i^2}{N_A W_B} e^{\frac{qV_{BE}}{kT}} \tag{7.4}$$

[5] For simplicity, it is assumed that the drawn length equals the effective length though in practice due to overlap, the effective channel length is less than the drawn length.

where $q = 1.6 \times 10^{-19} C$ is the electron charge, A_{BE} is the area of the emitter, D is the diffusion constant of the carrier in m^2/sec, $n_i = 1.5 \times 10^{16} m^{-3}$ is the intrinsic carrier concentration of silicon at 300 degrees kelvin, Q_B is the total doping in the base per emitter area in m^{-2}, and V_{BE} is the base-emitter voltage.

The transconductance of BJT under low-level injection has the following expression

$$g_m = \frac{qI_C}{kT}, \tag{7.5}$$

which shows that the transconductance varies linearly with the collector bias current. The transconductance of MOST, on the other hand, varies with the square root of the drain current. Therefore, at a given bias current, BJT has a higher transconductance compared to MOS. The high g_m / I ratio implies that for a fixed gain, the current drive of a BJT is always greater than that of a MOS transistor.

A useful figure of merit for high frequency design is the transition frequency f_T defined as the frequency at which the short-circuit current gain becomes unity. For sufficiently large current drive, f_T can be approximated by the following expression for MOST

$$f_T \approx \frac{1}{2\pi}\frac{g_m}{C_{gs}} = \begin{cases} \dfrac{3}{4\pi}\dfrac{\mu(V_{GS} - V_t)}{L^2}, & \text{long channel} \\ \dfrac{1}{4\pi}\dfrac{\mu E_{sat}}{L}, & \text{short channel} \end{cases} \tag{7.6}$$

In the long-channel regime, f_T varies inversely to the square of the channel length and as $V_{GS} - V_t$ becomes much larger than LE_{sat} f_T flattens out to a maximum value determined by the inverse of the channel length and the saturation velocity of the carrier, $0.5\mu E_{sat}$.

BJT has a similar expression for f_T as that of MOST where the channel length is replaced by the base width and $V_{GS} - V_t$ is replaced by the thermal voltage kTq^{-1}. When the current drive is low such that a BJT is operating under low-level injection and a MOST is operating in the long-channel region, the f_T of a BJT may actually be lower than that of a MOST since $V_{GS} - V_t$ is generally larger than the thermal voltage, typically around 26 mV. BJT has a similar behavior as MOST when driven with sufficient current such that the transistor performance saturates due to velocity saturation. In this case, the f_T of a BJT can be higher than that of MOST operating in the short-channel regime since W_B can be made lower than L, although lithography has improved over the years such that the performance gap is narrowing between MOST and BJT.

$$f_T \approx \frac{1}{2\pi\tau_F} = \begin{cases} \dfrac{\mu kT}{\pi q W_B^2}, & \text{low-level injection} \\ \dfrac{\mu E_{sat}}{4\pi W_B}, & \text{velocity saturation} \end{cases} \tag{7.7}$$

Table 7.1 summarizes numerous transistor parameters and values for a 0.25 μm CMOS process and a BJT process with a base width of 0.05 μm. It is worth noting that if (7.7) is applied, the transition frequency is estimated to be 178 GHz which is much higher than that listed in Table 7.1. The discrepancy results from the parasitic junction capacitances that begin to dominate as the base width and emitter area diminishes in size for the current BJT processes. These capacitances are neglected in (7.7). Empirically, the transition frequency is typically around 30 GHz for current BJT processes.

Table 7.1 List of transistor parameters.

Parameter		Definition	Typical Value
General	μ [6]	Carrier Mobility (bulk)	$5.5\times10^{-2}\ m^2/Vs$
		Carrier Mobility (channel)	$1.5\times10^{-2}\ m^2/Vs$
	E_{sat}	Saturation Electric Field	$4\times10^{6}\ V/m$
MOS (n-channel)	V_t	Threshold Voltage	0.7 V
	C_{ox}	Oxide Capacitance	3.5 $fF/\mu m^2$
	K_{MOS}	Flicker Noise Scale Factor	$4\times10^{-27}\ C^2m^{-2}$
	g_m	Transconductance[7]	$293\mu S/\mu m^2$
	J_{MOS}	Current Density[7]	$149\ \mu A/\mu m$
	f_T	Transition Frequency[7]	20-25 GHz
BJT (NPN)	D	Carrier Diffusion Constant	$1.4\times10^{-3}\ m^2/s$
	N_A	Carrier density in Base	$10^{23}\ m^{-3}$
	K_{BJT}	Flicker Noise Scale Factor	$10^{-25}\ A\cdot m^2$
	g_m	Transconductance[8]	$10\,mS/\mu m^2$
	J_{BJT}	Current Density[8]	$270\mu A/\mu m^2$
	f_T	Transition Frequency[8]	30 GHz

[6] Electron.

[7] Assumes 0.25-μm channel length and $V_{GS} - V_t = 0.8V$.

[8] Assumes 0.05-μm base width and $V_{be} = 0.8V$.

7.2 Noise

Noise figure is a key design parameter that has been discussed in Section 4.2 of Chapter 4. This chapter begins with a discussion of the various circuit noises encountered in an RF circuit that degrade noise figure. Other noise sources due to the environment have already been discussed in Chapter 2.

7.2.1 Thermal Noise

Thermal noise results from thermal agitation of charge carriers in conductor that results in noise current. Thermal noise is found in resistors or any conducting element such as the channel of a MOST in the on state. This type of noise can be modeled by a noise voltage source $\overline{v_n^2}$ or current source $\overline{i_n^2}$ as shown in Figure 7.2, where

$$\overline{v_n^2} = 4kTR\Delta f \tag{7.8}$$

$$\overline{i_n^2} = 4kT\frac{1}{R}\Delta f. \tag{7.9}$$

k is the Boltzmann's constant, T is the ambient temperature in kelvin, R is the resistance value of the conductor, and Δf is the measurement bandwidth [21][183]. The term $4kTR$ and $4kTR^{-1}$ can be viewed as the noise voltage and noise current density, respectively. Since the density is constant over frequency, thermal noise can be considered as white noise. This assumption holds true for at least the radio frequency.

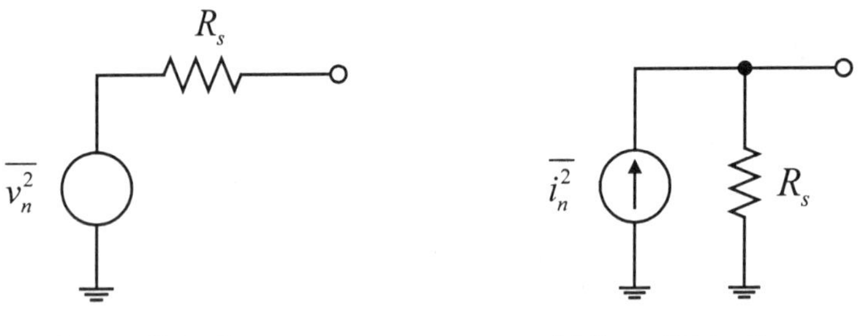

Figure 7.2 Thermal noise models.

Example 7.1 Consider a 1 $k\Omega$ resistor at 295 degrees kelvin. Then, according to (7.8), the RMS voltage per root Hz is

$$v_{rms} = \sqrt{4kTR} = \sqrt{4\times 1.38\times 10^{-23}\times 295\times 1000} = 4\,nV/\sqrt{Hz}.$$

If the BW is 1 MHz, then the total RMS voltage is 4 μV. Equivalently, the RMS current noise is 4 nA. □

Thermal noise is also found in both a BJT as well as a MOST. In a BJT, thermal noise arises from the base resistance r_b and in a MOST thermal noise arises from the channel resistance and has the following expression [184]

$$\overline{i_n^2} = 4kT\gamma g_{d0}\Delta f \tag{7.10}$$

where g_{d0} is the drain-source conductance at $V_{DS} = 0$. For a long channel devices, γ is 1 when $V_{DS} = 0$ and $2/3$ when $V_{DS} > V_{GS} - V_t$, i.e. when the transistor is in saturation. For short channel devices, γ is typically 2-3 times larger. Note, V_{DS}, V_{GS}, and V_t are all based on common naming convention for the drain-source voltage, gate-source voltage, and the threshold voltage of a MOST.

The drain-source conductance g_{d0} can also be expressed in terms of the transconductance g_m

$$g_{d0} = \frac{(1+\rho)^2}{1+0.5\rho} g_m = \frac{g_m}{\alpha} \tag{7.11}$$

Example 7.2 Consider a 0.25-micron n-channel MOST biased at $V_{GS} - V_t = 2V$ and with a drain current of 1 mA. Assume an ambient temperature of 300 degree K. Then, according to (7.2),

$$\rho = \frac{2}{0.25\times10^{-6}\cdot 4\times10^{6}} = 2.$$

According to (7.3), a channel width of 7.1 micron is required to support 1 mA drain current corresponding to a transconductance of

$$g_m = \frac{0.015\cdot 3.5\times10^{-3}\cdot 7.1\times10^{-6}\cdot 2}{0.25\times10^{-6}}\frac{(1+1)}{(1+2)^2} = 663\mu S.$$

Setting $\gamma = 4/3$ and applying (7.10) and (7.11), the RMS noise current due to channel conductance over a 1 MHz bandwidth is

$$i_{rms} = \sqrt{4\cdot 1.38\times10^{-23}\cdot 300\cdot\frac{4}{3}\frac{(1+2)^2}{1+1}663\times10^{-6}\cdot 1\times10^{6}} = 8.1nA. \ \square$$

Example 7.3 Consider a BJT transistor with a base resistance of 100Ω. According to (7.9), the RMS noise current over 1 MHz bandwidth is

$$i_{rms} = \sqrt{4\cdot 1.38\times10^{-23}\cdot 300\cdot\frac{1}{100}\cdot 1\times10^{6}} = 12.9\ nA.$$

Assuming a current gain of 10 at high frequency (2 GHz), the RMS current is 129 nA at the collector output and is nearly sixteen times that of the thermal noise in MOS transistor. Therefore, that base resistance tends to limit the noise performance of a BJT. □

7.2.2 Shot Noise

Shot noise occurs from the random nature in which charge carriers in a direct current transit across a potential barrier, such as a semiconductor junction. The noise current has a flat spectral density. Therefore, shot noise like thermal noise is white and has a RMS level that is proportional to the square root of the direct current I_{DC}

$$\overline{i_n^2} = 2qI_{DC}\Delta f \tag{7.12}$$

where q is the electric charge equal to [185] $1.6\times10^{-19}C$. In a BJT, shot noise exists at the base and collector while in the MOST shot noise occurs only at the gate due to the leakage current I_G.

Example 7.4 Consider a BJT with a 1-mA collector current. Then, according to (7.12), the RMS current per root Hz is

$$i_{rms} = \sqrt{2qI_{DC}} = \sqrt{2\times1.6\times10^{-19}\times10^{-3}} \approx 18pA/\sqrt{Hz}.$$

Over a bandwidth of 1 MHz, the total RMS current is 18 nA. A MOST, on the other hand, with a leakage current of 1 pA results in 0.6 $fA/\sqrt{Hz}$ noise current, which is much smaller than the shot noise of BJT even when referred to the output. □

7.2.3 1/f Noise

Section 4.4.3 discussed phase noise, which has a spectral density that varies inversely to frequency raised to the power n. Phase noise results from the spectral shaping due to the bandpass characteristic of a PLL and non-linear mixing of flicker noise, otherwise known as $1/f$ noise. Both BJT and MOST exhibit flicker noise and have the following general form

$$\overline{i_n^2} = K\frac{I_{DC}^n}{f}, \tag{7.13}$$

where I_{DC} is a direct current in the transistor and n is a constant that varies between 0.5 and 2 [183]. In a BJT and MOST, flicker noise occurs at the base and drain of the transistor, respectively. Therefore, I_{DC} in (7.13) corresponds to I_B for a BJT and I_D for a MOST.

It is widely accepted that the origin of flicker noise results from charge carriers trapped in surface defects that are released at random times. The trap and release mechanism is a random process that has a spectral density with energy concentrated in the low frequency. Since flicker noise results from surface defects, it can also be represented as a function of the surface area over which charge carriers must flow through. In fact, it has been shown that flicker noise varies inversely to the surface area. Alternative expressions for flicker noise based on this observation is shown below [186]:

BJT:
$$\overline{i_n^2} = \frac{K_{BJT}}{f}\frac{I_C}{A_j}\Delta f \tag{7.14}$$

MOST:
$$\overline{i_n^2} \approx \frac{K_{MOS}}{f}\frac{g_m^2}{A_g C_{ox}^2}\Delta f, \tag{7.15}$$

where A_j is the area of the junction, $A_g = WL$ is the area of the gate, and I_C is the collector current. The constants K_{BJT} and K_{MOS} are not well controlled for a given process but have typical values of $10^{-25}\ A \cdot m^2$ and $4 \times 10^{-27}\ C^2 m^{-2}$, respectively[9].

Example 7.5 Consider an n-channel MOST with the same conditions as in Example 7.2. To compute the RMS noise current from 1 Hz to 1 MHz, (7.15) is integrated over this frequency range to generate the following expression

$$i_{rms} \approx \sqrt{K_{MOS}\frac{g_m^2}{A_g C_{ox}^2}\ln\left(\frac{1\times 10^6}{1}\right)}$$

In Example 7.2, it has been determined that to sustain a bias current of 1 mA, $W = 7.1$ μm resulting in $A_g = 1.78\,\mu m^2$. Using this gate area, the RMS current is then 33.4 nA, which is much larger than the noise current induced by thermal noise. This implies that flicker noise can become a limiting factor in circuit noise performance when using MOST. The frequency at which the flicker noise density becomes less than that of thermal noise can be determined by setting (7.10) equal to (7.15) and solving for f. This frequency is also referred to as the corner frequency and is 1.2 MHz for this example. □

Example 7.6 Consider a BJT with the same conditions as in Example 7.3. Then, by applying (7.14), the RMS noise current due to flicker noise from 1 Hz to 1 MHz can be determined by

$$i_{rms} \approx \sqrt{K_{BJT}\frac{I_C}{A_j}\ln\left(\frac{1\times 10^6}{1}\right)}.$$

Using the current density of 270 μA/μm², a junction area of 3.7 μm² is needed to deliver 1 mA. The RMS current due to flicker noise then becomes 19.3 nA, which is smaller than that of MOST. The corner frequency is defined as the frequency at which the shot noise (7.12) equals the flicker noise (7.14). For this example, the corner frequency is 84 kHz for this example. □

7.2.4 Equivalent Input Noise Generators

Figure 7.3 summarizes the noise sources in BJT and MOST discussed in the previous section. The dominant noise sources in a BJT are the shot noise at the collector and the thermal noise generated by the base resistance. The shot noise and flicker noise

---

9 PMOS devices typically have a smaller normalization factor for flicker noise, e.g. $10^{-28} C^2 / m^2$ [186].

generated by the base current tend to be small due to the small base current. A MOST has even smaller shot noise current at its gate since the gate current is generated by the leakage current, which is on the order of 0.1-1 pA. In contrast to a BJT, a MOST does not have thermal noise due to parasitic resistance at its gate. It does however display thermal noise and flicker noise at its drain.

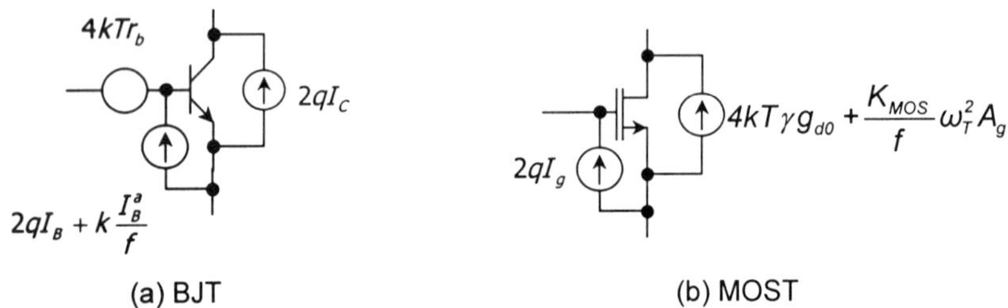

Figure 7.3 Noise sources in BJT and MOS transistors.

Referring to the definition of noise figure in (4.3), noise generated by the circuit is referred back to the input and compared against the available input noise. Therefore, to analyze the noise behavior of a circuit on the basis of noise figure, it is useful to refer all noise sources back to the input. Given a two-port model of a circuit, the noise sources within the two-port element can be referred back to the input in terms of a noise voltage source and a noise current source as shown in Figure 7.4. Both voltage and current noise sources are needed to take into account different input impedance. At low input impedance, voltage noise becomes dominant while at high input impedance current noise becomes dominant. The two-port noise model is precise if the two input noise generators are uncorrelated, a condition which is generally not true. When the input noise sources are correlated, this model generally predicts the worst-case performance.

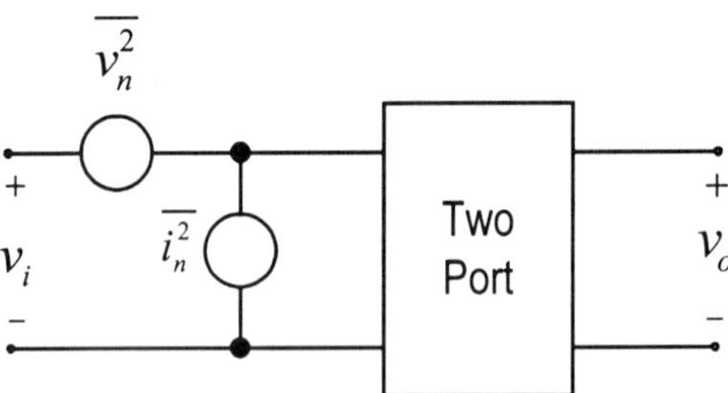

Figure 7.4 A two-port model with equivalent input noise sources.

The equivalent noise inputs for BJT and MOST are shown in Figure 7.5. The noise resistor $0.5g_m^{-1}\Delta f$ is the input referred voltage noise of the shot noise $2qI_C\Delta f$ at the collector output. The collector current can be referred to the input by dividing it by the square of the AC current gain, which can be approximated by ω_T^2/ω^2. Similarly, the noise current at the MOST output results in an equivalent input current noise scaled by ω_T^2/ω^2. The output current noise also results in an equivalent input voltage noise that consists of $0.8kT\delta g_{d0}^{-1}\Delta f$ and flicker noise $K_{MOS}\Delta f/(WLC_{ox}^2 f)$ [183]. A typical value for δ is 2γ [141].

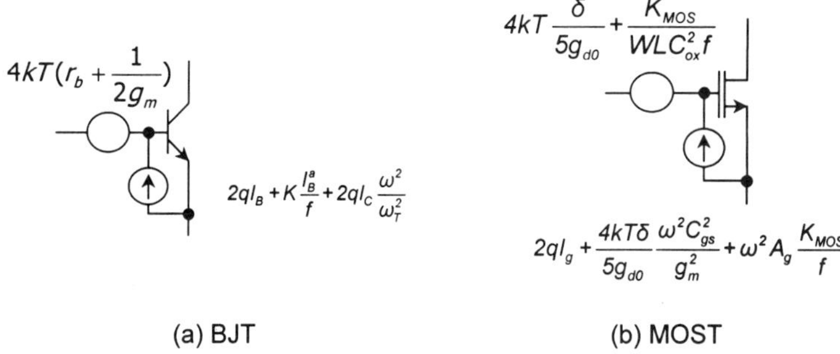

Figure 7.5 Equivalent input noise sources for BJT and MOS transistors.

Assume a bias current of 1 mA at the collector and drain for a BJT and MOST, respectively, a noise bandwidth of 1 MHz, and a carrier frequency of 1 GHz. Then, the input-referred noise levels are found as shown in Table 7.2. From this table, it can be observed that generally a BJT has a higher input current noise and a lower input voltage noise than a MOST. Therefore, if the input impedance is high then using MOST's results in better noise performance but if the input impedance is low then BJT outperforms MOST. Note that the large input voltage noise of MOST is due mainly to flicker noise and can be reduced by using a larger device.

Table 7.2 Input noise levels in BJT and MOST biased at 1 mA.

	Input Current Noise	Input Voltage Noise
BJT	11.2 nA	1.4 nV
MOST	61 pA	50 μV

7.3 Passive Elements

Passive elements such as resistors, capacitors, and inductors appear in all RF circuits. Good passives enable the design of high performance RF circuits. This is particularly true for inductors, which offers substantial improvements in noise performance otherwise not obtainable. This section summarizes key properties for each type of passive element and discusses the application of passives in RF circuitry.

7.3.1 Resistor

Active elements in RF circuits must be biased at a desired region of operation. Biasing circuitry often requires resistors to setup a certain voltage or current in the active element as shown in Figure 7.6a. Other applications of resistors include impedance matching (Figure 7.6b) and loading the output of the transistor (Figure 7.6c) to achieve a specified gain.

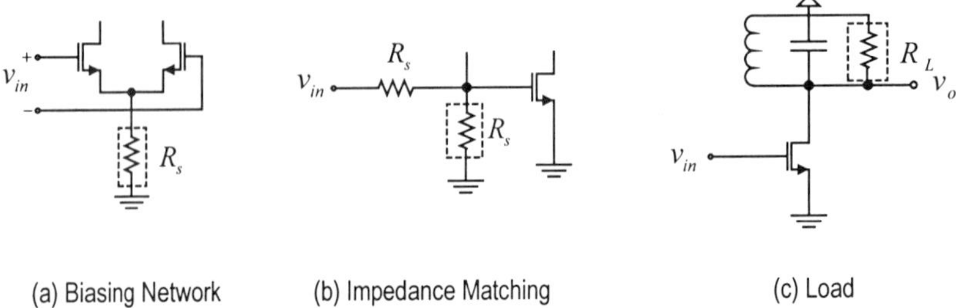

Figure 7.6 Use of resistors for biasing, impedance matching, and passive loads.

Since integrated resistors tend to have a poor tolerance (20-40%), a high temperature coefficient, and low ohmic values per square, the main design considerations are in designing the circuits such that resistors can be integrated and are insensitive to temperature variations and tolerances. Details of achieving these goals are outside the scope of this book but may be found in [141][183][186]. Table 7.3 summarizes characteristics of resistors commonly used in integrated circuit designs.

Table 7.3 Typical resistor characteristics.

Type	Resistance ($\Omega / square$)	Tolerance (%)	Temperature Coefficient (ppm/C°)
Aluminum	0.05	20	100
Silicided Polysilicon	5-10	20-40	1000
Well	1000-10000	50-80	3000-5000
Polysilicon	1500	1	100

7.3.2 Capacitor

Capacitors are useful for DC blocking between RF stages or at DC bias points as shown in Figure 7.7a. For high frequency circuits, capacitors are also essential to provide building blocks in impedance matching networks and tuned loads shown in Figure 7.7b-c. Table 7.4 summarizes characteristics of capacitor commonly used in integrated designs.

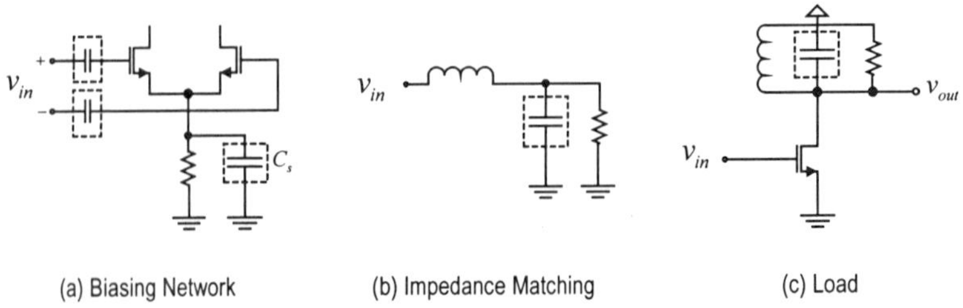

Figure 7.7 Use of capacitors in DC block, impedance matching, and tuned loads.

Table 7.4 Typical capacitor characteristics.

Type	Capacitance	Tolerance (%)	Temperature Coefficient (ppm/C°)
Metal Interconnect	0.1-0.5 fF/μm	10	30-50
Metal Area	0.05 fF/μm²	10	30-50
Poly-Poly	0.6 fF/μm²	1-5	20-30
MOS Gate	1-5 fF/μm²	5-10	20-30

7.3.3 Inductor

To operate in the GHz range, low noise and high gain can be achieved by the use of inductors in impedance matching networks and tuned loads. In contrast, integrated low frequency designs most often do not need to use inductors because they become too large to integrate. This section discusses two common means of implementing inductors for RF designs based on planar spirals and bondwires. The former is suitable for on-chip integration while the latter require additional provisions in packaging.

Planar spirals come in many variations [187]. Figure 7.8 shows two basic square spiral inductors. The inductance per area (nH/μm^2) of the spirals can be approximated by the following expression

$$\frac{L}{A_L} \approx \frac{\mu_0 \sqrt{A_L}}{8\times10^3 (D+W)^2}, \tag{7.16}$$

where $\mu_0 = 4\pi\times10^{-7}\ H/m$ is the free-space permeability, A_L is the area of the inductor in m^2, D is the distance between metals in meters, and W is the width of the metal in meters. Both D and W depend on the process design rule. Typical values are on the order of 1-3 μm. Expression (7.16) implies that $L \propto A_L^{1.5}$. Since an inductor is typically applied with a capacitor of value C to achieve resonance at $(LC)^{-1/2}$ and $C \propto A_C$, the area of an inductor varies inversely to the resonant frequency

$$A_L \propto \frac{1}{A_C^{2/3}\omega^{4/3}}, \tag{7.17}$$

where A_C is the area of the capacitor.

Example 7.7 Consider a spiral inductor of area 100 μm × 100 μm, D of 3 μm, and W of 3 μm. Then, according to (7.17), the inductance per area is 4.4×10^{-4} nH/μm^2 and the total inductance is 4.4 nH. □

One can also express inductance per length (nH/mm) shown below

$$\frac{L}{l} \approx \frac{\mu_0 \times 10^6}{8} \sqrt{\frac{l}{D+W}}, \tag{7.18}$$

where l is total length of the spiral inductor.

Example 7.8 Consider a spiral inductor of area 100 μm× 100 μm, D of 3 μm, and W of 3 μm. The total length of this inductor is approximately 1.7 mm determined by $l \approx A_L (D+W)^{-1}$. Then, according to (7.18), the inductance per length is 2.6 nH/mm. □

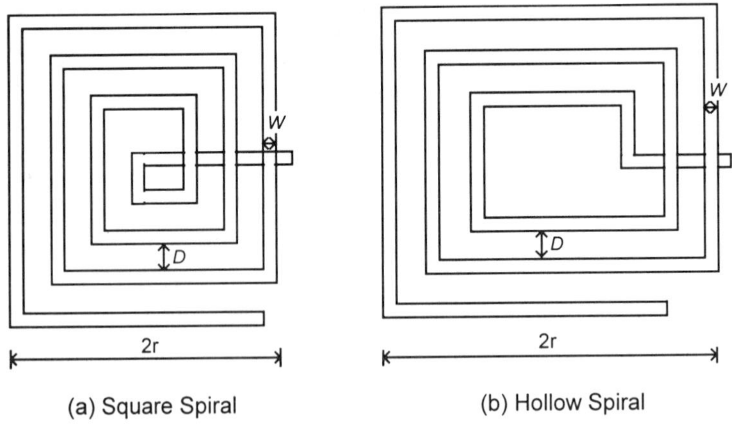

(a) Square Spiral (b) Hollow Spiral

Figure 7.8 Spiral Inductors.

While it might appear that obtaining inductance on the order of 10's nH are easily obtainable with on-chip spirals, the problem with on-chip spirals lies in the low quality factor when integrated on a lossy substrate such as silicon. The Q of the inductor at resonance can be approximated by

$$Q \approx \frac{\sqrt{L/C}}{r_s}. \tag{7.19}$$

The lossy substrate introduces capacitance and resistance paths beneath the spiral inductor. Parasitic capacitance to the substrate increases the value of C while the resistance paths in the substrate promote image currents that tend to decrease the value of L. Both of these effects decrease the quality factor. To make the matter worse, spirals can have significant series resistance due to skin loss at high frequency. Though it is possible to reduce the amount of series resistance by widening and thickening the metal lines, the lines cannot be widened without increasing parasitic capacitance to both the substrate and the cross-under feed point shown in Figure 7.8a. The thickness of the line is also limited by standard IC process to less than a micron.

Ignoring the effect of image current and substrate loss, the quality factor of a spiral inductor can be approximated by

$$Q \approx \frac{1-e^{-T/\delta}}{4}\left[\frac{4\mu_0\sigma^2 W^3}{nC_{ox}(W+D)^2}\right]^{1/4}, \tag{7.20}$$

where T is the thickness of the metal line, σ is the conductivity of the metal line (approximately $4\times10^7\,S/m$), n is the number of turns in the spiral, and δ is the skin depth. Expression (7.20) indicates that the metal width should be increased to offset the increase in series resistance due to skin effect. However, care should be taken when increasing the metal line width since the increase in parasitic capacitance could diminish the improvement substantially. While (7.20) seems to indicate that Q increases indefinitely with larger line width, in reality Q eventually decreases due to substrate loss. Also, to increase Q, the number of turns should be minimized for a given inductance value. A possible configuration to achieve this is shown in Figure 7.8b. Again while (7.20) indicates that Q decreases monotonically with increasing n, in reality Q degrades at lower number of turns due to substrate loss. Thus, an optimum n and W exist where the Q peaks.

Finally, whenever possible, the effective thickness of the metal line should be increased to mitigate the degradation in conductivity due to skin effect. It should be noted that even with appropriate optimizations, typical on-chip inductors on silicon substrate are limited to a quality factor of about ten [317].

Example 7.9 Consider again a spiral inductor of area 100 μm× 100 μm, D of 3 μm, and W of 3 μm. The number of turns in the inductor is approximately 8.33 determined by $n \approx 0.5\sqrt{A_L}\,(D+W)^{-1}$. Then, according to (7.20), the quality factor is

$$Q \approx \frac{1}{4}\left[\frac{4\cdot 4\pi\times10^{-7}\cdot\left(4\times10^{7}\right)^2\cdot\left(3\times10^{-6}\right)^3}{8.33\cdot 3.5\times10^{-3}\cdot\left(6\times10^{-6}\right)^2}\right]^{1/4} = 5.33. \;\square$$

High-Q inductors can be achieved using external components, such as bondwires or commercially available inductors in the 0402 form factor. Typical quality factors achieved using these types of inductors are in the range of 50-100. The inductance per length (nH/mm) of bondwires can be expressed by [188]

$$\frac{L}{l} = \frac{\mu_0\times10^6}{2\pi}\left[\ln\frac{2l}{R}-\frac{3}{4}\right], \tag{7.21}$$

where l is the length of the bondwire and R is the radius of the inductor. A typical value of R is 12.5 μm. Applying (7.21) and assuming a length of 1 mm, the inductance is approximately 1 nH/mm. The quality factor of the bondwire inductor is less prone to parasitic effects since the separation of bondwires from the die is typically large on the order of several mils and also the series resistance is lower due to the large conductor diameter. A rough approximation of the quality factor is shown below

$$Q_{wire} \approx \sqrt{4\pi f \mu_0 \sigma R^2}\left(\ln\frac{2l}{R}-\frac{3}{4}\right). \tag{7.22}$$

where f is operating frequency.

Example 7.10 Consider an operating frequency at 2 GHz and a 4-mm bondwire with a 12.5-μm radius. Then according to (7.22), the Q factor is

$$\left(4\pi\cdot 2\times 10^{9}\cdot 4\pi\times 10^{-7}\cdot 4\times 10^{7}\cdot\left(12.5\times 10^{-6}\right)^{2}\right)^{0.5}\left(\ln\frac{0.004}{12.5\times 10^{-6}}-\frac{3}{4}\right)\approx 70.5.$$

The computed Q is much larger than what is achieved in practice and therefore (7.22) serves only as a guideline rather than a design tool. In particular, it neglects the effect of resistive losses and parasitic capacitances. More accurate model must be developed or even better EM simulators should be used. □

7.4 Low Noise Amplifier

The received RF signal can be in the -100 dBm range, corresponding to about 3 μV of signal amplitude but at the same time this weak signal can have a strong interference as much as 80 dB higher in power. The receiver therefore must perform the down conversion with substantial gain and dynamic range, typically 100 dB or more. The high dynamic range also implies low noise as discussed in Section 4.3.5.

To achieve low noise corresponding to a low system noise figure, it has been shown in Section 4.2.1 that the NF of the first stage immediately after the antenna must be minimized and its gain should be maximized as much as possible. A low system NF directly translates to the amount of noise margin defined in (5.29) available in the system design. As its name implies, a low-noise amplifier (LNA) is an amplifier with a low NF and a high gain. It is ideally inserted immediately after the antenna to minimize the effect of noise on the rest of the system. However, in practice, it is sometimes preceded by a bandpass filter that helps to increase the receiver dynamic range, especially in presence of strong adjacent interference.

7.4.1 Noise Performance

A LNA can be modeled by an equivalent two port with input noise sources discussed in Section 7.2.4. Figure 7.9 shows a two-port model with its input driven by a source having impedance Z_s and a matching network which ideally serves to maximize the power while minimizing the noise at the LNA input. For maximum power match, the input impedance at the matching network should be set equal to

$$Z_{in} = Z_s^* \tag{7.23}$$

and for noise match, the output impedance Z_{out} of the matching network should be set such that [189]

$$|Z_{out}|^2 = R_{opt}^2 = \frac{\overline{v_n^2}}{\overline{i_n^2}}. \tag{7.24}$$

The basis for (7.24) can be interpreted intuitively in the following way. The equivalent input noise source can be viewed as a noise resistor of value $\sqrt{\overline{v_n^2} / \overline{i_n^2}}$. Minimizing the input noise would then be equivalent to matching the source resistance to the noise resistor value, according to (7.23), such that maximum noise power is delivered to the source rather than to the input of the two-port. However, (7.24) generally cannot be satisfied concurrently with (7.23). Therefore, in an actual design the matching network should be designed to achieve a balance between the amount of noise injected into the LNA and the amount of power delivered.

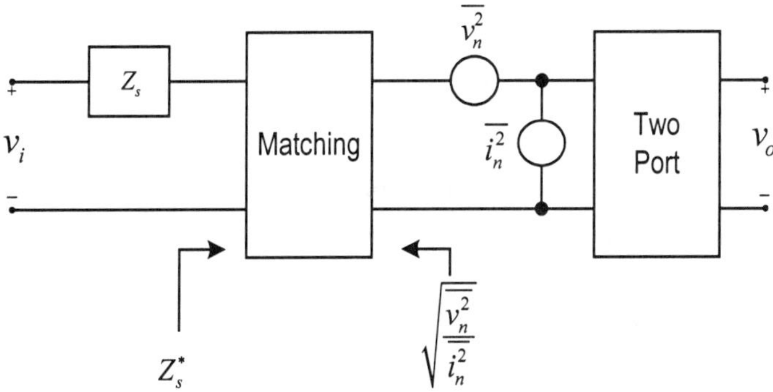

Figure 7.9 Matching requirements for a two-port model.

Expression (7.24) is now applied to derive the minimum noise figure for a MOST. The derivation can be easily extended to a BJT. Substituting the expressions for equivalent input noise sources from Figure 7.5 into (7.24) but ignoring the flicker noise and shot noise due to gate current, the optimum resistance can be determined

$$R_{opt} = \sqrt{\frac{\gamma}{\alpha\delta g_g g_m}} = \sqrt{\frac{5\gamma}{\delta}}\frac{1}{\alpha\omega C_{gs}}. \tag{7.25}$$

Based on (4.2), the minimum noise factor can then be derived by substituting $4kTR_{opt}$ for N_0 as shown below

$$F_{\min} = 1 + \frac{v_n^2 + i_n^2 R_{opt}^2}{4kTR_{opt}}, \tag{7.26}$$

where the numerator of the fraction represents the total input-referred noise. Substituting (7.25) into (7.26), the following expression is obtained for the minimum noise figure

$$F_{\min} = 1 + 2\sqrt{\frac{\gamma\delta}{5}}\frac{f}{f_T}. \quad (7.27)$$

Equality in (7.27) is achieves when the input noise sources are independent since if they are not independent then some of the noise contribution in (7.26) would have been counted multiple times. If noise correlation is taken into account, then (7.25) and (7.27) are scaled by $\sqrt{1-|c|^2}$, where c is the correlation coefficient between the noise sources [141].

Expression (7.27) indicates that the minimum noise figure is a function of three process dependent parameters γ, δ, and f_T. In particular, the noise performance of the circuit improves as the f_T of the transistor increases with smaller feature size. However, the benefit of increasing f_T is limited by short channel effects in sub-micron devices. Expression (7.27) also indicates that at a low operating frequency relative to f_T, a low noise figure can be obtained.

Table 7.5 shows the minimum achievable noise figure for MOST with $\gamma = 2$, $\delta = 4$, and $|c| = 0.4$. In practice, the actual noise figure is typically higher than those shown in Table 7.5 because flicker noise has not be taken into account and power match has not been considered. Flicker noise can contribute a substantial fraction of the noise figure for direct-conversion designs and a good power match may result in degradation in noise match. Therefore, LNA performance should be gauged in terms of both NF and input match s_{11}, which is a scattering parameter that measures the amount of reflected signal at the input of the circuit. Good LNA performance with current mature technology (e.g. 0.5 μm MOST) ranges from 2-3 dB with a –10 to –20 dB s_{11} and around 10 dB s_{21}.

Table 7.5 Minimum noise figure for MOST.

f/f_T	Noise Figure (dB) (Uncorrelated Noise)	Noise Figure (dB) (Correlated Noise)
0.1	0.98	0.91
0.2	1.78	1.65
0.4	3.04	2.85
0.8	4.81	4.56
1.0	5.48	5.21

In [141][190], an optimum source resistance is derived based on a constraint that takes into account both noise and power dissipation. The result is shown below

$$R_{opt} = \frac{2}{9\omega_0 C_{gs}}. \tag{7.28}$$

Comparing (7.28) to (7.25) and substituting typical values for α, δ, and γ, the power-optimized method results in a device size that is 5-10 times smaller than the noise-only optimization.

7.4.2 S-Parameters

The S-parameters shown in Table 7.6 are useful in general for high frequency circuits since it does not require an actual short circuit or an open circuit to obtain two-port circuit parameters. Rather, all four parameters shown in Table 7.6 can be determined by Impedance matching either the input or the output. With respect to a LNA, the input reflection coefficient s_{11} should be low (< -20 dB) to ensure that low power loss occurs at the LNA input. On the other hand, the forward gain s_{21} should be sufficiently large (> 10 dB) to lower the sensitivity of the system NF due to noise sources in later stages. For direct-conversion receivers, the reverse gain s_{12} should be low (< -40 dB) to reduce the amount of LO radiation.

Table 7.6 Summary of S-parameters.

S-Parameters	Description
s_{11}	Measures the amount of signal reflection at the input, i.e. input matching.
s_{21}	Measures the amount of input signal that gets to the output, i.e. forward gain.
s_{12}	Measures the amount of output signal that gets to the input, i.e. reverse gain.
s_{22}	Measures the amount of signal reflection at the output, i.e. output matching.

7.4.3 Linearity

While it is desirable to achieve both high linearity and low noise figure, it is generally not possible. For instance, to have a high system noise figure, the gain of the LNA should be high but since the amplifier is driven into non-linear region with smaller signal power at higher gain, the IIP3 decreases. To see the effects of distortion due to non-linear behavior of the amplifier, the transistor is modeled using (7.1).

Now consider a common-source amplifier with an equivalent load of R_L. Then, the output voltage varies around a bias point with the following amplitude

$$v_o = -\left(I_D - I_d\right) R_L, \tag{7.29}$$

where I_D corresponds to a bias current at $V_{GS} = V_{bias}$ and I_d corresponds to the total current through the transistor with $V_{gs} = V_{bias} + v_{in}$. Using power series expansion on (7.29), the coefficients β_1, β_2, and β_3 of (4.14) are determined to be

$$\beta_1 = -\frac{\mu C_{ox} W R_L}{L}\left(V_{bias} - V_t\right)\left(\frac{1+0.5\rho}{\left(1+\rho\right)^2}\right), \tag{7.30}$$

$$\beta_2 = -\frac{\mu C_{ox} W R_L}{2L}\frac{1}{\left(1+\rho\right)^3}, \tag{7.31}$$

$$\beta_3 = \frac{\mu C_{ox} W R_L}{2L}\frac{1}{\left(1+\rho\right)^4 L E_{sat}}. \tag{7.32}$$

Observe that coefficient β_1 is the small signal gain of the amplifier equal to the inverse of the product of transconductance (7.3) and R_L.

Expressions (7.30)-(7.32) can be applied to the intercept points derived in Chapter 4. For instance, these expressions are applied to obtain an expression for *IIP3* shown below in dBm

$$IIP3 = 10\log\left[\frac{4\rho\left(1+0.5\rho\right)}{3}\frac{\left(V_{sat} + LE_{sat}\right)^2}{R_s}\right] + 30, \tag{7.33}$$

where $V_{sat} = V_{bias} - V_t$. Expression (7.33) indicates that for a smaller feature size the amplifier becomes more linear with higher *IIP3*. This is because as the channel length decreases, the transistor transconductance becomes weakly dependent on its signal drive level. Moreover, *IIP3* increases with increasing V_{sat} since β_3 decreases rapidly with increasing V_{sat}. However, V_{sat} cannot be made arbitrarily large since it is constrained by the supply voltage which in turn limits the output swing of the transistor. Since the transistor moves out of active region when V_{ds} becomes less than V_{sat}, the output swing can be no greater than $V_{DD} - V_{sat}$. Thus, V_{sat} is limited to V_{DD} less the output swing.

Example 7.11 Consider a channel length of 0.25-μm, a source resistance of 50 Ω and V_{sat} of 0.1 V. Then, according to (7.33) the *IIP3* is

$$10\log\left[\frac{4\cdot 0.1\left(1+0.5\cdot 0.1\right)\left(0.1+1\right)^2}{3\cdot 50}\right] + 30 = 5.3\,dBm.$$

As v_{sat} increases to 1 V, the IIP3 increases to 22.0 dBm. □

7.4.4 Common-Source Configuration

The simplified two-port model shown in Figure 7.10 is useful in analyzing the performance of an RF circuit. The parameter r_o is the output resistance of a MOST looking into its drain. The output resistance is generally ignored given its large value (100's $k\Omega$).

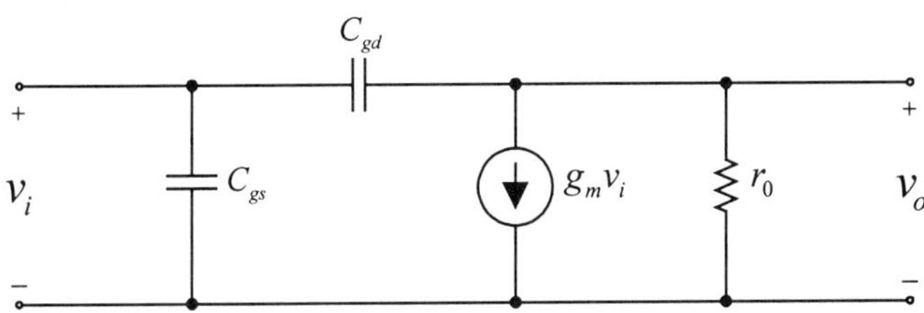

Figure 7.10 Two-port small signal model for MOST.

Consider now a common-source amplifier shown in Figure 7.11 for the LNA. Using the small signal model in Figure 7.10, it is not difficult to show that the input impedance of the amplifier with load impedance Z_L can be expressed by

$$Z_{in} \approx \frac{1}{s\left[\left(1+g_m Z_L\right)C_{gd} + C_{gs}\right]}. \tag{7.34}$$

At high frequency, the input of the amplifier appears to be a large capacitor with value $\left(1+g_m Z_L\right)C_{gd} + C_{gs}$. This capacitor has a large contribution from the overlap parasitic capacitance C_{gd} amplified by $1+g_m Z_L$ as a result of the Miller effect [183]. Because the Miller capacitance can be relatively large, the input impedance drops quickly as the frequency increases and it becomes more difficult to achieve a good power match with a 50 Ω source impedance at high frequency.

Example 7.12 Consider a DC gain of 50 corresponding to 17 dB for the common-source configuration. For simplicity, the load impedance is assumed to be real. Then, $1+g_m Z_L = 51$. The overlap capacitance for a small feature size can be a significant fraction of the gate-source capacitance. Assuming a ratio of 1:2 and using (7.34), the equivalent capacitance at the gate of the amplifier becomes approximately $26.5C_{gs}$. Based on (7.34) and assuming $Z_s = R_s$, the 3-dB bandwidth is

$$\omega_{3dB} \approx \frac{1}{26.5C_{gd}R_s}.$$

Substituting $C_{gs}R_s = 2/(9\omega_0)$ from (7.28), the 3-dB bandwidth becomes

$$\omega_{3dB} \approx 0.2\omega_0,$$

which implies the gain is substantially reduced at the RF center frequency ω_0. □

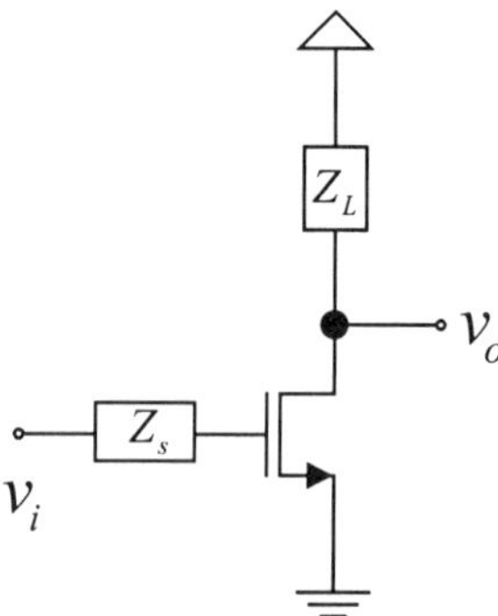

Figure 7.11 A common-source amplifier.

The output impedance of the amplifier experiences similar amplification on the overlap capacitance due to the Miller effect:

$$Z_{out} \approx \frac{1}{s\left(1+g_m Z_s\right)C_{gd}}. \tag{7.35}$$

where Z_s is the source impedance. When using a tuned load such as a RLC tank, the large effective output capacitance detunes the load and can be compensated by decreasing the inductor size. In practice, the size of the capacitance must be sufficiently larger than $\left(1+g_m R_s\right)C_{gd}$ to allow the circuit to become insensitive to unpredictable variations in the overlap capacitance and circuit transconductance. However, increasing the capacitance in the tuned load is undesirable because it causes a decrease in the bandwidth of the amplifier.

Another problem associated with the common-source amplifier is the potential for instability at a high bias current often needed to achieve good noise performance. The instability arises from a zero introduced by the overlap capacitance in the transfer function shown below

$$\frac{v_0}{v_i} = \frac{\left(sC_{gd} - g_m\right)Z_L}{\left(1+sC_{gd}Z_L\right)\left[1+s\left[\left(1+g_m Z_L\right)C_{gd}+C_{gs}\right]Z_s\right]}. \tag{7.36}$$

The zero unfortunately is in the right half of the s-plane. A positive zero tends to make the system less stable.

7.4.4.1 Inductive Degeneration

To address the problem with the low input impedance, inductive degeneration (Figure 7.12) is used to generate a resistive term in the input impedance shown as the $g_m L_s / C_{gs}$ term below

$$Z_{in} \approx sL_g + \frac{sL_s + \dfrac{1}{sC_{gs}} + \dfrac{g_m L_s}{C_{gs}}}{1 + \left(1 + g_m R_L\right)\dfrac{C_{gd}}{C_{gs}}} \tag{7.37}$$

The first term sL_g is a series inductor placed in series with the gate to cancel out the admittance due to the gate-source capacitor. Here, it is assumed that the tuned load is placed in resonance and therefore appears to be a pure resistive load R_L.

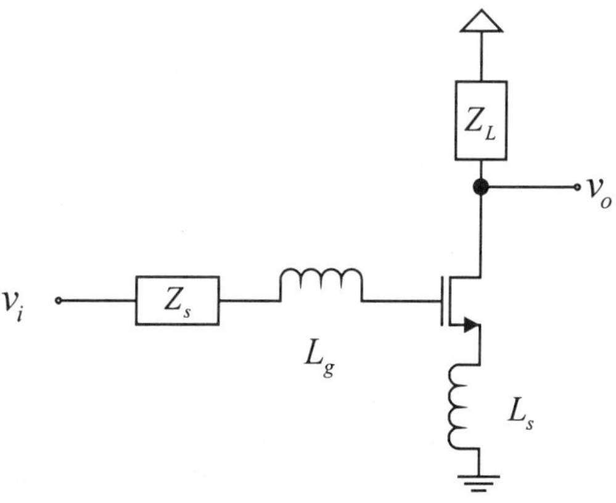

Figure 7.12 A common-source amplifier with inductive degeneration.

In contrast to using resistive impedance matching, using inductive degeneration results in less noise generation since the resistive component does not correspond to a physical resistor. To obtain input matching, the source inductor L_s is set to

$$L_s = \frac{R_s C_{gs}}{g_m}\left(1 + \left(1 + g_m R_L\right)\frac{C_{gd}}{C_{gs}}\right), \tag{7.38}$$

which is derived from (7.37). To obtain this pure resistive term at the input, the combined reactance seen at the input must cancelled out the input admittance. To achieve this cancellation, the gate inductance should be set to

$$L_g = \frac{1 - \omega_0^2 L_s C_{gs}}{\left[1 + \left(1 + g_m R_L\right)\dfrac{C_{gd}}{C_{gs}}\right] C_{gs}\omega_0^2}, \tag{7.39}$$

where ω_0 is the resonant frequency. At resonance, the voltage across C_{gs} is Q times greater than the input voltage, where Q is the quality factor of the series resonant structure of the input matching network. Using Table 4.2, the Q of the resonant structure is $\left[\omega_0 C_{gs}\left(R_s+\omega_T L_s\right)\right]^{-1}$. Therefore, the total gain of the amplifier is

$$\frac{v_0}{v_i} \approx -\frac{\omega_T R_L}{2\omega_0 R_s}, \tag{7.40}$$

which shows that the degradation in gain due to decrease in input impedance has been compensated by the input resonant matching network.

7.4.4.2 Cascode Configuration

While using inductive degeneration solves the input matching problem, the instability and detuning that occur due to the Miller effect still exist. To reduce the Miller effect but at the expense of reduced signal swing, a cascode stage can be inserted between the output of the common-source amplifier and the tuned load as shown in Figure 7.13. The impedance looking into the source of the cascode amplifier is simply g_{m2}^{-1} which has a low resistive value of 10's of ohms. The Miller capacitor at the input therefore is reduced to $\left(1+g_{m1}g_{m2}^{-1}\right)C_{gd1} \approx C_{gd1}$. At the output of the cascode amplifier, the overlap capacitor does not experience the Miller effect since the gate of the amplifier is at AC ground. Thus the tuned capacitor only has to be large enough to make the tank insensitive to C_{gd2}. Also, with a low impedance point at the output of the common-source amplifier, the instability caused by the zero associated with the overlap capacitance is greatly reduced. Finally, with an AC ground at the gate of the cascode amplifier, the output is decoupled from the input, giving a high reverse isolation to the cascode configuration. Given all of these advantages, most designs use the cascode configuration.

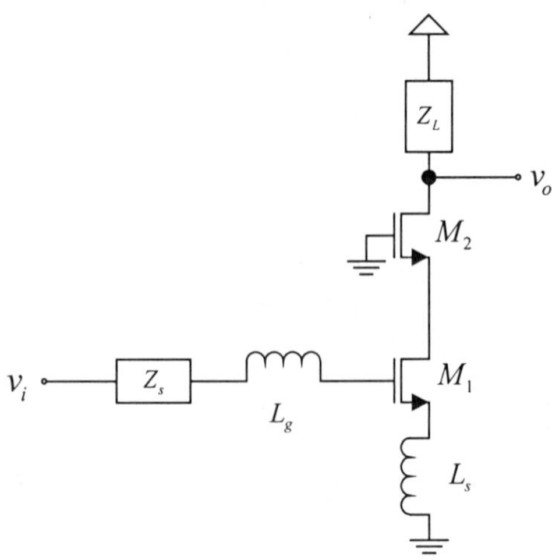

Figure 7.13 A common-source amplifier with a cascode output stage.

7.5 Mixer

Mixers are needed to either up-convert the transmitted signal or down-convert the received signal. In the process of frequency conversion, it must maintain good linearity since as discussed in Section 4.3.4, overall linearity of the receiver depends more so on the later stages rather than the earlier stages. The system noise performance, on the other hand, shows the opposite behavior. Nevertheless, it would still be important to maintain as low a noise figure as possible in the mixer such that noise performance is not greatly compromised for high linearity. This section examines the various design trade-offs associated with mixer circuits.

7.5.1 Mixer Performance

A Mixer is a three terminal device consisting of an RF port, a LO port, and an IF port as shown in Figure 7.14. The linearity and noise performance of the mixer are measured from the RF to the IF port while assuming a constant signal input at the LO port. The linearity measure for a mixer can either be *IIP3* or input compression point defined in Section 4.3. Having a good linearity in the mixer is critical to maintain an adequate linearity for the overall transceiver since mixers are positioned later in the receive chain, typically after the LNA. While it would be ideal to build a low-noise mixer that has good linearity and gain such that the LNA can be eliminated, in practice this is not possible to achieve in one stage. Given that the mixer is a non-linear device which frequency converts not only the desired signal but also its image, any noise in the image frequency also folds into the mixer output. Therefore, mixer's noise performance tends to be poor, making it necessary to break the amplification and frequency conversion in two stages where the LNA handles the amplification with sufficient gain and low NF. The high gain in the LNA compensates for the high NF of the mixer, typically in the range of 10-15 dB.

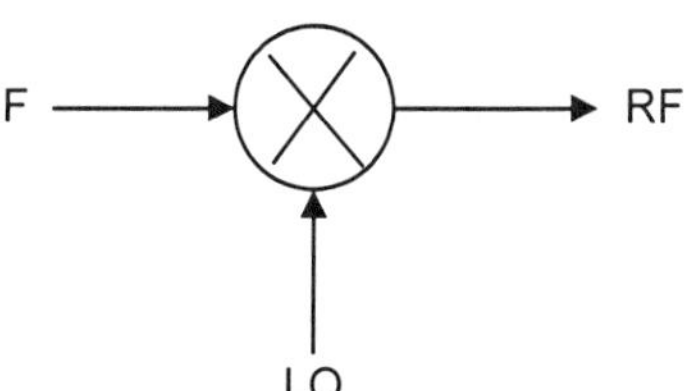

Figure 7.14 Mixer block diagram.

Moreover, because of the non-linear characteristics of a mixer, flicker noise of active devices and phase noise of the LO get modulated to the mixer output, further increasing its noise figure. Also, such noise increases the phase jitter in the transceiver output that degrades the error-rate performance. Therefore, a mixer should be designed to minimize the phase noise at its output.

Finally, mixer isolation among the different ports is essential to prevent spurs in both the transmitted and received signals. Isolation among the ports is an important design consideration for a mixer since the LO power is generally high and therefore can result in substantial signal coupling. As shown in Table 7.7, there are three main signals coupling that results in desensitization of the transceiver.

Table 7.7 Mixer coupling.

Coupling	Description (assuming a down-conversion mixer)
LO-IF	LO couples to the mixer output and causes interference if LO frequency is close to the mixer output.
LO-RF	1) LO couples to the mixer input and self-mixes to DC. This is undesirable for direct-conversion receivers. 2) LO couples to the mixer input and radiates out towards the antenna causing EMI to other transmissions.
IF-RF	The mixer IF port couples to the RF port.

7.5.2 Active Mixer

A single-balanced active mixer [191] is shown in Figure 7.15a where transistor M3 acts as a transconductor that converts the RF signal v_{RF} into a current of magnitude $g_{m3}v_{RF}\left(1+sg_{m3}L_s\right)^{-1}$. Transistors M1 and M2 act as current switches that steer the current depending on the polarity of the LO signal v_{LO}. If v_{LO} is positive then current is switched through M1 and if v_{LO} is negative then current is switched through M2. The overall effect of current steering is mixing the current through M1 with a square wave, which has a fundamental frequency at ω_{LO} with the associated odd harmonics.

The mixer output then consists of frequency components at $\{m\omega_{LO}, m\omega_{LO} \pm \omega_{RF}\}$. The desired output is either $\omega_{LO}+\omega_{RF}$ or $|\omega_{LO}-\omega_{RF}|$ depending on whether the upper sideband or lower sideband is desired. Note the LO feedthrough at $m\omega_{LO}$ is a result of the constant bias current in M3. The desired frequency component is obtained by properly filtering the mixer output, usually done by using a tuned load Z_L. Since the desired frequency is associated with the fundamental frequency component of the square wave, the conversion gain of the mixer can be determined to the first order by

$$G_c = \frac{2}{\pi}\frac{g_{m3}}{1+sg_{m3}L_s}, \tag{7.41}$$

where $2/\pi$ is half of the amplitude at the fundamental frequency. The feedthrough of the LO signal (i.e. $m\omega_{LO}$) to the mixer output can be reduced with a double balanced configuration shown in Figure 7.15b.

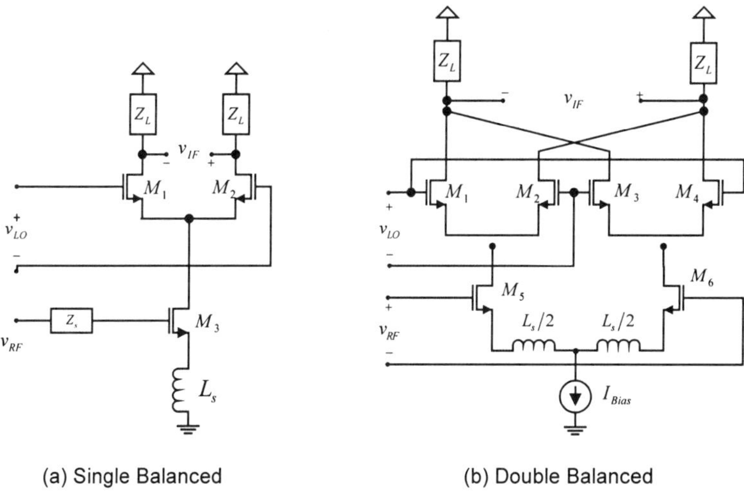

(a) Single Balanced

(b) Double Balanced

Figure 7.15 Active mixers.

Since the drain current in M3 is a function of $g_{m3}v_{RF}$, the mixer response can become non-linear as the RF signal level increases. To linearize the mixer, the source degeneration inductor is selected such that $\omega g_{m3}L_s >> 1$ thereby making the current independent of the transistor transconductance g_{m3}.

The noise of an active mixer is determined mostly by the NF of the transconductor and the switching speed of M2 and M3. Design approaches discussed for LNA NF may also be applied to the mixer to design for low noise. The LO should be designed to have sufficiently large drive and fast edges such that the time during which M2 and M3 are both on is reduced. This helps to minimize the contribution of noise from M2 and M3 at the mixer output. Also, since noise current from M3 is transferred to the mixer output through M2 and M3, flicker noise gets up converted in frequency and can cause significant degradation for low IF or direct-conversion architectures. To minimize the effect of flicker noise, the transistor size for M3 should be made as large as possible.

7.5.3 Passive Double-Balanced Mixer

Rather than switching currents as in active mixers, double-balanced passive mixers [192] switch voltages as shown in Figure 7.16. Such passive mixers offer high linearity since no transconductance amplifier is required. When M2 and M4 are on M1 and M3 are off, the mixer output is $2v_{RF}/\pi$, where $2/\pi$ arises from the square-wave spectral component at the fundamental frequency ω_{LO}. The amplitude at ω_{LO} is actually $4/\pi$ but is reduced by a half due to mixing two sinusoids.

A key design consideration for the passive double balanced mixer is to ensure that the transistors in the bridge must remain in the linear region of operation. Therefore, the gates of M1 to M4 should be driven at a high signal level. The passive mixer can potentially achieve better noise performance because there are fewer active elements and also no flicker noise due to lack of bias current. However, since M1-M4 provides a switching structure with no gain stage, the mixer offers no conversion gain but rather conversion loss and therefore worsens the NF. Based on the earlier discussion, the conversion gain of a passive mixer is

$$G_c = \frac{2}{\pi}. \tag{7.42}$$

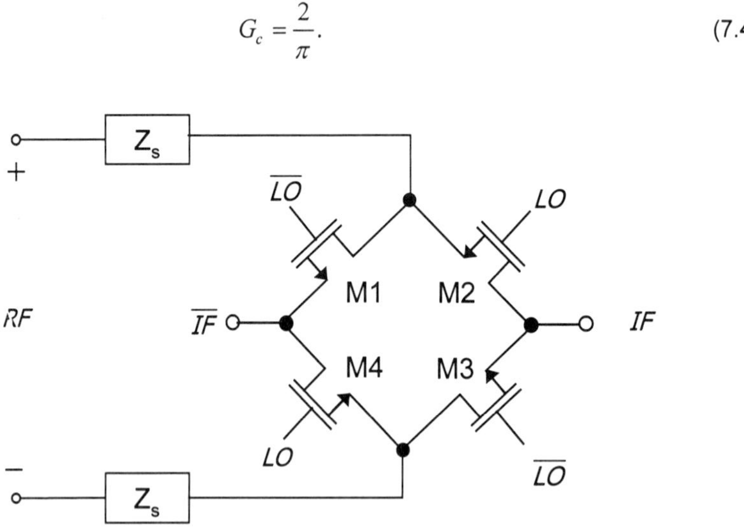

Figure 7.16 Passive double-balanced Mixer.

7.5.4 Sub-sampling Mixer

If an RF signal is sampled at a fraction of the carrier frequency, then the resulting spectral content at the sampler output contains aliases at multiples of the sampling rate F_s as shown in Figure 7.17. If the sub-sampling ratio M is chosen such that one of the alias bands centers around ω_{IF}, then a bandpass filter can be applied at the sampler output to obtain the desired signal output. Figure 7.17 shows the case when zero IF is used. M can be determined with the following expression

$$M = \frac{\omega_{RF} - \omega_{IF}}{2\pi F_s}. \tag{7.43}$$

Though seemingly simple and attractive, the sub-sampling mixer has noise enhancement problem due to noise folding from the alias bands into the desired signal band. The amount of noise enhancement equals M, which is the range of 10-100. Therefore, the noise figure of a sub-sampling mixer easily degrades by 10-20 dB.

Furthermore, any phase noise in the sampling clock directly affects the sampler output. Therefore, low-phase noise clock must be used but is also difficult to realize in practice. Finally, the sampling switch shown in Figure 7.17 does not have the accuracy required by the mixing function and therefore more complex sample-and-hold circuits should be used. While sub-sampling mixer [193] has noise problems, it does however offer high linearity and simplicity that would make it applicable to short-range applications in which having a high NF is not required. Such short-range applications include Bluetooth.

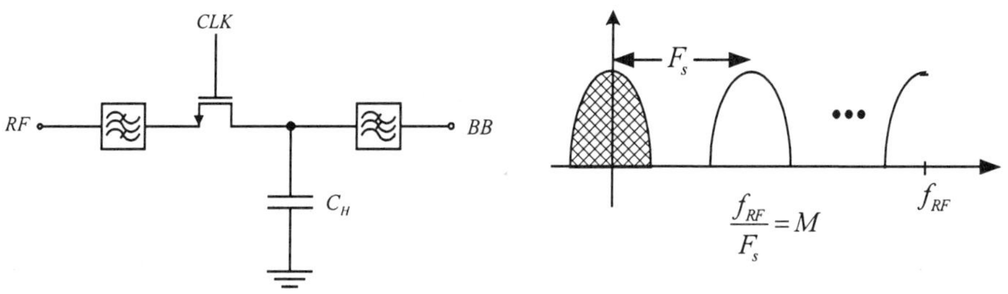

Figure 7.17 Sub-sampling mixer.

7.6 Synthesizer

Synthesizers generate the LO's at the transmitter as well as at the receiver. Synthesizers are implemented using direct frequency synthesis or indirect frequency synthesis. Direct synthesis based on direct-digital frequency synthesizer will be discussed in Chapter 8. This section concentrates on indirect frequency synthesis based on phase-locked loops. The basic theoretical background on PLL has been described in Section 2.5.2 of Chapter 2. Here its application to RF synthesizer designs is described.

System requirements that are pertinent to synthesizer designs are summarized in Table 7.8. Small channel spacing requires higher tuning accuracy and lower phase noise. Wide aggregate transmission bandwidth requires wide tuning range and linearity in the output frequency versus control voltage. High receiver sensitivity requires low phase noise to minimize reciprocal mixing. Time-division duplexing requires fast settling time (i.e. fast tracking) such that the receiver and transmitter can tune to the correct frequency without incurring excessive overhead. Use of frequency-hop to obtain diversity may also require fast settling time if the hop rate is high. Unfortunately, designing a synthesizer that meets high tuning accuracy, low phase noise, high tuning range, high linearity, and low settling time is difficult since the design parameters conflict with one another. However, with careful design choices, one can obtain a balanced trade-off to meet adequate system performance.

Table 7.8 Synthesizer design requirements.

System Requirement	Synthesizer Specification
Channel Spacing (W_c)	Tuning Accuracy (steady-state error). Phase Noise (spectral purity)
Available Bandwidth (W_A)	Tuning Range and linearity
Sensitivity (S)	Phase Noise
Duplexing	Settling Time (tracking speed)
Diversity via Frequency Hop	Setting Time

7.6.1 Synthesizer Parameters

Figure 7.18 shows a basic synthesizer block diagram implemented using a PLL. The crystal oscillator generates a low-frequency reference signal $v_{ref}(t)$ that has good frequency stability (e.g. 1-10 ppm) over temperature. The phase detector compares the phase of $v_{ref}(t)$ against the phase of the prescalar output. Often, a multiplier implements the phase detector by mixing $v_{ref}(t)$ with the prescalar output to generate an approximation of the phase error $-K_d(t\Delta\omega + \phi_{LO})/N$. The phase detector output contains higher order harmonics which are filtered out by the prefilter. To maintain stability, the prefilter should have a much wider bandwidth than the loop filter.

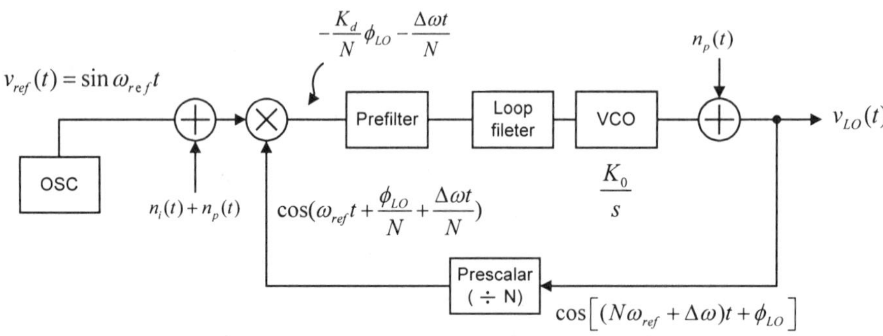

Figure 7.18 Block diagram of an integer-N synthesizer.

The phase error consists of a constant phase offset ϕ_{LO} and a frequency offset $\Delta\omega$, which could be slow time varying. The loop filter averages the phase error to generate a control signal to the voltage-controlled oscillator (VCO) and tunes the VCO output frequency until the average phase error converges to zero. The tuning of the VCO frequency occurs in a negative feedback loop where the VCO output is first scaled by a prescalar in the phase domain and then sent to the phase detector. Ideally when the PLL is locked, the frequency and phase of the prescalar output should equal to that of the crystal oscillator. In other words, when the PLL reaches steady state, the

VCO output should be at $N\omega_{ref}$. The synthesizer can therefore be tuned by changing the scale factor N over the tuning range of the VCO. This is essential for the transceiver to tune over multiple frequency channels in the given available bandwidth W_A.

The performance of the synthesizer implemented based on the architecture shown in Figure 7.18 depends predominantly on the choice of loop filter, the VCO gain K_0, measured in $rad \cdot s^{-1}V^{-1}$, and the phase-detector gain K_d, measured in $rad \cdot V^{-1}$. Two configurations of PLL can be used for this synthesizer. These configurations are often referred to as type-I and type-II loops, where the former uses a constant gain block for the loop filter while the latter uses a lag-lead network. Table 7.9 summarizes the characteristics of the two PLL configurations. Detailed discussions and derivations of the parameters listed in Table 7.9 can be found in Section 2.5.2.

Table 7.9 Phase-locked loop summary.

Coupling	Type I	Type II
Transfer Function	$N\dfrac{K_{DC}}{s+K_{DC}}$	$N\dfrac{K_\infty s + K_\infty \tau_2^{-1}}{s^2 + K_\infty s + K_\infty \tau_2^{-1}}$
Loop Filter Transfer Function	$F(0)$	$\dfrac{\tau_2 s + 1}{\tau_1 s}$
DC Loop Gain (K_{DC})	$K_d K_0 F(0) N^{-1}$	$K_{DC} \to \infty$
AC Loop Gain (K_∞)	$\dfrac{K_d K_0 F(\infty)}{N}$	$K_\infty = \dfrac{K_d K_0 \tau_2}{N\tau_1}$
Phase Error (θ_e)	$\theta_e \to 0$	$\theta_e \to 0$
Frequency Error (ω_e)	$\omega_e \to N\Delta\omega / K_{DC}$	$\omega_e \to 0$
Noise BW (B_L)	$0.25NK_{DC}$	$\dfrac{N\omega_n}{2}\left(\zeta + \dfrac{1}{4\zeta}\right)$
Phase Jitter $\left(\overline{\theta_0^2}\right)$ Due to Input White Noise	$\dfrac{NK_{DC}}{4\overline{\gamma} R_b}$	$\dfrac{N\omega_n}{2\overline{\gamma} R_b}\left(\zeta + \dfrac{1}{4\zeta}\right)$

Since the settling time and steady-state error of a PLL depends inversely on the loop gains, the larger K_0 and K_d lead to more accurate and faster tracking performance. In a type-I loop, the noise bandwidth scales with the DC loop gain. Therefore, even though lower frequency error can be obtained with a high DC loop gain, phase jitter increases due to the higher noise bandwidth. In contrast, type-II loop has two design parameters $\zeta = 0.5\sqrt{\tau_2 K_\infty}$ and $\omega_n = \sqrt{K_\infty \tau_2^{-1}}$ such that one can obtain a balanced tracking and noise performance. The tracking speed and noise bandwidth can be controlled by adjusting the resonant frequency ω_n and the damping factor ζ. In fact, it can be shown that the optimum noise bandwidth occurs at $\zeta = 0.5$ and has a value half that of the resonant frequency. Another advantage in using a type-II loop is that the steady-state error for both phase and frequency approaches zero. For the above reasons, type-II is almost always preferred over a type-I loop.

7.6.2 Fractional-N Synthesizer

The synthesizer architecture shown in Figure 7.18 is also referred to as an integer-N synthesizer because the output frequency equals *N* times the reference frequency. For small channel spacing W_c, the reference frequency must be small but the multiplying factor *N* must be large to cover a large available bandwidth W_A. With a large *N*, loop gains are decreased, resulting in worse tracking and steady-state error performance as well as potential for instability. Moreover, the output phase jitter increases as a function *N* and thus the noise performance also degrades as *N* is increased. Reducing *N*, therefore, enables lower phase jitter, faster tracking, and a more accurate steady-state response. However, often times, given the requirement for high tuning resolution, reduction in *N* cannot be achieved with an integer-N synthesizer.

In the above discussion, only white noise is considered. However, the effect of phase noise from the reference oscillator and VCO has not been considered. If these are also taken into account, the output noise spectral density can be described as follows

$$P_n(f) = \left[P_{wn}(f) + P_{ref}(f)\right]\left|H(f)\right|^2 + P_{VCO}(f)\left|1 - H(f)\right|^2 \tag{7.44}$$

where $P_{wn}(f)$ is the input white noise density, $P_{ref}(f)$ is the phase noise of the reference oscillator, $P_{VCO}(f)$ is the phase noise of the VCO, and $H(f)$ is the PLL transfer function. Expression (7.44) indicates that lowering the loop bandwidth reduces phase jitter due to the input white noise and the phase noise of the reference oscillator. However, a lower loop bandwidth also increases the phase jitter due to the phase noise of the VCO. With an integer-N synthesizer the loop bandwidth must be made smaller than the channel spacing to filter out harmonics of the reference oscillator. While this helps to reduce the phase jitter due to the first term in (7.44), the second term is inevitable enhanced, which compromises the overall jitter performance.

A different architecture approach based on the fractional-N synthesizer [194] offers a more balanced design. The fractional-N synthesizer has the same basic architecture as the integer-N synthesizer but with a different implementation of the prescalar. Rather than fixing the divide ratio, the divide ration in this case is changed between *N* and *N+K* with a duty cycle D. The average divide ratio then becomes

$$\overline{N} = (N+K)D + N(1-D) = N + KD. \tag{7.45}$$

Since D can be made less than unity, small channel spacing can be obtained with a relatively high reference frequency. Therefore, loop bandwidth can be made larger than the channel spacing without sacrificing spurious performance.

Example 7.13 Consider a reference frequency of 2 MHz and a carrier frequency of 1.8 GHz with a channel spacing of 200 kHz. N is chosen to be 900 such that the output of the synthesizer is centered on 1.8 GHz. To tune the output at a resolution of 200 kHz, D is set to 0.1. Different channels can be selected by adjusting K. □

It should be noted that a fractional-N synthesizer also has problems. In particular, since the instantaneous divide ratio changes with duty cycle D, low-frequency sidebands are generated that could increase the spurious levels at the synthesizer output. Fortunately, since the sidebands are introduced by a known mechanism, it can also be calibrated out.

7.6.3 Oscillators

Synthesizers consist of many components comprised of phase detector, loop filter, prescalar, and oscillators. Rather than focusing on the details of all components, we focus only on the oscillator which imposes fundamental limitations on the performance of a synthesizer. Readers interested in detailed discussions on other synthesizer components might find [195][196] useful.

The performance of Oscillators influences a number of important synthesizer parameters. These include spectral purity, linearity of tuning voltage versus output frequency, tuning range, gain, and tuning accuracy. Spectral purity is largely determined by the amount of phase noise generated by the oscillator. In particular, as discussed in 7.6.2, phase noise of both the reference oscillator and the VCO influences the phase jitter at the output of the synthesizer. Since the influence of reference oscillator's phase noise diminishes with a smaller loop bandwidth but that of the VCO increases, phase noise cannot be suppressed completely.

In general, reference oscillators should have low phase noise and high frequency stability such that the synthesizer performance is not substantially compromised even if a large loop bandwidth is used to suppress the VCO phase noise and to achieve fast tracking performance. While ideally the VCO noise should also be minimized but generally this is not possible with the high tuning range and gain requirement. High tuning range is essential to tune across a wide transmission band while high gain is needed for improved accuracy and tracking performance. The following sub-sections describe a few basic implementations of reference oscillators and VCO's.

7.6.3.1 Positive Feedback

Ideally, an oscillator generates a tone at the desired center frequency without any external input. Therefore, an oscillator is inherently unstable. There are many ways of implementing oscillators but they are all based on the principle of positive feedback to create the instability needed for oscillation. Figure 7.19 shows the general block diagram of a positive feedback loop where $A(s)$ is the forward gain and $\beta(s)$ is the feedback gain. The closed loop response $H_{cl}(s)$ is

$$H_{cl}(s)=\frac{A(s)}{1-\beta(s)A(s)}. \tag{7.46}$$

In contrast to a negative feedback loop, the denominator of (7.46) is $1-\beta(s)A(s)$ instead of $1+\beta(s)A(s)$. When the loop gain $A(s)\beta(s)$ becomes one, the close-loop response approaches infinity, implying an oscillatory output. More generally, to overcome losses in the feedback loop, the loop gain should be made greater than unity with zero phase at the desired frequency.

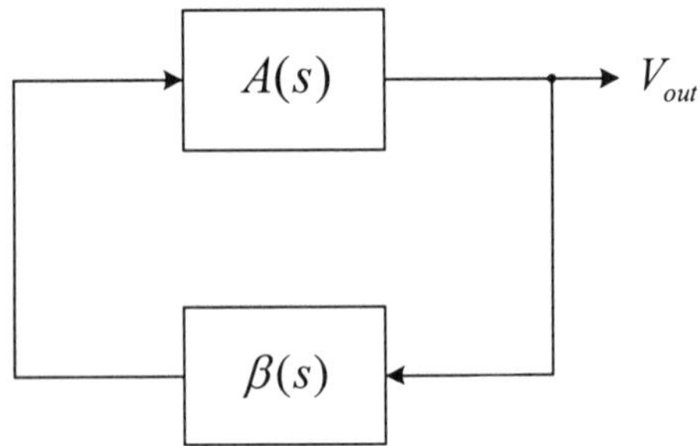

Figure 7.19 A positive feedback loop.

7.6.3.2 LC Oscillator

Since the synthesizer output is generally a pure tone with extremely narrow bandwidth, a tuned RLC tank is generally used for the feedback network $\beta(s)$. The RLC tank is given some excitation using an active device that pushes it into oscillation. Two configurations are typically used for the active device, common-source and common-gate amplifiers. To ensure positive feedback, the output of the common-source amplifier is fed back to the gate while the output of the common-gate amplifier is fed back to its source. Figure 7.20 shows the Colpitts and Hartley oscillators based on the common-gate configuration.

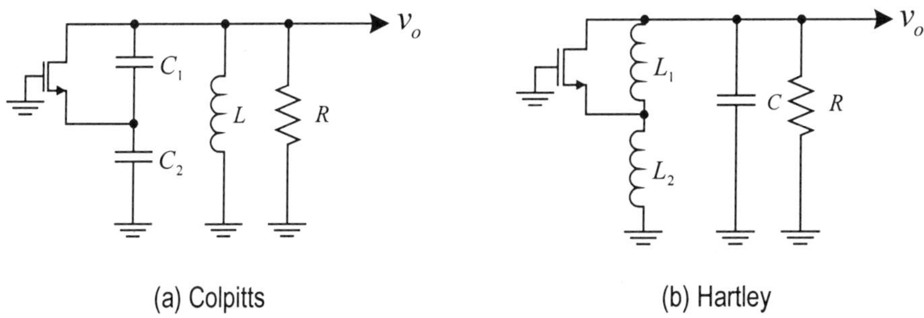

(a) Colpitts (b) Hartley

Figure 7.20 LC Oscillators.

Properties of these two oscillators are shown in Table 7.10. The transconductance of the common-gate amplifier must be sufficiently larger than C_2/C_1 for the Colpitts oscillator and L_1/L_2 for the Hartley to oscillator. The resonant frequency is determined largely by the inductor and capacitor values though there is a weak dependence on the series resistance r_s of the inductor. The series resistance has an adverse effect on the quality factor of the oscillators.

Table 7.10 Properties of Colpitts and Hartley oscillators.

	Colpitts	Hartley
Resonant Frequency	$\sqrt{1+\frac{r_s}{R}}\frac{1}{\sqrt{L\frac{C_1C_2}{C_1+C_2}}}$	$\sqrt{1+\frac{r_s}{R}}\frac{1}{\sqrt{C(L_1+L_2)}}$
Quality Factor	$\frac{\sqrt{1+\frac{r_s}{R}}}{\frac{1}{R}\sqrt{\frac{L(C_1+C_2)}{C_1C_2}}+r_s\sqrt{\frac{C_1C_2}{L(C_1+C_2)}}}$	$\frac{\sqrt{1+\frac{r_s}{R}}}{\frac{1}{R}\sqrt{\frac{L_1+L_2}{C}}+r_s\sqrt{\frac{C}{L_1+L_2}}}$
Oscillation Condition	$g_mR>\frac{C_2}{C_1}$	$g_mR>\frac{L_1}{L_2}$

Having a high-quality factor in the *RLC* tank is essential to achieve good spectral purity since phase noise varies inversely to the square of the quality factor as shown below

$$P_\phi(\omega)\propto\frac{\omega_0^2P_n(\omega)}{4Q^2\Delta\omega^2P_{sig}}. \tag{7.47}$$

where $P_n(\omega)$ is the PSD of the noise arising from the oscillator circuitry, $\Delta\omega$ is the frequency offset from the resonant frequency ω_0 and P_{sig} is the power of the oscillator output. A derivation of (7.47) is shown in Section 4.4.3 of Chapter 4. Unfortunately, as mentioned earlier, high-Q on-chip passives are not possible in standard silicon process due to the lossy substrate. Therefore, LC oscillators implemented using on-chip passive components result in poor phase noise performance. To obtain high spectral purity, high-Q external passives or resonators, such as dielectric resonators or crystals are used.

7.6.3.3 Crystal Oscillator

Crystals consist of piezoelectric material that releases charge when stressed mechanically. When driven with the right frequency, it can be driven into mechanical resonance and thus can be used as a resonator. Electrical equivalent model of a quartz crystal is shown in Figure 7.21. C_s and L_s form a series LC tank that models the transduction behavior of the crystal. The series resistance is undesirable but because of the large L_s and small C_s, quality factors on the order of 10^5 to 10^6 can still be obtained. C_p is a parasitic capacitor due to packaging of the crystal.

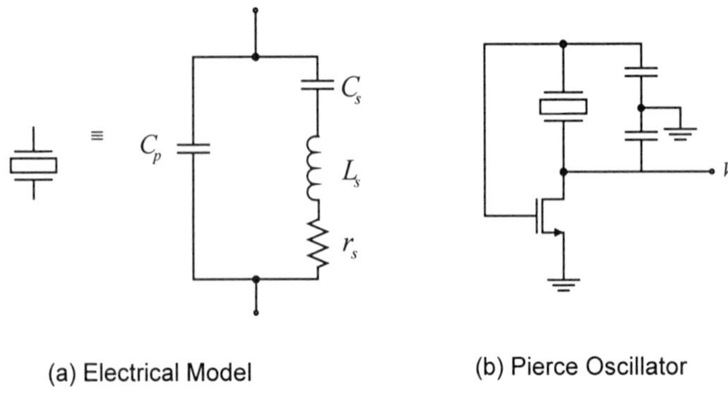

(a) Electrical Model (b) Pierce Oscillator

Figure 7.21 Crystal resonators.

Table 7.11 summarizes key properties of crystal resonators in both series mode and parallel mode. Because C_p is generally larger than C_s, the resonant frequency of the series mode and parallel mode are very close to one another. Therefore, significant peaking in the tank impedance can be obtained for high quality factor. Very good spectral purity can be obtained with a crystal resonator because of the high Quality factor of the crystal. Figure 7.21b shows a Pierce crystal oscillator which is essentially a Colpitts oscillator with the inductor replaced by a crystal. Such an oscillator has good spectral purity but small tuning range due to the closeness of the resonant frequencies in the two modes. Expression (7.48) approximates the tuning range of a Pierce crystal oscillator and has typical values of 10-50 ppm. High temperature stability can also be achieved on the order of 1-10 ppm/C°. Crystal oscillators are therefore ideal for use as the reference oscillator in the synthesizer to provide a frequency reference that has long-term stability and high spectral purity.

$$tuning\ range \approx \frac{C_1}{2C_p}\frac{1}{\sqrt{L_s C_s}} \tag{7.48}$$

One limiting factor with crystals is that frequencies are often limited in the range of 10's MHz since a crystal normally used for resonators is a bulk device with physical limitation on the minimum usable thickness. High frequency crystals are difficult to manufacture and thus expensive. Higher frequencies can be obtained via dielectric resonators or SAW resonators. Fortunately, since the reference frequency often is not larger than several 10's of MHz, the maximum frequency limitation is generally not an issue.

Table 7.11 Properties of crystal resonators.

	Series Mode	**Parallel Mode**
Resonant Frequency	$\frac{1}{\sqrt{L_s C_s}}$	$\frac{1}{\sqrt{L_s\left(\frac{C_p C_s}{C_p + C_s}\right)}}$
Quality Factor	$\frac{1}{r_s}\sqrt{\frac{L_s}{C_s}}$	$\frac{1}{r_s}\sqrt{\frac{L_s\left(C_p + C_s\right)}{C_p C_s}}$

7.6.3.4 Negative-Resistance LC Oscillator

Given that a good fixed reference can be built using a crystal oscillator, Let us now focus on the VCO required to tune the synthesizer over a wide frequency range. To tune the oscillator frequency, voltage-controlled capacitors, also known as varactors, can be used to vary the resonant frequency subject to a control voltage. A varactor has a capacitance that varies inversely to a control voltage raised to the power m as shown below

$$C\left(V_{cntl}\right) \approx \frac{K^m C_{j0}}{V_{cntl}^m} \tag{7.49}$$

where K, C_{j0} and m are constants that depend on the physical device. Generally, a reverse-biased junction capacitor is used.

When a varactor is used, the resonant frequency becomes

$$\omega_0 \approx \left(\frac{V_{cntl}}{K}\right)^{m/2}\frac{1}{\sqrt{LC_{j0}}}. \tag{7.50}$$

Since m tends to be process dependent and does not equal to 2, the synthesizer frequency does not vary linearly with the control voltage. In fact, larger tuning range intensifies non-linear behavior.

While low phase noise is desirable in a VCO, it is not the prevailing design consideration in comparison to a wide tuning range. Thus, most designs do not use voltage-controlled crystal oscillator (VCXO) due to its narrow tuning range. Rather, LC oscillators are used to obtain a wide tuning range. Since high Q in the LC components helps to reduce phase noise but resistive loss tends to lower the Q, LC oscillators use negative resistance impedance transformers to cancel out the undesirable resistance. Such measure is particularly important if the LC components are integrated on-chip. A voltage-controlled LC oscillator built based on this concept is shown in Figure 7.22. A PLL synthesizer tunes the oscillator frequency by adjusting the control voltage across the varactors.

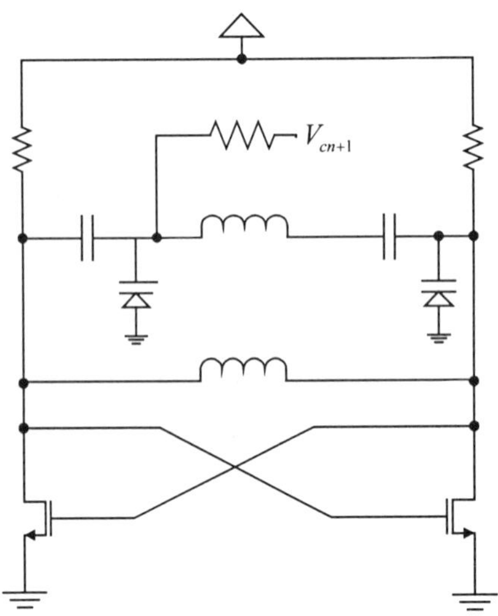

Figure 7.22 A negative-resistance LC oscillator.

7.6.3.5 Relaxation Ring Oscillator

For integrated transceivers, the inductors and capacitors in the LC tank are costly in area. Therefore, often solutions are sought that do not require a LC tank. An example of such oscillator is the relaxation ring oscillator shown in Figure 7.23 where each stage consists of a differential pair whose bias current is voltage controlled to modify the VCO output frequency. The output frequency is related to the number of stages by

$$\omega_0 = \frac{g_m}{2nC_t} \tag{7.51}$$

where g_m is the transconductance of the differential pair, n is the number of stages, C_t is the effective capacitance at each stage. If a cascode configuration is used then C_t approximately equals C_{gs}. Therefore, the maximum output frequency can be approximated by $0.5\omega_T / n$. To achieve high frequency, n should be minimized. Using a sub-micron BiCMOS process with a transition frequency of 20 GHz, oscillation frequency up to 6 GHz has been demonstrated [197].

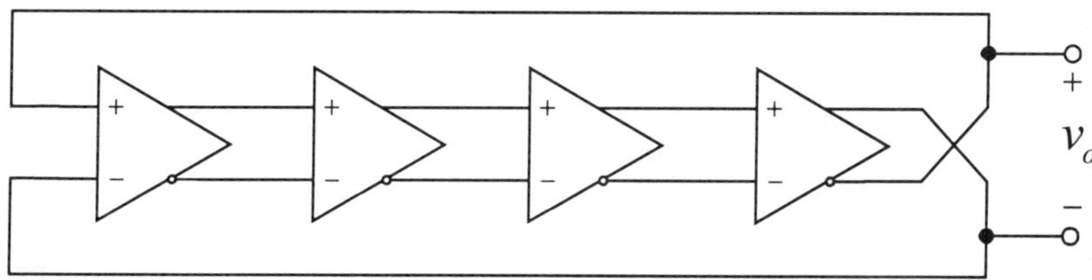

Figure 7.23 Relaxation ring oscillator.

Adjusting the bias current in each stage changes the transconductance and thereby varies the oscillator frequency. However, since the transconductance varies as square root of the bias current in CMOS, tuning behavior is non-linear though a wide tuning range is possible. Moreover, since transconductor is generally used to control the frequency, designing for high VCO gain K_0 often results in additional non-linear behavior in the frequency transfer curve. Finally, the main disadvantage with a ring oscillator lies in its poor spectral purity show below

$$P_\phi(\omega) = \frac{\omega_0^3}{\Delta\omega^2} \frac{nC_t P_n(\omega)}{4\pi R_L I_{bias}}, \tag{7.52}$$

where I_{bias} is the bias current in each stage, R_L is the load resistance in each stage, and $P_n(\omega)$ is the noise PSD due to the circuit element. A predominant noise source for MOST-based ring oscillator is the flicker noise. In contrast to a LC oscillator, there are no high-Q elements in a ring oscillator to help reduce the phase noise as evident in (7.52). Therefore, ring oscillators exhibit higher phase noise than a LC-based design.

7.6.3.6 Quadrature Generation

So far oscillators with a single output have been discussed but often both I and Q references are needed for I/Q up-down conversion. To meet this requirement, two popular approaches based on ring-oscillator and 90°-phase shifter are used [198]. Figure 7.24a shows a ring oscillator that has three non-inverting stages and one inverting stage. It can be shown that, outputs separated by two stages are in quadrature.

The phase shifter shown in Figure 7.24b splits the synthesizer output into two signals, one filtered by a RC lowpass filter and the other by a CR highpass filter. The phase difference $\Delta\phi$ and amplitude ratio ΔA between the Q and I channel can be expressed by

$$\Delta\phi = -\left(\tan^{-1}\omega_o\tau + \tan^{-1}\frac{1}{\omega_o\tau}\right) \qquad (7.53)$$

$$\Delta A = \omega_o\tau.$$

where $\tau = (RC)^{-1}$ is the time constant of the RC-CR network. If the time constant is set equal to $1/\omega_o$, then $\Delta\phi$ and ΔA become -0.5π and one, respectively. However, ideal phase and amplitude matching are limited over a narrow range of bandwidth and component mismatch.

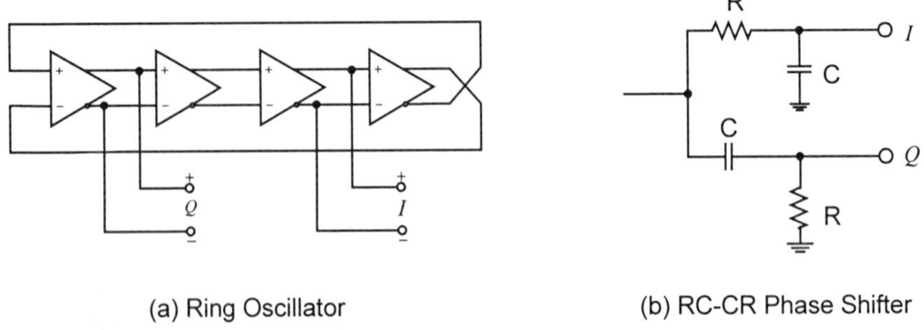

(a) Ring Oscillator (b) RC-CR Phase Shifter

Figure 7.24 Quadrature generation.

7.7 Power Amplifier

A power amplifier is perhaps the most important component in a transmitter power. For systems that require relatively high ranges (e.g. > 500 m), the power amplifier can consume half or more of the total radio power. The high power consumption arises from the need to transmit at relatively high power (e.g. 27 dBm) while maintaining a low out-of-band emission. The problem gets substantially worse for non-constant envelope modulation as discussed in Section 2.4.1.3 of Chapter 2. This section discusses a few common power-amplifier configurations and design trade-offs that affect the system performance.

7.7.1 Power-Amplifier Parameters

The main function of a power amplifier is to deliver sufficient transmission power to meet a specific range requirement and to maintain minimal distortion on the transmitted signal to achieve a low out-of-band emission. Three design parameters are discussed below that characterize the power efficiency and distortion of a power amplifier.

Power efficiency η_{PA} is defined as the ratio of the power delivered to the load P_{load} to the power delivered to the supply P_{supply}.

$$\eta_{PA} = \frac{P_{load}}{P_{supply}} \tag{7.54}$$

It measures the amount of power that gets delivered to the antenna with respect to the total amount of power dissipated by the supply. Ideal power efficiency is 100% which is not possible in practice due to losses and non-linearity in circuit components. Non-linearity forces backoff on the output power and therefore degrades efficiency.

Power-added efficiency (PAE) is defined as the ratio of amount of power provided by the power amplifier alone to that dissipated by the supply. PAE takes into account the power dissipated in driving the power amplifier whereas power efficiency defined in (7.54) does not. PAE can be expressed in terms of the amplifier power gain G and power efficiency as shown below

$$PAE = \frac{P_{load} - P_{in}}{P_{supply}} = \left(1 - \frac{1}{G}\right)\eta_{PA}. \tag{7.55}$$

According to (7.55), PAE is always less than the power efficiency. While power efficiency can be high but if the gain is low then the amplifier still has poor PAE.

The Input-referred compression P_{1-dB} defined by (4.17) and the input-referred third-order intercept $IIP3$ defined by (4.26) characterize the linearity of an amplifier. Input-referred distortion measures are particularly useful for input devices such as an LNA. On the other hand, for output device such as power amplifiers, output compression point is often used to measure their linearity. The 1-dB output compression point is defined as the product of the power gain and the input compression point. The output compression point provides an indication of the amount of output backoff needed to meet a given out-of-band emission requirement. High output backoff is required for modulations with a high peak-to-average power ratio to accommodate the large amplitude deviation that occurs from time to time. Some modulations, such as PSK, have a peak-to-average power ratio of unity but when filtered the power ratio becomes large. Therefore, filtered PSK requires a significant output backoff to reduce spectral regrowth as discussed in Section 2.4.5.2.

7.7.2 Class A

A general-block diagram of a power amplifier is shown in Figure 7.25. A common-source amplifier is used to achieve high power efficiency. The inductor L_{RFC} suppresses high frequency ripples in the supply due to the RF signal being amplified.

The DC-blocking capacitor C_{DCB} prevents AC perturbation on the bias point and AC couples the output to the impedance matching network. Because of the high power requirement, the impedance matching network performs a downward impedance transformation. To achieve high gain and reduced distortion, often a tuned LC load is used to resonate the output at the carrier frequency. The LC tank is in parallel with the output load Z_L, which models the antenna. For ease of discussion, the output load is assumed to be 50 Ω.

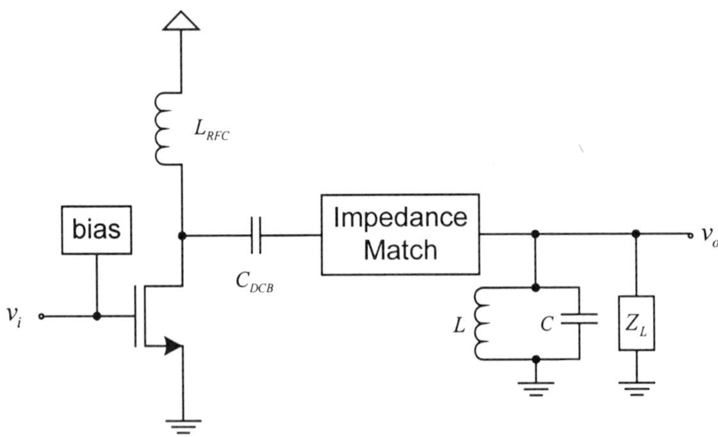

Figure 7.25 General circuit diagram of a power amplifier.

Class-A operation imposes a non-zero bias current I_{bias} in the transistor. The resulting drain current and voltage are shown in Figure 7.26a. Since the current flowing through the RF-choke inductor cannot change instantaneously, the drain voltage actually rises above that of the supply. The high peak voltage helps to increase the output swing and therefore the transmit power for a given supply voltage. The high peak voltage, however, places a high voltage stress on the amplifier which complicates thermal management. Also, since the transistor must not breakdown under the high voltage stress, sub-micron MOST with low breakdown voltages is not suitable to implement power amplifiers that must transmit at high power.

Power efficiency of a Class-A amplifier can be derived by making the following observations. The power delivered to the load is $P_L = 0.5V_o^2 R_{eq}^{-1}$, where V_o is the output current amplitude and R_{eq} is the equivalent load resistance after impedance transformation. The power consumed by the supply can be expressed by $V_o R_{eq}^{-1} V_{DD}$. Thus, the power efficiency for a Class-A amplifier becomes

$$\eta_{classA} = \frac{V_o}{2V_{DD}}. \tag{7.56}$$

To keep the transistor in saturation, the drain-source voltage v_{ds} must be greater than $V_{sat} = V_{GS} - V_t$ or alternatively expressed as

$$V_{sat} = \frac{I_d + \sqrt{I_d^2 + 2kI_d L^2 E_{sat}^2}}{kLE_{sat}}, \tag{7.57}$$

where $k = \mu C_{ox} W L^{-1}$. Thus, $|V_o|$ can be no greater than $|V_{DD} - V_{sat}|$. In addition, other second-order effects decrease V_o even more. For example, parasitic lead inductance of the IC package can induce voltage drop that further decreases V_o. Therefore, in practice, power efficiency of Class-A amplifiers is around 30-45%.

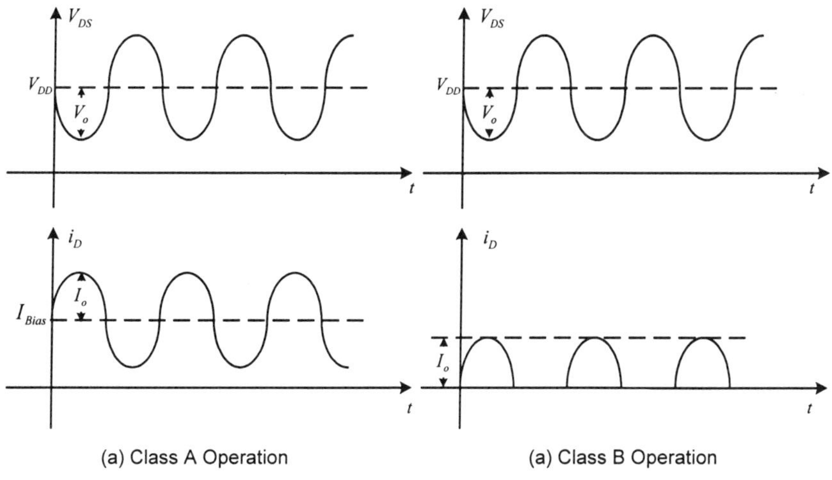

Figure 7.26 PA drain current and voltage waveforms.

Example 7.14 Consider a supply of 3.3 V and V_{sat} of 0.5 V. Then, according to (7.56), power efficiency is

$$0.5 \times \frac{3.3 - 0.5}{3.3} = 0.42. \ \square$$

In the active region, a Class-A amplifier has a voltage gain of $-g_m R_{eq}$. If a source resistance of R_s is assumed, then the power gain of the amplifier can be expressed as

$$G_{classA} = g_m^2 R_{eq} R_s, \tag{7.58}$$

where g_m is the transconductance of the transistor. The equivalent resistance is determined based on the transmit power as shown below

$$R_{eq} = \frac{V_o^2}{2P_{load}}. \tag{7.59}$$

The source resistance depends on the input impedance of the power amplifier and is not necessarily $50\,\Omega$. Generally, the gain is designed to be greater than 10 dB. Therefore, the PAE is at least 90 % of the power efficiency.

The output 1-dB compression point of the amplifier can be obtained by multiplying its ICP by $|\beta_1|^2 R_s R_{eq}^{-1}$ as shown below

$$|\beta_1|^2 R_s R_{eq}^{-1} P_{1-dB} = 10\log\left[433.84 P_{load}\frac{(1+0.5\rho)^3}{\rho}\right]. \tag{7.60}$$

Referring to (7.56) and (7.60), both the efficiency and output compression point increase as V_{sat} decreases. However, the opposite behavior is observed for the ICP and IIP3. Therefore, a trade-off exists between efficiency and linearity of the device. In particular, by applying (7.2) and (7.56), the following constraint on ρ subject to a given amplifier efficiency can be derived

$$\rho = \frac{(1-2\eta_{classA})V_{DD}}{LE_{sat}}. \tag{7.61}$$

While it is possible to increase the output compression point by increasing ρ and therefore maintain high linearity, the amplifier efficiency could be greatly reduced due to a limited supply voltage.

Example 7.15 Consider a supply of 3.3 V, V_{sat} of 0.5 V, 0.25-µm CMOS process, and a transmit power of 600 mW. Then, according to (7.60), the output compression point is

$$10\log\left[433.84\cdot 0.6\cdot\frac{(1+0.5\cdot 0.5)^3}{0.5}\right] = 30.1\,dBm.$$

Since the transmit power is 27.8 dBm and is about 2.3 dB higher than the output compression point, the power amplifier is kept in the linear region at the rated transmit power. According to (7.56), the amplifier efficiency is

$$\frac{3.3-0.5}{2\cdot 3.3} = 42\%.$$

According to (7.59), the equivalent resistor is then

$$\frac{(3.3-0.5)^2}{2\cdot 0.6} = 6.5\,\Omega.$$

The corresponding bias current is

$$\frac{3.3-0.5}{6.5}=430.8\ mA.$$

According to (7.1), the transistor width in 0.25-μm CMOS is

$$W=2I_d\frac{LE_{sat}+V_{sat}}{\mu C_{ox}V_{sat}^2E_{sat}}$$
$$=2\cdot 0.4308\cdot\frac{0.25\times 10^{-6}\cdot 4\times 10^6+0.5}{0.015\cdot 0.0035\cdot(0.5)^2\cdot 4\times 10^6}=24.6\ mm.$$

According to (7.3), the corresponding transconductance is approximately

$$\sqrt{\frac{2\cdot 0.015\cdot 0.0035\cdot 0.0246\cdot 0.4308}{0.25\times 10^{-6}}\frac{1+0.5\cdot 0.5}{(1+0.5)^{1.5}}}=1.44\,S$$

By applying (7.58) and assuming a source impedance of 50 Ω, the power gain of the amplifier is

$$1.44^2\cdot 6.5\cdot 50=28.3\,dB.$$

The input 1-dB compression point is then 1.8 dBm. □

7.7.3 Class B

Class-A operation requires a non-zero quiescent current, implying that the amplifier will dissipate power in the absence of an input to the amplifier. The efficiency of a Class-A amplifier at quiescence can therefore be quite poor. Furthermore, the quiescent current contributes a non-zero power dissipation at all times in the supply, resulting in a maximum possible efficiency of only 50%. In contrast, a Class-B amplifier is designed to have essentially zero bias current at quiescent and dissipates current only half of the signal cycle as shown in Figure 7.26b. The amplifier in fact has the same configuration as the Class A amplifier shown in Figure 7.25 but in this case the bias point sets the transistor in cut-off mode half of the signal cycle.

The average current through the supply is I_o/π and has an amplitude of $0.5I_o$ at the fundamental frequency, where I_o is the amplitude of the current waveform. Therefore, the power dissipated by the load is $0.125I_o^2R_{eq}$ and the power dissipated in the supply is I_oV_{DD}/π. The amplifier efficiency then becomes

$$\eta_{class\,B}=\frac{\pi V_o}{4V_{DD}}, \tag{7.62}$$

which is 0.5π times higher than that of a Class-A amplifier.

Despite the large distortion generated when the transistor cuts off, the output voltage is still approximately sinusoidal due to the tuned load that resonates at the carrier frequency. However, when the transmitted signal contains data modulation, the distortion due to transistor cut off becomes worse, especially, for non-constant envelope modulation. This crossover distortion can be reduced with a push-pull configuration where a PMOS is used to supply current to the load when the NMOS cuts off on the negative input cycle. Some amount of biasing current may be needed to keep the crossover distortion low. Other configurations of amplifiers are possible such as Class C in which current flows during a portion of the signal cycle defined by 2Φ, where Φ is also known as the conduction angle. Class-B amplifier is a special case of Class C with $\Phi = 0.5\pi$.

Due to the 0.5π conduction angle, the power gain of a Class B amplifier is a factor of four lower than that of a Class-A amplifier as shown below

$$G_{class\ B} = 0.25 g_m^2 R_{eq} R_s . \tag{7.63}$$

The peak voltage and current on the transistor is the same as that of class A. The compression point is more difficult to derive since it depends on the crossover distortion in addition to the non-linear characteristic of a transistor.

7.7.4 Class F

Class A and B amplifiers are considered to be linear amplifiers since the output power is a linear function of the input power. We now consider non-linear amplifiers where the output power no longer has a linear dependence on the input power. Such amplifiers are overdriven to generate a square wave that has the same frequency as the carrier. Tuned tank circuits at the output of the amplifier then select the fundamental frequency of the square wave to produce the desired signal for transmission.

Figure 7.27 shows a Class-F [199][200] power amplifier using a quarterwave transmission line between the tuned load and the output of the transistor. The quarterwave transmission line has the following impedance transformation property

$$Z_{in} = \frac{Z_o^2}{Z_L}, \tag{7.64}$$

where Z_{in} is the input impedance, Z_o is the characteristic impedance, and Z_L is the load impedance. At the carrier frequency, the input impedance equals Z_o if $R_L = Z_o$. At odd harmonics, the input impedance appears to be an open circuit since the load impedance is a short due to the LC tank. At even harmonics, on the other hand, the input impedance appears as a short circuit since the transmission line behaves like a halfwave transmission line at even harmonics of the carrier frequency. By suppressing the even harmonics of the drain current, the quarterwave transmission line effectively generates a square voltage at the output of the transistor. The tuned element selects the fundamental frequency of the square wave to produce the desired sinusoid at the carrier frequency. While the Class F amplifier can easily amplify a tone, substantial spectral regrowth could result for modulated waveforms, in particular those with a non-constant signal envelope.

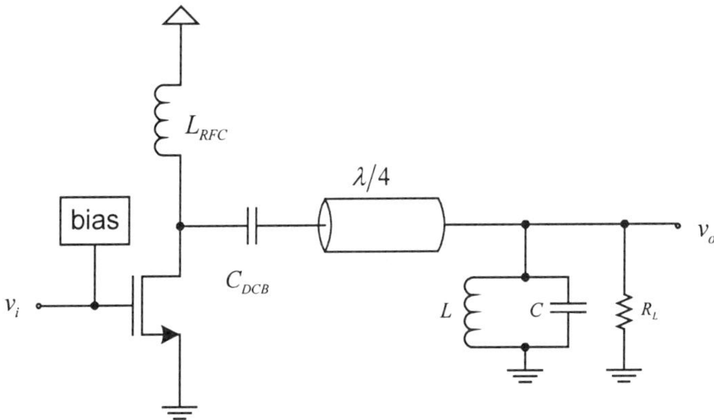

Figure 7.27 A Class F power amplifier.

The efficiency of a Class-F amplifier can be derived by computing the transistor drain current and voltage. Due to clipping, the drain voltage is a square wave centered at V_{DD} with a peak-to-peak amplitude of $2V_{DD}$. The fundamental component of the square wave has an amplitude of $4\pi^{-1}V_{DD}$. The peak drain current is therefore equal to the peak-to-peak amplifier output voltage divided by R_{eq}, i.e. $8V_{DD}\pi^{-1}/R_{eq}$. The drain voltage and current have the same phase relationship and characteristics as that of the Class-B amplifier shown in Figure 7.26b. We may conclude then that expression (7.62) also applies to a Class-F amplifier. Substituting $V_o = 4\pi^{-1}V_{DD}$ into (7.62), a maximum efficiency of 100% can be obtained.

The power gain of the class-F amplifier is not easily computed since the output power of the transistor does not depend linearly on its input power. In general, the transistor should be sized large enough to support the peak current requirement and should be driven with sufficiently large voltage level to provide fast transition, e.g. 10% of a carrier cycle. A larger voltage level means that the power gain can be lower than that of a linear amplifier for a given bias current.

CHAPTER 8

Digital Transceiver Design

Unlike the RF design flow, a digital design flow relies heavily on computer-aided design (CAD) tools to automate the various design steps. A typical design flow for a digital transceiver is shown in Figure 8.1. At the top of the design process, system requirements are derived via detailed system and network simulations. The system requirements guide the selection of communication algorithms and network protocols. The communication algorithms have been discussed in Chapter 2 and 0 but a discussion of the network protocols has not been discussed in detail. Instead, readers interested in networking are referred to [201][202].

In the algorithm design stage, floating point simulation is performed to assess trade-offs in computation complexity and performance. The algorithm design guides the selection of a suitable architecture for implementation on digital hardware. Numerous CAD tools are available to facilitate the implementation process. These tools require as inputs a netlist description of the design containing information about the underlying building blocks and how the blocks are interconnected. Standardized hardware description languages (HDL) have been created to describe the design netlist. VHDL [203] and Verilog [204] are two popular HDL being used in most of the current designs.

In general, the HDL representation is at a module level, where a module is defined as a functional block such as a DDFS, an adder, a multiplier, or a register. The module itself may be described at a behavioral or structural level. The former involves representing the module at a functional level and the latter involves representing the module at a block or gate level. Given a module library, a top-level netlist can be written to implement a particular architecture using the module library components as the underlying building blocks for the design. The top-level HDL netlist is validated in simulation and finally synthesized to a gate-level netlist targeted for a specific technology. A place-and-route tool then takes the gate-level HDL and generates an IC layout based on components from the target technology library. Often, delay information from the layout is extracted and back annotated into the HDL design.

With more precise delay information, the design can be re-verified in a post-synthesis simulation to ensure that all the timing rules are met. Finally after detailed post-synthesis simulations, the IC is fabricated and packaged for testing.

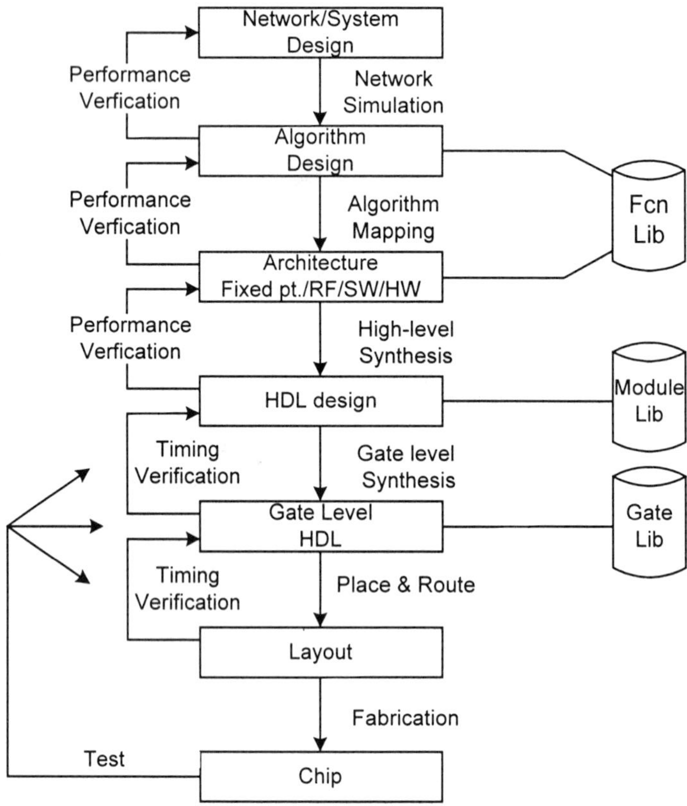

Figure 8.1 Design flow for the digital transceiver design.

Unlike RF IC designs which require extensive numbers of test runs to fine tune the layout and circuit design to meet the target design specifications, a digital design has a very high success rate of first-time working silicon, especially, when well characterized device or gate-level simulation models are used. Often, emulations may also be employed to obtain functional verification at the gate level so that the likelihood of a working chip is further enhanced.

Simulations performed at each step of the design process validate the design's performance and functionality. Accurate verification often requires extraction of design parameters, such as delays, that are back annotated into the HDL netlist to achieve more precise simulation results. However, further into the design process, it becomes increasingly difficult to verify a design without requiring huge amounts of computation resources. Therefore, compact test vectors are often times created to validate the design at the IC level.

Unfortunately, current simulation tools are not sophisticated enough to provide precise circuit-level verification of a single-chip digital radio that includes digital circuitry as well as sensitive RF circuitry. In particular, it is difficult with finite computation resources to simulate the effects of coupling between the digital circuits and the sensitive RF circuits. The difficulty arises due to the lack of a good model for the digital-RF cross coupling and the long simulation time needed to validate the entire design at the circuit level. Often, precision is sacrificed by modeling the digital circuits as equivalent capacitors running at the associated clocking frequencies. Due to the difficulty in validation of a digital radio IC, several iterations of test chips are usually needed before obtaining a working prototype.

When considering just the digital transceiver, the key design challenge lies in deriving a suitable architecture that can be implemented efficiently in digital hardware to meet a given set of system and algorithm constraints while being small, low cost, and if needed low power. To meet the above challenge, a designer should focus on the design choices made during the first three design stages: namely, the system, algorithm, and architecture designs. A careful choice of system parameters leads to the greatest reduction in power, complexity, and cost. An appropriate choice of algorithm for radio functions leads to the next level of power, complexity, and cost reduction. The requirements derived during the system design stage are used to guide the selection of algorithms that are then mapped to an architecture appropriate for hardware implementation. A poor architecture design can negate the efficiency gained from the system and algorithm level optimizations. In Chapter 6, numerous digital radio architectures have been described along with detailed discussions on the associated performance trade-offs. This chapter describes the architecture design of common digital transceiver building blocks that can be used to implement many of the algorithms described in Chapters 2 and 3. These building blocks also form the basis for the digital transceiver architectures mentioned in Chapter 6. Circuit level considerations when dealing with having both digital logic and sensitive RF circuits on the same substrate are discussed later in Chapter 9. Also, details of digital circuit and logic designs will not be described. Instead, interested readers are referred to [205][206].

8.1 Design Considerations

The design of a digital transceiver involves numerous techniques in the area of computer arithmetic. Specifically, computer arithmetic forms the fundamental theory of number representation and arithmetic for computations in the digital domain. This section covers some background materials in these two areas that are fundamental to digital transceiver designs.

8.1.1 Performance Metric

Before delving into the specifics of computer arithmetic, several commonly used performance metrics for digital transceivers are described below. These performance metrics are useful in analyzing the performance trade-offs associated with different architecture choices.

MIPS or millions of instructions per second measures the performance of general-purpose processors, micro-controllers, and digital-signal processors (DSP). Although the MIPS rating varies quite a bit depending on the benchmark, the metric is widely used because of its simplicity. To allow a fair comparison between different processors and controllers, several industry standard benchmarks have been created. A commonly used standard is the Dhrystone MIPS obtained by running the Dhrystone benchmark [207]. Today, the most widely used embedded microprocessor for portable applications is based on the ARM architecture developed by Advanced RISC Machine [208]. An ARM-based processor may be integrated as part of the digital transceiver to handle control functions, such as power control, man-machine interface, framing, and buffering of data. Current digital cellular handsets require embedded processors with low MIPS ratings such as the 40 MIPS ARM9 core.

MOPS or millions of operations per second measures the performance of signal processors. For high performance transceivers, the computational complexity may be measured in terms of Giga operations per second or GOPS or 1000 MOPS. The MOPS rating differs from the MIPS rating in the definition of operations versus instructions. An operation for a signal processor usually refers to an arithmetic computation or a memory access while an instruction refers to a combination of operations. A common operation in a digital transceiver is the multiply-and-accumulate (MAC). A signal processor is designed to handle the MAC operation efficiently, at least one MAC operation per clock cycle. On the other hand, a general-purpose processor typically requires many instructions per clock cycle to execute a MAC. Thus, a digital transceiver that requires a 100 MOPS signal processor can require as much as ten times more MIPS to execute in a general-purpose processor. This may change, however, as embedded microprocessors begin to include special hardware accelerators or instruction sets to improve their efficiency in executing signal-processing functions. Note that general signal processors require additional MOPS for software control. An application-specific processor or custom transceiver, on the other hand, has minimal control overhead and thus provides the best performance but with limited programmability.

Transistor density measures the area efficiency of a digital design in units of number of transistors per square mm. Other metrics can also be used to measure area complexity. These metrics include the absolute area in square mm, the number of transistors, the number of gates, or gate density. Gates generally refer to simple logics such as NAND, NOR, and XOR and consists of four transistors. Current digital transceivers can contain millions of transistors on a 100 square mm. die. Since a design can have a small transistor count but take up a large area, the transistor density metric provides a more accurate assessment of area utilization. Current digital transceivers have a transistor density that typically ranges between one to ten thousand transistors per square mm.

Power Efficiency measures the amount of power dissipated per MIPS or MOPS in units of mW/MOPS mW/MIPS. Equivalently, power efficiency can be viewed as the amount of energy consumed per operation or instruction. Current transceivers dissipate between 0.01 nJ to 1 nJ per operation. The former is typical for a fixed datapath design whereas the latter is typical for a programmable design. While new programmable technology is claiming power efficiency similar to that of a fixed design, one might still pay a power penalty for applications that require high computation complexity. Since a programmable processor typically achieves a lower MOPS as compared to a fixed design, an implementation could require multiple processor chips that would dissipate more power in contrast to a single-chip solution based on a fixed datapath design.

Figure of Merit (FOM) is a combined metric that incorporates the transistor density, power efficiency, and speed constraints to measure the efficiency of a design. A possible FOM measure is shown below

$$FOM = \frac{\text{Transistor density}}{\text{Power efficiency}} R_b, \tag{8.1}$$

having units of $\frac{transistors}{mm^2} \cdot \frac{MOPS}{mW} \cdot \frac{bits}{s}$. A high value indicates a design that is compact and power efficient for a given throughput.

8.1.2 Number Systems

A digital system operates on bits with binary values, i.e. a logical '0' or a logical '1'. A DAC converts real-world signals to discrete digital words of m bits for processing in the digital domain. Many number systems are available to represent the discrete digital values. This section describes three number systems commonly used in a digital transceiver design.

8.1.2.1 Two's Complement

Consider an analog waveform x with values ranging from x_l to x_h. A two's complement number representation of this number with Δ_x precision requires m bits as shown below

$$m = \left\lceil \log_2 \left(1 + \frac{x_h - x_l}{\Delta_x} \right) \right\rceil. \tag{8.2}$$

The two's complement number $\mathbf{u}$ can be represented as a vector of m elements $[u_{m-1}, u_{m-2}, \cdots, u_0]$, where u_{m-1} is the most-significant bit (MSB) and u_0 is the least significant bit (LSB). The relationship of **u** to the discrete analog value $x_{2's}$ can be determined using

$$x_{2's} = -2^{m-1}u_{m-1} + \sum_{i=0}^{m-2} 2^i u_i. \tag{8.3}$$

Because of the negative sign associated with the MSB, u_{m-1} is also known as the sign bit. In (8.3), the decimal point follows the LSB. Therefore, (8.3) covers only integer numbers. To represent fractional numbers, the decimal point can be shifted left k bit positions toward the MSB by dividing (8.3) by 2^k where $k \le m-1$.

Example 8.1 Consider a set of discrete values $\{-4,-3,-2,-1,0,1,2,3\}$, where $x_l = -4$, $x_h = 3$, and $\Delta_x = 1$. By applying (8.2), it can be shown that $m = 3$ or 3 bits are required to represent the set of values in two's complement representation. Using (8.3), the set of integer values given above may be represented by the following mapping:

$u_2u_1u_0$	Integer	Fractional
0 0 0	0	0
0 0 1	1	0.25
0 1 0	2	0.5
0 1 1	3	0.75
1 0 0	–4	–1
1 0 1	–3	–0.75
1 1 0	–2	–0.5
1 1 1	–1	–0.25

In the above table, an additional column has been inserted to illustrate fractional representation where $x_l = -1$, $x_h = 0.75$, and $\Delta_x = 0.25$. Also, it can be seen that a negative number can be derived from the corresponding positive number by simply inverting all the bits and adding a '1' at the LSB. □

8.1.2.2 One's Complement

The one's complement representation $x_{1's}$ is related to the two's complement representation by the following relationship

$$x_{1's} = \begin{cases} x_{2's}, & x_{2's} \ge 0 \\ x_{2's} + LSB, & x_{2,s} < 0 \end{cases}. \tag{8.4}$$

Expression (8.4) implies that negation in one's complement can be achieved by simply inverting all the bits without adding a '1' at the LSB. Therein lies the main advantage of one's complement since negation does not require an m-bit adder rather it requires only a bank of m inverters. Adding a '1' at the LSB position not only requires more hardware but also induces a longer delay because the '1' at the LSB position must propagate across the m-bit adder to obtain the final result. However, a one's complement representation does have two main disadvantages. First, negative numbers have a bias of one LSB and second a zero value cannot be uniquely represented.

Example 8.2 Consider the table in Example 8.1. Using (8.4), it can shown that the corresponding one's complement representation has the following mapping:

$u_2u_1u_0$	Integer	Fractional
0 0 0	0	0
0 0 1	1	0.25
0 1 0	2	0.5
0 1 1	3	0.75
1 0 0	−3	−0.75
1 0 1	−2	−0.5
1 1 0	−1	−0.25
1 1 1	0	0

□

8.1.2.3 Signed Magnitude

The signed magnitude representation x_{sm} is related to the two's complement representation by the following relationship

$$x_{sm} = \begin{cases} x_{2's}, & x_{2's} \geq 0 \\ \left(1 - 2u_{m-1}\right)\left(x_{2's} + 2^{m-1}\right), & x_{2,s} < 0 \end{cases}. \tag{8.5}$$

Expression (8.4) implies that negation in signed magnitude can be achieved by simply inverting the MSB or the sign bit, even simpler than one's complement. In this case, only one inverter is needed. However, additions in a sign-magnitude representation require more complex circuitry than additions in two's complement. This observation can be inferred from (8.3) which shows only one expression for both positive and negative numbers where as both (8.4) and (8.5) show different computations needed for negative numbers.

Example 8.3 Consider the table in Example 8.1. Using (8.5), the following mapping can be derived for the sign-magnitude representation:

$u_2u_1u_0$	Integer	Fractional
0 0 0	0	0
0 0 1	1	0.25
0 1 0	2	0.5
0 1 1	3	0.75
1 0 0	0	0
1 0 1	−1	−0.25
1 1 0	−2	−0.5
1 1 1	−3	−0.75

□

8.1.3 Arithmetic Functions

Given a number system, computations can be performed using digital representations. In digital transceivers, the predominant computations involve multiplies and adds. Figure 8.2a-b show the block-level representations of an $m \times n$ multiplier a 2-input m-bit adder, respectively. The outputs of the multiplier and adder are respectively $m+n$ and $m+2$ bits for full precision. However, full precision is rarely used since it is costly in complexity, especially for cascaded stages. Reduction in the number of output bits is achieved either by truncation or rounding, both of which result in quantization errors. In the former, the quantization error has a bias or mean whereas the latter has zero mean. Rounding requires a conditional addition at the LSB position whereas truncation requires no additional logic. Depending on usage, rounding may be necessary such that the result has a quantization error with zero mean. To prevent overflow or underflow from occurring due to partial precision, saturation logic is implemented that limits the outputs to a maximum or minimum value.

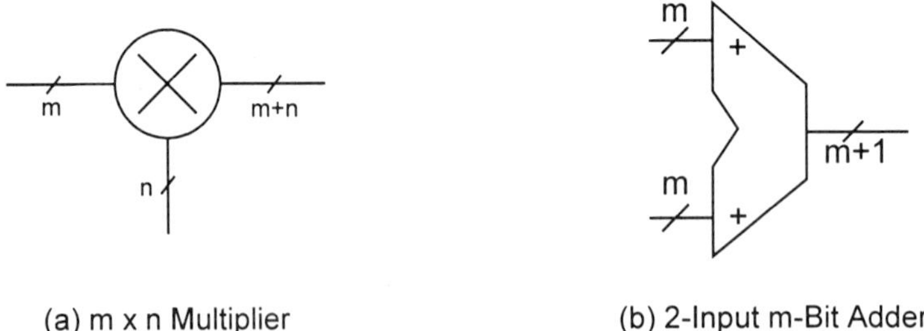

(a) m x n Multiplier (b) 2-Input m-Bit Adder

Figure 8.2 Symbols for basic arithmetic units.

8.1.3.1 Half/Full Adder

An $m \times n$ multiplier can actually be viewed as m rows of n-bit adders or n rows of m-bit adders or an array of $m \times n$ one-bit adders. Thus, the adder is in fact a primitive block from which a multiplier may be constructed. A one-bit adder can either be a half adder (HA) or full adder (FA). A HA adds two one-bit numbers and generates a sum bit s and a carryout bit c_0 whereas a FA adds three one-bit numbers. Figure 8.3 shows the block diagram, truth table, and gate-level diagram for a FA and HA. The truth table can be understood by viewing the one-bit adder as a counter that enumerates the number of '1's at its input and outputs a sum and a carry with weights of one and two, respectively. In other words, $s+2c_o$ represents the number of '1's that appear at the inputs of the adder.

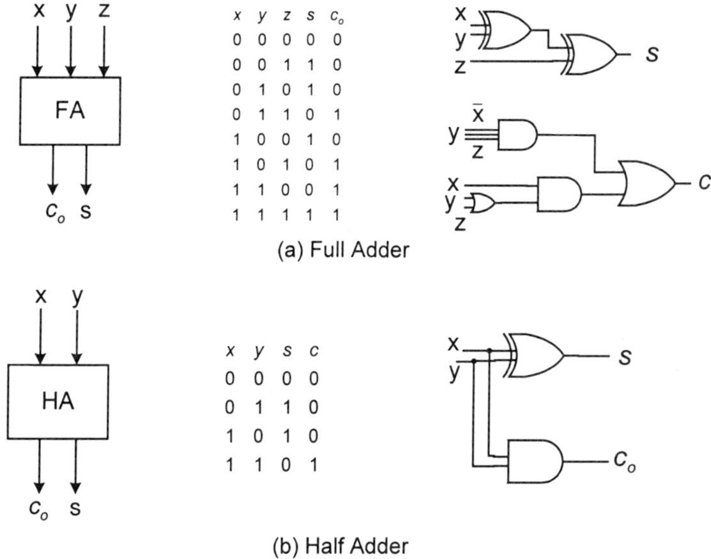

Figure 8.3 Bit-level adders.

8.1.3.2 Carry-Save Adder

Often, an algorithm requires the addition of multiple numbers. When more than two numbers need to be added, a carry-save adder array offers an efficient solution [209]. A carry-save adder preserves the carry and sum outputs of each FA in the array. Although it might appear that keeping the carry output for each bit position in the array is wasteful of computation resources since an extra output word is required, this approach offers an improvement in power and speed performance. In particular, consider adding k N-bit numbers. Then, the number of FA stages required can be approximated by

$$\left\lceil \frac{7}{4}\left(\log_2 k - 1\right) \right\rceil. \tag{8.6}$$

The overall delay is simply the expression in (8.6) multiplied by the FA delay t_{FA}. Since (8.6) is independent of N, a carry-save adder offers a high-speed implementation to add multiple numbers. Its suitability for low-power designs will also become clear in Section 8.1.7.

Figure 8.4 shows the carry-save adder array for adding five 3-b numbers, denoted by **v**, **w**, **x**, **y**, **z**.. The array output consists of a 4-b sum and a 3-b carry. Next section shows how the carry and sum outputs are combined to form a single sum value.

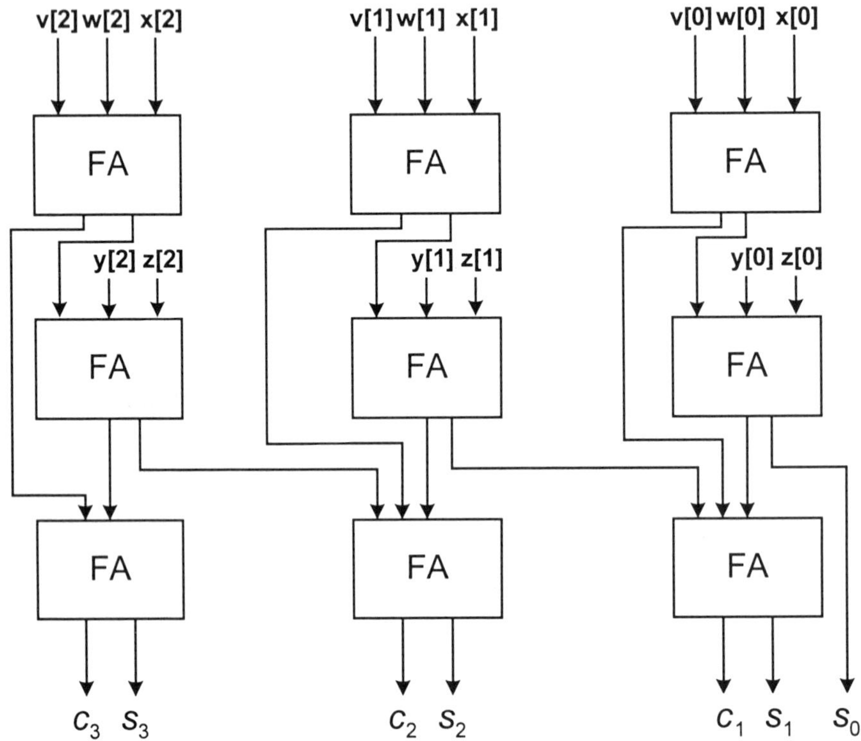

Figure 8.4 A carry-save adder array for adding five 3-b numbers.

8.1.3.3 Carry-Ripple Adder

A carry-save adder produces sum and carry outputs. To obtain a single word that represents the final sum based on the carry-save adder outputs, a carry-ripple adder can be used. A carry-ripple adder takes two numbers and generates a digital word that represents their sum. Consider two numbers **x** and **y**. Let their sum and carry be denoted by **s** and **c**. It can be inferred that the carry-out bit of the jth bit position has the same weight as the sum of the $(j+1)^{th}$ bit position. The jth c_o can thus be fed into the FA as a carry-in c_i at the $(j+1)^{th}$ bit position. A FA with its associated circuitry for each bit position is also known as a bit slice. An *N*-bit carry-ripple adder is formed by cascading N FA's with the c_o of each bit slice feeding into the c_i of the next bit slice in an ascending order toward the MSB as shown in Figure 8.5.

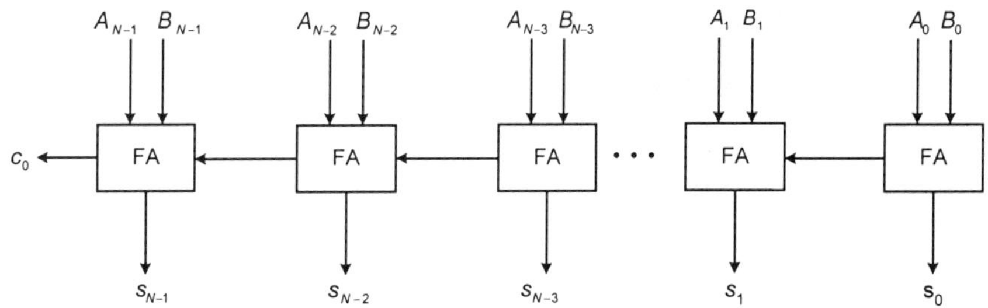

Figure 8.5 An N-bit carry-ripple adder.

Due to the rippling of the carry signal from LSB to MSB, a sum cannot be computed until after N FA delays. Thus, for a large wordlength, the delay introduced by the ripple adder may be too large to meet the speed requirement. Also, as the carry signal ripples across the bit slices, transients are induced at each bit slice that increases dynamic power dissipation. Several architectures have been developed to overcome the long carry-ripple delay. These include the carry-select adder [210], the carry look-ahead adder [211], and the conditional-sum adder [212]. All of these approaches tries to pre-compute ahead of time the results of different carry input combinations such that the carry does not need to propagate across the entire N bit slices before the sum can be computed. Pipelining techniques discussed in the next section also offers a means to speed up a carry-ripple addition.

Example 8.4 Consider the carry-save adder in Figure 8.4. To compute the final sum based on the carry and sum outputs of the carry-save array, a ripple adder can be used and is shown below.

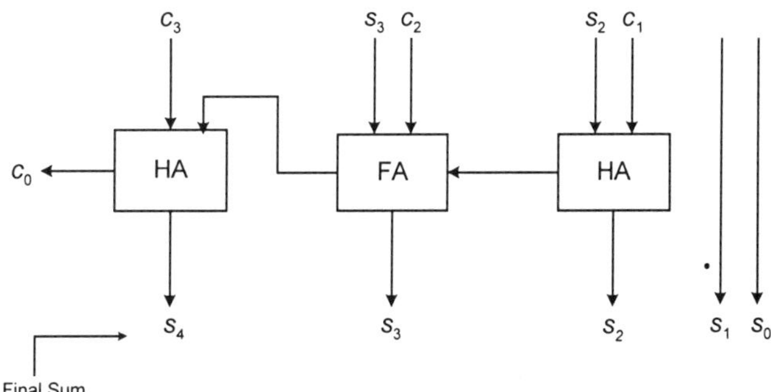

Figure 8.6 A carry-ripple adder that generates the final sum for the carry-save array in Figure 8.4.

Now consider using a carry-ripple adder to add k N-bit numbers. It is easy to show that the number of stages required is

$$\lceil \log_2 k \rceil \tag{8.7}$$

and that the total adder delay is

$$\left(N + \lceil \log_2 k \rceil - 1\right) t_{FA}. \tag{8.8}$$

If the final carry-ripple addition is taken into account then the total delay using a carry-save array becomes comparable to that of (8.8). Where then is the speed advantage of the carry-save method? To answer this question, let us consider a high-speed application that requires the addition of k N-bit numbers in less than the delay of a FA. If a carry-ripple adder is used then all k ripple adders in the array must be pipelined down to the bit level. On the other hand, in the carry-save implementation, pipelining is required only between the carry-save stages and in the final carry-ripple adder. □

8.1.4 Pipelining

Pipelining is a simple yet effective technique that can be applied to speed up computation and also to reduce dynamic power dissipation. Consider a computation that requires three sub-blocks with delays of τ_1, τ_2, and τ_3 as shown in Figure 8.7a. The total delay is then $\tau_1 + \tau_2 + \tau_3$ and the resulting throughput is $\left(\tau_1 + \tau_2 + \tau_3\right)^{-1}$. Pipelining introduces a register at between each of the three sub-blocks as shown in Figure 8.7b. The throughput now depends only on the block that has the maximum delay, in other words, $1/\max\{\tau_1, \tau_2, \tau_3\}$. The overall delay increases to $3\max\{\tau_1, \tau_2, \tau_3\}$. In a pipelined system, this delay is also referred to as latency since it is the amount of time required before a valid output is available after the circuit initially turns on. The throughput is improved since $\max\{\tau_1, \tau_2, \tau_3\}$ is always less than $\tau_1 + \tau_2 + \tau_3$. Pipelining can also be applied in a much finer grain down to the transistor level rather than just block boundaries.

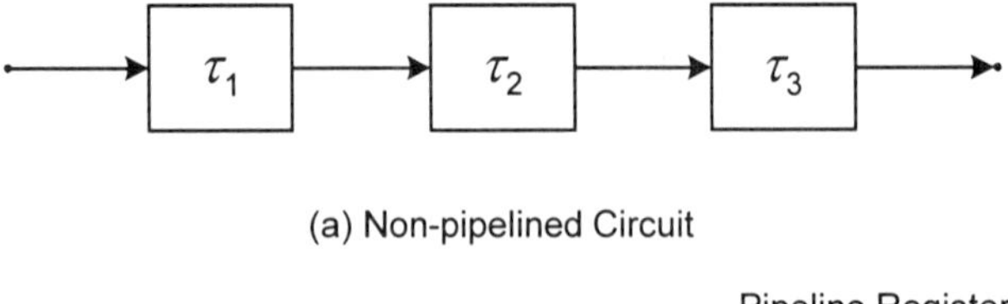

(a) Non-pipelined Circuit

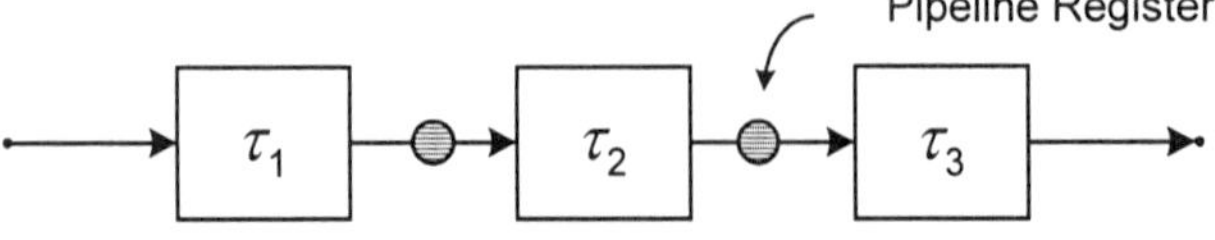

(b) Pipelined Circuit

Figure 8.7 Pipelining.

Example 8.5 Consider a 3-bit carry-ripple adder. Assume that each FA has a worst-case delay of 1 ns and that the system requires the adder to run at 1 GHz. To achieve this speed, pipelining should occur at the bit-slice boundary as shown in Figure 8.8. Note that de-skewing registers are needed at the inputs and outputs of the adder to ensure that the pipelined samples are aligned in time. □

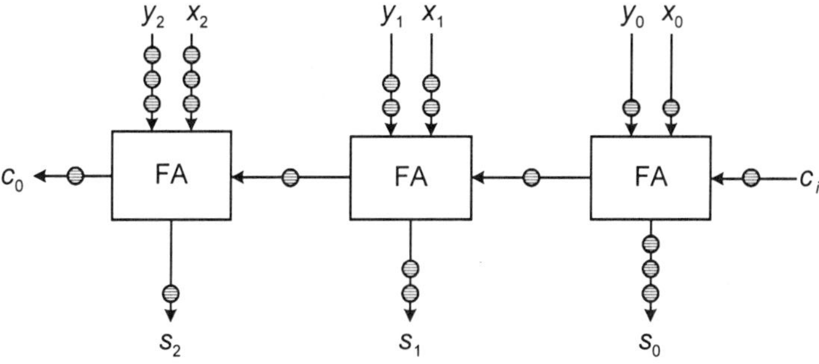

Figure 8.8 A bit-level pipelined 3-b carry-ripple adder.

8.1.5 Serial Vs. Parallel

A computation can often be partitioned into multiple sub-tasks which are computed in parallel or serially as shown in Figure 8.9. A parallel architecture demultiplexes the input stream into N sub-streams, each of which is processed by a processing element (PE). This assumes that the overall task can be mapped to N sub-tasks that are run in parallel, each on a PE. The results of the PE's are then multiplexed back into a single output. Since the input stream has been partitioned into N sub-streams, the rate at which each PE must run is N times slower than the input clock rate. Similarly, the output stream runs faster than the PE clock rate but depending on the application the output rate may differ from that of the input rate.

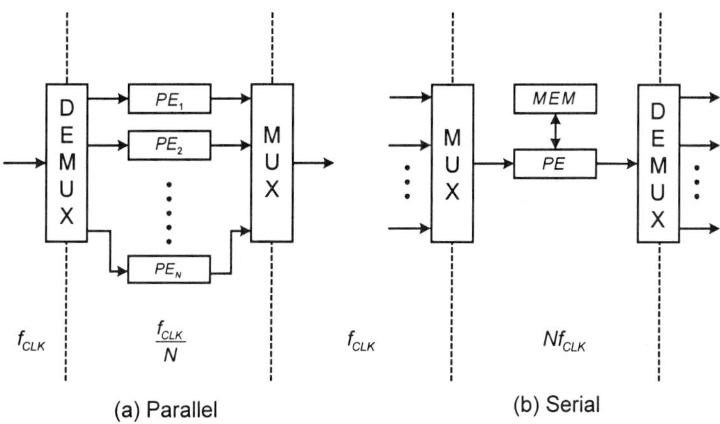

Figure 8.9 Two methods of computing.

A serial approach, on the other hand, maps the sub-tasks onto a single PE which is time-shared among all the sub-tasks. Shared memory elements are needed to store the results from each of the sub-tasks until all have completed execution and their results are ready to be sent to the output. Unlike the parallel approach, in the serial case, the PE runs at N times the input rate.

The parallel approach lends well to high-speed designs since only the DEMUX and MUX must run at high speed while the individual PE's run at a much lower clocking rate. However, it requires more complexity circuitry since N PE's are needed. In contrast, the serial approach has the advantage in lower complexity since the sub-tasks are time shared on a single PE. This assumes that the shared memory and associated controls do not become an excessive overhead. To reach the low complexity, a serial approach trades off speed performance. Since the PE must run at a high clocking rate, it becomes more difficult to support high-speed operation using the serial architecture. Examples of parallel and serial architectures will be discussed in Sections 8.2.6.2 and 8.2.6.3.

8.1.6 Programmable Vs. Dedicated Datapath

Current digital transceivers are either implemented as dedicated hardware or programmable DSP's and sometimes both, where a dedicated hardware accelerator accompanies a DSP to improve the performance of certain types of computation. While both a programmable DSP and dedicated hardware are implemented as integrated circuits, the distinction between them lies in the ability to reprogram the hardware to run an arbitrary algorithm. Dedicated hardware possesses no versatility to reprogram itself whereas a programmable DSP can be reprogrammed with a high-level language such as C. While complete reprogrammability is a desirable feature, it is achieved at the cost of lower power efficiency. For example, the most advanced DSP from TI can operate at 1 GOPS but dissipates a few Watts, which corresponds to power efficiency on the order of 1 μW/OP. A dedicated solution can easily achieve power efficiency several orders of magnitude better than a programmable solution. The inferior power performance arises from the unnecessary overhead needed to reprogram the PE's in a programmable architecture. This overhead can be reduced somewhat by writing in assembly language though this introduces other problems such as inefficient code reuse and prolonged design time.

8.1.7 Low Power Design

Power dissipation is an important metric for portable applications. This section discusses several guidelines for the design of a low power system. An in-depth survey of low-power designs can be found in [213].

Power dissipation in a CMOS circuit can be expressed by the following equation:

$$P_{total} = p_t C_L V_{dd} \Delta V \cdot f_{clk} + I_{SC} V_{dd} + I_{static} V_{dd} + I_{leakage} V_{dd}, \quad (8.9)$$

where p_t is the activity factor, C_L is the load capacitance, ΔV is the voltage swing, V_{dd} is the supply voltage, f_{clk} is the clocking rate, I_{SC} is the short-circuit current, and $I_{leakage}$ is the leakage current. The first term in (8.9) represents dynamic power dissipation that results from voltage switching on parasitic capacitors in CMOS circuits. Figure 8.10a illustrates the source of dynamic power dissipation in a CMOS inverter which is driving a load of C_L. The power dissipated by the supply is used to charge C_L from zero to V_{dd} over the duration of a clock cycle. This results in the $CV_{dd}\Delta V \cdot f_{clk}$ term in (8.9) with $\Delta V = V_{dd}$. The additional scale factor p_t accounts for logic transitions since unless a voltage transition occurs on C_L, the capacitor does not dissipate any power.

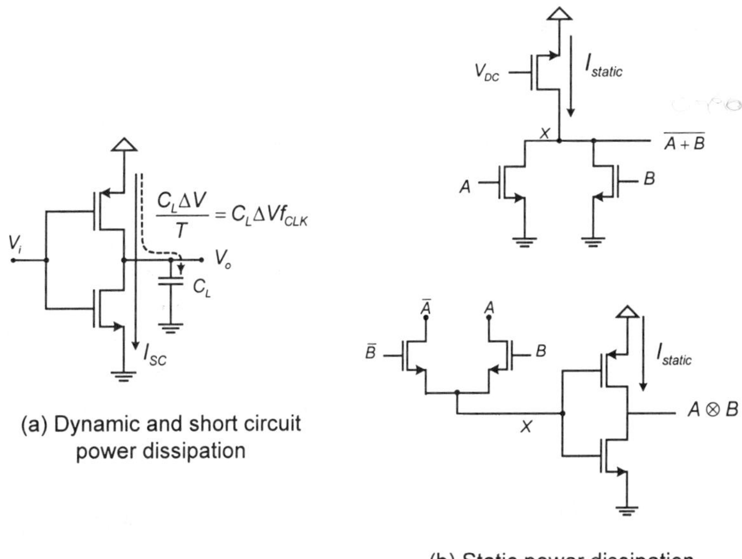

Figure 8.10 Sources of power dissipation in CMOS circuits.

The second term in (8.9) models the power dissipation due to the short-circuit current that flows from the supply to ground during the time when both transistors are turned on as shown in Figure 8.10a. Generally, short-circuit currents can be minimized by appropriately sizing transistors to have equal rise and fall times [214]. In a well-designed circuit, the contribution from short-circuit currents can be considered negligible.

The third term in (8.9) accounts for the static current that results from internal nodes not fully charged to the same level as the supply voltage, typical in circuits that use pseudo NMOS/PMOS or transmission gates. As shown in Figure 8.10b, the maximum voltage in the intermediate node X stays between ground and V_{dd}. Therefore, the NMOS gates do not completely shut off, resulting in a static current flowing from the supply to ground.

Finally, leakage currents arise from subthreshold effects when the supply voltage is close to the threshold voltage of the device or when the threshold voltage itself is close to zero volts. Power dissipation due to leakage current, the fourth term in (8.9), can be minimized by leaving an adequate margin on the threshold voltage so that the device can be driven to the cutoff state.

In most cases, the dynamic power dissipation is the predominant source of power dissipation in a digital design. Several useful guidelines are described below to minimize dynamic power dissipation. The techniques listed are by far not comprehensive but are most relevant to the design of digital transceivers.

Scale the supply voltage: The first term in (8.9) has a quadratic dependence on the supply voltage. Clearly then, to reduce power dissipation, one would try to scale the supply voltage down as much as possible subject to technology constraints. However, since speed is inversely proportional to the supply voltage, voltage scaling can slow down the circuit below the specified operational requirement. Two techniques exist that circumvent this problem. The first method, discussed in Section 8.1.5, utilizes parallel computations so that the speed requirement of individual PE's are reduced sufficiently to allow voltage scaling. However, there is still a lower limit imposed by the noise margin of CMOS circuits and the amount of overhead needed to implement the parallel architecture. The second method utilizes pipelining, described in Section 8.1.4, to provide sufficient margin in speed performance as the supply voltage is scaled down. Noise margin and circuit overheads again limit the minimum supply voltage that can be achieved through pipelining.

Minimize capacitance: Besides voltage scaling, techniques that reduces parasitic capacitances can provide lower power dissipation since according to (8.9), dynamic power scales linearly with capacitance. Several methods are possible. At the architecture level, one could try to minimize the number of bits required to implement various computations within the digital transceiver while keeping the quantization noise down to an acceptable level. Also, integration of all digital functions on a single chip eliminates power-hungry line drivers. At the circuit level, a judicious choice of circuit style and transistor sizing provide further reduction in power. An example of circuit level design is the use of circuits that require fewer transistors, such as pass transistor logic, or less loading on global signals such as clock lines. In the latter case, synchronous designs may have a large loading on clock lines and therefore may require power-hungry buffers to drive the clock lines on chip. By using dynamic register circuits such as TSPC, clock loading can be reduced to help minimize power dissipation.

Minimize spurious transients: Ideally during each clock cycle, it is desirable to have at most a single logic transition. However, in practice, multiple transitions occur due to transient behavior arising from unequal propagation delay of multiple signal paths. Signal transients create additional switching activity and therefore increase power dissipation. To minimize spurious transients, one can introduce enable lines or pipeline registers to limit the logic depth such that transients do not propagate extensively in a given design. Cascaded structures should also be avoided since transients tend to propagate along the length of the structure. An example of this is the carry-ripple adder. Finally, dynamic logic should be used since internal nodes are precharged to a known state. Therefore, during the evaluation phase only one transition occurs at each node in the circuit.

Minimize activity: Since dynamic power dissipation scales linearly with p_t, methods that reduce p_t also reduce dynamic power dissipation. While p_t can be reduced by virtue of circuit design, the greatest power saving derives from activity-driven power down or smart power management. An example of activity-driven power down has been described in [215] where unused taps in a digital filter are disabled dynamically depending on the input SNR. In a broader sense, power management applies to RF circuits as well; examples are power control and DTX described in Chapter 2. Finally, a suitable choice of number representation can result in a lower activity level. For example, when a signal varies between positive and negative values, it may prove useful to use the sign-magnitude representation rather than two's complement.

8.2 Communication sub-blocks

Referring to Figure 8.1, to implement the transceiver architecture in HDL, a module library is needed to provide a modular implementation of the transceiver architecture. This section discusses some basic modules blocks useful in constructing a digital transceiver. Often, these modules are used to build higher-level functional blocks to achieve a hierarchical design. These higher-level functional blocks might include a modulator/demodulator, a spread-Spectrum encoder, synchronization loops, and diversity receivers.

8.2.1 Accumulate and Dump

An accumulate-and-dump (A&D) is a widely used in digital transceivers as a digital implementation of the analog correlator discussed throughout Chapter 2. An analog correlator performs the following operation

$$y(kT) = \frac{1}{T}\int_{(k-1)T}^{kT} x(t)dt, \tag{8.10}$$

where $x(t)$ is the input and $y(kT)$ is the output sampled at time kT. Because the integrator must reset after each integration window of duration T, (8.10) is often referred to as an integrate-and-dump (I&D).

A direct digital implementation of I&D can be derived in a straightforward manner as shown below

$$y_k = \frac{1}{N} \sum_{i=(k-1)N}^{kN} x_i, \tag{8.11}$$

where $x_i = x(i\Delta_t)$ is the digitized version of $x(t)$ sampled at a rate of Δ_t^{-1} and $y_k = y(kT)$ is the sampled version of $y(t)$ at a decimated rate of $(N\Delta_t)^{-1}$. Note that $N = T / \Delta_t$ and Δ_t is the sampling interval or inverse of the sampling rate. Since the integrator has been replaced by an accumulator, (8.11) is referred to as an A&D block. Comparing (8.11) to (2.13), it can be seen that an A&D can also be viewed as a matched filter when the transmitted pulse shape is a square pulse of duration $N\Delta_t$.

A digital implementation of (8.11) is shown in Figure 8.11a where the running sum has been implemented with an accumulator and the dump function is obtained with a control line *DCK* which enables the output registers and resets the accumulator register at the Nth cycle of the accumulation clock *ACK*. The associated timing diagrams are shown in Figure 8.11b.

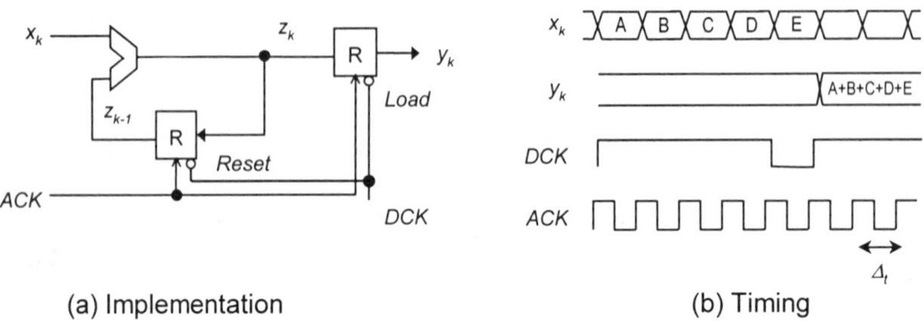

Figure 8.11 An accumulate-and-dump block.

8.2.2 Constant Scaling

Constant scaling is required in many places within the digital transceiver. Although it can be implemented with a multiplier with one of its inputs held at a constant value, it is more efficient to implement it using shifts and adds. Take for example a constant scale factor 0.875 which corresponds to 0.111 in fractional two's complement representation. Using the fact that a shift by one bit to the right of the decimal point corresponds to scaling by 0.5, 0.111 can then be represented by shifts and adds. To see this, 0.111 is decomposed into three terms shown below

$$0.111 = 0.001 + 0.010 + 0.100.$$

Now consider a number x scaled by 0.111. The scaling can be implemented with three shifts and an add as shown in Figure 8.12,

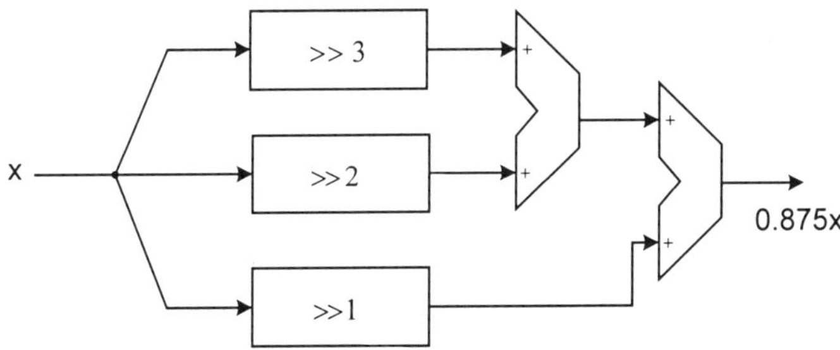

Figure 8.12 A shift-and-add implementation of $0.875x$.

where $>> B$ denotes a shift to the right by B bit positions. The complexity has been reduced from a full-fledged multiplier to two two-input adders since the shifts cost no additional hardware resources.

The complexity of multiplication by a constant can be further reduced by using a canonic-signed digit (CSD) representation, where each bit position is allowed to hold three values {-1, 0, 1} rather than just {0, 1} as in conventional radix two representations. Using CSD, 0.111 can now be represented more efficiently by $\bar{1}.001$ where a $\bar{1}$ represents –1. The number of shifts has reduced from three to two, implying that an adder has been eliminated. While this may not seem to be a major hardware reduction when the number of bits is three, the reduction becomes more significant with larger wordlengths. Later in Section 8.2.6, CSD is exploited to dramatically reduce the complexity of fixed-coefficient digital filters.

8.2.3 Loop Filter

As discussed in Section 2.5.2.1, the performance of a phase or timing recovery loop depends largely on the choice of loop filter and the associated loop gain and time constants. Most commercial application can be accommodated by either a zero-order or first-order loop filter. A zero-order loop filter is a simple gain block with a gain K that results in a first-order feedback loop. A first-order loop filter, on the other hand, has a pole and a zero with time constants τ_1 and τ_2, respectively. When placed in a feedback loop, a first-order loop filter results in a second-order loop. The reader may wish to refer back to Chapter 2 for a review of synchronization loops.

Consider a zero-order loop filter with a constant gain K. This loop filter can be easily implemented with hardwired shifts and adds as discussed in the previous section. In contrast, a digital first-order loop filter is much more complex. An analog first-order loop filter has the following transfer function

$$F(s) = \frac{\tau_2 s + 1}{\tau_1 s}. \tag{8.12}$$

By applying the bilinear transform [216], a digital version of (8.12) is obtained and has the following z-domain transfer function

$$F(z) = c_1 + \frac{c_2}{1 - z^{-1}}. \tag{8.13}$$

The corresponding digital implementation is shown in Figure 8.13, where $c_1 = (2\tau_2 - \Delta_t)/\tau_1$ and $c_2 = \Delta_t/\tau_1$. The z^{-1} operator is implemented by a register and the gain blocks can be implemented by simple shifts and adds.

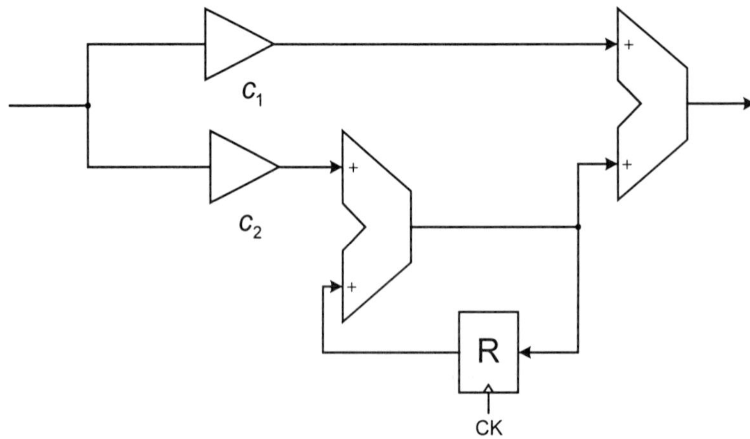

Figure 8.13 A first-order digital loop filter.

8.2.4 Numerically Controlled Oscillator

A numerically controlled oscillator (NCO) is an essential component in a timing-recovery loop. It is the digital counterpart of an analog voltage controlled clock oscillator. Figure 8.14 shows its architecture, which grossly resembles an accumulator with the exception that there is some addition circuitry at the input of the accumulator. The one-bit quantizer at the output requires no additional hardware since it merely pulls the MSB to the output. The control word W_{cntl} is an unsigned number with N bit resolution and the accumulator is clocked by CK at a rate of f_{NCO}. The gain and adder blocks at the input convert a signed two's complement phase-detector output to an unsigned number ranging from 0 to $2\alpha_0$, where $\alpha_0 f_{NCO}$ is the free running output frequency generated by the NCO. The input register, clocked by *UPD*, prevents the NCO from being updated too quickly. The typical update rate is chosen to be no greater than the free-running output frequency.

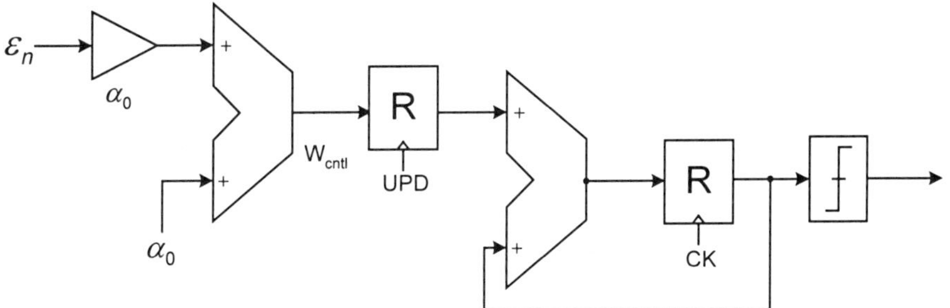

Figure 8.14 A numerically controlled oscillator.

The output frequency can be controlled by changing W_{cntl}. For example, if the accumulator content is initially zero, then having a $W_{cntl} = 2^{-N}$ will generate an output frequency of $2^{-N} f_{NCO}$ since it takes 2^N NCO clock cycles before the accumulator overflows. By the same token, if $W_{cntl} = 2^{-N+1}$ then the time needed for the NCO to overflow reduces by a factor of two and therefore the output frequency doubles. More generally, the output frequency of an NCO is related to the control word by the following expression

$$f_{out} = W_{cntl} f_{NCO}. \tag{8.14}$$

Note that if the phase error is zero then $W_{cntl} = \alpha_0$. Therefore, the output frequency becomes $\alpha_0 f_{NCO}$, which is equal to the free-running frequency normally set equal to the desired system clock rate. The parameter α_0 is made small (e.g. 1/8) to keep the timing jitter down to an acceptable value. Therefore, the accumulator must be clocked at many times (e.g. 8 times) the system clock rate generated at its output. For example, to generate a 50-MHz system clock, the NCO needs to operate at 400 MHz if $\alpha_0 = 1/8$.

Furthermore, for tracking small frequency offsets, the wordlength of the NCO should be made large enough to achieve adequate frequency resolution Δ_f of

$$\Delta_f = \frac{f_{NCO}}{2^N}. \tag{8.15}$$

For a large wordlength (e.g. 32 bits), high-speed operation (e.g. 400 MHz) becomes difficult even with sub-micron CMOS technology. To accommodate for the high-speed and high frequency resolution, a pipelined architecture can be used. Pipelining a computation block with feedback must be done at the at the bit level [217] as shown below for the 4-b case in Figure 8.15.

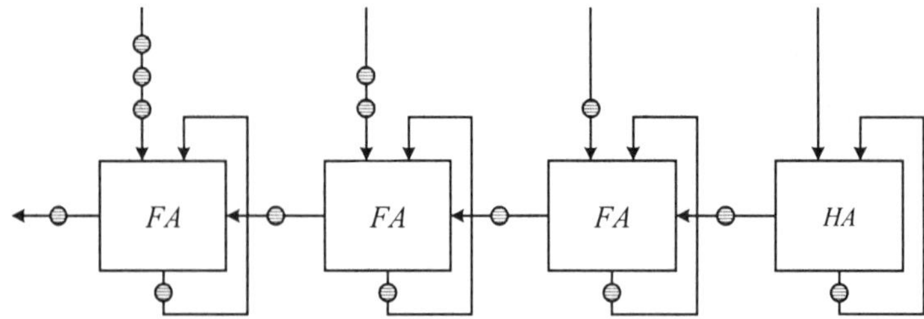

Figure 8.15 A pipelined 4-b numerically controlled oscillator.

8.2.5 Direct Digital Frequency Synthesizer

The NCO generates a square wave whose frequency is set by the control word and allows the timing recovery loop to recover symbol, chip, or hop timing by tuning its frequency. An NCO, however, cannot be used for phase recovery because of the high harmonic content in the square wave. To provide a clean carrier reference, an NCO must be modified to generate sinusoidal waveforms. An NCO with sinusoidal outputs is commonly referred to as a direct-digital frequency synthesizer (DDFS) shown in Figure 8.16.

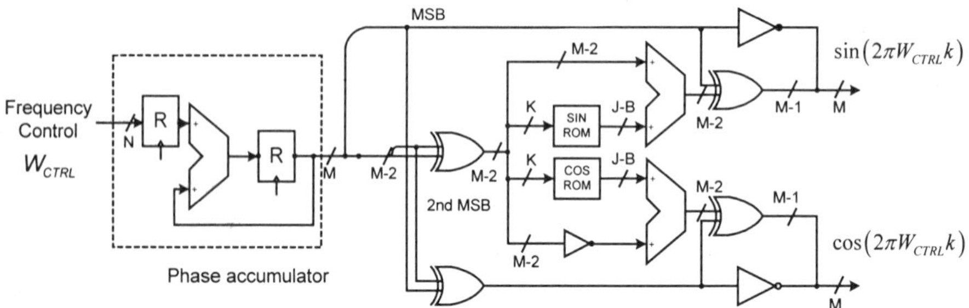

Figure 8.16 A direct-digital frequency synthesizer

The DDFS has the same update register and accumulator as in an NCO. In particular, this circuit block is also referred to as a phase accumulator since it generates phase advances at every clock cycle with increments specified by the frequency control word W_{cntl}. Like the NCO, the frequency resolution depends on the wordlength N. To save hardware, the output of the phase accumulator is often truncated to M bits but at a cost of decreased spectral purity of the generated waveforms. The amount of truncation that can be tolerated depends on the desired spectral purity.

A brute force approach to generate a sinusoidal waveform is to store a complete cycle of the sine wave in a ROM and use the phase accumulator output to address the ROM contents. This requires $2^M(J-B)$ bits of ROM, where J is the full precision ROM output wordlength and B is the number of truncated bits at the output of the ROM.

The truncation to *J-B* bits saves hardware but again introduces degradation in the spectral purity of the DDFS output. For an M of 16 bits and *J-B* of 10 bits, the ROM size is 655360 bits which incur significant complexity. However, if the symmetry of the sine wave is exploited, the ROM size can be reduced to $2^K(J-B)$ where $K < M$. For instance, one can exploit the quarter-wave symmetry to just store a quarter of the sine wave, shown as the shaded region in Figure 8.17a. The ROM size immediately reduces by a factor of four. Addressing in this case has a simple implementation based on taking the XOR of the *M-2* LSB's of the phase accumulator output with the 2nd MSB of the same output signal as shown in Figure 8.16.

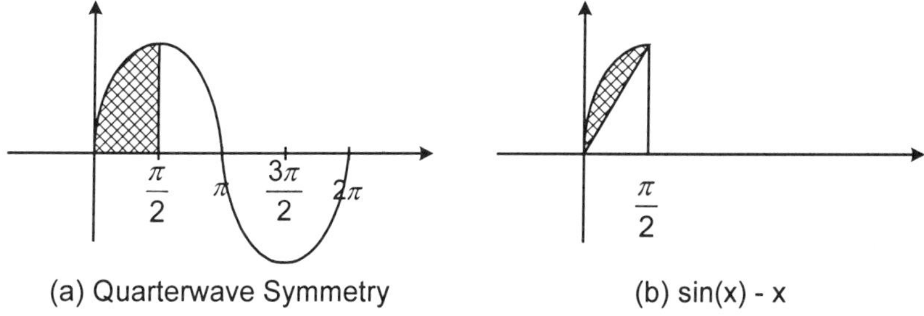

Figure 8.17 Reducing ROM size by utilizing quarterwave symmetry.

Further reduction in ROM size can be achieved by partitioning one large ROM into smaller ones, whose outputs are then summed to obtain the desired waveform. A direct method to partition the ROM is to express the sine wave as a sum of more elementary basis functions. For instance, using a power series, a sine wave can be expressed as the sum of odd-powered polynomials. Unfortunately, the series has an infinite number of terms and cannot be implemented with a ROM of finite size. To circumvent this problem, the first few terms can be used while the rest can be expressed in terms of the difference between the sine function and the first few terms. A simple example of this is to express a sine function as

$$\sin x = x + (\sin x - x). \tag{8.16}$$

Graphically, (8.16) is depicted in Figure 8.17b, where the second term is shown as the shaded region. (8.16) offers two advantages: 1) the second term has a lower dynamic range requirement and therefore fewer bits are needed per ROM entry and 2) the first term requires no storage since it is precisely the phase accumulator output from 0 to $\pi/2$. Implementation of (8.16) is illustrated in Figure 8.16, where the SINE ROM output is added to the address line. Note that since the phase of a cosine leads the sine function by 0.5π, an inverter is applied to the address line before summing with the output of the COSINE ROM and an XOR of the MSB and 2nd MSB of the phase accumulator output is used to provide a control signal to obtain a signed 2's complement representation at the DDFS cosine output.

Even more reduction can be achieved by taking more terms. Similar techniques based on coarse/fine ROM partition [218][219] show significant reduction in ROM size. For example, a 12×12 ROM can be partitioned into an 8×9 and an 8×3 ROM, resulting in a 16 times reduction.

At this point, the DDFS generates rectified sine and cosine waveforms since the outputs are unsigned. The bit-parallel XOR and inverter shown at the outputs of the DDFS in Figure 8.16 convert the rectified waveforms into full-swing sinusoids by using the MSB of the phase accumulator output.

A DDFS has many advantages over an analog VCO, including 1) low spurious response (< -100 dBc), 2) high tuning range (50% of clock rate), 3) virtually no phase noise except that induced by the crystal reference, 4) perfect quadrature generation, 5) process independence, and 6) high frequency resolution. It is important to realize that the performance limitation in the DDFS lies mainly in power dissipation at high frequency and the availability of precision clock crystal oscillators. The first issue results from the linear dependence of the DDFS output frequency on clock rate. A high frequency output (e.g. 1 GHz) requires at least a 2 GHz sampling rate, making it extremely difficult to implement in CMOS and with low power. Several high speed but high power DDFS implementations have been reported in [220] and [221]. The former reports CMOS implementation with a maximum output frequency of 400 MHz and the latter reports a GaAs IC with a maximum output frequency of 512 MHz. The second issue is more subtle in that the phase noise and frequency stability of a DDFS depends heavily on the crystal oscillator.

8.2.6 Transmit and Receive (Matched) Filters

In Section 2.1.2, pulse-shaping filters are used to achieve high bandwidth efficiency and maintain good SNR performance with minimal ISI. The difficulty with ideal pulse-shapes is that they have infinite time duration, implying non-causal behavior which makes the pulse-shape unrealizable in practice. To overcome this problem, ideal pulse-shapes are often truncated and windowed appropriately to maintain good frequency response with a well defined cut-off and high stopband attenuation of 50-70 dB as shown in Figure 8.18. Many filter design algorithms [222] and tools (e.g. Matlab) are available to take a frequency-response requirement and determine a set of filter coefficients $\{h_i\}$ that meets the filter specification. Filter requirements may be specified in terms of passband ripple, stopband ripple, passband gain, and stopband attenuation.

There are two types of filters: finite-impulse response (FIR) and infinite-impulse response (IIR). FIR filters have finite-duration impulse responses and have the advantages of being numerically stable and having linear phase. IIR filters, on the other hand, have an infinite-duration impulse response, cannot achieve linear phase, and tend to have numerical stability problems. The main advantage of an IIR filter though is that it generally requires fewer coefficients to implement a sharp transition and high stopband attenuation in its frequency response.

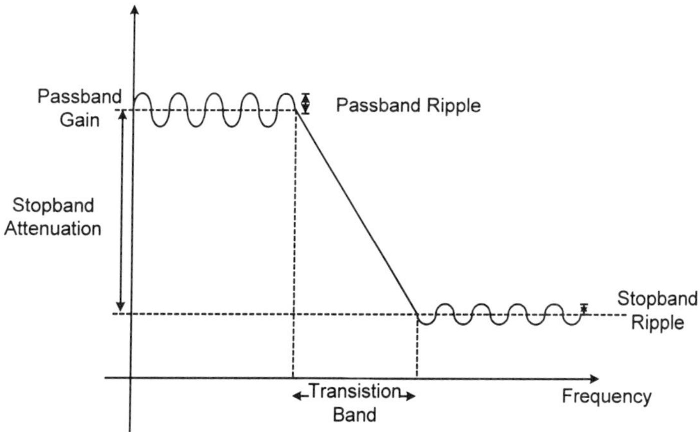

Figure 8.18 Filter requirement.

Once the filter coefficients have been determined, an architecture is designed to efficiently implement the filter, namely expression (2.13). This section describes several architectures for implementing a digital filter given the filter response, focusing primarily on the FIR filter.

8.2.6.1 Direct Form

The output y_n of an FIR filter can be computed based on (2.13). Since each transmitted symbol is pulse shaped, the impulse response $h(t)$ should be over-sampled. Denoting an over-sampled symbol by $\xi_i^{'}$ and the over-sampling ratio by M, the output of the FIR filter can be expressed as

$$y_n = h_0\xi_n^{'} + h_1\xi_{n-1}^{'} + \cdots + h_M\xi_{n-M}^{'}, \tag{8.17}$$

where $\{h_i\}$ are the filter coefficients of the over-sampled $h(t)$ and $\{\xi_i^{'}\}$ are the over-sampled input samples. The over-sampled input symbol can also be represented as an interpolated signal shown below

$$\xi_n^{'} = \sum_{i=-\infty}^{\infty} \xi_n \delta_{n-iM}. \tag{8.18}$$

A direct implementation of (8.17) is shown in Figure 8.19. The delay line formed by the M-bit shift register generates the delay samples $\{\xi_i^{'}\}$ that are tapped off into fixed gain blocks whose outputs are then summed to form the output sample. Rather than using full-fledged multipliers, the fixed gain blocks can be implemented by using the CSD number system mentioned in Section 8.2.2. Since there can be multiple CSD representations for a given number, the one with the minimum number of non-zero CSD digits is desired to minimize the number of additions needed. Algorithms are available to achieve optimal CSD coefficient assignments [223].

The main advantage of a direct-form architecture is that each delayed input signal sees a constant load. Thus, it is not necessary to insert power-hungry buffers at the filter input. However, the direct-form architecture does require an M+*1* input adder. Therefore, it is less suited for high-speed operation since a deeply pipelined carry-save adder array followed by a carry-ripple adder may be required.

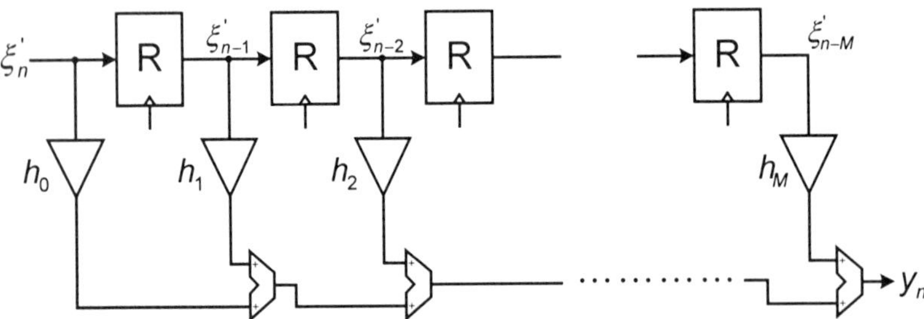

Figure 8.19 A direct-form implementation of an FIR filter.

8.2.6.2 Transpose Direct Form

An alternative to the direct-form architecture is the transpose direct form shown in Figure 8.20. It is straightforward to show that the transpose form and the direct form have the an equivalent filter response. The advantage of the transpose form lies in its inherent pipelined structure, where each adder is separated from the other adders by a register. Therefore, it is ideal for a high-speed and/or low power implementation [224][225]. For best performance, a carry-save adder can be used for each tap with dual registers to pipeline both the carry and save outputs. The filter output is obtained by using a carry-ripple adder at the final tap to generate the true sum. Despite its advantages, the transpose form does require a well-designed input buffer stage since the input must drive a fairly large load consisting of $M+1$ filter taps.

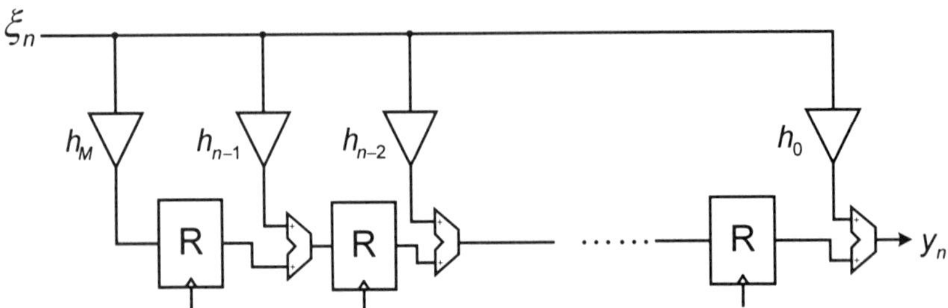

Figure 8.20 A transpose direct-form implementation of an FIR filter.

8.2.6.3 Time Shared

For systems where speed and power are not an issue but rather low complexity is desired, a time-shared architecture shown in Figure 8.21 may be used. In the time-shared architecture, each filter tap is time shared to achieve an M-fold hardware saving on adders and constant gain blocks. However, some overhead is required for the control logic, the multiplexer, the real multiplier, and the coefficient ROM [226]. Also, the MAC unit and the coefficient ROM must operate at an M-fold increase in clock rate. It should be noted that the register and multiplexer can be replaced by a dual-port memory with circular addressing for a more compact layout.

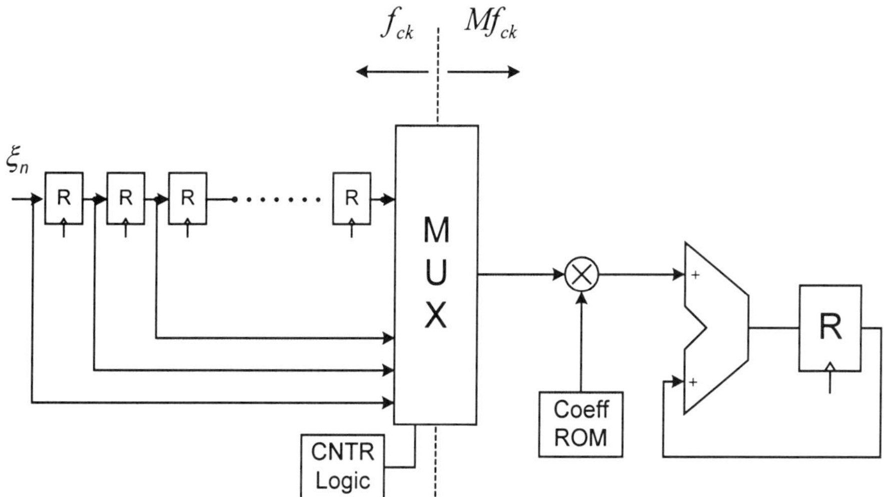

Figure 8.21 A time-shared direct-form architecture.

8.2.6.4 Linear Phase

Not all FIR filters are linear phase, only those that satisfy the following symmetry constraint

$$h_n = \pm h_{M-1-n}. \tag{8.19}$$

A filter satisfying (8.19) with and without a sign inversion is also referred to as being anti-symmetric and symmetric, respectively. Because of (8.19), only half of the coefficients are unique and therefore half of the filter taps can be reused to reduce filter complexity as shown in Figure 8.22 for a symmetric filter with M even.

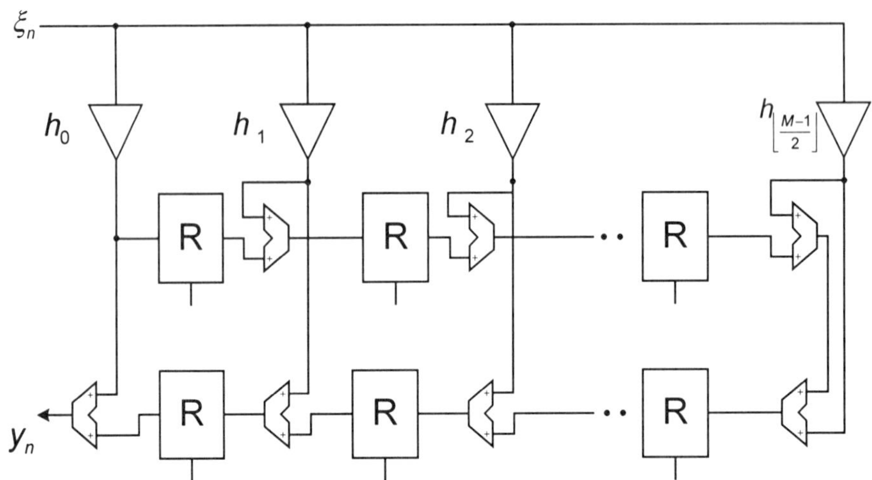

Figure 8.22 A transpose direct-form linear-phase symmetric FIR filter.

8.2.7 Multirate Filters

A digital transceiver generally needs to process multirate signals as discussed in Chapter 6, especially for digital IF and digital RF architectures. Multirate processing requires two key elements: 1) an interpolation filter for up sampling and 2) a decimation filter for down sampling.

8.2.7.1 Interpolation Filter

Often a signal with a sampling rate F_s needs to be sampled to a higher sampling rate of $\beta_I F_s$, where β_I is the interpolation ratio. The interpolation process involves inserting '0's in between each sample while not changing the frequency spectrum of the baseband signal. Figure 8.23a shows the frequency spectrum of the interpolated signal with images that are centered on multiples of F_s. The spectrum has not changed around DC but images have been introduced centered on $(m\beta_I+1)F_s \cdots (m\beta_I+\beta_I-1)F_s$ with $m=0,1,\cdots$. These images distort the interpolated signal and must be filtered out. However, due to the high sampling rate, such filters require too many taps as discussed in Section 6.2.1. To reduce the complexity of the image-reject filter, the filtering could be combined with interpolation and performed in multiple stages as shown in Figure 8.23b. At each stage, the input is interpolated by a factor N_I followed by a $0.5N_I$ band filter, which rejects undesired images. The bandwidth is normalized to $N_I F_s$. By performing the filtering in stages, the filter complexity reduces from $O(\beta_I)$ to $O(N_I \log_{N_I} \beta_I)$.

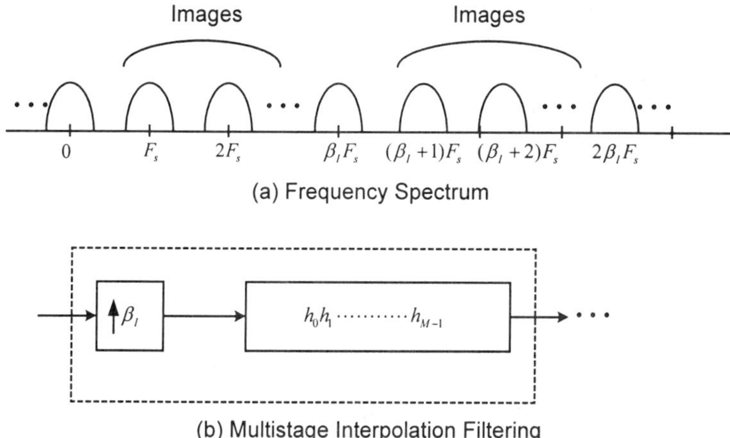

Figure 8.23 Interpolation.

A direct implementation of interpolation requires the filter to operate at β_I times the original sampling rate. A polyphase filter helps to reduce the speed requirement by partitioning the filter into β_I sections with coefficients indicated in Figure 8.24. The output of each sub-filter is multiplexed to form the final filter output. The only high-speed block in a polyphase design is the multiplexer while the sub-filters can operate at the original sampling rate. Since the number of filter taps M may not be multiples of β_I, m should be set to $\lceil M/\beta_I \rceil$ and the filter coefficients with an index greater than M-1 should be set to zero. While the above discussion assumes the interpolation is performed in one stage, a polyphase filter structure easily applies to a multistage implementation as well.

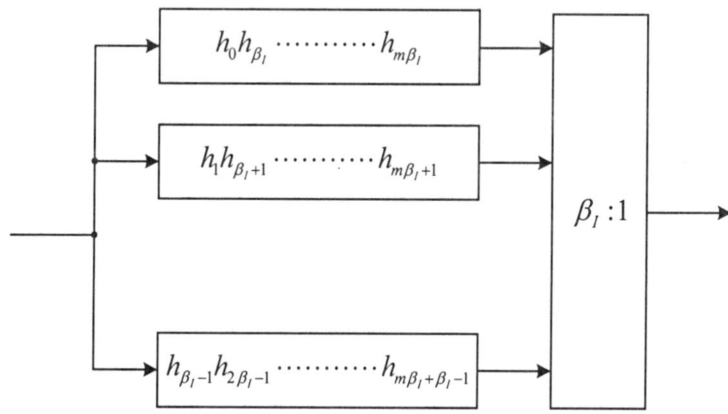

Figure 8.24 Polyphase Interpolation filter architecture.

For applications in which the signal bandwidth is much less than the interpolated sampling rate, a simple filter structure based on cascaded integrator comb (CIC) filters may be sufficient [227]. The CIC filters provide extremely steep rejection at multiples of F_s but with only over a narrow frequency range, which is fine when the interpolation factor is high. An Lth order CIC filter has the following transfer function

$$H(z) = \left(\frac{1 - z^{-\beta_I}}{1 - z^{-1}} \right)^L, \tag{8.20}$$

which can be implemented as a cascade of L comb filters. Each comb filter takes the running sum of its current input and $\beta_I - 1$ delayed samples. When used as the post-filter in an interpolator, the cascaded structure can be simplified by moving the interpolator in between a cascade of comb filters and integrators as shown in Figure 8.25. In this case, each comb filter has a memory of one rather than β_I.

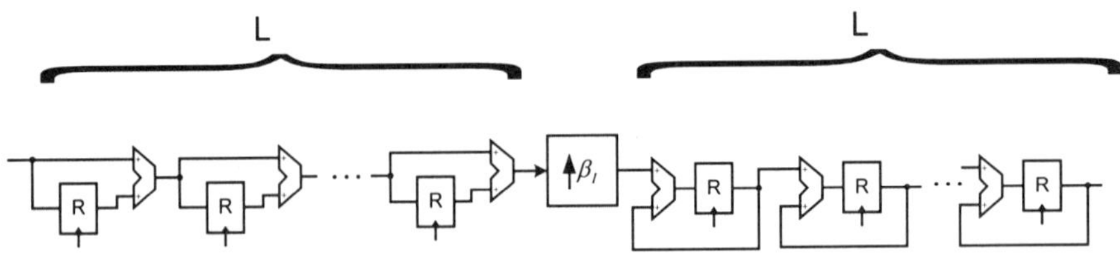

Figure 8.25 An Lth order CIC filter used for interpolation.

The CIC filter has the following frequency response

$$H\left(e^{j2\pi f \Delta_t}\right) = \left| \frac{\sin\left(\pi f \beta_I \Delta_t\right)}{\sin\left(\pi f \Delta_t\right)} \right|^L \tag{8.21}$$

over the frequency range between 0 and Δ_t^{-1} with nulls at multiples of $\left(\beta_I \Delta_t\right)^{-1}$. With interpolation the sampling interval is $\left(\beta_I F_s\right)^{-1}$. An example response for $\beta_I = 4$ is shown in Figure 8.26 for L = 1. By increasing the order of the CIC, more attenuation can be obtained in the image band.

8.2.7.2 Decimation Filter

In contrast to the interpolation process, decimation requires deletion of samples at regular intervals to reduce the sampling rate of the signal. Decimation has applications in digital transceivers to down sample the received signal, which is often sampled at a rate much higher than the Nyquist rate. To reduce the speed requirements in the baseband processing, decimation is used to convert the high sample rate received signal to a rate closer to two times the Nyquist rate. Rather than having image problems, decimation can experience aliasing problems due to down sampling in which samples containing information are discarded.

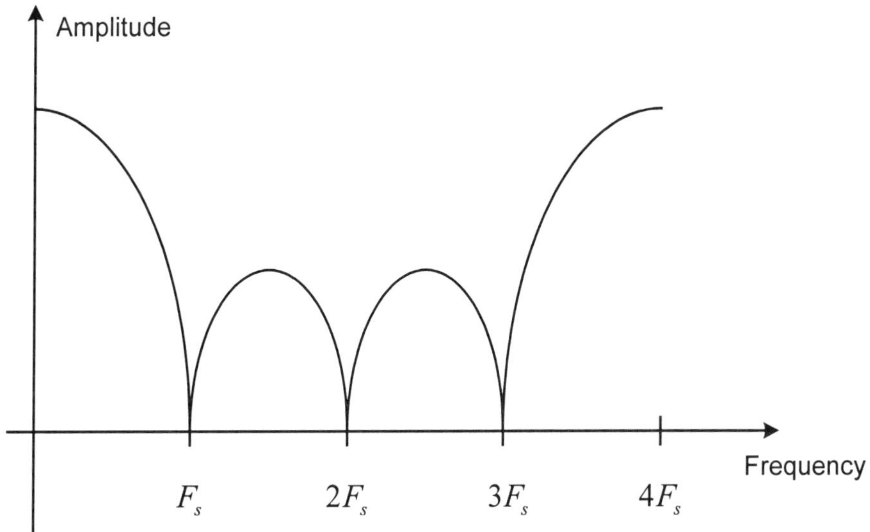

Figure 8.26 A CIC filter with $\beta_I = 4$ and L = 1.

Figure 8.27a shows the frequency response of the original signals and the decimated signal. Since decimation may be viewed as sampling of the analog waveform by a sampling rate F_s / β_D, where β_D is the decimation factor, frequency components within $mF_s/\beta_D \pm W_s$ will fold into the desired signal band for $m = 1, 2, \cdots \beta_D - 1$. To prevent aliasing, a $0.5\beta_D^{-1}$ band filter should be used prior to decimation although with a high decimation factor, a $1/\beta_D$ CIC may be sufficient. The filter bandwidth is normalized to F_s.

Similar to interpolation, a multistage filter structure may be used to reduce the size of the pre-filter for large decimation ratios as shown in Figure 8.27b. By performing the filtering in stages, the filter complexity reduces from $O(\beta_D)$ to $O(N_D \log_{N_D} \beta_D)$. In [228][229], the multistage architecture has been implemented in a recirculating architecture whereby outputs are fed back or recirculated to the input for further interpolation or decimation. In other words, a single stage may be time shared to achieve low complexity. Retimed architectures [226][230][231] are also possible but will not be discussed here.

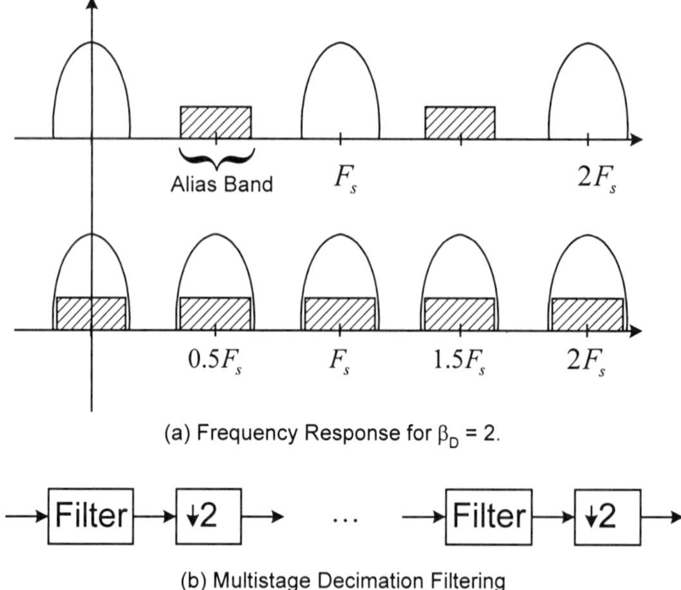

(a) Frequency Response for $\beta_D = 2$.

(b) Multistage Decimation Filtering

Figure 8.27 Decimation.

Each stage of the decimation filter can be effectively implemented using either a CIC filter similar to the one discussed in the previous section or a polyphase structure shown in Figure 8.28. The polyphase filter performs a serial to parallel conversion on the input samples so that the β_D parallel outputs are filtered by β_D sub-filters. In this case, a one-stage decimation is assumed. The outputs of the sub-filters are then summed to form the decimated output. In this case, the only high-speed component is the input multiplexer. The sub-filters operate at a lower frequency of F_s / β_D. Since M may not be multiples of β_D, m should be set to $\lceil M / \beta_D \rceil$ and the filter coefficients with index greater than M-1 should be set to zero.

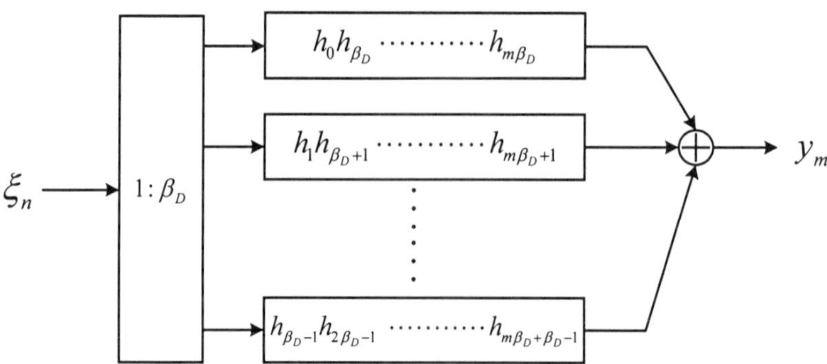

Figure 8.28 Decimation stage using a polyphase filter.

8.3 Modulator/Demodulator

A modulator/demodulator, or modem for short, constitutes a baseline digital transceiver. The modulator converts a serial bit stream into the transmitted symbols with values corresponding to the points on the signal constellation. The demodulator recovers the transmitted symbols in the presence of noise. Using the maximum likelihood (ML) criterion the average error probability of the received signal is minimized. A demodulator employing the ML criterion is also known as an ML receiver or a minimum distance receiver. The theoretical basis for digital modulation and ML reception has been discussed in Chapter 2.

8.3.1 BPSK

BPSK is the simplest form of digital modulation and has good energy efficiency as defined by (2.43). Its bandwidth efficiency, however, is at a nominal 1bps/Hz. Because of its simplicity, it has been widely used in many wireless systems.

8.3.1.1 Coherent Modem

A digital IF BPSK modem is shown in Figure 8.29. The modulator consists of an interpolator followed by a transmit filter whose output is then up-converted using a DDFS. The output sampling rate of the filter must be at least twice $\frac{\omega_{IF}}{2\pi}+W_s$ to satisfy the Nyquist criteria. Therefore, the interpolation factor β_I becomes $2+\frac{\omega_{IF}}{\pi W_s}$. The transmit filter is generally implemented using the square-root raised cosine pulse shape discussed in Section 2.1.2 and the interpolation is implemented using any of the techniques described in the previous section.

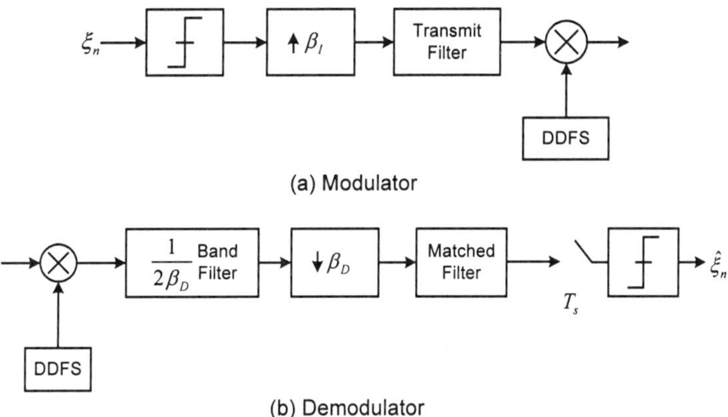

Figure 8.29 A coherent BPSK modem architecture.

The demodulator consists of a digital down converter using the DDFS and a multiplier. The down-converted signal is then decimated to 2-4 times the Nyquist rate and processed by a matched filter followed by a sampler clocked at the symbol rate. The sampler output forms the soft-decision which consists of both noise and the transmitted signal. The purpose of the matched filter is to maximize the SNR at the output of the sampler. To recover the actual symbol value a slicer is used to make a hard decision based on the ML criteria as described in Section 2.3. For BPSK, the constellation points lie on $\pm\sqrt{E_s}$. Therefore, the slicer has a threshold set at zero. The demodulator shown in Figure 8.29b assumes that the DDFS is locked in carrier frequency and phase and that timing of the sampler has been synchronized. Circuits that implements carrier phase and timing recovery will be discussed later in Section 8.5.

8.3.1.2 Differentially Coherent Modem

PSK inherently has phase ambiguity as discussed in Section 2.4.1.1. To overcome the phase ambiguity problem, differential encoding is used to encode transmitted symbols in phase transitions rather than in absolute phase values. For BPSK, the differential encoder and decoder can be derived by referring to the tables shown in Figure 8.30 based on expression (2.119).

ϕ_k	$\hat{\phi}_{k-1}$	$\hat{\phi}_k$
0	0	0
0	π	π
π	0	π
π	π	0

$\hat{\phi}_{k-1}$	$\hat{\phi}_k$	ϕ_k
0	0	0
0	π	π
π	0	π
π	π	0

Figure 8.30 Phase transition table for differential encoding and decoding.

In BPSK, each phase ϕ_k represents a symbol and if phases 0 and π are represented respectively by '0' and '1', then coding and decoding the differential phase $\hat{\phi}_k$ can be easily implemented using an XOR and a register as shown in Figure 8.31. A differentially coherent modem has the same general architecture as the coherent modem shown in Figure 8.29, except a differential encoder is placed at the input of the modulator and a differential decoder is placed at the output of the slicer.

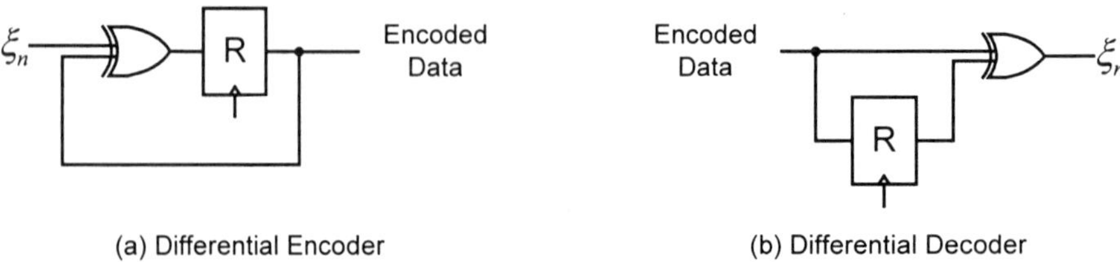

Figure 8.31 A differential encoder and decoder.

8.3.2 QPSK

QPSK achieves higher energy and bandwidth efficiency with respect to BPSK but it generally has twice the complexity, making it more difficult to integrate on an IC. However, with advances in IC technology, more systems, such as IEEE 802.11, are using QPSK that can now be easily integrated on a single chip [232][233].

8.3.2.1 Coherent Modem

Figure 8.32 shows a coherent QPSK modem architecture. Because QPSK has both in-phase and quadrature components, the modulator consists of two interpolation filters, one for the I-channel and one for the Q-channel. Inputs to each channel are generated from a 1:2 demultiplexer with a '0' mapped to +1 and a '1' to −1 at the output of the demultiplexer. The interpolated outputs are up converted using a quadrature DDFS, which generates both cosine and sine signals. The up converted in-phase and quadrature signals are then summed to form a bandpass QPSK signal according to (2.2).

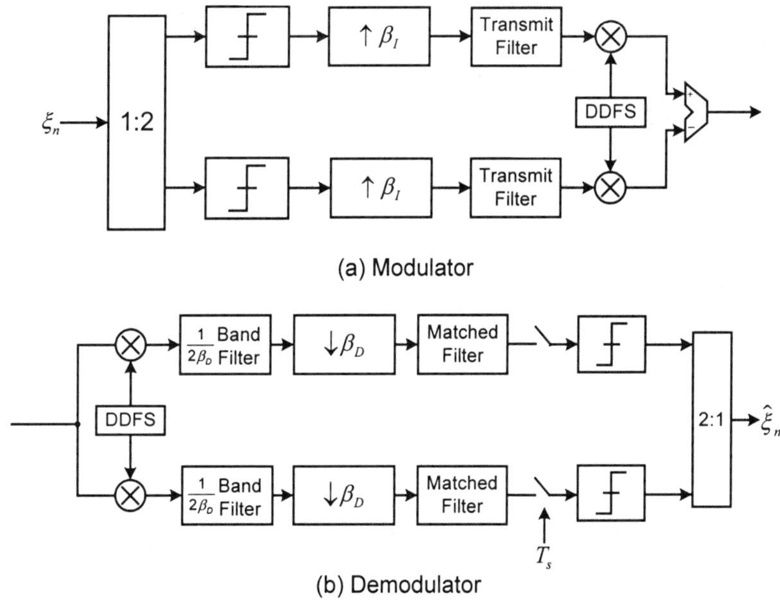

Figure 8.32 A coherent QPSK modem architecture.

The QPSK demodulator resembles that used for BPSK except it contains two channels, one to recover the I-channel data and the other to recover the Q-channel data. The down converted baseband signal is generated by mixing the received input with the cosine and sine outputs from a quadrature DDFS. The hard-decision data are then multiplexed to form the decoded bit sequence. Since the two constellation points along either channels are $\pm\sqrt{0.5E_s}$, the hard decision thresholds is set at zero in both channels.

8.3.2.2 Differentially Coherent Modem

The derivation of the differential encoder for QPSK is more involved because there are more phases that results in a larger phase transition table. Specifically, the transition table for QPSK has sixteen entries in contrast to four entries for BPSK. Also, QPSK requires two bits to represent its phase versus one bit in BPSK. While an increase in phase transitions and symbol size make the derivation of a differential encoder/decoder more involved, the derivation follows from that illustrated for BPSK. It is left as an exercise to show that with Gray coding, the encoder and decoder have the implementation shown in Figure 8.33.

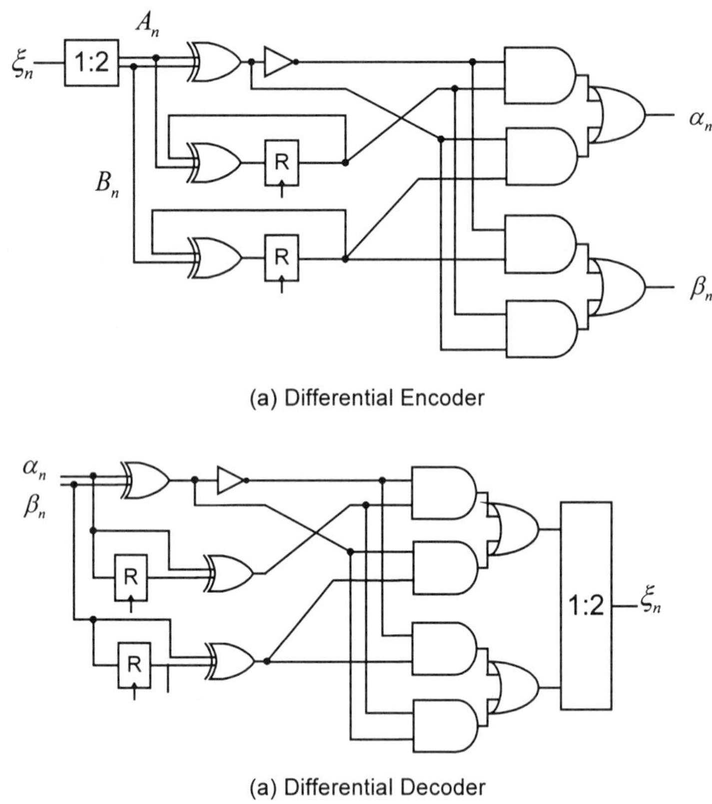

Figure 8.33 A differential encoder and decoder for QPSK.

The encoder is placed after the 1:2 demultiplexer in the modulator and after the slicers in the coherent demodulator.

8.3.2.3 Offset QPSK

PSK has the problem of spectral regrowth due to non-linearity in the power amplifier. OQPSK has the advantage of needing less backoff in the power amplifier as discussed in Section 2.4.1.3. An OQPSK modem can be implemented with very little modification to the QPSK modem shown in Figure 8.32. Modification in the transmitter involves adding a half-symbol delay in the Q-channel at the output of the 2:1 demultiplexer. In the receiver, the Q-channel matched filter is sampled with a half symbol lag relative to the sampling clock of the I channel.

8.3.3 QAM

QPSK is actually a special case of QAM when the constellation size is four. In general, QAM constellation points vary in both phase and amplitude and can have a wide range of geometry. Compared to PSK, high order QAM has higher energy efficiency. QAM is being considered for some of the third-generation systems to enable high-speed transmissions in indoor and static environments.

8.3.3.1 Coherent Modem

Figure 8.34 illustrates a coherent modem architecture for QAM. The QAM modem differs from the QPSK modem mainly in the symbol mapping at the modulator and the hard-decision logic at the demodulator. The input bit stream at the modulator is demultiplexed into an N-bit word which addresses the stored symbol values in the I/Q ROM's. The stored symbol values depend on the geometry of the QAM constellation. In Figure 8.34b, a symmetric constellation is assumed such that the hard-decision logic at each channel inverts the soft-decision value to $N/2$ information bits that are multiplexed to form a single serial output.

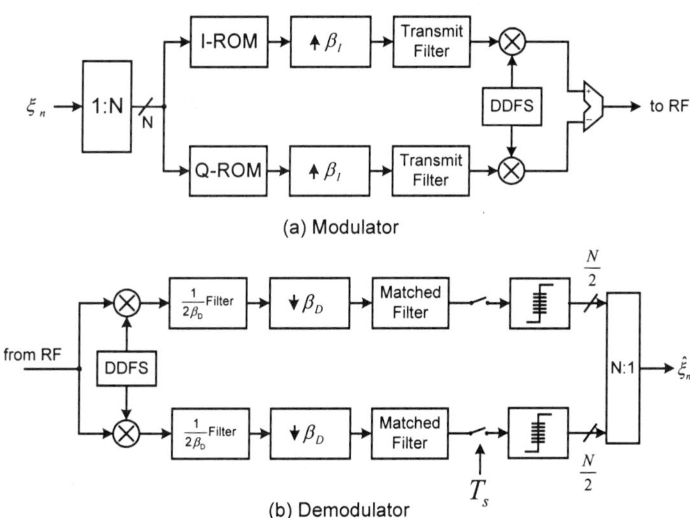

Figure 8.34 A coherent QAM modem architecture.

8.3.3.2 Differential Encoding

Since QAM still uses phase information, it also exhibits phase ambiguity. At first glance, it may seem difficult to apply differential encoding to QAM due to the amplitude variations in QAM constellation points. However, if the constellation points are partitioned into four groups, one for each quadrant in the I-Q plane, then differential encoding can be applied to the individual groups. In practice, this is achieved by differentially encode the MSB of the *N*-bit word that is used to label each constellation point. Thus, the differential encoding and decoding used for QPSK can be directly applied to QAM.

8.3.4 FSK

FSK can have a very low E_b / N_0 requirement but poor bandwidth efficiency. FSK is ideal for low data rate systems, such as paging, or for applications in which bandwidth is not an issue.

8.3.4.1 Modulator

A FSK modulator can either be double sideband or single sideband as shown in Figure 8.35. A double sideband modulator demultiplexes the input bit sequence into an *N*-bit work that addresses a ROM storing the symbol values, which in the case of FSK are different frequencies. The ROM outputs appropriate frequency control words to the DDFS that generates a sinusoid with the modulation frequencies corresponding to the transmitted symbols. The modulation is double sideband since for every modulation frequency f_i there is a negative counterpart. When the baseband signal is up-converted to IF or RF, the transmission bandwidth becomes approximately $2f_i$.

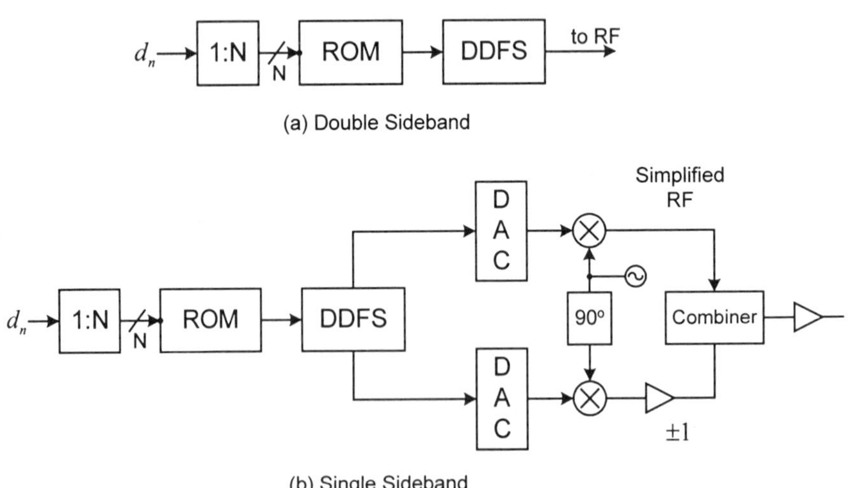

Figure 8.35 An FSK modulator.

A single sideband modulator, on the other hand, has no negative frequency components and therefore results in a two times saving on bandwidth. While all real signals have negative frequency components, complex signals can be created that exhibit only positive or negative frequency contents. Thus, instead of representing an FSK signal by $\cos(2\pi f_i t)$, one may represent it using complex notation shown below

$$e^{j2\pi f_i t} = \cos(2\pi f_i t) + j\sin(2\pi f_i t).$$

The frequency spectrum of the real signal has two impulses, one at f_i and the other at $-f_i$ but the complex signal has only one impulse at f_i. Since real-world signals are not complex, the complex representation is only used as a tool to process the signal. Eventually, it is converted to a real signal by taking its real part.

For the single sideband modulator, the complex representation is used to represent the tone modulation and the up conversion; i.e. $e^{j2\pi f_i t} \times e^{j2\pi f_o t}$, where f_o is the carrier frequency. The final output, however, is obtained by taking the real part of the above complex multiplication as shown below

$$\begin{aligned} R\left\{e^{j2\pi f_i t} \times e^{j2\pi f_o t}\right\} &= \cos\left[2\pi\left(f_i + f_o\right)t\right] \\ &= \cos(2\pi f_i t)\cos(2\pi f_o t) - \sin(2\pi f_i t)\sin(2\pi f_o t). \end{aligned} \tag{8.22}$$

The frequency spectrum of (8.22) has frequency components only from the upper sideband $\pm(f_o + f_i)$ whereas with double sideband modulation lower sideband components $\pm(f_o - f_i)$ are also present. Figure 8.35b shows a single sideband modulator that generates the upper or the lower sideband depending on the sign of the unity gain block before the combiner. A sign inversion generates the upper sideband.

8.3.4.2 Envelope Detection Demodulator

Although a coherent demodulation can be done for FSK, it is typically not used because of the complex hardware needed for carrier phase recovery. Non-coherent demodulators sidestep the need for phase recovery and offer a simpler solution. A non-coherent demodulator does however incur a doubling of bandwidth since to maintain orthogonality each tone must be separated by the symbol rate whereas coherent modulation requires a tone separation of only half the symbol rate. Moreover, a non-coherent demodulator incurs a 3-dB degradation in SNR compared to a coherent demodulator.

Figure 8.36 shows a non-coherent demodulator based on envelope detection. The envelope correlation between each tone and the received signal is compared to determine the tone that generates the maximum correlation energy. The envelope detector for each correlator is implemented based on the following observation

$$\int_{T_s} \left|r(t)e^{-j2\pi f_i t}\right|^2 dt = \int_{T_s} \left[r_I\cos(\omega_i t) + r_Q\sin(\omega_i t)\right]^2 + \left[r_Q\cos(\omega_i t) - r_I\sin(\omega_i t)\right]^2 dt \tag{8.23}$$

where $r(t)$ is the received baseband signal and r_I and r_Q are its in-phase and quadrature components, respectively. The integration over the symbol period is performed using the A&D. From the envelope detector outputs, a comparison block determines the symbol corresponding to the tone that generated the maximum correlation energy.

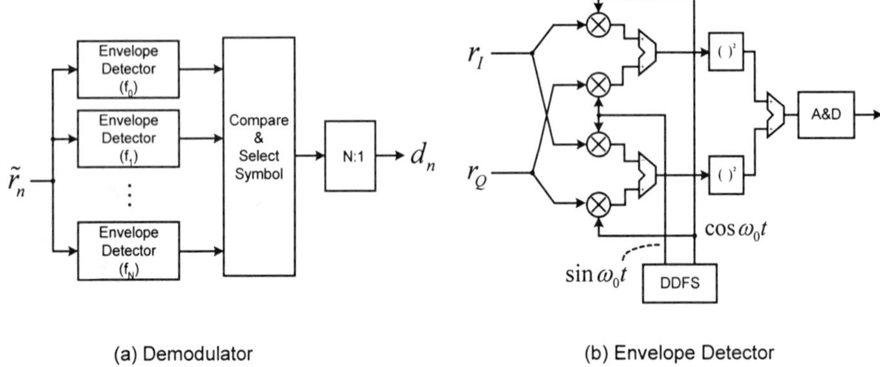

Figure 8.36 A non-coherent FSK demodulator using envelope detection.

8.3.5 GMSK

GMSK is a special case of CPM discussed in Section 2.4.2. GMSK exhibits lower spectral regrowth in the presence of non-linear amplification. Because of its insensitivity to non-linear amplification, it is currently used in several commercial systems, such as GSM.

8.3.5.1 Modulator

Equation (2.125) describes the mapping of transmitted symbols onto a continuous phase waveform $\phi(t;\xi_k,\sigma_k)$. The modulated RF waveform is simply

$$s(t)=\cos\left[\phi(t;\xi_k,\sigma_k)\right]\cos(\omega_o t)-\sin\left[\phi(t;\xi_k,\sigma_k)\right]\sin(\omega_o t) \tag{8.24}$$

and can be directly implemented using the modulator architecture shown in Figure 8.37a. The input bit sequence is converted to an N-bit symbol value using a 1:N demultiplexer whose output is interpolated and shaped by a Gaussian filter described by (2.136). The output of the filter, scaled by the modulation index h, forms the frequency pulse shape that is fed to a quadrature DDFS which generates the baseband in-phase and quadrature components of (8.24). The baseband signals are then converted to analog signals, up-converted in quadrature, and combined to produce the modulated RF signal. Note that it is possible to perform all of (8.24) digitally with a single DDFS. However, the DDFS must then operate at several times the RF frequency, which is generally considered too costly and power consuming.

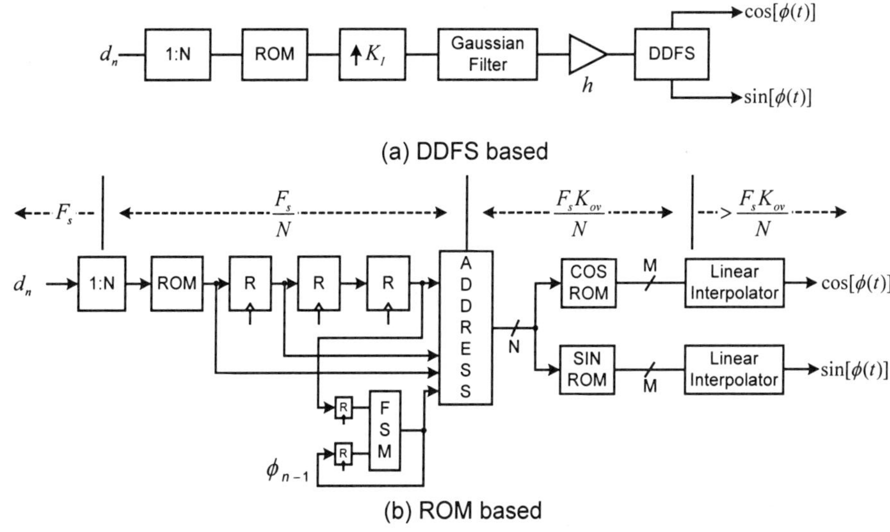

Figure 8.37 GMSK modulator architectures.

Figure 8.37b shows an alternative modulator architecture where the baseband in-phase and quadrature waveforms have been stored in two ROM's. The address to the ROM's is formed by using the L most recent input symbols $\xi_n, \xi_n, \cdots, \xi_{n-L+1}$ and the phase states are computed recursively using $\phi_n = \phi_{n-1} + \pi h \xi_{n-L+1}$. Since the baseband waveforms need a reasonably fine resolution, multiple samples should be generated for each modulator state to capture fine details of the waveforms. To achieve this, the address generator is clocked at K_{ov} times the symbol rate. The ROM size depends on the over-sampling ratio K_{ov}, the symbol size 2^N, the number of phase states ρ, the correlation length L of the phase pulse, and the ROM word size N_{ROM} as shown below:

$$\text{ROM size} = K_{ov} \rho 2^{NL} N_{ROM} \tag{8.25}$$

For a large symbol size or correlation length, the ROM may become too large to implement. To help reduce the ROM size, a linear interpolator can be used to reduce the over-sampling ratio.

Example 8.6 Consider the GMSK example in Section 2.4.2.2 which uses modulation parameters similar to that of GSM: namely, $BT = 0.3$, $L = 2$, $\rho = 4$, and $N = 1$. Assume that the oversampling ratio is 8 and that each ROM entry has 10 bits. Then, according to (8.25), the ROM size for either the in-phase or quadrature component becomes

$$8 \times 4 \times 2^2 \times 10 = 1280 \text{ bits},$$

a reasonable size. □

8.3.5.2 Viterbi Decoding

The optimum demodulator for GMSK or general CPM is based on Viterbi decoding. Referring to (2.125), $\phi(t;\xi_n,\sigma_n)$ can be expressed as the sum of the phase resulting from the L most recent symbols and the phase resulting from all other symbols. The former is defined as the phase branch ϑ_n and the latter the phase state φ_n. There are 2^{NL} phase branches and ρ phase states. Each of the possible phase branches and phase states is indexed by $\vartheta_n(i)$ and $\varphi_n(j)$, respectively, where $i=1,2,\cdots 2^{NL}$ and $j=1,2,\cdots \rho$. In general, there are $\rho 2^{NL}$ possible phases for a given time index n and each possible phase value can be represented by $\phi_n(l)$, where $l=1,2,\cdots,\rho 2^{NL}$ and

$$\phi_n(l)=\vartheta_n(i)+\varphi_n(j). \tag{8.26}$$

Figure 8.38a shows the architecture of a CPM demodulator based on Viterbi decoding. The received baseband signal at time index n represented by $r_n = r_{I_n} + jr_{Q_n}$ is used to compute all the $\rho 2^{NL}$ possible branch metrics $\lambda_n(l)$ based on (2.129) using stored values of the phase branches and phase states, i.e. (8.26). Figure 8.38b shows details on the branch metric computations for a given index l. The Viterbi detector based on (2.130) processes the $\rho 2^{NL}$ branch metrics and determines the optimum symbol after a decoding delay D. Architectures for Viterbi detectors will be discussed later in Section 8.7.3. For the theoretical background on Viterbi decoding, the reader should refer to Section 2.6.3.2 in Chapter 2.

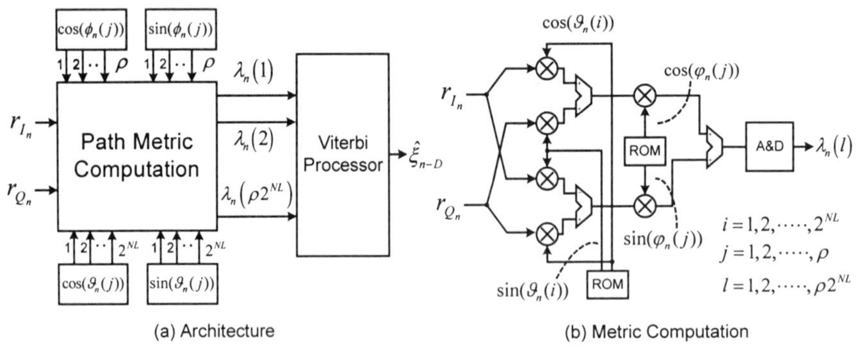

Figure 8.38 A CPM demodulator based on Viterbi decoding.

Example 8.7 Consider the GMSK system in Example 8.6. The ROM size required to store the phase state and phase branches is small. For the phase state, the ROM size is $4\times2\times10=80$ bits, assuming a 10-bit ROM word size. This also holds for the phase branches. However, sixteen computation elements (Figure 8.38b) are required for the branch metric computation. With a parallel implementation, the complexity due to the metric computations could be high but can certainly be accommodated with current CMOS technology. However, with more phase states and longer correlation lengths, the complexity can increase substantially making it more difficult to implement on a single chip. □

8.4 Spread-Spectrum Modulation

Generating a spread-spectrum waveform can be viewed as a type of modulation in that either the phase or frequency of the carrier is modulated by the spread-spectrum code. Spread-spectrum modulation spreads the spectrum of the original signal to provide frequency, time, and path diversity. This section describes two main types of spread-spectrum modulations based on direct-sequence and frequency-hop.

8.4.1 Direct-Sequence

The most common modulation formats used with direct-sequence spread spectrum are BPSK and QPSK since higher-order modulation results in a reduction in processing gain as discussed in Section 3.3.2. Thus, this section focuses only on direct-sequence spreading using BPSK or QPSK. Moreover, for illustration purposes, the digital baseband architecture is used. However, the technique can easily be applied to the digital IF and digital RF architectures as well.

8.4.1.1 BPSK Spread

Expression (3.2) captures the essence of direct-sequence encoding. Figure 8.39 illustrates the block diagrams of spread-spectrum encoders for quadrature and binary modulations. If the quadrature code $c_Q(t)$ is not used then the encoding is also referred to as BPSK spread. The PNGEN block produces the in-phase code signal $c_I(t)$ at the chip rate. The code signal can be any given PN-codes listed in Table 3.2. Since I_n and Q_n are both binary signals an XOR can be used in place of a multiplier. The spread signal is converted to an NRZ waveform and filtered for transmission by the RF front-end.

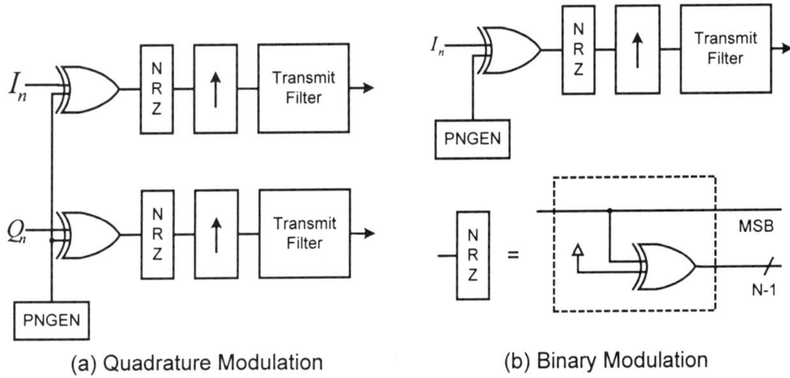

Figure 8.39 BPSK spreading.

8.4.1.2 QPSK Spread

In contrast to BPSK spread, QPSK spread employs both the in-phase and quadrature code signals as shown in Figure 8.40. The two codes are generally from the same family and are chosen with minimum cross-correlation. In general, QPSK spread provides additional protection against interference due to the additional degree of freedom obtained by the quadrature code.

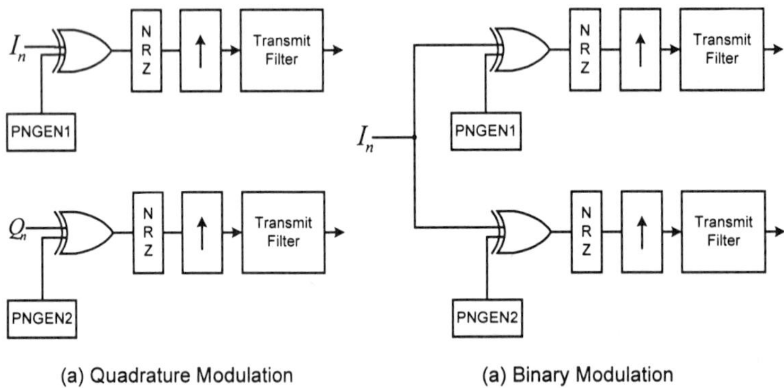

Figure 8.40 QPSK spreading.

8.4.1.3 Code Generators

A direct-sequence spreader requires the PNGEN block to produce PN-code running at the chip rate. Several codes have been discussed in 0, including the M-, Gold, Kasami, and Walsh sequences. All of these codes can be implemented using linear feedback shift register (LFSR) type circuits. LFSR implementation is illustrated for the M-sequence and the Gold sequence.

M-sequence can be generated by dividing two finite-field polynomials $I(z)$ and $C(z)$, also known as the initial-condition polynomial and the characteristic polynomial. In particular, we consider polynomials over the binary field GF(2). If $C(z)$ is a primitive polynomial, defined in Section 2.6.1.3, then with an arbitrary non-zero $I(z)$, the polynomial division generates the M-sequence. $C(z)$ has degree m, where $m = \log_2\left(1+N_c\right)$ and N_c is the period of the M-sequence. Also, the constant term is always non-zero in $C(z)$. Thus, $C(z)$ has the following general form

$$C\left(z\right) = z^m + c_{m-1}z^{m-1} + \cdots + c_1 z + 1. \tag{8.27}$$

The polynomial division can be implemented with a simple LFSR circuit shown in Figure 8.41, where the multiplication and addition in GF(2) are done with AND and XOR gates, respectively. Each tap of the LFSR corresponds to the coefficient of $C(z)$ and the internal states of the registers correspond to the coefficients of $I(z)$. An intuitive explanation of the LFSR implementation is that the polynomial division can be viewed as digital filtering of $I(z)$ by an infinite-impulse response (IIR) filter with transfer function $1/C(z)$ with all arithmetic done in GF(2). Note that for a fixed coefficient generator, a '0' tap corresponds to an open circuit and a '1' tap corresponds to a hardwired connection. Table 8.1 lists primitive polynomials of various degrees with c_0 always being '1'. More elaborate tables are available in [114].

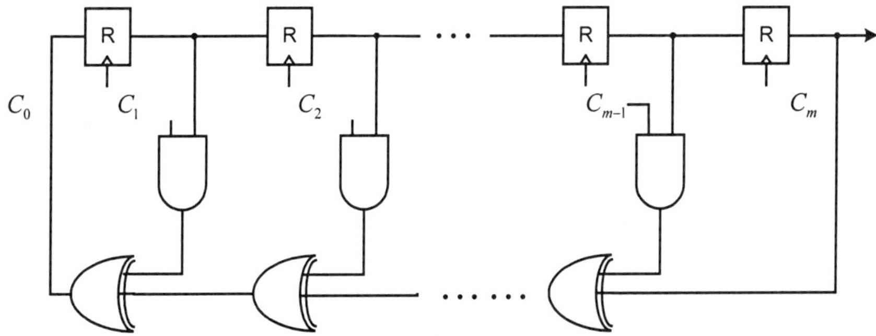

Figure 8.41 An LFSR implementation of an M-sequence generator.

Table 8.1 Some examples of primitive polynomials in GF(2).

N_c	m	Primitive Polynomial
7	3	x^2+x+1
15	4	x^4+x+1
31	5	x^5+x^2+1
63	6	x^6+x+1
127	7	x^7+x^3+1
255	8	$x^8+x^4+x^3+x^2+1$
511	9	x^9+x^4+1
1023	10	$x^{10}+x^3+1$

Gold sequence generators consist of two M-sequence generators with certain constraints on the two characteristic polynomials as defined in [122]. The outputs of the two M-sequence generators are XOR'ed to produce the Gold sequence as shown in Figure 8.42. It should be noted here that, given the filter analogy, a transpose architecture also exists for the LFSR implementation and can be obtained by the following simple rules:

1) Reverse the signal direction.
2) Replace all XOR's by shorts.
3) Replace all shorts by XOR's.
4) Obtain the coefficients c'_k for the transpose architecture using $c'_k = c_{m-k}$.

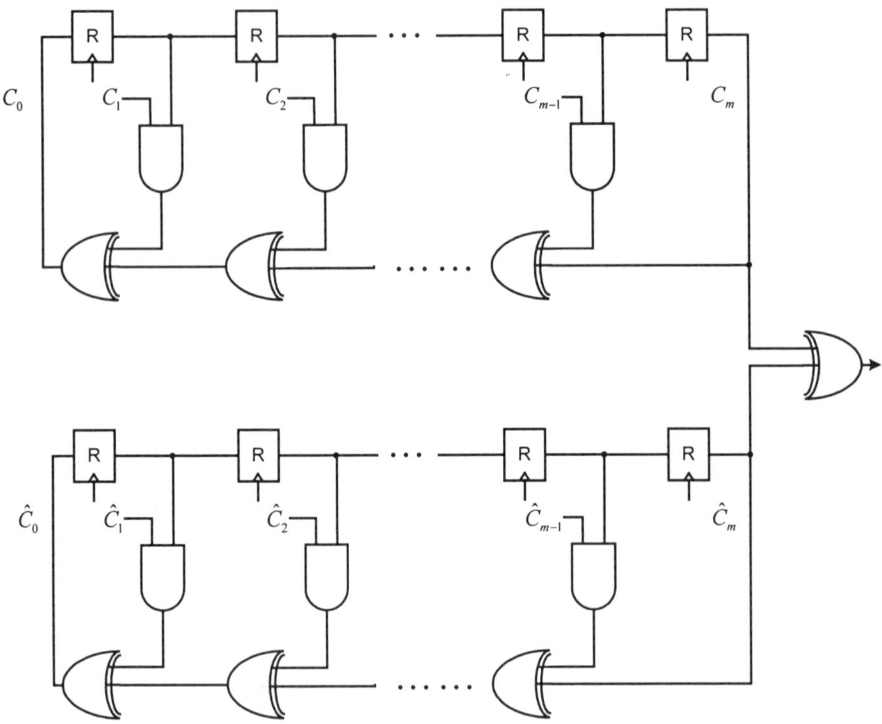

Figure 8.42 An LFSR implement of the Gold sequence.

8.4.2 Frequency Hop

Frequency-hop spread-spectrum has been introduced in Section 3.3 of 0. While it is generally true that frequency-hop spread-spectrum uses MFSK as the data modulation, other modulations such as MPSK or MQAM can also be used, in particular for slow-hop systems. For the sake of illustration, this section focuses on the specific case of frequency-hop with FSK data modulation.

8.4.2.1 Digital IF

Figure 8.43 shows the architecture of an MFSK frequency-hop digital IF transceiver, which resembles the MFSK transceiver shown in Figure 8.35b with the exception that an additional control word is added at the input of the DDFS. The additional control word corresponds to the hopping frequency, whose value is set such that the DDFS outputs a signal at IF. A quadrature DDFS is used to allow single sideband modulation and in this case the upper sideband is selected using the quadrature RF up converter. Note that the RF frequency in this case is fixed. This architecture has been used in [180][289] to implement a fully integrated RF transceiver.

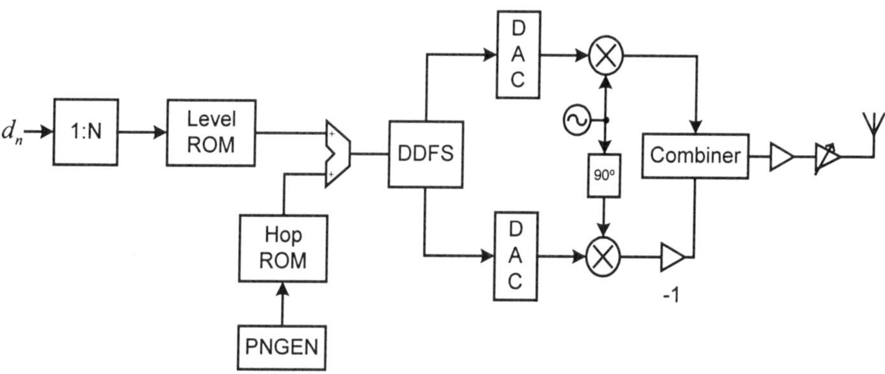

Figure 8.43 A digital IF MFSK FH transceiver architecture.

8.4.2.2 Digital RF

Instead of hopping at IF, it is possible to hop at RF with a digital RF architecture using more costly technology such as GaAs. The digital RF architecture eliminates the need for a quadrature DDFS and a quadrature RF up converter. Instead, a single DDFS can be used followed by a DAC to generate the RF signal directly as shown in Figure 8.44. Despite its apparent simplicity, the design requires high-speed circuits up to several GHz for both the DDFS and the DAC.

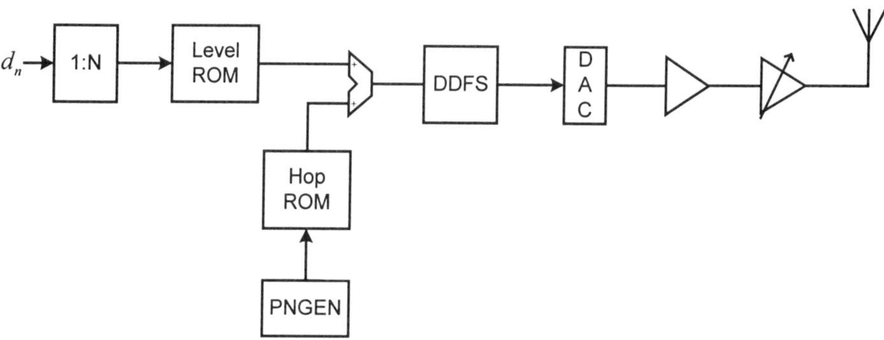

Figure 8.44 A digital RF MFSK FH transceiver architecture.

8.5 Synchronization

Both spread-spectrum and narrowband modulated waveforms require synchronization to achieve adequate performance. A thorough discussion of synchronization techniques and issues has been described in Section 2.5. This section focuses on the architecture and implementation issues associated with synchronization.

8.5.1 Digital Phase-Locked Loop

A PLL can be analyzed using a linear model shown in Figure 2.51. For digital implementations, this linear model can still be used to design the various loop parameters. However, some modifications are needed. In particular, Z-transforms must be used in place of Laplace transforms and in some cases the bilinear transform is applied to convert the analog blocks to digital blocks [138].

The transfer function of a linearized digital PLL (DPLL) is shown below

$$H(z) = \frac{K_d F(z) N(z)}{1 + K_d F(z) N(z)}, \tag{8.28}$$

where $F(z)$ and $N(z)$ are the transfer functions of the loop filter and the DDFS or NCO, respectively. One can see that (8.28) and (2.142) have essentially the same form if the term $K_0 s^{-1}$ is replaced by $N(s)$ in (2.142). Applying the bilinear transform as shown in Section 8.2.2, $F(z)$ can be derived from its analog counterpart $F(s)$. The Z-transform of a DDFS or NCO can be derived directly by treating it as essentially an accumulator, where

$$N(z) = \frac{K_0 z^{-1}}{1 - z^{-1}}. \tag{8.29}$$

Figure 8.45 shows the DPLL architecture that implements (8.28). Specific implementations of the Loop filter, DDFS, and NCO have been discussed in Section 8.2. Similar to an analog PLL, the phase detector implementation largely determines the tracking and noise performance of the DPLL.

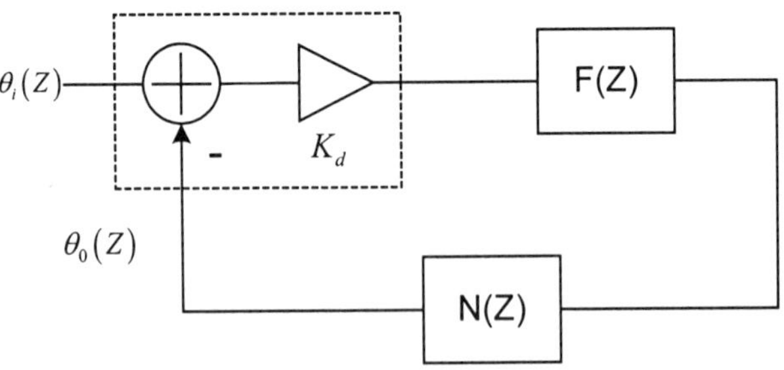

Figure 8.45 A digital phase-locked loop.

Since the performance of an analog PLL has already been thoroughly characterized, it would be useful to extend these results for a digital implementation. In particular, the DPLL parameters such as the loop filter coefficients, phase detector gain, and NCO/DDFS gain should be related to the analog loop parameters such as the damping factor and the resonant frequency. The derivation using a second-order loop filter is illustrated as an example. First, (8.28) is expanded by substituting $F(z)$ and $N(z)$ with (8.13) and (8.29), respectively. The resulting expression has the following form

$$H(z)=\frac{K_oK_d\left[(c_1+c_2)z^{-1}-c_1z^{-2}\right]}{1+\left[K_oK_d(c_1+c_2)-2\right]z^{-1}+(1-K_oK_dc_1)z^{-2}}. \tag{8.30}$$

The goal is to relate the various design parameters such as c_1 and c_2 to the damping factor and resonant frequency corresponding to an analog design with the transfer function shown in (2.148). By applying the bilinear transformation to (2.148), a digital approximation is obtained as shown below

$$H_d(z)=\frac{\left[4\zeta\omega_n\Delta_t+(\omega_n\Delta_t)^2\right]+2(\omega_n\Delta_t)^2z^{-1}+\left[(\omega_n\Delta_t)^2-4\zeta\omega_n\Delta_t\right]z^{-2}}{\left[4+4\zeta\omega_n\Delta_t+(\omega_n\Delta_t)^2\right]+\left[2(\omega_n\Delta_t)^2-8\right]z^{-1}+\left[4-4\zeta\omega_n\Delta_t+(\omega_n\Delta_t)^2\right]z^{-2}}. \tag{8.31}$$

Then by equating the denominator polynomials in (8.30) and (8.31), c_1 and c_2 are determined to be

$$c_1=\frac{1}{K_oK_d}\frac{8\zeta\omega_n\Delta_t}{4+4\zeta\omega_n\Delta_t+(\omega_n\Delta_t)^2} \tag{8.32}$$

$$c_2=\frac{1}{K_oK_d}\frac{4(\omega_n\Delta_t)^2}{4+4\zeta\omega_n\Delta_t+(\omega_n\Delta_t)^2}. \tag{8.33}$$

From (8.32) and (8.33), the loop filter coefficients can be determined after the sampling interval Δ_t, the loop natural frequency ω_n, and the damping factor ζ are specified. The phase detector gain K_d is specific to the particular application of the DPLL. For both the DDFS and NCO, K_o can be expressed by $2\pi\alpha_0$, where α_0 is the scale factor that sets the free running frequency.

8.5.2 Applications of DPLL

The above results can be applied to the synchronization loops listed below.

The **Costas loop** can be implemented digitally by using digital processing blocks in place of the analog elements shown in Figure 2.49 and Figure 2.54. In particular, the quadrature-phase VCO is implemented by a quadrature DDFS and the I&D is implemented by an A&D dumped every symbol time. Since only the sign bit is needed by the Costas loop for BPSK and QAM, the multiplier in the phase detector can be replaced by a bit-parallel XOR with one of its inputs tied to the sign bit. This performs a one's complement inversion instead of a two's complement inversion. Therefore, an LSB error is introduced when the number being inverted is negative, though this error is negligible with a sufficient bit resolution. The Costas loop for

MPSK is more complex due to the need for real multipliers in the phase detector to handle the multi-level quantized outputs from the slicer. It can be shown that for QAM and MPSK, the phase detector gains are respectively $0.5A$ and $0.5A^2$, where A is the amplitude of the transmitted waveform.

The **Phase Derotator** shown in Figure 2.55 is designed to recover the phase of the down-converted baseband signal. Therefore, it is most suited for a digital baseband receiver. The phase derotator can be implemented with a quadrature DDFS followed by a complex multiplier, where the DDFS generates the phase signal needed to derotate the received signal and the complex multiplier performs the actual derotation. The phase detector has the same structure as that used in the Costas loop for QAM and MPSK.

The Symbol Timing Recovery Loop has two basic implementations based on either a coherent or a non-coherent early-late gate correlator, shown in Figure 2.58 and Figure 2.61, respectively. In both implementations, the I&D's are again replaced by an A&D dumped at the symbol rate. The rectifier can be implemented with a bit-parallel XOR block with one of the bits tied to the MSB of the input signal. The output is an unsigned number with N-1 bits, where N is the input wordlength. Again, this is based on one's complement inversion and has an LSB error for negative inputs. The half-symbol delay can be achieved with a clock generation circuit that clocks the register at twice the symbol rate. In both coherent and non-coherent designs, the phase detector gain is $A'T_s^{-1}\Delta_t$, where A' is the amplitude of the input signal to the EL gate, T_s is the symbol duration or dump interval, and Δ_t is the input sampling rate.

The **Direct-Sequence Code Tracking Loop** differs from the symbol timing recovery loop mainly in the early and late gate correlator processing. In direct-sequence code tracking, the received signal is correlated with an early and late copy of a locally generated PN code. The result of the late correlation is subtracted from that of the early correlation to form the error signal as shown in Figure 3.17. On the other hand, the symbol timing recovery loop has no PN-code modulation. Therefore, the correlation is done by simply gating the received signal with an early and late clock signals as shown in Figure 2.58 and Figure 2.61. Similar to the symbol timing recovery loop, the PN-code tracking loop may be either coherent or non-coherent and requires the use of a coherent or non-coherent correlator as discussed in Section 8.5.3. In both types of tracking loop, the phase detector gain is $2A'\left(1+N_c^{-1}\right)/m$, where m is either 1 or 2 for a half-chip or full-chip delay, respectively.

8.5.3 PN Acquisition Loop

Before a spread-spectrum signal can be tracked, it must first be acquired by the PN acquisition loop. There are two basic types of PN acquisition loops: serial and parallel (matched filter). Also, the loop may be either coherent or non-coherent. Since Section 3.5 has already discussed in detail the algorithm and architecture trade-offs of PN-acquisition loops, the following discussion focuses only on their implementations.

The same guidelines described in the previous section on implementation of tracking loops can also be applied for the PN-acquisition loop. In particular, in the correlator blocks shown in Figure 3.14 and Figure 3.15, the multipliers can again be implemented by a bit-parallel XOR and the summation block can be implemented by an A&D. As an illustration, let's consider the serial and parallel correlators and the case in which all signals are real. Then, the correlators can be implemented as shown in Figure 8.46. The control needed for the search procedure is usually specified in VHDL as a finite-state machine (FSM) and is synthesized automatically using a standard-cell library. Though for high-speed designs, often the timing control is implemented manually.

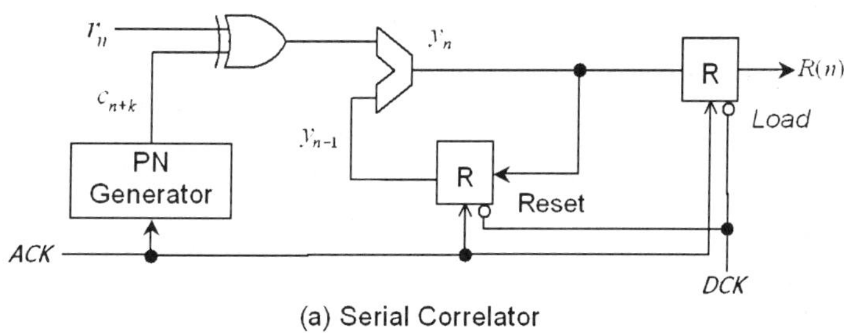

(a) Serial Correlator

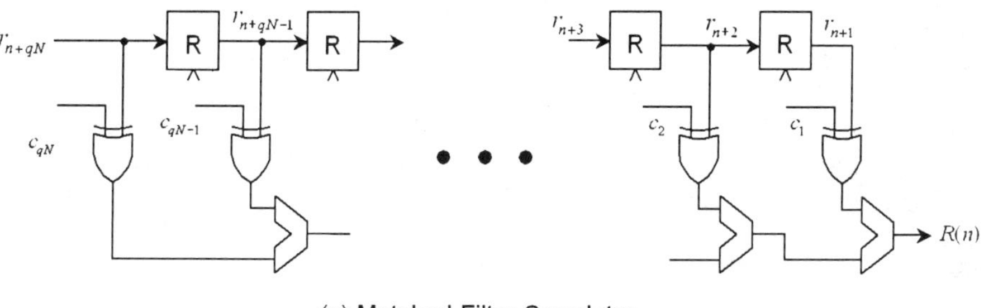

(a) Matched Filter Correlator

Figure 8.46 Correlator implementations.

8.5.4 Frequency Acquisition

Frequency offset arises from Doppler shift and crystal oscillator stability. To compensate for the frequency offset, the receiver must first acquire the frequency, especially for a large offset. Depending on the application, the Doppler shift can be up to several kHz. For instance, satellite links due to high orbital velocity exhibit a Doppler frequency of several kHz whereas digital cellular links exhibit a much lower Doppler shift of a few hundred Hz. In the latter case, the oscillator frequency stability dominates the frequency offset since a low-cost oscillator is accurate only to within

10-30 ppm, which translates to a 10-30 kHz offset at 1 GHz. When the total offset is too large for a DPLL to handle with adequate noise and tracking performance, a frequency estimator becomes necessary to aid the DPLL in acquiring the frequency.

A robust frequency estimator has been described in (2.153) and its implementation is shown in Figure 8.47. The implementation of (2.153) is straightforward. The implementation consists of a delay line, a complex multiplier, A&D, and a ROM to store values of the inverse tangent. Techniques to reduce the ROM size discussed in Section 8.2.5 can also be applied to this case.

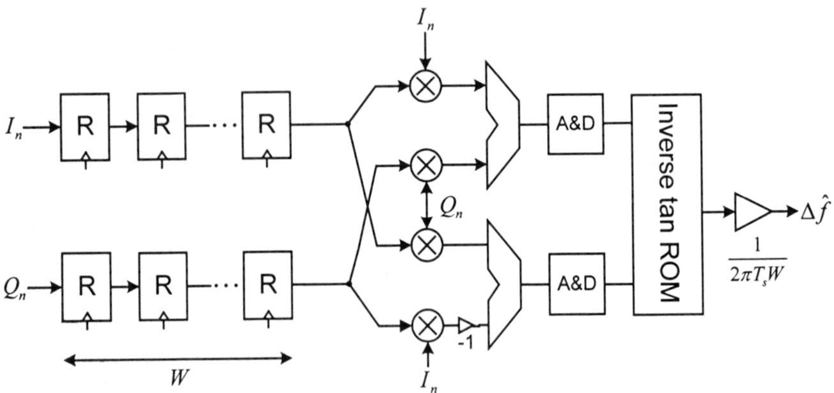

Figure 8.47 Frequency estimator.

8.6 Path Diversity

Thus far, the previous sections have described the implementation of three fundamental blocks in a digital transceiver: namely, the modem, the spread-spectrum modulator, and synchronization loops. This section describes the implementation of advanced functional blocks used to achieve path diversity, including the Rake receiver, MLSE, FFE, and DFE.

8.6.1 Rake Receiver

A Rake receiver can simultaneously demodulate K_D dominant paths in a multipath channel and optimally combine the soft-decision outputs to achieve diversity gain. As shown in Figure 3.7, the high-level architecture consists of K_D demodulators, also referred to as fingers. Each demodulator, also known as a finger, independently acquires and tracks a multipath in the channel. To facilitate the acquisition and tracking, a searcher is implemented to find the K_D dominant multipaths and assigns the initial timing of the multipaths to the despreading and tracking circuit.

8.6.1.1 Pilot Generation

The searcher locks on to the transmitted code phases corresponding to the multipath delays to determine the initial timing for each finger. To lock on to the code phases, the searcher relies on the correlation peak computed by the acquisition circuit using the locally generated PN codes. If data modulation is present, the hard-decision data is used as well. Due to the non-zero acquisition time, the searcher requires a finite amount of time to acquire the phases before demodulation can occur in each finger. There are two general methods to acquire the code phases. The first method allocates the time in the form of a preamble for every packet or message sent over the air. The preamble is set long enough to ensure that the searcher acquires the code phases with high probability. The preamble is also sometimes referred to as a pilot symbol and is generally unmodulated so that acquisition is achieved without the need for decision feedback. The method based on pilot symbols is most appropriate for packet-switched systems such as IEEE 802.11 and third-generation cellular systems that support packet switching.

The second method is based on a continuous pilot whereby a pilot is transmitted continuously over a common control channel as shown in Figure 8.48. A different code $c_{pI} + jc_{pQ}$ is allocated to the pilot to isolate it from the user code $c_I + jc_Q$. To mitigate degradations due to multiple access interference, the pilot is assigned slightly higher power (i.e. $\alpha_p = 1.2$) than the user codes. The pilot code is also designed to have good correlation properties so that its correlation value can be easily distinguishable in high noise or interference environments. Therefore, the pilot code is usually chosen to have a long period, such as 32768 in the IS-95 system. Searching through a long code quickly can require a substantial amount of hardware and dissipate significant amounts of power. Fortunately, for a circuit-switched system, a long search delay even on the order of seconds may be tolerable. Therefore, a low-complexity serial architecture can be used. If fast acquisition is needed but with lower power, a hybrid architecture that exploits both serial and parallel acquisition may be used [132].

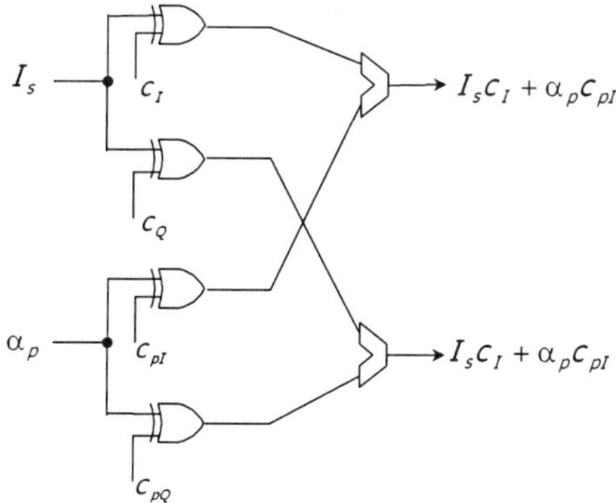

Figure 8.48 A continuous pilot channel encoder.

8.6.1.2 Rake Finger Processing

The pilot-based searcher initializes timing for the despreading circuits in each Rake finger as shown in Figure 8.49. Using the locally generated PN sequence, a non-coherent despreader extracts the pilot from the received signal. The despreader shown in Figure 8.49 has an architecture similar to that of a non-coherent correlator but does not have the decision feedback and squaring logic. The outputs of the despreader still contain residual phase from the down conversion process. The residual phase $\Delta\phi_k$ and the multipath fading amplitude α_i are estimated using an averaging circuit based on A&D's with a dump cycle of N_p symbols. Two constant gain blocks remove the pilot amplitude from the channel estimates. The estimated phase and amplitude form two normalizing factors $\alpha_i \cos\Delta\phi_k$ and $\alpha_i \sin\Delta\phi_k$ that are used to derotate the received signal. The derotated signals are then despread using the user codes. Because of the approximate orthogonality between the pilot and user codes, the averaged despread signal contains the in-phase and quadrature symbols scaled by α_i^2. The scaled outputs are combined to form the final soft-decisions. The derotated signals are also fed to a PN-tracking loop in each finger to track the timing variations caused by the Doppler spread of the channel.

While the Rake finger shown in Figure 8.49 is relatively complex, current CMOS technology has enabled the integration of many Rake fingers on a single chip. For instance, [234] reported a CDMA digital transceiver for IS-95 that integrated four Rake fingers along with many other transceiver functions on a 111 mm^2 chip.

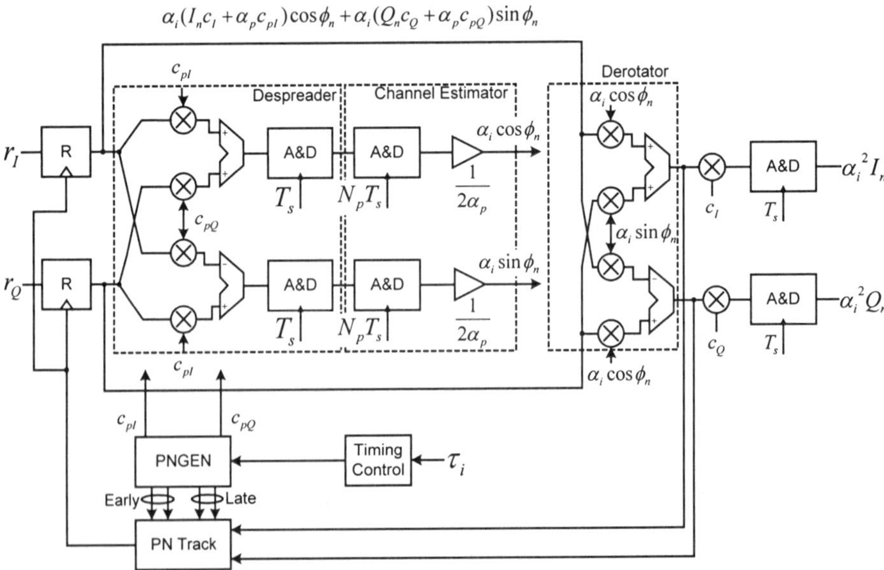

Figure 8.49 A Rake finger.

8.6.2 MLSE (Viterbi Equalizer)

A multipath channel can be modeled as a time varying filter whose taps have values equal to the fading amplitudes. For instance, if a channel has L_c dominant paths, then this channel can be modeled by an L_c+1-tap FIR filter with a memory of L_c-1. Each tap of the filter is time varying. Details of this model have been discussed in Section 2.2.6.1 of Chapter 2. Given a multipath channel, diversity reception can be achieved by using an MLSE, also known as the Viterbi equalizer, which searches through all the possible states of the channel and determines the sequence with the minimum error. See Section 2.7.2.5 of Chapter 2 for the theoretical basis of MLSE detection.

8.6.2.1 Branch Metric

A multipath channel imposes a certain structure on the received signal similar to the trellis of a CPM signal. In this case, the state for the kth symbol time interval in the trellis corresponds to the L_c-1 previous symbols and can be denoted by $\sigma_k=(\xi_{k-1},\xi_{k-2},\cdots,\xi_{k-L_c+1})$. Just as in CPM, the state transitions depend only on the previous state and the current symbol ξ_k. A transition from state σ_k at time k to state σ_{k+1} at time k+1 is represented by (σ_k,σ_{k+1}). Each transition $\lambda_k(\sigma_k,\sigma_{k+1})$ has a branch metric computed based on the magnitude square of the error between the received baseband signal $\tilde{r}_k$ and the local version $\tilde{s}_k$ reconstructed from the estimated CIR coefficients $\left\{\hat{h}_j\right\}$.

Let $\hat{h}_j$ be represented in the complex notation $\hat{h}_j=\hat{h}_{Ij}+j\hat{h}_{Qj}$. Then, the reconstructed signal is simply

$$\tilde{s}_k=\sum_{j=0}^{L_c-1}\left[\hat{h}_{Ij}I_{k-j}-\hat{h}_{Qj}Q_{k-j}\right]+j\left[\hat{h}_{Ij}Q_{k-j}+\hat{h}_{Qj}I_{k-j}\right] \tag{8.34}$$

where $\left\{I_j\right\}$ and $\left\{Q_j\right\}$ are all the possible symbols that have been transmitted during L_c symbol durations. The branch metric is then

$$\lambda_k\left(\sigma_k,\sigma_{k+1}\right)=-\left[r_{Ik}-\sum_{j=0}^{L_c-1}\left(\hat{h}_{Ij}I_{k-j}-\hat{h}_{Qj}Q_{k-j}\right)\right]^2-\left[r_{Qk}-\sum_{j=0}^{L_c-1}\left[\hat{h}_{Ij}Q_{k-j}+\hat{h}_{Qj}I_{k-j}\right]\right]^2, \tag{8.35}$$

where the negative sign is needed for the Viterbi decoder to find the symbol sequence that minimizes the squared error. Figure 8.50 shows an architecture that implements (8.35) to generate one of the M^{L_c} branch metrics used by the Viterbi decoder where M is the number of levels used to represent a symbol. The rest of the branch metrics can be generated using the same circuit in a time-shared fashion or using M^{L_c} replicas of the circuit in a parallel fashion.

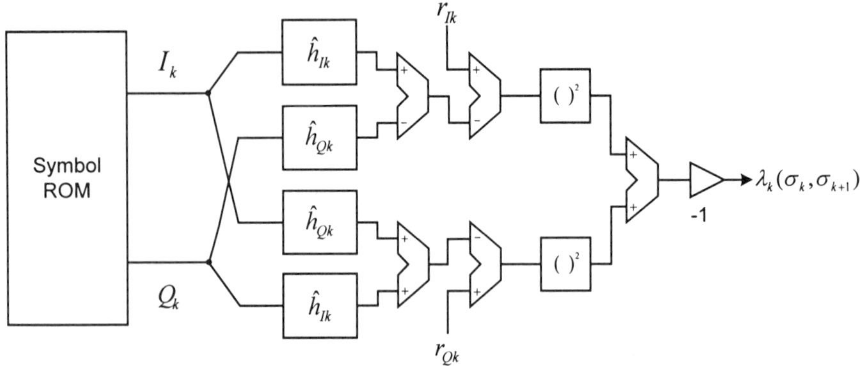

Figure 8.50 Branch metric computation for MLSE.

8.6.2.2 Pilot-based Channel Estimator

The performance of a Viterbi equalizer depends critically on the accuracy of the estimates for the CIR coefficients. A pilot channel proves useful to obtain the CIR estimates and as in the Rake receiver either a continuous or a symbol-based pilot may be used. In the following discussions, a symbol-based pilot is considered whereby a preamble with good correlation properties is selected so that the channel coefficients can be determined by applying (2.196). In general, a PN-code is used for the preamble and the challenge is to design it such that the auto-correlation behaves like a delta function even at the boundaries of the preamble. Also, the code should have a short enough period so that multiple periods can be included in the preamble of length N. Estimation SNR is improved by using the averaging described in Section 2.7.7.1 and network overhead is reduced by keeping N small.

An implementation of a pilot-based channel estimator is shown in Figure 8.51. The preamble consists of periodic repetitions of a PN code $c(t)$. The complex matched filter in Figure 8.51 generates the cross-correlation of the received in-phase and quadrature signals with $c(t)$. Because of the good auto-correlation property of $c(t)$, the output of the matched filter closely approximates the estimate of the complex CIR, $\hat{h}_{Ik} + j\hat{h}_{Qk}$. The additional processing surrounding the matched filter identifies the major peaks of the CIR and also averages these peaks to achieve a high estimation SNR. The peak detection circuitry first rectifies the matched filter outputs and combines the in-phase and quadrature components to give an indication of the magnitude of the CIR. When the magnitude exceeds a certain threshold, the accumulation of the channel impulses is triggered. The purpose of the accumulation is to identify the portion of the CIR that contains the most energy. When the accumulated output begins to flatten out and crosses a second threshold, the matched filter has captured the region with the most energy. Within this region, the matched filter outputs are averaged by an A&D before stored into memory for use in the branch metric computation.

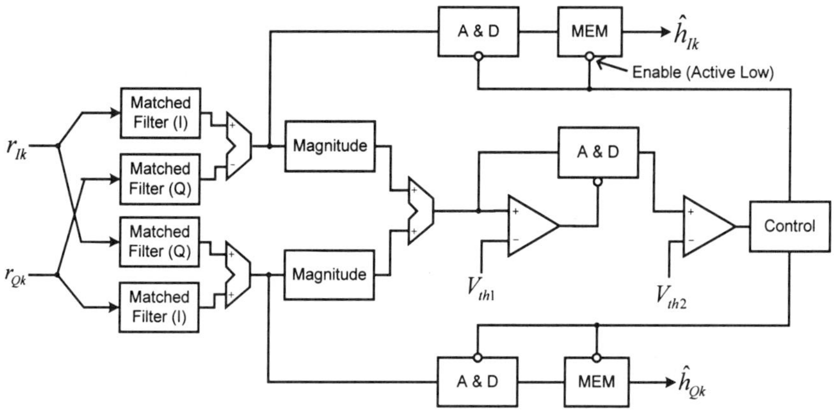

Figure 8.51 A pilot-based channel estimator.

8.6.3 Feed-Forward Equalizer

A feed-forward equalizer (FFE) inverts the channel impulse response such that the channel appears to be frequency-flat after equalization. Sections 2.7.2.2 and 2.7.2.3 described the theoretical basis of the FFE. Figure 8.52 depicts its implementation based on a digital complex FIR filter with programmable coefficients. The filter coefficients need to be programmable to allow updates of the filter coefficients as the channel changes.

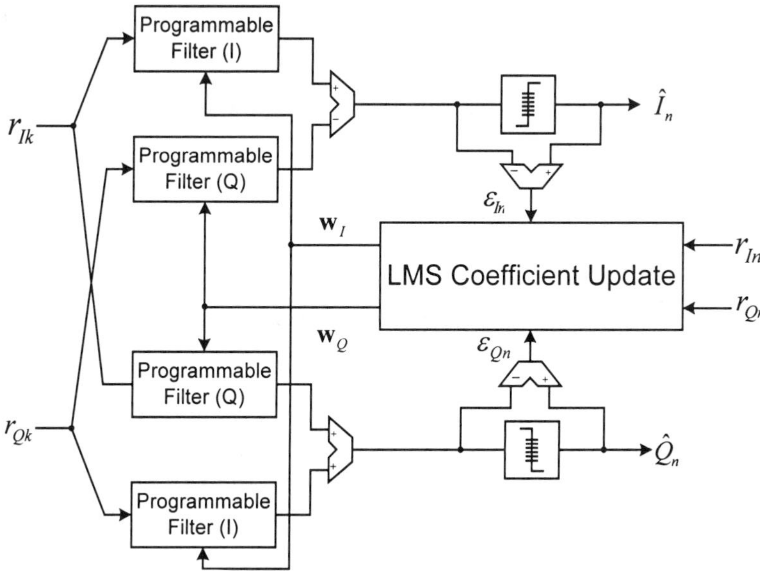

Figure 8.52 An FFE architecture.

A programmable filter based on the transpose direct form is shown in Figure 8.53. The FFE consists of a complex N-tap programmable filter that consists of four N-tap real filters, each with an implementation shown in Figure 8.53. The LMS coefficient update block generates the filter coefficients using the errors derived from the hard and soft decisions at the filter outputs according to (2.204) and (2.205).

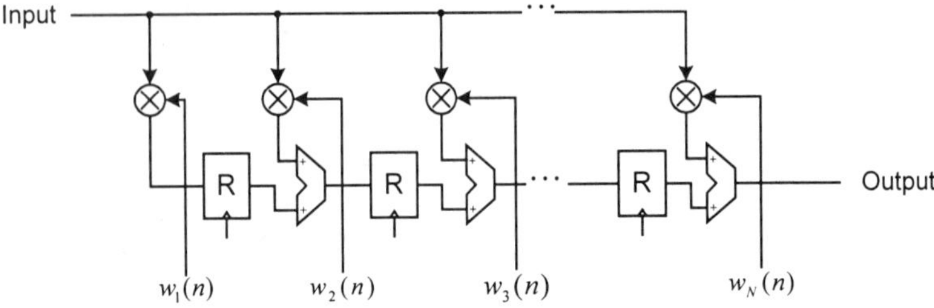

Figure 8.53 The transpose direct-form implementation of a programmable FIR filter.

Figure 8.54 shows the implementation of the LMS update for the kth tap of the filter at time index n. The two registers delay the baseband signal at each tap to align the LMS inputs with the coefficients as dictated by (2.204). The time alignment is not needed with the direct-form architecture. Also, the registers are not required for the first tap. The baseband signals are multiplied with the error signals using a complex multiplier. In many implementations [235][236], only the sign bits of the baseband signals are used to reduce the complexity of the multiplier. The accumulator implements the recursion in (2.204) and since any overflow or underflow may cause a persistent error in the estimated coefficients, saturation logic is used to prevent error accumulation. Generally, because of the iterative process it is important to keep high resolution in the accumulator and truncate only at the accumulator outputs. An example would be to use 32 bits internal for the accumulator and truncate to 16 bits at its output. The LMS step size μ is often implemented as simple shifts by using powers of two scale factors.

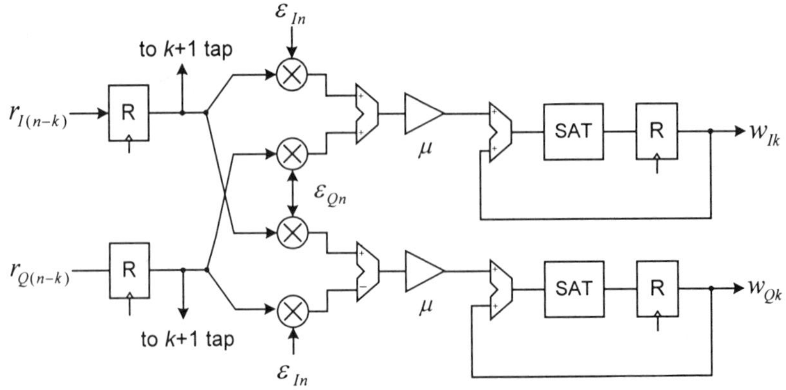

Figure 8.54 LMS update for the kth tap of the FFE.

8.6.4 Decision-Feedback Equalizer

Compared to the FFE, a decision-feedback equalizer has more robust performance in reducing residual ISI and noise enhancement as discussed in Section 2.7.2.4. Figure 8.55 shows an implementation of the DFE based on (2.186). The FFE block has K_1+1 taps and has the same architecture as that shown in Figure 8.53. The output of the FFE is summed with the output from the DFE block to eliminate residual ISI from the output of the FFE. The DFE consists of a complex K_2-tap programmable filter whose inputs are the hard-decisions, $\hat{I}_n$ and $\hat{Q}_n$. The complex filter consists of four real $K2$-tap filters implemented in the transpose direct-form shown in Figure 8.53. The LMS update block with an equivalent architecture as the one shown in Figure 8.54 generates the coefficients to the complex DFE filter based on the error signals ε_{In} and ε_{Qn} as well as $\hat{I}_n$ and $\hat{Q}_n$. In contrast to the FFE coefficient update, the baseband inputs to the LMS update block have been replaced by the hard-decision outputs.

Highly integrated DFE's have been reported in [237][238][239][240] for applications in digital TV transmissions as well as cable and DSL modems. For instance, [237] reports a four tap complex DFE CMOS chip that achieves up to 60 Mbaud or 480 Mbps with 256 QAM in 1.2 μm CMOS technology. To accommodate large delay spreads that span beyond the four filter taps, several of these chips could be cascaded. Later in Chapter 9, a QAM transceiver is described that uses DFE together with spatial diversity to achieve 64 Mbaud data rate over the wireless link.

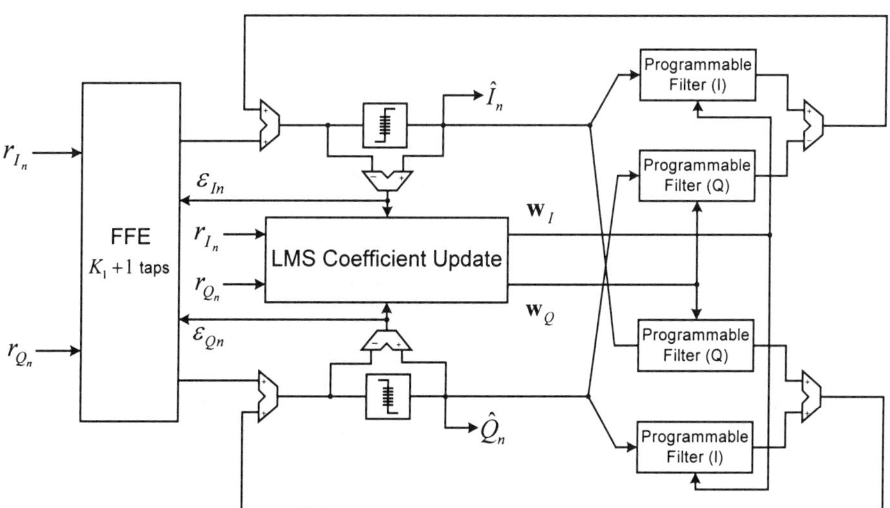

Figure 8.55 The transpose direct-form implementation of a DFE.

8.7 Time Diversity

While path diversity effectively combines energy from the various multipath components at the receiver, time diversity combines energy from amplitude fading that occurs at different points in time. Such fading occurs in a frequency-flat or time-selective fading channel as discussed in Section 2.2.6.2. Also, as discussed in Section 2.7.4, two time-diversity techniques can be used to combat time-selective fading: namely, frequency hop with interleaving and FEC with interleaving. This section describes implementations of the FEC and the interleaver/de-interleaver.

8.7.1 Interleaver/De-interleaver

A block interleaver can be viewed as a matrix operation where the input data is written one row at a time into the matrix and then read out column wise to obtain the desired time interleaving in the transmitted bits. As shown in Figure 2.74, the interleaver matrix has m rows and Sd columns with each matrix position containing one bit. At the receiver a de-interleaver inverts the interleaving process by performing the same operation but using a matrix that has Sd rows and m columns, i.e. the transpose of the interleaver matrix.

An efficient implementation of the matrix operation can be achieved using random-access memory. At the transmitter, the effective size of the memory is $m \times Sd$ bits while it is $Sd \times m$ bits at the receiver. A serial to parallel conversion demultiplexes the serial stream into a word that is written into the memory one row at a time. At the transmitter the serial stream is converted into an Sd-bit word while at the receiver the serial stream is converted into an m-bit word. The column-wise readout requires non-conventional column decoding to allow the columns to be read out one at a time.

Conventional memory designs typically organize a rectangular memory array of $z \times y$ cells into a square array of $n \times n$ cells. Let's assume without loss of generality that $y = 2z$. The $z \times 2z$ array can be broken down into two equal segments, each with size $z \times z$. Two columns of the same position from each of the memory segments are decoded using a 2:1 MUX or if the memories have tri-state outputs, the memory outputs can be shorted.

In contrast, the column decoding logic in an interleaver or a de-interleaver requires a z:1 MUX for each of the two segments. Although decoding can be implemented directly for a $z \times 2z$ matrix as is, the decoding logic tends to be more inefficient. By breaking the memory into smaller pieces, the decoding logic becomes much more manageable. Also, with smaller memory partitions, unused partitions can be shut down using the chip select (CE) to obtain power savings. Figure 8.56 shows a block interleaver of size $z \times 2z$ but can be easily extended to an interleaver of arbitrary size.

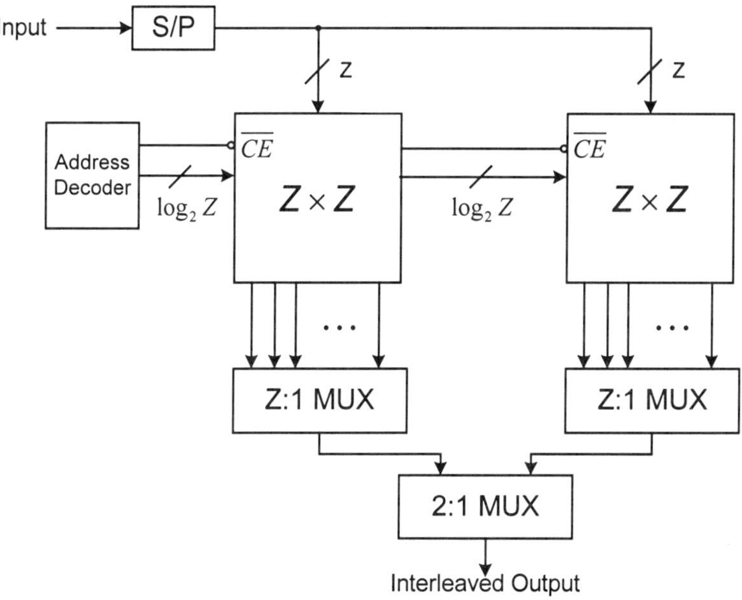

Figure 8.56 A $z \times 2z$ block interleaver implementation.

8.7.2 Viterbi Encoder

Convolutional codes are widely used in current systems together with interleaving to achieve diversity gain in a time-selective channel. A generic encoder is shown in Figure 2.62 and consists of *KN* generator polynomials in GF(2), which can also be viewed as digital filters using GF(2) operations. Each tap then can be implemented as a short or open circuit, corresponding to a "1" or a "0" in the polynomial coefficient. Each polynomial power corresponds to the number of registers in cascade. The summation can be implemented using a modulo-2 addition, i.e. an XOR gate.

The following discussion use the rate 1/3 constraint length 3 convolutional encoder discussed in Section 2.6.3 as an illustration of the general guidelines for the implementation of Viterbi encoding and decoding. The three generators required for performing this encoding are $\mathbf{g}_{11} = [100]$, $\mathbf{g}_{12} = [110]$, and $\mathbf{g}_{13} = [100]$ and can be implemented based on binary field arithmetic as shown in Figure 8.57. Although there are three polynomials, only two registers are needed since they can be time shared by the three generators. Taps are not connected in positions that correspond to '0's in the generators and connected in positions that correspond to '1's in the generators. The tap outputs are added to form three signals that are multiplexed onto a single serial output stream running at three times the input data rate. The states diagram shown in Figure 8.57 corresponds to the register content $[s_1 s_0]$, where s_0 is the output of the register closest to the input of the encoder. Thus, new inputs are shifted into s_0 or the LSB of the state label $[s_1 s_0]$. A dashed and solid line indicates an input '1' or an input '0', respectively.

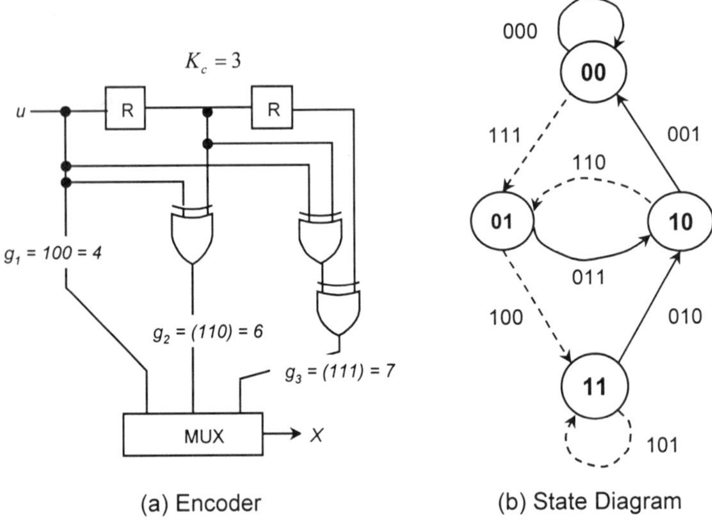

Figure 8.57 A rate 1/3 constraint length 3 convolutional encoder.

8.7.3 Viterbi Decoder

The Viterbi decoder offers an optimal means to compute solutions for several important communication functions, including CPM demodulation, MLSE, and decoding of convolutional codes. The key equations (2.131) that govern Viterbi decoding comprise computations of the branch metric $\lambda_l(\sigma_{l-1}, \sigma_l)$, the path metric $\mu_l(\sigma_l)$, and the survivor paths $\Gamma_l(\sigma_l)$. Figure 8.58 shows a general block diagram of a Viterbi decoder that implements the recursion described in (2.131) with an add-compare-select unit together with the necessary memory elements to store the branch metrics, the path metrics, and the survivor paths. While read-only memory is adequate to store the branch metrics, random-access memories are required for both the path metrics and the survivor metrics. The *Branch ROM* block corresponds to the ROM that stores all possible branch metrics according to (2.129), (2.172), (2.173), or (2.188) depending on the application.

At the lth step of the decoding process, branch metrics are computed by the branch metric unit (BMU) for the set of state transitions from σ_{l-1} to σ_l. All branch metrics which belong to the same state at the lth step are added to the path metric from their originating states. The maximum sum is used to update the path metric for state σ_l. For each ACS computation, the required path metrics must be retrieved from path memory and the updated metric for each state is stored back into the path memory. At each step of the recursion, the survivor path associated with each state is also updated after all path metrics have been computed.

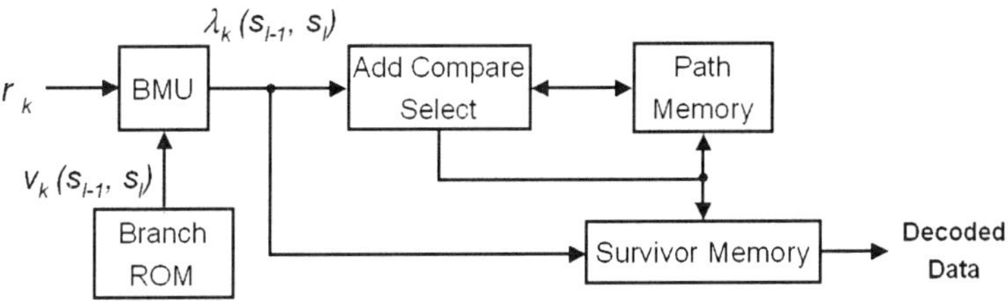

Figure 8.58 General block diagram of a Viterbi decoder.

The Viterbi decoder shown in Figure 8.58 can be implemented using many architectures [241][242][243][244][245][246][247][248][249][250][251]. Two basic architectures are based on the state parallel and state sequential approaches. For illustrative purposes, the rate 1/3 code is used as an example. In this case, there are four states and two transitions per state as shown in Figure 2.63. The state parallel architecture simply implements the trellis by directly mapping each state to a processing element (PE) shown in Figure 8.59a. Each processing element contains an ACS and survivor memory as shown in Figure 8.60b. The BMU and path memory are not shown but are implemented as global resources shared by all four PE's.

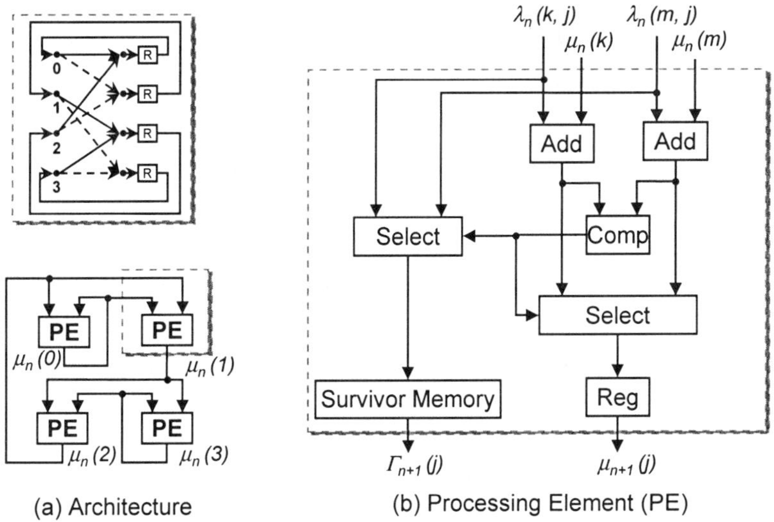

Figure 8.59 A state-parallel Viterbi decoder.

In contrast, the state sequential architecture is a time-shared version of the parallel architecture, whereby multiple states are mapped on to a single PE. Figure 8.60a shows the case where all four states are mapped to two PE's, also known as a butterfly unit because of the resemblance between its trellis diagram and the butterfly of an FFT. Figure 8.60b shows the implementation of a PE for the sequential architecture. The complexity of a state-sequential architecture remains approximately the same regardless of the number of states. However, the processing speed of the PE's increases linearly with the number of states. The opposite holds for the state-parallel architecture, where its complexity rises linearly to the number of states but its processing speed remains fixed. Also, the state-parallel architecture tends to consume greater routing resources in an integrated circuit, sometimes as much as 50% [246]. When voltage scaling is applied to the state-parallel architecture, Viterbi decoding can be achieved with extremely low power – on the order of 41.7 nW/state/kbps [250]. With careful design, [248] also reports a low power (0.65 nW/state/kbps) but a much lower complexity design based on the state-sequential architecture. The state-parallel architecture achieves a lower power at the expense of higher complexity with 37,500 transistors/state versus 254 transistors/state in the state-sequential design.

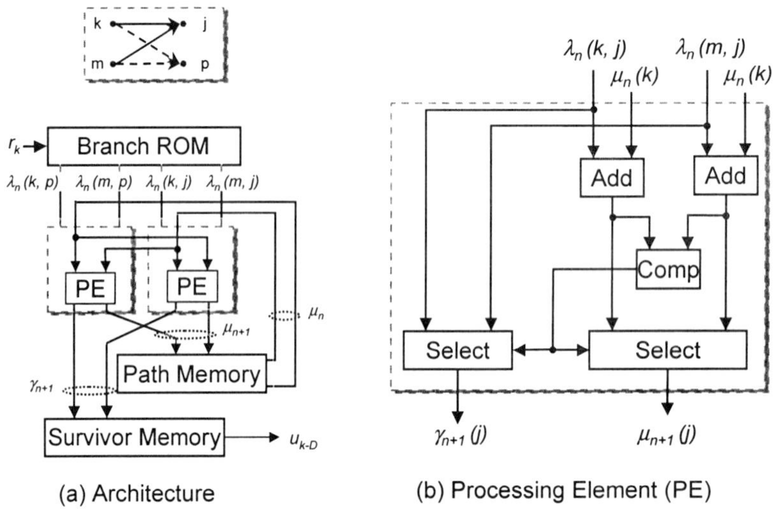

Figure 8.60 A state-sequential Viterbi decoder.

8.7.3.1 BMU for Decoding of Convolutional Codes

The BMU block provides a common resource in both the state-parallel and state sequential architectures. The BMU computes the branch metric at the lth time index based on (2.172) for hard-decision decoding and (2.173) for soft-decision decoding. Implementations of the two cases are illustrated in Figure 8.61. In the hard-decision case, the branch ROM stores all 2^{K+K_c-1} possible *N*-bit codewords for a rate *K/N* convolutional code. With hard-decision decoding, the decisions are made on the received symbols before they are used by the BMU to compute the Hamming distance of the received codeword relative to each of the possible codewords.

In Figure 8.61a, the hard-decisions are shown as y_{lk} where k indicates the kth bit of the demodulated codeword at the lth time index. The Hamming distance is computed by taking the bitwise XOR of the demodulated codeword and the stored codeword, and summing the XOR outputs. In contrast, a soft-decision BMU determine the branch metric based on (2.173), which is equivalent to taking the inner product of the soft-decision symbol with the codewords stored in the branch ROM. Since the soft-decision values are no longer binary, it is necessary to convert the branch ROM contents from binary to NRZ waveforms, i.e. ±1. Figure 8.61b shows the implementation for the inner product whereby the XOR's in Figure 8.61a have been replaced by real multipliers and the outputs of the branch ROM are converted to NRZ by a simple circuit shown in the inset of Figure 8.39b.

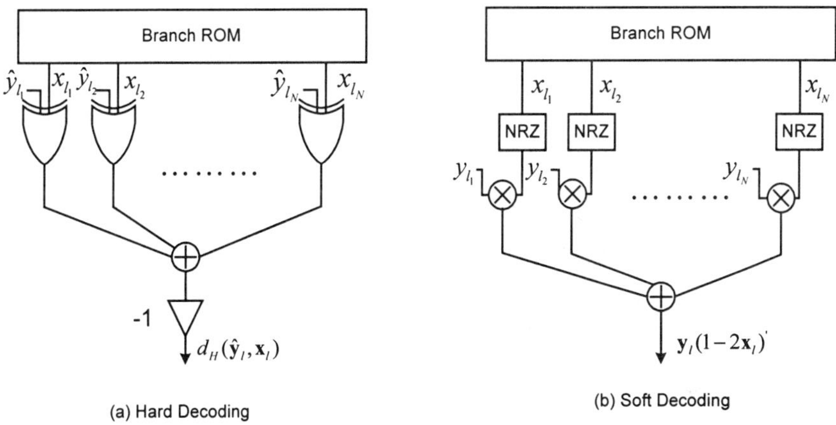

Figure 8.61 Branch metric unit for convolutional decoder.

8.7.3.2 Path Memory

The design of path memory for the state-parallel architecture is straightforward. Each PE has a local memory consisting of a single register that stores the current path metric. In contrast, the state-sequential architecture requires a single memory resource that stores all 2^{K_c-1} path metrics. At each cycle, the PE reads path metrics from memory and writes back the updated path metrics. If static addressing is used to dedicate a memory location for each path metric, then data erasure results from writing over memory locations of old path metrics that are still needed by other states in subsequent cycles. For instance, during the first cycle for the 1/3 convolutional code shown in Figure 2.63, states S_0 and S_1 are updated by using the path metrics associated with states S_0 and S_2 in the ACS computation. If each state has a fixed memory location, then the new metrics are written into memory locations associated with S_0 and S_1. During the second cycle, the computation of new path metrics for S_2 and S_3 would then result in an error since S_1 no longer holds the old path metric but the updated metric derived during the first cycle.

To overcome the data erasure problem, the most direct method is to implement double buffering as shown in Figure 8.62a whereby the new metrics are stored in a temporary memory $t_0 - t_3$ until all metrics have been computed. Then, the new metrics are written into the main memory unit $m_0 - m_3$ for use in the next iteration. However, double buffering increases memory complexity and memory accesses and results in a larger silicon area and power dissipation. Therefore, the alternative approach based on in-place memory read/write [252] is preferred and is shown in Figure 8.62b. This method requires dynamic addressing whereby at the end of each cycle the updated metrics are written into locations that are no longer needed in future cycles. For example, in the first cycle, new metrics for S_0 and S_1 are written into memory locations m_0 and m_2, respectively. In the second cycle, new metrics for S_2 and S_3 are written into m_1 and m_3, respectively. In the next iteration, the memory accesses are back to the natural order. The access pattern repeats every other iteration. The same principle can be extended to convolutional codes with arbitrary code rate and constraint lengths.

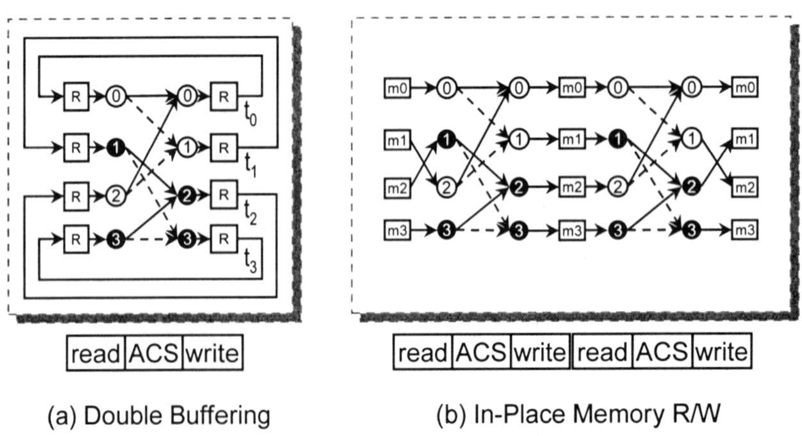

Figure 8.62 Path memory architectures.

8.7.3.3 Survivor Memory

At the end of each iteration, the survivor path associated with each state is updated. Each survivor path consists of the state transitions in the trellis diagram that led to the maximum path metric. A straightforward method of performing the update is based on register exchange in which the survivor path is updated by appending the current decoded symbol to the survivor path from the previous state. Figure 8.63 shows the application of register exchange [245] to the rate 1/3 convolutional code example. The numbers annotated along the edges of the trellis in Figure 8.63a represent the branch metrics and the sequence of bits in Figure 8.63b represents the decoded data symbols.

There are four memory locations as shown in Figure 8.63b, each corresponding to a state. During an iteration of the Viterbi algorithm, the survivor path from the previous state that led to the largest path metric is selected and written into a shift register. The decoded data symbol is then shifted into the shift register and the content of the shift register is then written back into the survivor memory. Data decoding can be extracted from the state with the largest path metric by simply reading out the register content. Figure 8.63b shows that by the end of the fourth cycle, the maximum likelihood path corresponding to state S_0 with a total path metric of 2.5 contains the following decoded data – 0001.

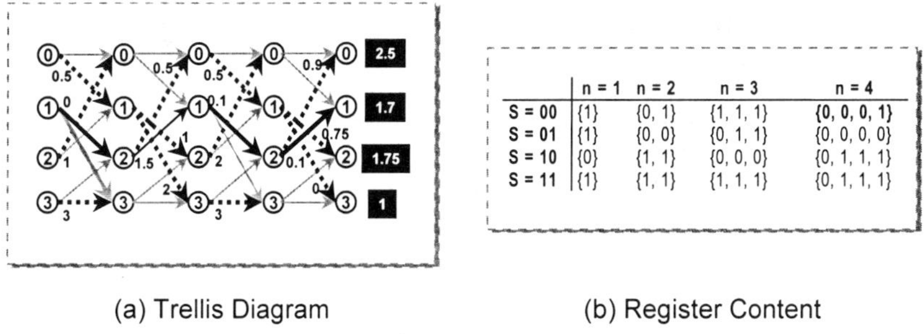

	n = 1	n = 2	n = 3	n = 4
S = 00	{1}	{0, 1}	{1, 1, 1}	**{0, 0, 0, 1}**
S = 01	{1}	{0, 0}	{0, 1, 1}	{0, 0, 0, 0}
S = 10	{0}	{1, 1}	{0, 0, 0}	{0, 1, 1, 1}
S = 11	{1}	{1, 1}	{1, 1, 1}	{0, 1, 1, 1}

(a) Trellis Diagram (b) Register Content

Figure 8.63 Register exchange applied to the rate 1/3 convolutional code example. Thick lines are the survivor paths for each state. The dashed and solid lines correspond to a decoded data of 1 and 0, respectively.

Figure 8.64 shows the architecture for the register exchange scheme. Survivor paths from state *k* and state *m* are read from the survivor memory and selected based on the ACS result. The selected path is appended to the current symbol and written back into the survivor memory at a location corresponding to state *j*. The architecture shown in Figure 8.64 works for convolutional codes with $K = 1$ but can be easily extended to arbitrary *K*.

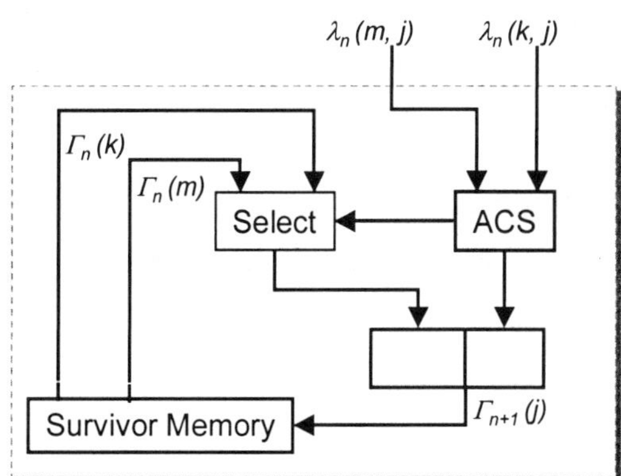

Figure 8.64 The register exchange architecture.

An alternative to the register exchange architecture is based on traceback discussed in Section 2.6.3.2. In contrast to register exchange, the architecture for traceback shown in Figure 8.65 only stores the previous state associated with the state transition that resulted in the largest path metric. If soft decoding is desired, then the path metrics associated with the survivor path can also be stored.

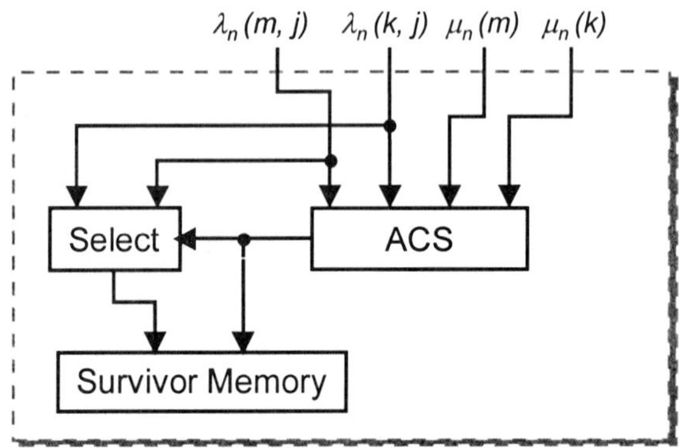

Figure 8.65 The traceback architecture.

The traceback memory organization is shown in Figure 8.66a for the 1/3 convolutional code example. During the lth iteration, the label of the previous state σ_{l-1} associated with the optimum state transition is written into memory for each σ_l. In other words, each cell in the memory array stores a pointer to the previous state. Tracing back through the memory structure shown in Figure 8.66a can be done with a circular addressing scheme shown in Figure 8.66b where two pointers are kept, a TB pointer and a WR pointer [253]. The former points to the memory location at which traceback begins while the latter points to the location where the current update is written. The update and traceback operations are performed for each state and can be done concurrently if a dual port RAM is used. The TB pointer is incremented in the clockwise direction while the WR pointer is incremented in the counterclockwise direction. The highlighted cell in Figure 8.66b represents the cell containing the decoded data symbol.

	n = 1	n = 2	n = 3	n = 4
S = 00	{10, 1.0}	{10, 0.5}	{10, 2.0}	{10, 1.0}
S = 01	**{00, 0.5}**	**{10, 1.5}**	**{00, 0.5}**	**{10, 0.1}**
S = 10	{01, 0.0}	{01, 1.0}	{01, 0.1}	{01, 0.7}
S = 11	{11, 3.0}	{01, 2.0}	{11, 3.0}	{01, 0.0}

(a) Memory Content

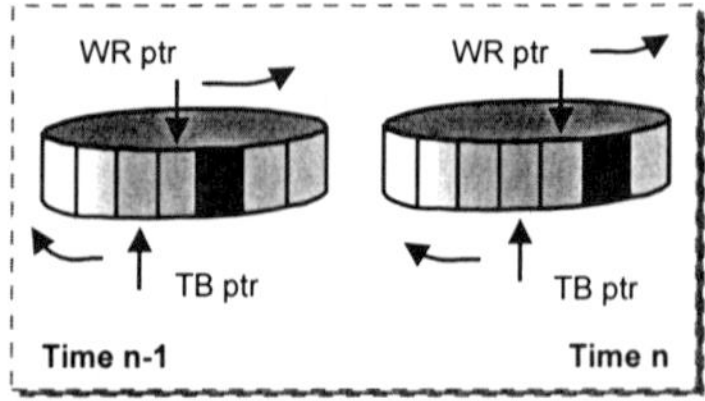

(b) Circular Addressing

Figure 8.66 Traceback applied to rate 1/3 convolutional code example.

Table 8.2 tabulates the required complexity of the register exchange and traceback schemes when used for block decoding or sliding-window decoding. The complexity is measured in terms of the number of memory bits and the number of bits accessed per symbol decoded. For a review of block decoding and sliding-window decoding, refer to Section 2.6.3.2.

Table 8.2 Comparison between the register exchange and traceback architectures.

	Block Decoding		Sliding-Window Decoding	
	Complexity	Memory Access	Complexity	Memory Access
Register Exchange	$K2^{K_c-1}$	$\frac{3}{2}K(B-2)2^{K_c-1}$	$\frac{KN_d2^{K_c-1}}{B}$	$2KN_d2^{K_c-1}B^{-1}\times (B-1.5-0.25N_d)$
Traceback	$(K_c-1)2^{K_c-1}$	$(1+2^{K_c-1})\times (K_c-1)$	$(N_d+1)\times \frac{(K_c-1)}{B}2^{K_c-1}$	$\left[B+\frac{(B+1)N_d-N_d^2}{2^{K_c-1}}\right]\times \frac{(K_c-1)2^{K_c-1}}{B}$

Generally speaking, the traceback technique requires $(K_c-1)/K$ times more memory than the register exchange architecture. However, the traceback approach requires much fewer memory accesses per symbol decoded. For instance, when block decoding is used, traceback reduces the amount of memory accesses by a factor of approximately $1.5KB/(K_c-1)$. Similarly, when sliding window is used, traceback reduces memory accesses by a factor of $2KN_d/(K_c-1)$, assuming a large constraint length and $N_d << B$. Because of the large saving in memory accesses, traceback is usually preferred over the register exchange architecture.

8.8 Spatial Diversity

This chapter concludes with a discussion of transceivers that support spatial diversity. Such transceivers are becoming increasing important in emerging systems, especially in the area of broadband wireless. Although many space diversity techniques exist, only a few of these are practical.

8.8.1 Selection Diversity

A practical implementation of spatial diversity is based on selection diversity discussed in Section 2.7.1. A receiver architecture based on selection diversity is shown in Figure 8.67. The receiver consists of two independent digital radios whose output strengths are measured by a power detector. The output strength is also referred to as the received signal strength indicator (RSSI). The two RSSI's are compared and the soft-decision output of the transceiver having the larger RSSI is used to generate the demodulator outputs, $\hat{I}_k$ and $\hat{Q}_k$. The selection-diversity architecture shown in Figure 8.67 has two independent radios but can be easily extended to an arbitrary number of radios. Ultimately, antenna size and radio complexity limit the maximum number of radios that can be implemented.

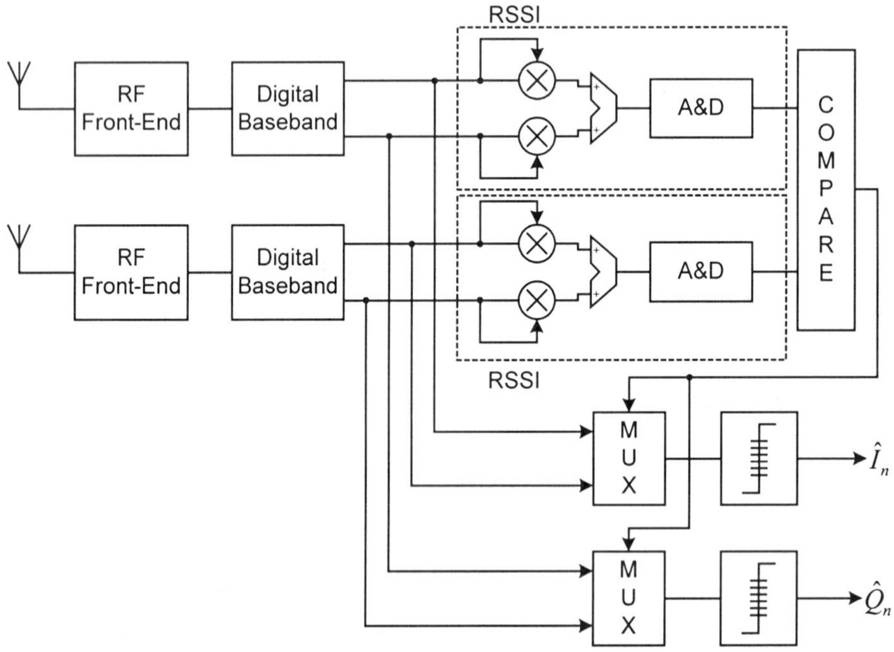

Figure 8.67 A digital receiver with selection diversity.

8.8.2 Digital Beamforming

Selection combining results in a lower diversity gain compared to MRC. If higher performance is needed, MRC should be implemented at the cost of increased complexity. Digital beamforming offers an efficient means to implement spatial diversity with MRC as shown in Figure 8.68. The digital beamforming receiver consists of K_D RF front-ends followed by K_D complex multipliers whose outputs are demodulated by K_D independent digital transceivers. Soft decisions from these transceivers are summed to form two combined I and Q soft decisions which together with the associated hard-decisions are fed to a LMS update block to generate the K_D

complex weights used at the outputs of the RF front-ends. Note that $\mathbf{w}_i$ and $\mathbf{r}_i$ represents the complex weight and received signal from the ith RF front-end.

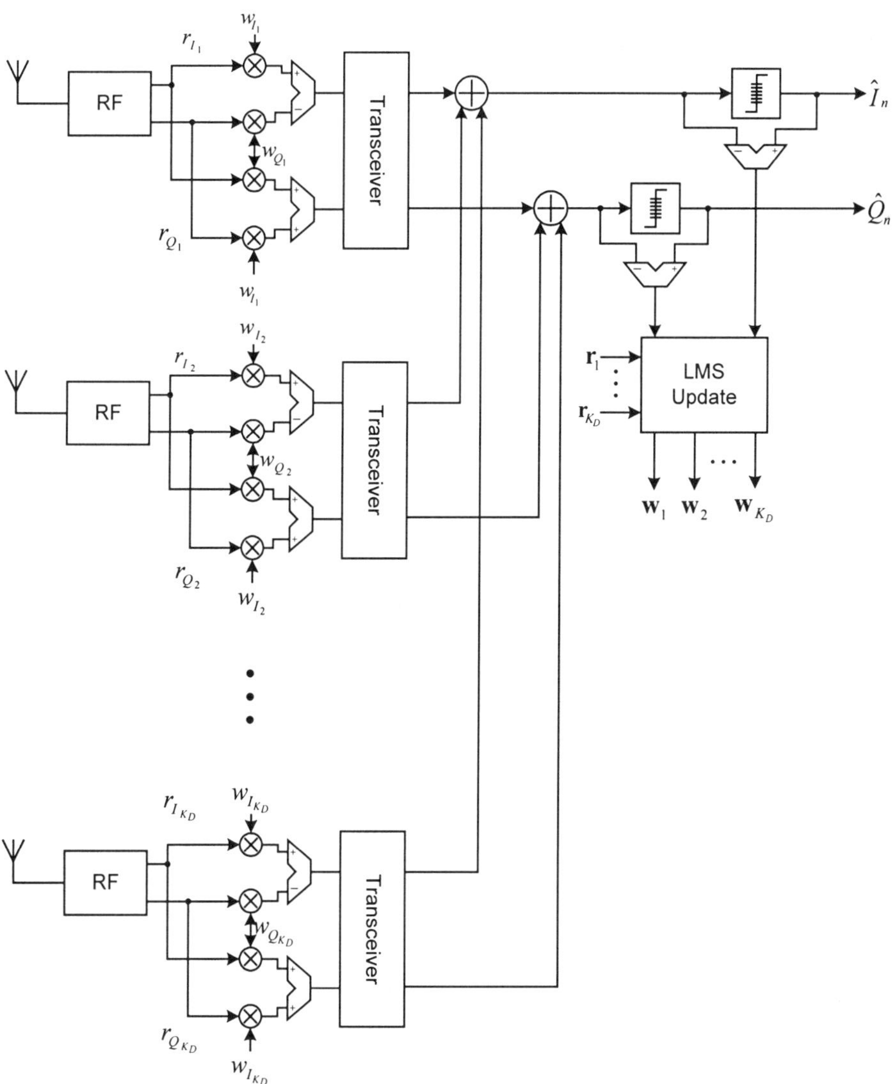

Figure 8.68 A diversity receiver with digital beamforming.

CHAPTER 9

Case Studies

The fundamental theory behind digital transmission has been around since the early 1900's. For instance, the well-known Shannon's capacity bound that placed a theoretical constraint on the capacity of digital systems was developed in the 1940's. Similarly, spread-spectrum system concepts were also developed around that time. However, until about ten years ago, all mainstream communications such as television, broadcast radio, and mobile radios have used analog transmissions. Moreover, they have been implemented using analog circuits.

In the past ten years, systems based on digital transmissions have begun to penetrate traditional systems based on analog transmissions. Examples are digital cellular systems, ISM-band spread-spectrum cordless phones, and digital television broadcast. Moreover, new systems based on digital transmission are proliferating at an astounding rate. Some of the new and emerging systems include third-generation digital cellular systems, wireless local area networks based on the IEEE 802.11 and HiperLAN standards, wireless local loops, HomeRF, and Bluetooth.

Much of the digital communications revolution owes its success to the availability of low-cost integrated circuits technology that enabled the integration of many discrete components and complex functions on a single chip. It should be noted that while the transmission is digital, the radio transceiver can still be implemented using analog circuits. In fact, some, though few, analog transceivers have been reported to implement digital transmission and reception. Although more power and area efficient, analog transceivers usually cannot fully exploit many of the sophisticated signal processing described in Chapter 2 and 0 to mitigate the degradations encountered in the wireless environment. Therefore, high-performance radio transceivers for digital systems require both an RF section and a digital section, partitioned appropriately depending on the system requirement as discussed in Chapter 6. The designs of key functional blocks for both the RF section and the digital section have been described in Chapter 7 and Chapter 8, respectively.

This chapter presents case studies of digital radio transceiver IC designs for several mainstream commercial and experimental digital systems. In particular, a complete top-down design is developed for GSM. Then examples of integrated chip-sets are described for DECT digital cordless system, indoor PCS, 64-Mbps indoor WLAN, and InfoPad [306], a 2-Mbps CDMA-based information access system. The IC examples illustrated in the case studies are based primarily on CMOS because it is felt that in the near-term, CMOS technology with its cost advantage and high circuit density offers a promising solution for low-cost, highly integrated digital radio transceivers that require complex digital signal processing. There are several issues related to obtain a fully integrated radio IC. Integration issues will be discussed toward the end of this chapter, at which point an example of a commercial single-chip radio for spread-spectrum digital cordless phone is also described.

9.1 Global System for Mobile Communications (GSM)

GSM is a digital cellular radio system which began its development in 1982. The European Posts and Telecommunications Conference (CEPT) initiated the development of GSM to provide a pan-European cellular system. In 1988, the standardization process was transferred to the European Telecommunications Standards Institute (ETSI), back then a newly created organization. Commercial launch of GSM began in 1992 to provide service in Europe. Within a year, 39 countries worldwide showed official support for GSM and one million subscribers had signed up for service, mostly in Europe. By 2000, GSM has captured about 315 million subscribers worldwide and has deployed systems in more than 130 countries. Initially, the main service carried by GSM is for mobile voice communications. Recently, with the increase in demand for data services, in particular wireless Internet access, GSM is being upgraded to support high-speed data services such as general packet radio service (GPRS), high-speed circuit-switched data (HSCSD), and enhanced data rate for global evolution (EDGE). This section provides a system overview of the basic GSM system, a rationalization of the radio specifications for GSM radio transceivers, and several examples of GSM transceiver IC's.

9.1.1 System Overview

Digital cellular system consists of cells, each controlled by a basestation as shown in Figure 5.7. Mobile units within a cell communicate through the basestation which routes their calls to the destination through the backbone network. As a mobile user moves across its cell boundary into another cell, an on-going call is handed off to the basestation controlling the new cell. The boundary of a cell is defined by its coverage. For a high-capacity system, it is desirable to have small cells, i.e. small coverage with low transmission power. Small cells often conflict with the requirement to support high-speed mobile operation since the overhead to handle mobile handoffs becomes excessive. More detailed discussions related to cellular systems can be found in Sections 5.1-5.2.

For improved capacity, cells are arranged in frequency reuse plans, some of which are shown in Figure 5.8. Conventional analog systems typically deploy cells with a reuse factor of seven, i.e. seven cells per reuse cluster. With interference diversity techniques such as frequency hopping and sectored antenna, a reuse factor of 3-4 may be obtained as shown in Figure 5.8. In practice, more complex reuse patterns are used to fully utilize frequency hopping and sector antenna. Interested readers are referred to [254] for more details.

GSM supports both voice and data services, summarized in Table 9.1. The FER is measured based on frame errors incurred in a typical urban channel with a mobile speed of 50 km/hr, i.e. the TU50 channel. Recently, improved GSM systems introduced new packet-switched services (e.g. GPRS) for high-speed data up to 64 kbps. Soon in a couple of years, data services will be available at similar quality with 384 kbps bitrate based on EDGE. Ultimately, evolving to a third-generation system based on Universal Mobile Telecommunication System (UMTS), high-speed access can be achieved at a data rate of 2 Mbps in quasi-static environments.

Table 9.1 Standard GSM services (TU50 model).

Services	**Frame-error rate (%)**
Full-rate speech (13 kbps)	6-9.6
9.6-kbps data	0.778
4.8-kbps data	0.0112
2.4-kbps data	0.00126

The second-generation GSM system has been designed to deliver high-quality voice services to mobile users at vehicular speeds up to 250 km/hr over different environments classified as typical urban (TU), rural (RA), hilly terrain (HT), and equalization test (EQ). Each channel model specified in the standard assumes both time-selective and frequency selective fading. Table 9.2 illustrates the 6-tap model of TU, RA, and HT channels. The Time Column indicates the occurrence in time at which the channel exhibits a certain RMS amplitude measured relative to the main peak. The EQ-channel is not listed but has a simple flat response of 0 dB at 3.2 μs intervals. A more precise model with 12-taps CIR is also available in the standard [255].

Table 9.2 GSM channel models.

TU		**RA**		**HT**	
Time (s)	**Amp. (dB)**	**Time (μs)**	**Amp. (dB)**	**Time (μs)**	**Amp. (dB)**
0.0	-3.0	0.0	0.0	0.0	0.0
0.2	0.0	0.1	-4.0	0.1	-1.5
0.5	-2.0	0.2	-8.0	0.3	-4.5
1.6	-6.0	0.3	-12.0	0.5	-7.5
2.3	-8.0	0.4	-16.0	15.0	-8.0
5.0	-10.0	0.5	-20.0	17.2	-17.7

Time-selective fading due to mobility introduces time variation on each of the taps and can be modeled by applying a Doppler filter to the individual taps. Except for the main peak in the HT model, a Doppler filter based on the power spectrum described in (2.88) is used to model Rayleigh-type fading. The main peak in the HT exhibits Rician fading and thus a slightly different Doppler filter is used [255].

9.1.1.1 Frequency Allocation

Several frequency bands have been allocated for GSM and depending on the band, GSM is referred to with different names. For instance, GSM operating in the 1.9 GHz band is referred to as PCS-1900. The various frequency bands are listed in Table 9.3. Note that FDD is used with a duplexing spacing of 45 MHz, 95 MHz, and 80 MHz for GSM, DCS-1800, and PCS-1900, respectively. However, to relax the requirement on the duplexer, GSM also uses TDD whereby the mobile unit transmits three time slots after receiving a transmission from the basestation.

Table 9.3 GSM frequency bands.

System	Frequency Bands (MHz)	
	Down Link	Up Link
GSM	935-960	890-915
Extended GSM	925-960	880-915
DCS-1800	1805-1880	1710-1785
PCS-1900	1930-1990	1850-1910

GSM uses both FDMA and TDMA for medium access. The set of frequency channels available for transmission in each cell depends on the cell planning. However, the total number of frequency channels available for use is fixed by the band allocation. Given that the channel spacing is 200 kHz, there are 125, 175, 375, and 300 channels available for GSM, E-GSM, DCS-1800, and PCS-1900, respectively. Usually some of the channels along the band edges are used to form guard bands between GSM and other services operating in adjacent frequency bands.

9.1.1.2 Transmit Power Level

GSM offers several classes of transmit power levels depending on the type of terminal equipment and cell size as shown in Table 9.4. For instance, GSM class-1 terminal equipment is intended to be a mobile car phone that has plenty of power supply and therefore can afford to transmit at high power to achieve a long range, suitable for large cell. The BS for a class 1 system also transmits at a high power to cover the large cell. In contrast, a class-5 terminal has a limited power supply to support portable applications and therefore it transmits at a substantially smaller power level. Up-banded GSM systems (e.g. DCS-1800) support even less transmit power for micro-cellular operations. All GSM classes support power control and typically adjust power in 2 dB steps.

Table 9.4 GSM transmit power level.

System	Maximum Transmit Power Level	
	Down Link	Up Link
GSM (class 1)	55 dBm	43 dBm
GSM (class 5)	43 dBm	29 dBm
DCS-1800 (class 1)	43 dBm	30 dBm
DCS-1800 (class 2)	40 dBm	24 dBm

9.1.1.3 TDMA Frame Structure

GSM organizes time slots into frames, multiframes, superframes, and hyperframes as shown in Figure 9.1 [256]. Each time slot is approximately 577 μs in duration and each frame, consisting of eight time slots, is 4.615 ms in duration. A multiframe consists of a group of frames. GSM supports two types of multiframes: a 26-multiframe for traffic channels and a 51-multiframe for control channels. A 26-multiframe consists of 26 frames and has a duration of 120 ms whereas a 51-multiframe consists of 51 frames and has a duration of 235.365 ms. A superframe consists of $26 \cdot 51 = 1326$ frames and has a duration of about 6.12 sec. A hyperframe consists of 2048 superframes and has a duration of 3 hours 28 minutes and 53.76 seconds.

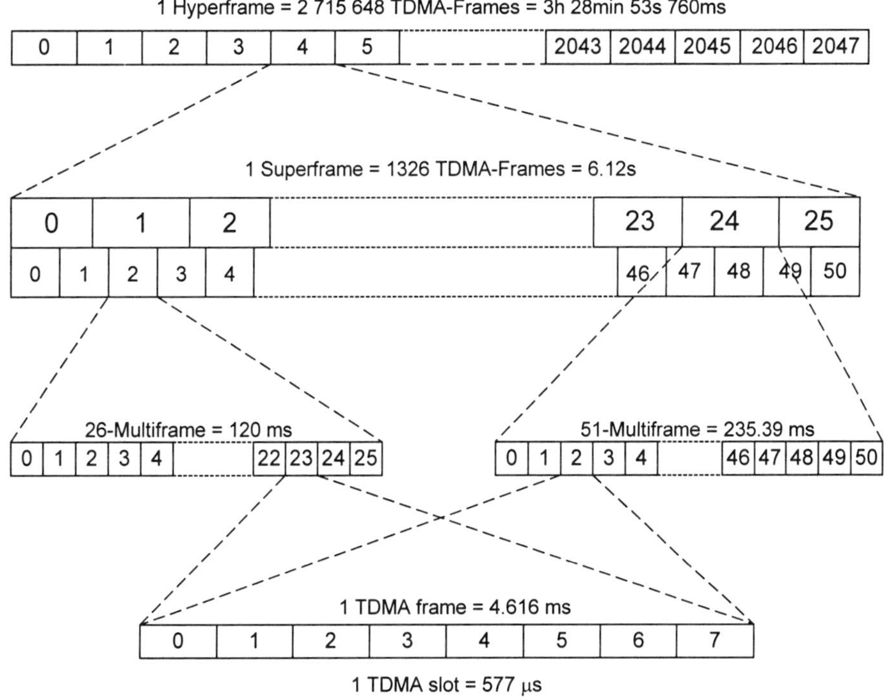

Figure 9.1 GSM TDMA frame structure.

GSM defines several types of frames, also known as logical channels, to convey system as well as user information for setting up a call, maintaining a call, and disconnecting a call. These logical channels and their associated functions are briefly described in Table 9.5 [256]. Altogether, FCCH, SCH, and BCCH make up the broadcast channel (BCH) on which a basestation broadcasts system data that are needed for network entry. RACH, AGCH, and PCH make up the common control channel (CCCH) on which a mobile unit can establish a call with the network. SDCCH, SACCH, and FACCH make up the dedicated control channel (DCCH) on which a mobile unit obtains dedicated out-of-band signaling channels for maintaining an existing link as well as for call setup.

Table 9.5 GSM TDMA logical channels.

Logical Channels	**Function**
TCH	*Traffic channel* carries user data: speech, 9.6 kbps data, etc.
FCCH	*Frequency control channel* carries the beacon used by a mobile unit to find the basestation (BS) and to synchronize its local frequency references.
SCH	*Synchronization channel* carries the training sequence required to demodulate information transmitted by the BS.
BCCH	*Broadcast control channel* carries system parameters needed by a mobile unit to access the network.
RACH	*Random access channel* carries requests from a mobile unit to obtain a dedicated link to the network.
AGCH	*Access grant channel* carries channel access information sent by the BS to a mobile unit in response to the RACH message.
PCH	*Paging channel* carries messages sent by the BS to alert the mobile unit of an incoming call.
SDCCH	*Stand-alone dedicated control channel* carries control information required for call setup.
SACCH	*Slow-associated control channel* carries system data for routine link maintenance, e.g. power control.
FACCH	*Fast-associated control channel* carries system data for time critical link maintenance, e.g. handoff.

Depending on the capacity need of the system, GSM arranges the logical channels into its frame structure in seven different combinations (I-VII). Each slot may be assigned to any one of the seven combinations. Combinations I, IV, VI, and VII are most commonly used for a high-capacity network.

Frame combination I, shown in Figure 9.2, contains 24 TCH frames (T), one SACCH frame (S), and one idle frame (I) [256]. Each frame contains eight time slots numbered from zero to seven. If the system assigns a time slot with combination I, then this time slot will carry speech data in the first 24 frames and control information in the 25th frame. During the 26th frame, marked as idle, the mobile unit may choose to perform various housekeeping functions, such as monitoring the power level of beacons from neighboring cells. Combination I assumes full rate source coding of the speech information at 13 kbps. GSM also defines combinations II and III to support half-rate speech coding at 6.5 kbps to achieve a higher capacity.

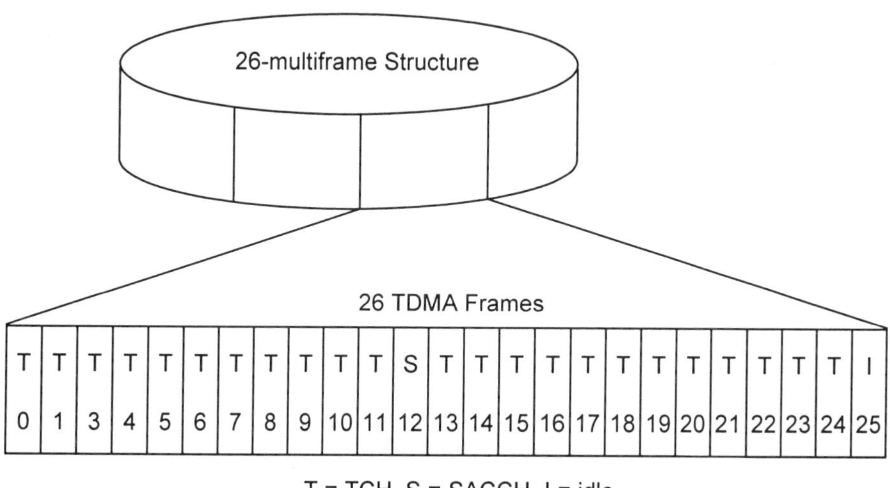

Figure 9.2 GSM frame combination I using the 26-multiframe structure.

Frame combination IV defines the occurrence of FACCH, SCH, BCCH, and CCCH channels as shown in Figure 9.3 [256]. Note that CCCH channels in the forward link consist of AGCH and PCH channels whereas in the reverse link CCCH channels consist only of RACH channels. GSM always assigns combination IV to the 0th slot of the beacon channel, which is designed to have a higher power spectral density than other frequency channels. The mobile unit detects the beacon channel based on its stronger signal strength. Once detected, the mobile unit acquires the frequency by estimating the frequency offset using information contained in the FACCH channel. Upon acquiring the frequency, the mobile unit synchronizes to the TDMA frame using the SCH channel. It is then able to demodulate parameters specific to the serving basestation via the BCCH channels. If the mobile user desires to initiate a call, it would then send a request through the RACH channel to the basestation. In response, the basestation acknowledges and, if bandwidth is available, grants the mobile user access to a SDCCH channel for further call setup procedures, such as authentication and connection setup. The basestation may also call the mobile unit via the PCH channels.

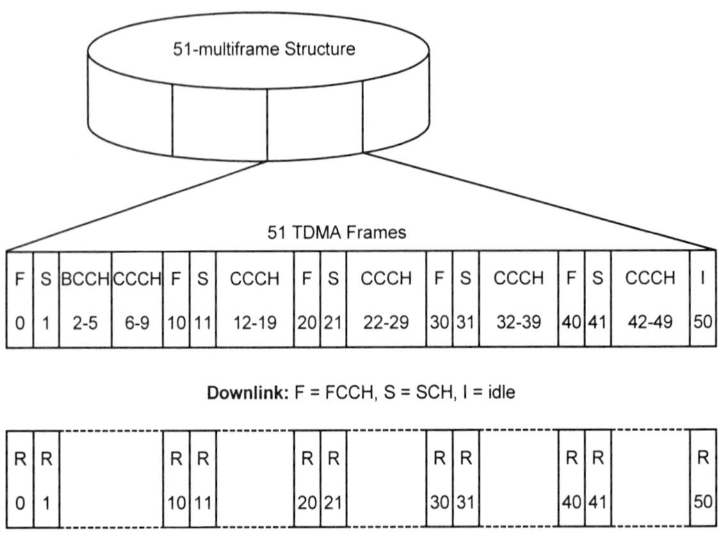

Uplink: R = RACH

Figure 9.3 GSM frame combination IV using the 51-multiframe structure.

In cases where call intensity is high, more control channels may be needed to support the large number of calls. GSM designates combination VI to accommodate such a situation. Combination VI has a structure similar to combination IV but occurs in the 2nd, 4th, and 6th time slots. To support low call intensity, GSM assigns combination V that consists of control channels available through combination IV as well as SDCCH channels required for call setup. If combinations IV and/or VI are used, GSM assigns the SDCCH channels through combination VII shown in Figure 9.4 [256].

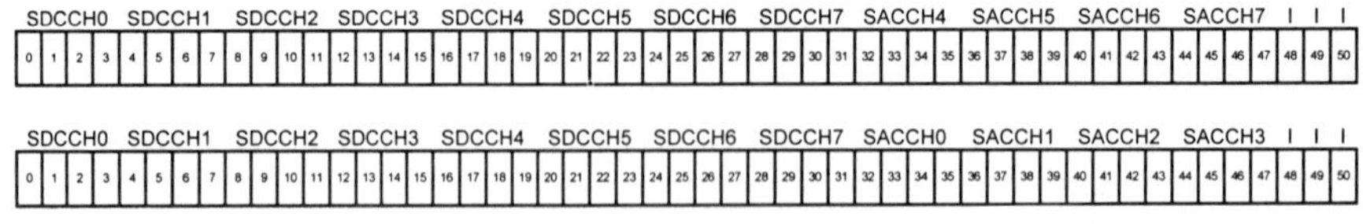

Downlink: SDCCH/8 + SACCH/8, I = idle

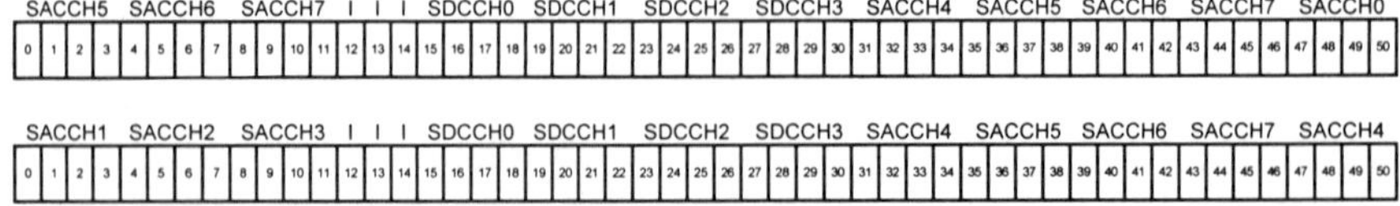

Uplink: SDCCH/8 + SACCH/8, I = idle

Figure 9.4 GSM frame combination VII using the 51-multiframe structure.

9.1.1.4 Speech Frames

GSM uses a digital codec to compress the speech data to improved system capacity. The digital codec is based on the RPE-LTP, which stands for regular pulse excitation and long-term prediction. The codec processes a 20-ms speech frame at 8 ksamples/s and 13 bits/sample. The equivalent input bitrate to the codec is 104 kbps. The codec compresses the digital speech information to 13 kbps or 260 bits per 20-ms speech frame. Since some parts of the encoded speech data require higher degree of reliability than other parts, unequal error protection is used to encode the speech data. In particular, the speech data is partitioned into three main groups: Class Ia, Class Ib, and Class II as shown in Figure 9.5. Class Ia bits are most sensitive to bit errors and are therefore encoded with a stronger code based on the concatenation of an inner block code and an outer convolutional code with rate 1/2 and constraint length five. Class Ib bits are less sensitive and are coded only with the convolutional code whereas Class II bits, being the least sensitive, are not coded at all. Altogether the channel encoded speech frame consists of 456 bits, corresponding to an equivalent code rate of 0.57.

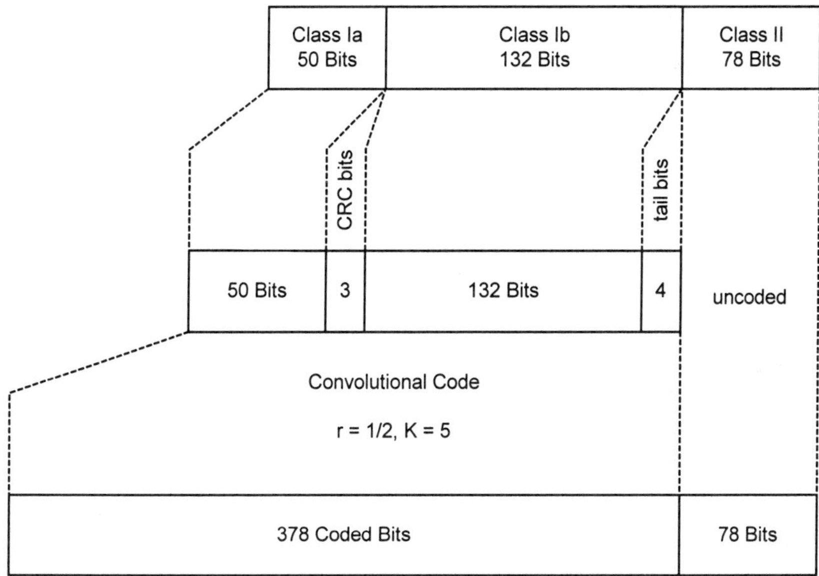

Figure 9.5 Format of a speech frame in GSM.

9.1.1.5 Burst Structure

Each of the logical channels maps into a burst structure that defines a bit-level transmission of the data. Table 9.6 describes the various burst structures defined in GSM. The format of the burst structures supports the particular function of the corresponding logical channel.

Table 9.6 GSM TDMA bursts.

Burst Type	Mapping to Logical Channels
NB	TCH, DCCH, BCCH, AGCH, and PCH are mapped to the *Normal Burst.*
FCB	FACCH is mapped to the *Frequency Control Burst.*
SB	SCH is mapped to the *Synchronization Burst.*
RAB	RACH is mapped to the *Random Access Burst.*

As an illustration, Figure 9.6 shows detailed format of the normal burst structure used for the TCH channel. The normal burst consists of an 8.25-bit guard-time field (GP), a 26-b training-sequence field, a 114-b data field, two 1-b stealing bits, and two 2-b tail-bit fields (T). In total, there are 156.25 bits per normal burst. With a burst duration of 577 μs, the corresponding bit duration is 3.692 μs. The guard-time field provides a 30.5-μs buffer for the power amplifier to settle as it ramps up and down between transmit and receive modes. The training-sequence field provides the bit sequence required to train the equalizer. By setting the stealing bits, the system can from time to time steal bandwidth from existing traffic channels to carry time-critical system function. The tail bits are used to add additional buffers for the power amplifier to ramp up/down and to provide the tail bits needed for synchronous demodulation and Viterbi decoding. The data field carries the user and system data.

T	Coded Speech Data	S	Training Sequence	S	Coded Speech Data	T	GT
3	57	1	26	1	57	3	8.25

S = stealing flag, T = tail bits, GT = guard time

Figure 9.6 Format of the normal burst consisting of 156.25 bits.

Depending on the type of traffic being carried in the TCH channel, i.e. speech or data, the data field is encoded differently. For speech frames, the data field is encoded using unequal error protection as discussed in the previous subsection. The 456 encoded bits of the speech frame are partitioned into eight 57-bit blocks. Blocks from two consecutive frames are odd-even interleaved onto eight TCH frames. The two blocks of speech data forms the 114-b data field of the normal burst.

9.1.1.6 System Capacity Design

Chapter 5 presented guidelines for a top-down design approach that determines the high-level system parameters needed to constrain the underlying circuit components. In particular, detailed steps are shown in Figure 5.11 for TDMA and CDMA systems. This section illustrates the design trade-offs captured by the heuristic rules developed in Chapter 5 with application to GSM.

To setup the analysis, relevant system constraints are specified as shown in Table 9.7 for GSM. For the purpose of this example, the standard GSM system operating in the 900 MHz band is considered. Also, to keep it simple, we consider only the up link,

Class 5 operation, and speech traffic. The BER of 10^{-3} listed in Table 9.7 is consistent with the GSM specification of 0.1% FER for a static channel. Substituting an FER of 0.1% into (2.31), a bound on BER is obtained as shown below

$$2\times10^{-5} \le P_b \le 10^{-3},$$

where the upper bound is the FER and the lower bound is $FER/50$ since there are 50 Class Ia bits. Note that the Class Ia error rate is equivalent to the FER of speech frames. In the following discussions, a worst-case BER of 10^{-3} is assumed. It will be left as an exercise for the reader to determine the system and link parameters given the best-case BER of 2×10^{-5}. The difference is approximately 2 dB in the required SIR.

Table 9.7 System constraints for standard GSM.

System Constraints	Value
Environment	See Table 9.2.
Available Bandwidth	25 MHz per direction
Carrier Frequency	890-915 MHz (up link)
Max. Transmit Power	29 dBm (Class 5, up link)
BER for Speech Data	10^{-3}

9.1.1.7 System Loading

According to Figure 5.11, the first step in computing the system capacity is to determine the traffic load. In practice, traffic load is determined based on statistics collected by the system. To facilitate discussion, a few assumptions are made to arrive at the traffic load. The first assumption is that traffic load varies in proportion to population density. Given that the penetration of cellular service is increasing rapidly, this is a reasonable assumption. The second assumption is that on average in a given hour, a portion of the cellular subscribers will make at least one call.

Given these two assumptions, the traffic load is estimated for a large metropolitan area such as Los Angeles which has a total population of 3.68 million over 1306 km^2, equivalent to a population density of 2818 people per km^2. Then, with a cell radius of 10 km, a total of 442389 subscribers must be served in a cell, assuming 50% penetration. If 0.8 % of the subscribers make two calls in a given hour with an average duration of 3 minutes, then the average traffic load is 354 Erlangs. The above scenario corresponds to a call arrival rate of 118 calls/minute. The call arrival rate encountered in practice varies with the time of day and locale. This particular scenario is chosen only for illustrative purposes. Using (5.15), approximately 355 channels are required to support a traffic load of 354 Erlangs.

9.1.1.8 Required E_b/N_0 and SIR

The required E_b/N_0 or $\overline{\gamma}_{req}$ can be determined by inverting the Q function for a given BER as shown below

$$\gamma_{req} = \frac{1}{2}\left[Q^{-1}(0.001)\right]^2 = 4.8 \text{ or } 6.8\,dB,$$

where a BER of 10^{-3} is used for speech data. The above is an estimate of the required E_b / N_0 assuming BPSK or QPSK.

Next, the required SIR is determined using (5.17) as shown below

$$SIR_{req} \le 4.8 \cdot 1.5 + \sqrt{1.5 \cdot 4.8 \cdot 3.16} = 12 \text{ or } 10.8\ dB$$

where a bandwidth efficiency of 1.5 bps/Hz, a noise margin (i.e. system noise figure) of 3.16, and γ_{req} of 4.8 are used. The computed result shows that the required SIR should not exceed 10.8 dB. To achieve high capacity, the SIR requirement may be lowered provided that appropriate interference diversity schemes are employed to accommodate the increase in interference. In GSM, interference diversity is implemented by frequency hopping, whereby each frame may be hopped over 64 different frequencies with an effective hop rate of 216.7 hops/sec. Since GSM interleaves a speech frame over eight TDMA frames, errors due to frequency collisions are spread over multiple frames. With coding, these errors are corrected at the receiver. Refer to Section 2.7.4 for a review of interleaving and coding. The following discussion assumes the worst case in which no frequency-hop is used and that the system operates with a SIR of 10.8 dB.

9.1.1.9 Network Efficiency

Sufficient information has been computed to plot the bounds on the number of available channels N_{ch} as a function of the network efficiency η_{net} defined as the product $r\eta_{TDMA}\eta_{FDMA}\eta_{sig}\eta_W$. Figure 9.7 shows a plot of the lower and upper bounds on N_{ch} as specified by (5.18). The plot assumes the following: $W_A = 25\,MHz$, $R_b = 13\,kbps$, $K_I = 6$, $SIR_{req} = 10.8\,dB$, and $n = 3.5$. Given that the network must support a loading of 350 Erlangs and requires 355 physical channels, Figure 9.7 shows that a network efficiency anywhere between 0.185 and 0.811 will meet this loading requirement.

The lower bound corresponds to a system that operates at the maximum SIR constrained by γ_{req} and γ_M defined in (5.17). If the required SIR is lowered through interference diversity for a fixed network efficiency, then more physical channels become available at a lower reuse factor as shown in Figure 9.7. With sufficient interference diversity, the SIR can be lowered until the upper bound is met at which point the reuse factor becomes one. GSM provides interference diversity through frequency hopping, DTX (i.e. voice activity), power control, and antenna sectorization. With interference diversity such as sectored antennas and spatial multiplexing, it is possible to further reduce the reuse factor below one.

A trade-off exists between network efficiency and the amount of interference diversity required by the system. For instance as depicted in Figure 9.7, with 355 channels to meet the lower bound requires an efficient network (i.e. $\eta_{net} = 0.811$) but with little interference diversity. On the other hand, to meet the upper bound requires lower network efficiency (i.e. $\eta_{net} = 0.185$) but with more interference diversity. The need for interference diversity arises from the reduction in the number of physical channels available for user traffic due to a decrease in network efficiency.

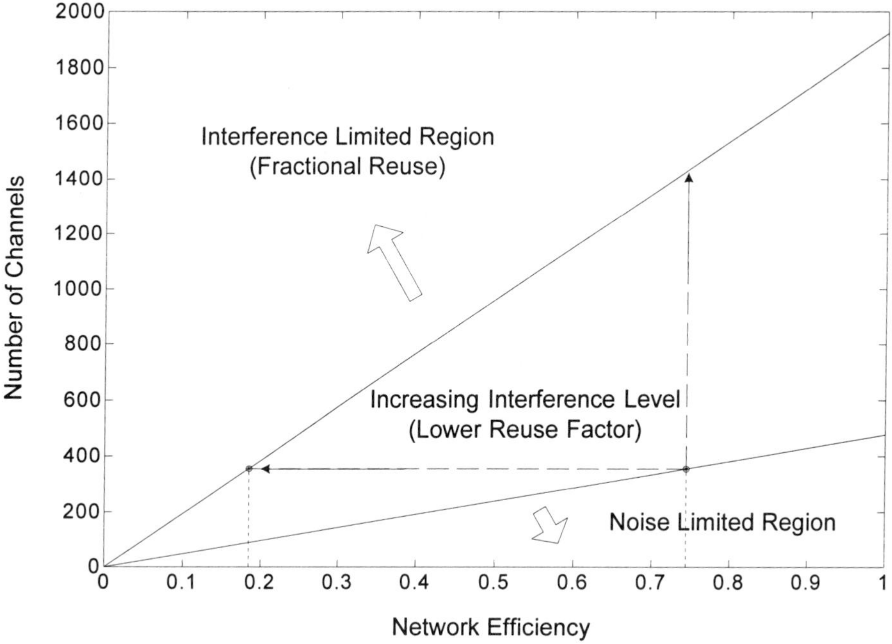

Figure 9.7 Number of available channels as a function of network efficiency.

In the current example, no interference diversity is assumed and thus an efficiency of 0.811 is needed to support 355 physical channels. However, in practice, such a high efficiency is difficult to implement especially in a mobile channel that exhibits both frequency-selective and time selective fading. Frequency-selective fading requires equalization that results in overhead, typically 25%, and time-selective fading requires FEC that also incurs a fair amount of overhead, typically 50%. Moreover, TDMA itself introduces overhead needed for guard time, controls, and out-of-band signaling. Therefore, the target network efficiency should be lowered. A reasonable compromise is to design the system to operate in between the two bounds, e.g. $\eta_{net} = 0.5$.

The network efficiency derived above places an overall constraint that guides the system design. In particular, the system should be designed so that the efficiency factors result in a product that meets the selected value of η_{net}. While extensive simulations must be performed to determine all the system parameters such that the network efficiency is met, one could use the set of guidelines described in Chapter 5 to arrive a set of estimated values to gain insight on the overall system trade-offs without delving into laborious iterations needed for a detailed system design. Table 9.8 shows the estimated as well as actual values of the various efficiency factors for GSM. The network efficiency based on the estimated values is 30 % which is off by 20% from the design target. The main sources of discrepancy result from the better TDMA and coding efficiencies that are actually achieved in GSM.

The predominant source of efficiency loss results from the coding efficiency. Using the unequal-error protection discussed in Section 9.1.1.4, GSM achieves a code rate of 57 % for the speech channel. The next largest efficiency loss is due to TDMA overheads that consist of the 42.25 bits used for training sequence, tail bits, and guard time as well as the 2/26 TDMA frames used for system management. Since GSM also supports multiple frequencies, it is a hybrid TDMA/FDMA system and one frequency channel at the lower band edge is used as a guard band, resulting in a relatively small efficiency loss of 0.8 %. Similarly, a frequency channel may be used for signaling purposes, resulting in another 0.8 % loss in efficiency. To achieve a high energy efficiency with low out-of-band emission due to non-linear amplification, GSM uses binary GMSK modulation that has a bandwidth efficiency of 1.35 bps/Hz which is in range of the 1-2 bps/Hz needed for an energy efficient system. The resulting overall network efficiency is approximately 51 %, close to the design target set based on Figure 9.7.

Table 9.8 Efficiency factors for GSM.

System Constraints	Estimated Value	Actual Value
Code Rate r	0.5	$\frac{260}{456} \approx 0.5702$
TDMA Efficiency η_{TDMA}	0.5	$\frac{114}{156.25}\frac{24}{26} \approx 0.6735$
FDMA Efficiency η_{FDMA}	0.9	$\frac{124}{125} = 0.992$
Signaling Efficiency η_{sig}	0.9	$\frac{123}{124} \approx 0.9919$
Bandwidth Efficiency η_W	1.5 bps/Hz	1.35 bps/Hz

9.1.1.10 Channel Spacing

To obtain an estimate of the channel spacing, (5.21) can be applied. The lower bound requires a value for $r\eta_W\eta_{TDMA}$. If detailed simulations have not been performed yet, the estimated value shown in Table 9.8 can be used to obtain $r\eta_W\eta_{TDMA}$. Rationales behind the estimated values have been discussed in Chapter 2 and Chapter 5. Rationales for the estimated values of r, η_W, and η_{TDMA} are summarized here.

First consider the code rate parameter. For a system that requires high capacity, any loss due to low coding efficiency is detrimental. Based on performance of commonly used convolutional codes listed in Table 2.9, a reasonable compromise in efficiency and coding gain is to select a code rate of 50%. If higher efficiency is desired, block codes, such as Reed-Solomon and BCH codes shown in Table 2.8 can be used. As a conservative estimate, a code rate of 50% is assumed in Table 9.8. Next, for an energy efficient system, a bandwidth efficiency of 1-2 bps/Hz should be used. A reasonable choice would be an average value of 1.5 bps/Hz. Finally, the TDMA efficiency is more difficult to estimate since it depends on many parameters, such as the number of control slots for maintaining the link, length of the training sequence, synchronization preamble, and guard time. Moreover, values for these parameters depend on channel conditions, transmission range, system architecture, and services being provided. However, it is reasonable to postulate that the system is poorly designed if it has to expend more than 50% of the time slots to perform control related functions. Thus, a conservative estimate of TDMA efficiency is 50%. Using these estimated values, the projected value for $r\eta_W\eta_{TDMA}$ becomes 37.5 %, which is about 1.4 times less than that actually achieved in GSM as shown in Table 9.8.

Using the conservative estimate of 0.375 for $r\eta_W\eta_{TDMA}$, the channel spacing is bounded by

$$\frac{13,000}{0.375} \approx 35\times10^3\ Hz < W_c \leq 25\times10^6\ Hz.$$

The precise value of channel spacing depends largely on the amount of equalization needed to mitigate the ISI introduced by the channel. The channel impulse response should be sampled with sufficient resolution to provide an adequate number of equalizer taps needed to resolve the individual multipath components. When MLSE is used, a large number of CIR samples can result in large loss in E_b/N_0 when the channel is minimum distance as shown in Table 2.10. However, if the CIR is sampled with too few samples, then the multipath components are not adequately resolved, resulting in a low diversity gain. Typically, a rule to thumb is to over-sample the CIR by a factor of ten. Referring to Table 2.10, the performance loss can get worse with too many CIR samples. A reasonable compromise might be to take five samples. GSM is designed to operate in a frequency-selective channel with a worst-case delay spread of 17.2 μs as specified in the HT channel. By taking five samples per CIR and assuming a bandwidth efficiency of 1.5 bps/Hz, a channel spacing of 194 kHz is obtained. The estimated spacing is close to the 200 kHz spacing actually used in GSM.

9.1.1.11 Number of Time Slots and Reuse Factor

Given the above derivations, one can now determine the number of time slots per frequency channel and the frequency reuse factor. The number of time slots is determined by applying (5.19) with $r = 0.5$, $\eta_W = 1.5$, $\eta_{TDMA} = 0.5$, $W_c = 210\,kHz$, and $R_b = 13\,kbps$, as shown below

$$N_{slot} = \left\lceil \frac{194\times 10^3}{13\times 10^3} \cdot 0.5 \cdot 1.5 \cdot 0.5 \right\rceil = 6\,slots.$$

The above estimate is 25% off from the actual value of eight slots used in GSM. The 25% difference results from the imprecise initial values selected for the code rate, bandwidth efficiency, and TDMA efficiency. More precision could be obtained with further iteration in the design process.

Next the reuse factor can be determined by applying (5.20) with $\eta_{FDMA} = 0.9$, $\eta_{sig} = 0.9$, $W_A = 25\times 10^6$, $W_c = 194\times 10^3$, $N_{slot} = 6$, and $N_{ch} = 355$. The result is shown below

$$K = \frac{25\times 10^6}{194\times 10^3}\frac{6}{355}0.9\cdot 0.9 = 1.8.$$

This is slightly less than the reuse factor of 3-4 implemented in GSM. According to (5.5), the SIR has to be lowered to 5 dB, which means an interference diversity gain of 5.8 dB is needed to achieve the required SIR of 10.8 dB derived earlier in Section 9.1.1.8. Such diversity gain can be accomplished through a combination of frequency hop, DTX, power control, and spatial multiplexing. Alternatively, the cell size can be decreased to reduce the traffic load per cell and thus lowering the number of required channels. Yet another possibility is to increase the frequency allocation to allow for addition frequency channels. For instance, EGSM and DCS1800/PCS1900 allow respectively 50 and 250 channels in addition to the 125 allocated for GSM.

9.1.1.12 Summary

Table 9.9 summarizes the results from the system capacity analysis derived thus far for GSM. The estimated system parameters can be used in the link design to constrain the implementation specifications for a GSM radio. Again it should be emphasize that the analytical process just illustrated is not meant to provide a precise solution but is meant to provide a general guideline for arriving at a good initial estimate for the system parameters and to obtain insights on the overall design. In an actual design detailed simulations must be performed to refine the design parameters in an iterative process depicted in Figure 5.15.

Table 9.9 Summary of system capacity analysis.

System Constraints	Estimated Value	Actual Value
TDMA Efficiency η_{TDMA}	0.5	0.6735
FDMA Efficiency η_{FDMA}	0.9	0.992
Signaling Efficiency η_{sig}	0.9	0.9919
Bandwidth Efficiency η_W	1.5 bps/Hz	1.35 bps/Hz
Code Rate r	0.5	0.5702
Number of Time Slots N_{slot}	6	8
Channel Spacing W_c	194 kHz	200 kHz
Reuse Factor K	1.8	2-4
Required E_b / N_0 $(\bar{\gamma}_{req})$	6.8 dB	7.29 dB
Required SIR $\left(SIR_{req}\right)$	10.8 dB[10]	10.80 dB [76]

9.1.2 Link Design

Link design determines values for link-related parameters, such as, the diversity gain required to meet the required E_b / N_0 and SIR determined from the system capacity analysis in the previous section. For a review of the steps associated with link design, referred to Section 5.5 of Chapter 5.

9.1.2.1 Link Compensation Factor

The link compensation factor, shown in (5.33), measures the excess gain required to counter losses due to noise, fading, distortion, and interference. A large compensation factor may result in excessively high transmit power and diversity gain. Expression (5.37) shows the break point that balances performance and complexity. The break point is a function of γ_M, $\bar{\gamma}_{req}$, and η_W. Substituting $\gamma_M = 3.16$, $\bar{\gamma}_{req} = 4.8$, and $\eta_W = 1.5$ into (5.37), the following breakpoint for the compensation factor is obtained

$$\alpha_\gamma^{(c)} = 3.16\left(1+\sqrt{\frac{4.8\cdot 1.5}{3.16}}\right) = 7.9.$$

The effective E_b / N_0 then becomes 15.8 dB.

The behavior of the effective E_b / N_0 as the required SIR varies can be characterized by using (5.28). Figure 9.8 plots γ versus SIR_{req} with $\gamma_M = 5\ dB$ and $\eta_W = 1.35$. The predicted behavior based on (5.28) is shown as the dashed curve while the actual values are shown as the solid curve. The actual values are determined based on

[10] Assuming an interference diversity gain of 6.2 dB.

analysis described in [52][76]. Although the estimated γ based on the simplified model (5.28) digresses substantially from the actual value for SIR less than 9 dB, the results are closely matched at SIR values greater than 9 dB. The deviation at low SIR results from the non-AWGN behavior of the interference signal.

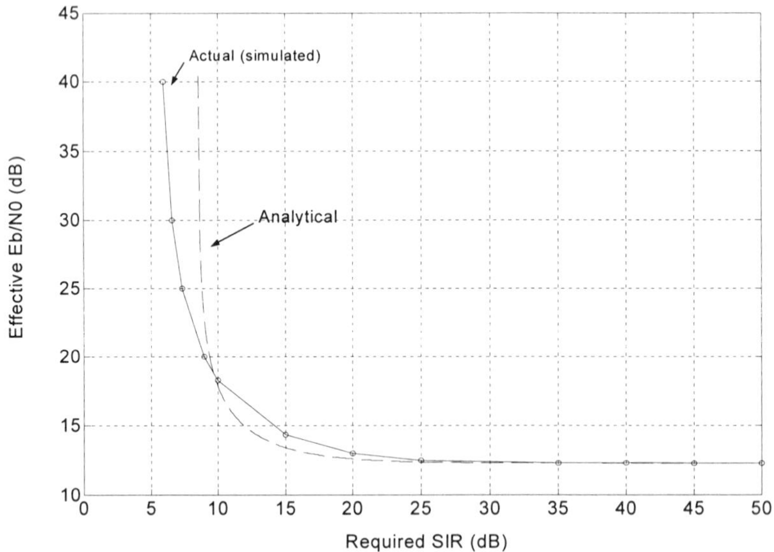

Figure 9.8 Effective E_b / N_0 Versus the required SIR.

9.1.2.2 Interference Diversity

GSM has several provisions for interference diversity, including DTX, power control, and slow frequency hopping. The amount of interference diversity gain and processing gain required is constrained by (5.38). Continuing with the analysis, it has been determined earlier that to obtain a reuse factor of 1.8, the SIR is 5 dB or 3.16. Applying (5.38) and assuming SDR/SIR of ten, the following is obtained for $G_d^{(1)}G_p$

$$G_d^{(1)}G_p = 10\log\left(\frac{1.5\cdot 4.8}{1-3.16/7.9}\frac{1+10}{10\cdot 3.16}\right) = 6.2\ dB.$$

To establish an initial estimate of the processing gain, the worst case is assumed where the processing gain compensates for all the loss in SIR. Based on (3.6), the required processing can be estimated as shown below

$$G_p = \frac{N_f W_c \eta_{TDMA}}{R_b N_{slot}}. \tag{9.1}$$

The number of hopping frequencies can then be determined using parameters that have already been estimated

$$N_f = \frac{G_p R_b N_{slot}}{W_c \eta_{TDMA}} = \frac{4.2 \cdot 13 \times 10^3 \cdot 6}{194 \times 10^3 \cdot 0.5} = 3.4.$$

In an actual system, both power control and DTX are applied to compensate for the SNR loss, Unfortunately, the interference diversity gain contributed from the power control and DTX cannot be easily modeled due to the bursty nature of the interference. In contrast, for a CDMA system interference diversity gain can be modeled as additive noise since CDMA provides a much more effective spreading of the interference energy over the signal transmission bandwidth.

9.1.2.3 Link Losses

Path loss and fading cause power loss in the transmitted signal as discussed in Sections 2.2.3-2.2.6 of Chapter 2. According to (5.30), the power loss may be represented by link margin L_M, path loss L_p, and fading loss L_{fading}. The link margin depends on the standard deviation of the lognormal fading in the channel. Referring to Table 2.3, for an urban environment, the mean standard deviation is approximately 8 dB. For 90% coverage, the link margin is approximately 1.7 times the standard deviation or 13.6 dB.

The path loss can be determined by a number of models, some of which have been described in Sections 2.2.3-2.2.4. For the current example, the partitioned model described by (2.67) is used. For distance less than 1 km, free space propagation is assumed and for distances greater or equal to 1 km, a path loss exponential of 3.5 is assume. The path loss is computed below for a cell radius of 10 km and the worst-case carrier frequency of 890 MHz used in the uplink:

$$L_p = 20\log\left(\frac{4\pi \cdot 1000}{0.3371}\right) + 35\log\left(\frac{10000}{1000}\right) = 126.4\ dB.$$

The fading loss has two dominant components, a part due to Rayleigh fading and a part due to frequency-selective fading. The loss due to Rayleigh fading can be estimated by applying (2.49) as shown below

$$Rayleigh\ Loss \approx 10\log_{10}\left(e^{4.8} - 1\right) - 10\log_{10} 4.8 \approx 14\ dB,$$

where a value of 4.8 has been substituted for $\overline{\gamma}_n$ in (2.49) for a BER of 10^{-3}. If the actual value of 5.36 is applied, then the loss is 16 dB, two dB higher than the estimated value.

The loss due to frequency selective fading can be estimated by applying the worst-case degradations listed in Table 2.10 for MLSE, which can achieve the best performance among all the equalizers for TDMA-based transmissions. The transmission environment is assumed to be the TU channel with a 5-μs delay spread. With a symbol rate of 200 kHz, the number of taps in the estimated channel response is roughly 1-2 taps. An MLSE theoretically achieves zero loss when the delay spread is less than three taps. Therefore, for a TU channel, the loss due to frequency-selective fading may be neglected provided that an MLSE is used at the receiver. While adequate for initial analysis of system performance, non-ideal effects that degrade the SNR should be accounted for in detailed designs. The non-ideal effects include imperfect channel estimates and channel dynamics.

9.1.2.4 Link Signal Gains

The link parameters estimated thus far are summarized in Table 9.10. By applying (5.39), the required signal gain can be obtained as shown below

$$\begin{aligned} P_{TX} G_{RX} G_{TX} G_d^{(s)} &= \alpha_\gamma \bar{\gamma}_{req} L_{fading} L_p L_M N_0 R_b \eta_{TDMA}^{-1} N_{slot} \\ &= 10\log(7.9) + 6.8 + 14 + 126.4 + 13.6 - 174 + 10\log(13000 \cdot 6 / 0.5) \\ &\approx 47.7\ dBm. \end{aligned}$$

For the uplink, the transmitter antenna is at the handset. Due to size constraint, a simple omnidirectional whip antenna is used. Under free-space propagation condition, a whip antenna can have gains greater than 0 dB. However, in practice loss is incurred due to interaction of the antenna with the handset package as well as the human body. Often, the gain can actually dip below 0 dB. For the purpose of discussion, the transmitter antenna is assumed to have a nominal gain of 0 dB. The basestation, on the other hand, has a lot more room for bigger antenna fixtures that provide a higher antenna gain. Generally, linear arrays are used for each sector of the cell. A much higher antenna gain can be achieved with the array, on the order of 10 dB.

Referring to Table 9.4, a Class-5 GSM handset can transmit up to 29 dBm. With a total average antenna gain of 10 dB, the required diversity gain then becomes 11.4 dB

$$G_d^{(s)} = 47.7 - 29 - 10 = 8.7\,dB.$$

Table 9.10 summarizes the link parameters that have been derived for GSM with a 10-km cell size. Comparing the estimated channel bitrate of 156 kbps, the actual channel bitrate for GSM is $13 \times 8 / 0.6735 = 154\,kbps.$ Note that the channel bitrate is distinct from the channel symbol rate which includes the overhead due to coding. The channel symbol rate for GSM is 270 kbps.

Table 9.10 Summary of link parameters for GSM with a 10 km cell size.

Link Parameter	Estimated Value	Actual Value
Compensation Factor $\left(\alpha_{\gamma}^{(c)}\right)$	7.9	8.0
Required Eb/N0 $\left(\overline{\gamma}_{req}\right)$	6.8 dB	7.3 dB
Path Loss $\left(L_p\right)$	126.4 dB	126.4 dB
Fading Loss $\left(L_f\right)$	14 dB	16 dB
Fading Margin $\left(L_M\right)$	13.6 dB	13.6 dB
Channel Bitrate $\left(R_{channel}\right)$	154 kbps	155 kbps
Interference Diversity Gain $\left(G_d^{(i)}\right)$	6.2 dB	0-2 dB
Signal Diversity Gain $\left(G_d^{(s)}\right)$	8.7 dB	11.2 dB
Processing Gain	6.2 dB	0-2 dB

9.1.3 Radio Specification

The radio specification can now be determined for a GSM mobile station based on the system and link parameters determined in the previous sections.

9.1.3.1 RF Specification

The main parameter that constrains RF transmission is the out-of-band emission, which should be small enough so that the amount of distortion incurred on adjacent channels does not appreciably degrade the link performance. To enforce a maximum tolerable out-of-band emission, GSM specifies a transmission spectral mask [255] as shown in Figure 9.9. The mask does not exhibit a sharp transition to ease the implementation of the pulse shaping at the transmitter. In GSM, GMSK is used with a Gaussian frequency pulse having a time-bandwidth product (BT) of 0.3. A more abrupt transition would have required a smaller time-bandwidth product and thus a larger transmit filter and a more complex MLSE. The downside of having a gradual transition is the increased interference at the adjacent and alternate channels, especially when the channel spacing is made as small as possible to improve spectral efficiency. Fortunately, high frequency components of the interfering signal have less deleterious effect on the desired signal. Thus, even with substantial level of out-of-band emission 100-200 kHz offset from the carrier, the amount of interference still results in adequate receiver performance.

The SDR due to out-of-band emission from an adjacent channel can be estimated by

$$SDR_{adj.\,chan.} = \frac{3\ln 10}{2\times 10^5}\frac{f_2 - f_1}{10^{-0.03 f_1} - 10^{-0.015(f_1+f_2)}} \tag{9.2}$$

where f_1 and f_2 are respectively the lower and upper frequency range over which interference incurs notable degradation. In the adjacent channel, the range of frequencies with the greatest impact lies between 150 kHz and 250 kHz as shown in Figure 9.9. Substituting $f_2 = 250\,kHz$ and $f_1 = 150\,kHz$ in (9.2), the resulting SDR becomes 20.5 dB, approximately 10 dB greater than the required SIR when no interference diversity is used. The computed result is consistent with the GSM reference sensitivity test for adjacent channel rejection whereby an interferer is injected at a power level 9 dB above that of the desired signal [255].

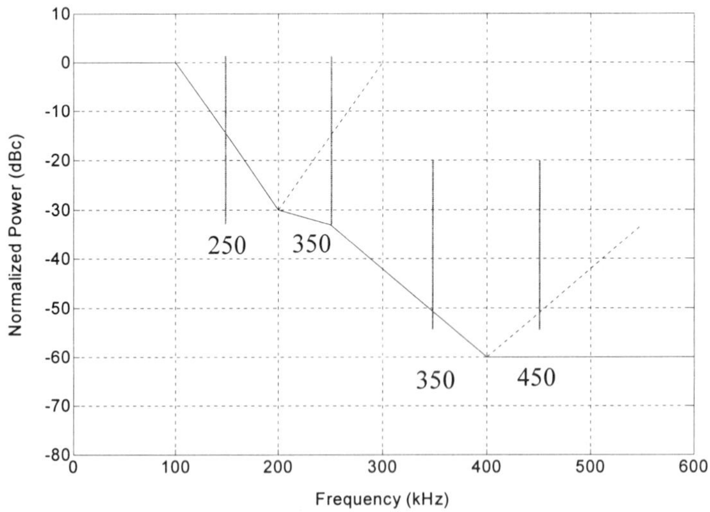

Figure 9.9 Spectral mask for GSM transmission.

The discussions above also hold for the alternate channel interference centered 400 kHz offset from the carrier of the transmitted signal. The SDR due to out-of-band emission from an alternate channel can be estimated by

$$SDR_{adj.\,chan.} = \frac{9\ln 10}{10^{4.2}} \frac{f_2 - f_1}{10^{-0.018 f_1} - 10^{-0.009(f_1+f_2)}}. \tag{9.3}$$

The out-of-band emission from 350 kHz to 450 kHz injects interference into the alternate frequency channel as shown in Figure 9.9. Using (9.3) with $f_2 = 450\,kHz$ and $f_1 = 350\,kHz$, the SDR at the alternate channel is 54.8 dB. The large SDR allows a higher transmit power for signals on alternate channels. GSM specifies a test with an alternate interferer transmitting at a power level that is 41 dB above that of the desired signal [255]. The 54.8 dB SDR provides a reasonable margin for the interference signal that is 41 dB above the desired signal power level.

The RF receiver performance is primarily specified by the NF, sensitivity, SFDR, IIP3, and phase noise. The system NF constrains the NF and gain of the LNA as discussed in Chapter 4 and Chapter 7. The LNA must have a couple of dB lower than the system NF and with sufficient gain to offset losses in the receiver front-end as well as to compensate for the high noise figure of the following mixer stage. For a system NF of 5 dB, a reasonable LNA NF is 2-3 dB with a gain of 10-20 dB depending on the NF and gain of the following mixer stage.

The receiver sensitivity is generally specified for noise-only condition based on (4.12). In GSM, GMSK is used with a BT of 0.3 and modulation index of 0.5. Based on the analysis in Section 2.4.2.2, the minimum distance can be approximated by $\sqrt{3.46E_b}$ or more precisely $\sqrt{3.56E_b}$ according to [52]. This minimum distance results in $\bar{\gamma}_{req}$ of 7.3 dB. For the traffic channel, GSM also uses a rate 1/2 convolutional code with a constraint length of five. This code provides a coding gain of approximately 3 dB as shown in Table 2.9. . Thus, the receiver sensitivity can be determined as follows

$$S = 7.3 - 3 + 10\log 154000 + 5 - 174 = -112.8\,dBm.$$

Compared to the GSM specification of –102 dBm for sensitivity on the portable handset, the computed sensitivity is 10.8-dB lower which provides additional margin for implementation losses.

Sensitivity measures the mobile handset's performance with weak received signal strength. To ensure that the mobile handset can also cope with large signal levels, GSM specifies that the handset must accommodate up to –15 dBm of received signal power [255] at a BER of 10^{-3}. Using the actual reference sensitivity of –102 dBm, the required dynamic range becomes 87 dB, which is reasonable for a mobile wireless channel. According to (4.33), the SFDR sets the requirement for the IIP3 of the receiver as shown below

$$\begin{aligned} IIP3 &= 1.5 \cdot 10\log SFDR + 10\log S \\ &= 1.5 \cdot 87 - 102 = 28.5\,dBm. \end{aligned}$$

If instead the estimated sensitivity of –112.8 dBm is used, the IIP3 becomes 33.9 dBm. However, such a high IIP3 is impossible to achieve given the high sensitivity requirement. A more realistic constraint on IIP3 can be based on (4.27). Since the GSM specification of maximum input power level does not assume adjacent channel interferers, α_1 and α_2 may be set to one. Assuming a SDR of 20 dB, which is 9.2 dB above the required SIR, the required IIP3 becomes

$$IIP3 \geq P_{\max} + 10, \tag{9.4}$$

where $P_{\max}$ is the maximum allowed input power level. Since GSM specifies a maximum power level of –15 dBm, (9.4) indicates that IIP3 should be at least –5 dBm.

GSM also specifies blocking signals with the characteristics shown in Table 9.11. Given two frequencies f_1 and f_2 separated by $|f_1 - f_2|$, IMD will be generated at $f_2 + |f_1 - f_2|$ and $f_1 - |f_1 - f_2|$ where $f_1 < f_2$. Expression (4.27) can be applied to determined the required IIP3 to accommodate the IMD introduced by blocker signals. Take for example, blockers at 943.4 MHz and 944.2 MHz, separated by 800 kHz, and a desired signal at 945 MHz. GSM specifies that the receiver sensitivity level is 3 dB above the reference sensitivity level when no blockers are present. Using the specified reference sensitivity of –102 dBm, the input receive power is then –99 dBm. Substituting –99 dBm for P_{in} in (4.27) and noting that the two blockers have equal power level of –43 dBm, the following constraint for IIP3 in dBm can be determined

$$IIP3 \geq \frac{1}{2}SDR + P_{in} + \alpha_1 + \frac{1}{2}\alpha_2. \tag{9.5}$$

Since both blockers are at a power level 56 dB above the received signal level, i.e. $\alpha_1 = \alpha_2 = 56\ dB$. Again, it is assumed that the SDR should be about 10 dB above the required SIR, i.e. 20 dB. Then, the required IIP3 becomes –5 dBm, consistent with the constraint derived from (9.4). A slightly tighter constraint results if the estimated receiver sensitivity of –112.8 dBm is used. It can be shown that in this case the required IIP3 is greater than 0.4 dBm.

Table 9.11 GSM blocking characteristics [255].

Blocker Frequency (MHz) f_0 = frequency of desired in-band signal f = frequency of in-band blocker signal f' = frequency of out-of-band blocker signal	Blocker Power (dBm)
$0.6 \leq \lvert f - f_0 \rvert \leq 1.6$	-43
$1.6 \leq \lvert f - f_0 \rvert < 3.0$	-33
$3.0 \leq \lvert f - f_0 \rvert$	-23
$980 \leq f' \leq 12700$ $0.1 \leq f' \leq 915$	0

This section ends with a discussion of the specification for phase noise in the RF transceiver. Recall from Section 4.4.3.2, phase noise causes spectral leakage into the adjacent channels. GSM specifies constraints on the allowable spurious emission over a span of 45 MHz offset from 915 MHz as shown in Table 9.12

The GSM specification may be rationalized by computing the amount of phase noise allowed so that the spectral leakage caused by a strong blocker can still be tolerated at the receiver. As an approximation, to achieve 10 dB SDR, the phase noise in (dBc/Hz) at a particular frequency offset from the carrier can be expressed by

$$\Phi(\Delta f) \approx -P_{blocker} + S - 10\log W_c - 10. \quad (9.6)$$

Phase noise corresponding to the specification in Table 9.11 is computed using (9.6) and listed in Table 9.13. Note that reciprocal mixing occurs at the receiver and constrains the receiver LO phase noise in a similar fashion.

Table 9.12 Allowable spurious emission [266].

Frequency Range (MHz)	Emission (dBc/Hz)
0 – 0.1	-53
0.2	-83
0.4 – 1.8	-113
1.8 – 3	-121
3 – 6	-123
6 – 10	-129
10 – 20	-150
20 – 45	-162

The close-in phase noise (i.e. less than 200 kHz offset) is constrained by the phase jitter tolerance of the demodulator. In GSM, the RMS phase jitter is specified to be 5 degrees [255]. The close-in phase noise of about –120 dBc/Hz at 100 kHz offset results in an RMS phase error of less than 5 degrees. See Example 4.13. The specifications listed in Table 9.13 forms a more stringent requirement than that of Table 9.12.

Table 9.13 Phase noise obtained from blocker and jitter requirements.

$\lvert\Delta f\rvert$ (kHz)	$P_{blocker}$ (dBm)	Phase Noise (dBc/Hz)
600	-43	-129 dBc/Hz
1600	-33	-139 dBc/Hz
3000	-23	-149 dBc/Hz

9.1.3.2 Analog-Digital Interface Specification

The sampling rate of the ADC depends on the architecture of the radio implementation. Section 5.6.2 and Chapter 6 are devoted to this topic. In this section, the baseband digital radio architecture is considered for GSM. The ADC sampling rate can vary between $2.5W_c$ (i.e. 500 kHz) and $2.5W_A$ (i.e. 62.5 MHz). An oversampling factor of 2.5 is chosen to satisfy the Nyquist Criteria. In general, having a higher oversampling ratio above 2.5 improves the performance of clock recovery at the expense of high-speed processing and complex digital circuitry. Another important trade-off is in the choice of digitizing only one channel or more than one channel up to the entire band. Digitizing only one channel lowers the speed requirement on the digital circuits. However, this approach requires a sharp high-Q analog anti-aliasing filter and precludes flexibility in demodulation multiple channels which is particularly useful for basestations.

Similar trade-offs exist for the sampling rate of the DAC. In this case, the sampling rate depends primarily on the selectivity of the image-reject filter. In general, oversampling ratios of 2-8 are used.

The ADC resolution depends largely on two factors: 1) the SFDR and 2) the required SIR. The SFDR for GSM is 87 dB for static channels, which translates to 15 bits. The number of bits can be reduced by allowing the AGC to handle a portion of the SFDR. For instance, an AGC with 40-dB dynamic range cuts down the ADC resolution to 8 bits. With an increasing AGC gain, the second factor may begin to dominate since the ADC resolution must provide a SNR greater than the required SIR. In GSM, the required SIR has been estimated to be about 11 dB, which corresponds to at least two bits. To minimize the implementation loss with too few bits, 4-6 bits are typically used for binary modulations [138]. Likewise, the DAC resolution should be made large enough to reduce the effect of signal distortion due to finite precision. If the SDR is set to 20 dB higher than the required SIR, at least five bits are required. Typically a few bits are added to accommodate the implementation loss.

9.1.3.3 Digital Transceiver Specification

In an energy-efficient system, it has been determined in Section 2.1.6 that candidate modulations should have bandwidth efficiency of 2.3 bps/Hz. Referring to Figure 2.12, since the efficiency is relatively flat near the optimum point of 2.3 bps/Hz, modulations with bandwidth efficiencies between 1-3 bps/Hz are also good candidates.

Besides energy efficiency, spectral regrowth due to non-linear amplification should also be considered when selecting a modulation. Sections 2.4.1.3, 2.4.1.4, and 2.4.2 described OQPSK, $\pi/4$-QPSK, and CPM that are more robust to non-linear amplification as compared to PSK and QAM. With lower modulation index and larger correlation interval, CPM can be designed with a low E_b/N_0 requirement as well as high bandwidth efficiency. However, more complex receivers using Viterbi equalizers with many states may be needed. A compromise is to use a relatively short correlation interval and a high modulation index. In GSM, GMSK is used with BT = 0.3, corresponding to a correlation interval of three, and a modulation index of 0.5 [257]. Section 2.4.5.2 showed the superior performance of GMSK when compared against $\pi/4$-QPSK and OQPSK subject to non-linear amplification. This modulation achieves E_b/N_0 within 0.5 dB of BPSK, lowest required for binary modulation, and a bandwidth efficiency of 1.35 bps/Hz.

GSM uses a frame size of 4.615 ms. There are two reasons behind this choice for frame size. The first reason is to meet the low delay requirement for speech services, which is 100-200 ms one way, depending on the line condition [258]. More specifically, the delay through the public land mobile network (PLMN) has been constrained to less than 90 ms [258]. While 4.615 ms seems short compared to 90 ms, the interleaver delay and transcoding delay in the network still make it difficult to meet the 90 ms delay requirement. The second reason is to accommodate a high mobile speed, typically 100 km/hr. According to (2.89), the coherence time can be expressed in terms of the mobile speed v and carrier frequency f_c as shown below

$$\tau_c = \frac{9c}{16\pi f_c v}. \tag{9.7}$$

GSM specifies a mobile speed up to 250 km/hr, which results in a coherence time of 846 μs at a carrier frequency of 915 MHz. To ensure that independent fades affect individual bursts that are transmitted across the channel, the burst duration should be less than the coherence time. Each frame contains eight TDMA slots or bursts and each burst has a duration of 577 μs less than that determined by (9.7). Having independent fades on each burst enables an efficient implementation of interleavers.

Interleaving combined with coding provide diversity gain for the received signals. As shown in Table 9.10, the required signal diversity gain is approximated 11.2 dB for GSM. Diversity is required to combat both time-selective and frequency selective fading. To combat time-selective fading, GSM employs time diversity in the form of convolutional encoding with interleaving. The design constraints on convolutional encoding and interleaving have been discussed in Section 2.6.3 and Section 2.7.4. Coding with interleaving is particularly effective against mobile fading channels where the channel exhibits burstiness in the received signal strength. In GSM, the channel is specified for both highly mobile (e.g. 100-300 km/hr.) as well as slowly mobile (e.g. 3 km/hr.) scenarios.

Let us first consider the highly mobile channel, using 100 km/hr. as an example. At 100 km/hr, the maximum Doppler shift is 85 Hz at 915 MHz. The fading rate and duration can be determined using (2.85) and (2.86) assuming $\rho = 0.21$, which is consistent with a fading margin of 13.6 dB. The resulting values are 42.8 fades/sec and 1 ms, respectively.

The depth of the interleaver m can be lower bounded by applying (2.190) and using $\tau(0.21) \approx 1\ ms$ as shown below

$$m > \frac{\tau(\rho)}{r\lfloor 0.5(d_{\min} - 1)\rfloor dT_b} \tag{9.8}$$

where $t = \lfloor 0.5(d_{\min} - 1)\rfloor$ is the highest number of correctable errors obtainable, d in this case is the number of time slots between each transmission of speech data, and $d_{\min}$ is the minimum distance or the free distance of the convolutional code. In GSM, a constraint length five rate 1/2 code is used with $d_{\min} = 8$ [72]. See Table 2.9. GSM also uses an 8-slot structure per TDMA frame as shown in Figure 9.1. Therefore, d = 8. Substituting these values into (9.8), the interleaver depth is constrained to be greater than 22 coded symbols as shown below

$$m > \frac{1000}{0.5 \cdot \lfloor 0.5(8-1) \rfloor \cdot 8 \cdot 3.69} = 22.6. \tag{9.9}$$

GSM actually uses an interleaving depth of 114 bits which is about five times greater than lower bound (9.9). The larger interleaving depth helps to combat a longer fading duration which occurs at lower mobility.

Constraint (2.189) on the span of the interleaver should be modified slightly for convolutional encoding. For block codes, the span of the interleaver should be large enough such that errors in any one row of the interleaver, corresponding to a code word, should be less than the number of correctable errors. Since convolutional codes have an infinite block size, the above constraint cannot be directly applied. However, note that if the number of errors exceeds the error correcting capability of the code (i.e. $\lfloor 0.5(d_{\min} - 1) \rfloor$) within the decoding depth, the decoder may make an incorrect decision. Therefore, a reasonable rule is to constrain the span S of the interleaver such that the above situation does not occur for a given K_c and $d_{\min}$ as shown below

$$S \geq \frac{5K_c}{\left\lfloor \frac{d_{\min} - 1}{2} \right\rfloor}, \tag{9.10}$$

where K_c is the constraint length of the code. Equality in (9.10) holds when the depth of the interleaver is designed so that only one error occurs per row of the interleaver matrix. Constraint (9.10) can be combined with (2.191) to further restrict the upper limit of the interleaver span as shown below

$$\frac{5K_c}{\left\lfloor \frac{d_{\min} - 1}{2} \right\rfloor} \leq S \leq \frac{N^{-1}(\rho)}{rT_b md}. \tag{9.11}$$

Substituting 42.8 fades/sec for the fading rate $N(\rho)$ and numeric values for other parameters associated with GSM, the span is restricted to the range prescribed below

$$\frac{5 \cdot 5}{\left\lfloor \frac{8-1}{2} \right\rfloor} \leq S \leq \frac{23300}{0.5 \cdot 3.69 \cdot 22.6 \cdot 8}$$

$$\Rightarrow \quad 8.33 \leq S \leq 69.8.$$

GSM actually uses a span of eight, which is close to the lower bound of 8.33 shown above.

Ultimately, the maximum tolerable delay of speech service limits the size of the interleaver. To see that GSM satisfies the delay requirement, we need to understand in more detail the interleaving method used by GSM [72] shown in Figure 9.10. A frame of 456 coded speech bits is segmented into eight blocks, each holding 57 bits. Two levels of interleaving are performed. The first level of interleaving is referred to as inter-burst interleaving, where bits from the eight blocks are block interleaved as shown in Figure 9.10a. The eight blocks may be viewed as eight columns in an interleaver matrix. Thus, the block interleaver has a span of eight. In contrast to standard block interleaving, GSM disperses bits within a burst in a pseudo-random fashion to mitigate the degradation due to correlated error patterns that have a period equal to the depth of the interleaver.

The second level of interleaving is referred to as block-diagonal interleaving which involves three speech frames as shown in Figure 9.10b. The first four blocks in the nth frame are interleaved with the last four blocks in the n-1 frame while the last four blocks in the nth frame are interleaved with the first four blocks of the n+1 frame. Moreover, two 57-bit blocks are interleaved in an even-odd pattern to form eight blocks of interleaved data each containing 114 bits. Each of the 114 bits of data forms the data part of a normal burst.

The interleaver incurs a transmission delay of 36.9 ms since each speech frame is interleaved over eight TDMA frames, each with a duration of 4.615 ms. The RPE-LTP vocoder [259] used in GSM incurs a delay of one speech frame time or 20 ms. Therefore, the total speech delay is then 56.9 ms and is well within the 90 ms limit. The remaining 30 ms is used to perform transcoding in the basestation interface and the speech decoding and de-interleaving in the mobile handset [258].

The performance obtained though time diversity in a mobile fading channel can be approximated by (2.194). For GSM, the achieved diversity gain is computed below using $\bar{\gamma}_n = 5.4$ and $d_{\min} = 8$:

$$G_d^{(s)} \approx 10\log_{10} 0.5(e^{5.4} - 1) - 10\log_{10}(e^{2\cdot 5.4/8} - 1) \approx 16dB.$$

Note that hard-decision is assumed; therefore, $d_{\min}$ in (2.194) is replaced by $0.5d_{\min}$. The resulting diversity gain is greater than the estimated value of 9-11 dB in the link design.

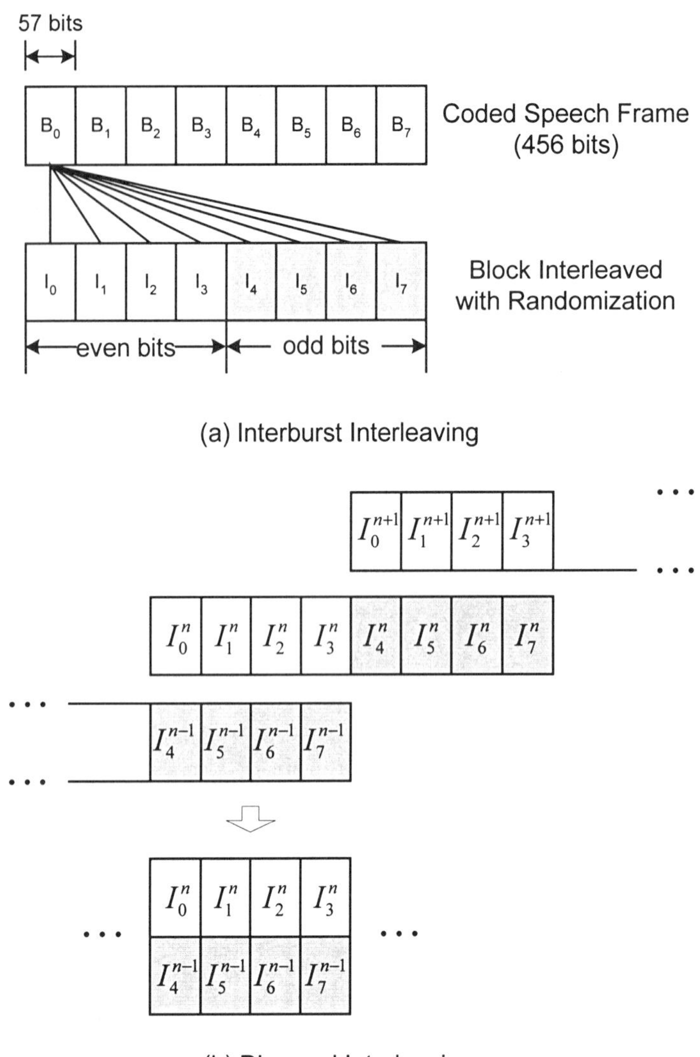

Figure 9.10 GSM interleaver for the speech channel.

While time-diversity can achieve substantial gains at high mobility, it performs poorly in slowly mobile environments, for instance, at pedestrian speeds of 3 km./hr. At 915 MHz, the maximum Doppler shift for a mobile speed of 3 km/hr is 2.5 Hz. Substituting $f_m = 2.5$ and $\rho = 0.21$ into (2.85) and (2.86), the fading rate and duration become 1.3 fades/s and 34.3 ms, respectively. Because of the long fading duration, the required interleaver depth becomes at least 775 coded symbols, approximately 33 times larger than the case at 100 km/hr. The total memory requirement of the interleaver is mS and can be constrained by using (9.8) and (9.11) as shown below

$$\frac{5K_c\tau(\rho)}{rdT_b\lfloor 0.5(d_{\min}-1)\rfloor^2} \le mS \le \frac{N^{-1}(\rho)}{rT_bd}. \tag{9.12}$$

Assuming that the TDMA frame structure remains at eight slots per TDMA frame, the memory requirement lies within the following range

$$\frac{5\cdot 5\cdot 34300}{0.5\cdot 8\cdot 3.69\lfloor 0.5(8-1)\rfloor^2} \le mS \le \frac{769000}{0.5\cdot 3.69\cdot 8}$$

$$\Rightarrow \quad 6455 \le mS \le 52100.$$

The memory requirement can range between 7 and 57 times larger than that of the interleaver at 100 km/hr. The large interleaver size requires more memory storage as well as additional processing, both of which are undesirable. More importantly, the large interleaver size incurs a delay of $mSdT_b$ which ranges between 191 ms and 1.5 s. Such a large delay is unacceptable for speech.

To meet the delay requirement, GSM does not use a larger interleaver. Therefore, with time-diversity alone, the performance degrades substantially at slow mobile speeds as shown in Figure 9.11. For instance, the 3 km/hr TU channel displays an error floor that is above the required BER of 0.1% whereas for high-speed operation of 50 km/hr and 100 km/hr, the system meets the BER requirement with a reasonable E_b/N_0 of about 13 dB.

To combat deep fades in a slowly mobile channel, frequency diversity can be employed as discussed in Section 2.7.4.2. GSM employs slow FH with one hop per TDMA frame, i.e. 216.6 hops/s with a maximum of 64 hopping frequencies. With FH, the performance over TU3 channel becomes comparable to that obtained at TU50 and TU100 channels. As expected, the performance gain due to frequency diversity diminishes for high-speed operation.

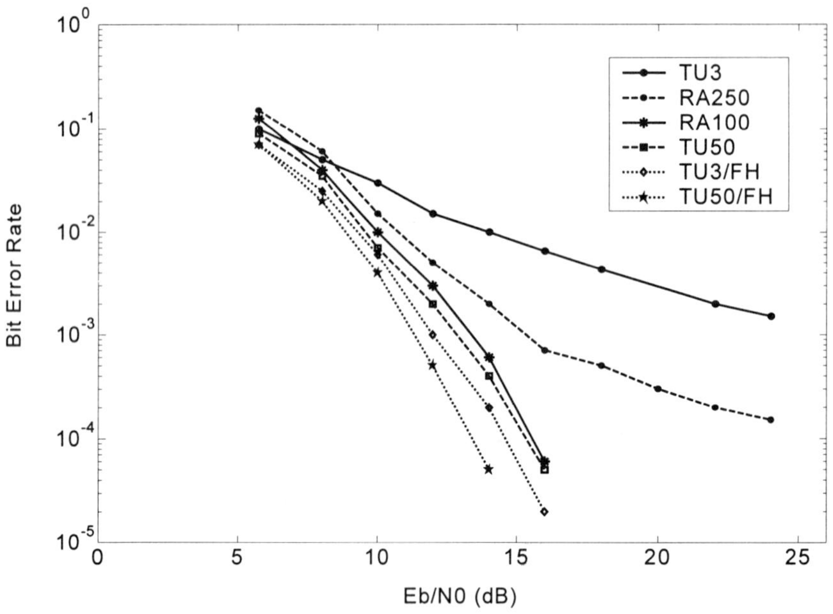

Figure 9.11 BER performance of Class I speech data [260].

In interference-limited environment, FH also offers interference diversity by providing processing gain to the system. According to discussions in Section 9.1.2.2, the required processing gain is estimated to be 5.8 dB. Using actual GSM parameters, the processing gain is even less, between 0-2 dB as shown in Table 9.10. In practice, SFH may be used as directed by the basestation to mitigate interference due to heavy traffic loading. It should be emphasized again that at slow moving speeds, having SFH is crucial to meet the required BER performance.

In designing GSM, several channels have been considered and are listed in Table 9.2. As discussed in Section 9.1.2.3, A trade-off exists between having too few samples on the CIR and therefore suffering a loss in diversity gain versus in having too many samples when the channel exhibit few multipaths. In the latter case, the channel introduces a small $d_{\min}$ in the trellis that results in a higher E_b/N_0 requirement. This trade-off is shown in Figure 9.12, where the BER performance of a 32-state MLSE is characterized for different delay separation between two Rayleigh faded paths. Since the channel bitrate is 270.8 kbps, the multipath resolution is about 3.7 μs. Figure 9.12 shows that as the separation between the two paths diminishes, the two paths can no longer be resolved and the BER performance degrades. On the other hand, as the path separation becomes larger than about five times the multipath resolution, i.e. 18.5 μs, the performance also degrades since the MLSE does not have sufficient memory to capture both paths. Optimum baseband processing would require the modem to adapt the number of states depending on the channel condition to obtain the best result.

This section concludes with a discussion of the burst structure and efficiency of TDMA. GSM burst structure for the speech channel is shown in Figure 9.6. According to (5.2), TDMA efficiency depends on the number of overhead bits in a TDMA time slot and the number of time slots used for control purposes. The three main sources of overhead bits are the bits used for training the equalizer, the bits used for synchronizing the frame, and guard time needed for power amplifier ramp up/down and variability in the propagation delay. Our initial estimate of TDMA efficient is 50%, close to the actual efficiency of 67.35%. In GSM, 26 bits are used to train the equalizer for the speech channel as shown in Figure 9.6. Section 2.7.7.1 discussed the trade-offs related to the training sequence length needed to obtain a good channel estimate. The problem is to find a training sequence that has good autocorrelation properties yet has a short duration such that multiple periods can be used to obtain an estimate with high SNR. Also, the duration of the training sequence should be at least equal to the maximum delay spread of the channel, in this case, 18 μs or 5 bits. Through extensive computer search, the 26-bit training sequence specified in GSM meets this requirement as well as good autocorrelation property.

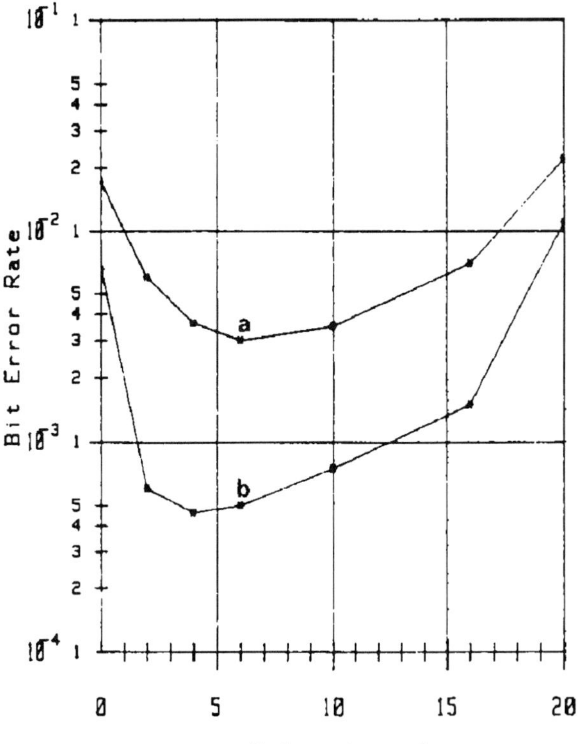

Figure 9.12 BER performance of MLSE as a function of delay spread; from D'Avella [261], © 1989 IEEE.

The GSM speech service requires an initial call setup to establish synchronization but does not require repeated synchronization for each TDMA burst. The synchronization overhead per TDMA burst is thus negligibly small. The initial synchronization is established in the mobile unit by tuning to the SCH channel transmitted by the basestation. The SCH frame is part of the 51-multiframe shown in Figure 9.3. The SCH contains a 64-bit preamble that allows the mobile unit to establish frame synchronization through a correlation process discussed in Section 2.5.5. Details of the burst structure for GSM can be found in [256]. Note that the SCH frame repeats every 10 frames or 46.16 ms. Therefore, a mobile unit can resynchronize to the basestation as needed every 46.16 ms, which should be more than sufficient when a reasonably accurate crystal oscillator is used in the mobile handset.

The above holds as long as the transceiver can estimate the frequency offset between the transmitter and receiver. Frequency offset results from oscillator frequency drift and Doppler shift. In GSM, the mobile unit estimates the frequency offset during the FCCH frame in which the basestation transmits a beacon. In Section 2.5.4, frequency estimation techniques have been described that can be applied for this purpose. Once estimated, the mobile tracks the residual frequency offset and if necessary will re-acquire the FCCH when the frequency error becomes large.

Taking into account the oscillator frequency stability Δ and the maximum Doppler shift f_m, the required update time T_{update} for the system to maintain synchronization can be approximated by

$$T_{update} < \left(1 + \frac{1}{\Delta + \dfrac{f_m}{f_0}}\right)\alpha T_s, \tag{9.13}$$

where α is the maximum tolerable timing offset measured in fraction of a symbol time and Δ is the frequency offset of the crystal oscillator in ppm.

Example 9.1 Consider $\alpha = 0.25$, a reasonable assumption for partial response modulation since the pulse shape extends over several symbols. At 100 km/hr, the maximum Doppler shift is 83 Hz at 915 MHz. The symbol duration for GSM is 3.69 μs. A readily available clock oscillator typically has a stability of 30 ppm. Therefore, by applying (9.13), the update time is 30.7 ms which is smaller than the repetition time of a SCH frame. However, if the offset is acquired to within 50 Hz then the update time relaxes to 16.9 seconds. □

Finally, 30 μs guard time is needed to allow sufficient time for the power amplifier to ramp up and down without generating excessive amount of switching transients that can raise the level of ACI beyond an acceptable level. A portion of this time is also used to accommodate the variability in propagation delay between the basestation and mobile unit. However, with a cell size that can be as large as 35 km, the allocated guard time seems to be too small to accommodate for the worst case roundtrip delay of 233 ms. GSM reduces the required guard time by time alignment [256] whereby the basestation estimates the delay to the mobile unit and advances the transmission in time to compensate for the propagation delay. The propagation delay is measured using the random access channel (RACH) in the uplink shown in Figure 9.3. The RACH burst has a longer guard time of 68.25 symbols or 251.8 μs to accommodate for the worst-case round trip propagation delay.

9.1.4 RF Transceiver

Current RF transceiver chips for GSM are all based on bipolar and GaAs technology, the latter of which is widely used for power amplifiers. Advances in technology have enabled high level of integration for the RF transceivers used in GSM. Typical transceiver chip sets consist of an RF chip, an IF chip, and one or two synthesizer chips. The complete RF front-end usually requires about 200 other discrete components, including: off-chip filters, passives for biasing and/or matching, crystal, TX/RX switch, power amplifier, and a duplexer [262]. As discussed in Chapter 6, some of these components, such as filters and passives, may be eliminated with direct-conversion or low-IF architectures. However, even with the passives and filters integrated on chip, 10-50 external components are still required in the near future. These include a power amplifier for high power applications, a crystal reference, a TX/RX switch, a duplexer, and discrete passives for biasing and matching.

In the early 1990's, [263][166] reported a transceiver chip-set based on the direct-conversion architecture. The chip-set consists of an RF direct up/down converter and a baseband analog IC. The RF chip contains only a pair of quadrature mixers needed for up/down conversion, a RC-CR phase splitter, and a pre-amplifier for the external PA. A high-gain (17 dB) external LNA is used to compensate for the high noise figure (18 dB) of the mixer whose noise performance had to be sacrificed to achieve the high dynamic range needed in a direct-conversion architecture. Moreover, since the architecture provides no isolation between the LO and the received RF carrier, the receiver suffers from DC offset problems. To reduce the offset, the baseband IC implements DC offset cancellation though no details of the technique was given. The chip-set is implemented in 9 GHz silicon bipolar process and dissipates 25 mA and 45 mA in RX and TX modes, respectively.

Later in the mid 1990's, more integrated transceivers have been reported. Reference [264] reports a two-chip solution that implements a high-IF architecture. The two chips, fabricated in 13 GHz bipolar process, consist of an RF chip and an IF chip with an IF frequency of 400 MHz. The RF and IF chips occupy 8.3 mm^2 and 10 mm^2, respectively. The chips operate from 2.7-5.5V supply and dissipates 51 mA in receive mode and 105 mA in transmit mode. The VCO's are not integrated on chip.

Reference [175] presented an integrated solution where the VCO's are integrated on-chip using a 12 GHz bipolar process. The architecture also differs from [264] in that the transmitter uses direct conversion and the receiver uses a lower IF at 71 MHz.

The power in transmit and receive mode is respectively 65 mA and 57 mA with 2.7 V supply. The direct up-conversion architecture contributes to the lower transmit power compared to [264] since direct up-conversion requires less hardware.

The main problem with direct up-conversion is that the high power transmit signal can couple back to the synthesizer and pull the VCO out of lock. VCO pulling is especially severe for direct up-conversion since the VCO is running at the same frequency as that of the carrier. To overcome VCO pulling, the design in [175] employs an offset VCO that generates the carrier frequency by mixing the outputs of two PLL's, each running at a frequency different from that of the carrier as shown in Figure 9.13. A frequency selection circuit (i.e. LO Filter) at the output of the mixer selects one of the mixing products, either $|f_1 - f_2|$ or $f_1 + f_2$. Since the VCO's are not running at the carrier frequency, pulling by the transmitted signal is reduced substantially. In [175], one of the PLL is fixed to lock at 117 MHz and the other PLL tunes over 996-1032 MHz.

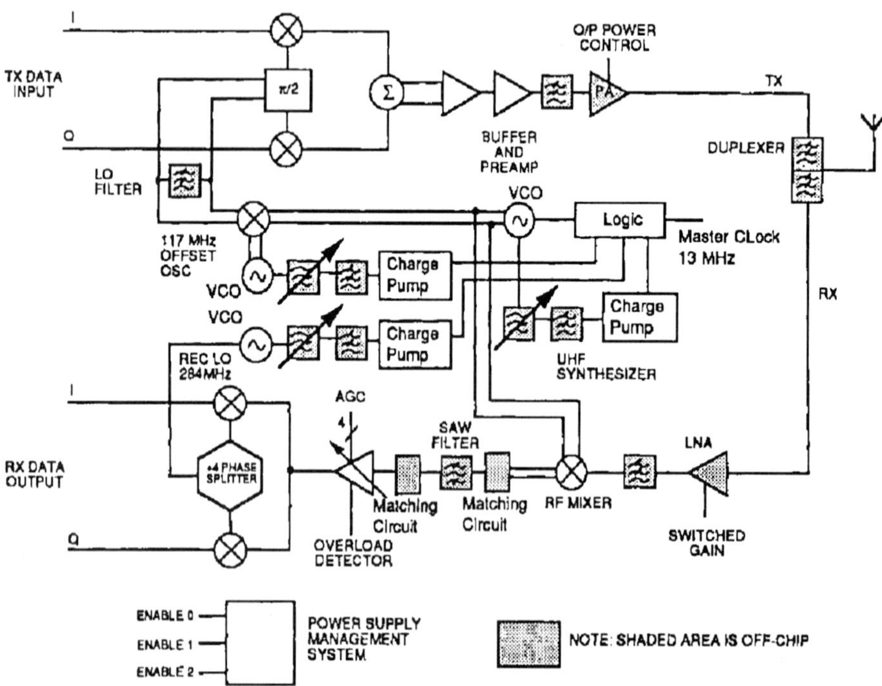

Figure 9.13 Offset VCO in a GSM transceiver; Stetzler [175], © 1995 IEEE.

Recently, [265] reported a 2.7-V RF transceiver IC that also uses direct up-conversion but by directly modulating the VCO with the baseband data. At the receiver, it uses dual IF conversion with 225 MHz and 45 MHz IF frequencies. The transceiver chip requires only a few external filters for the dual IF stages and a dual synthesizer. Due to the isolation achieved with the offset VCO, the authors claim that the duplexer and transmit filter can be replaced by a simple diode switch. This chip is implemented in 15 GHz bipolar process with similar current dissipation as in [175] and with a total area of 13 mm^2.

In general, these bipolar chip-sets have been designed for a reference sensitivity of –105 dBm, IIP3 of –3 to 1 dBm, a phase noise of –163 dBc/Hz at the receive band, and a dynamic range of 80-100 dB. The above specifications agrees with those derived in Section 9.1.3. All chip-sets require external filters, power amplifier, discrete passives, and crystals. Some require external synthesizer and duplexer as well.

With the drive toward highly integrated solution, current efforts in radio ICs are focused on low-IF and direct-conversion implementations. Due to the high performance requirement of GSM, more work has been reported that are based on the low-IF architecture and little on the direct-conversion architecture, though recently an impressive direct-conversion GSM radio chip-set has been announced in [262]. This direct-conversion IC consists of a dual-band direct-conversion RF IC in bipolar technology and a single-chip CMOS baseband IC that supports GPRS as well as Bluetooth.

While current commercial GSM RF IC's are still implemented in bipolar technology, many prototype RF transceivers have been implemented in CMOS. Two such examples will be described in the following sub-sections.

9.1.4.1 CMOS Dual-IF GSM Transceiver

This section is based on results reported in [266], which described a 0.25-μm CMOS 2.5V GSM transceiver using the dual-IF architecture shown in Figure 9.14. The shaded blocks indicate external components and double lines indicate differential signals. In addition to external filters, a power amplifier, and a duplexer, this transceiver IC also requires a dual synthesizer to generate the RX LO of 996-1057 MHz and the TX LO of 880-915 MHz. The transmitter converts the baseband signal in one stage. Therefore, if the VCO were on chip, an offset VCO should be used. The receiver uses a 71 MHz IF and requires a filter with sharp transition and high stopband attenuation to eliminate the close-in image. In this case, a duplexer and an inter-stage filter combined gave 70 dB of image suppression. However, the external filters introduce a high insertion loss that degrades the system NF. For instance, the duplexer contributes a loss of 3.2 dB or 35% of the system noise figure.

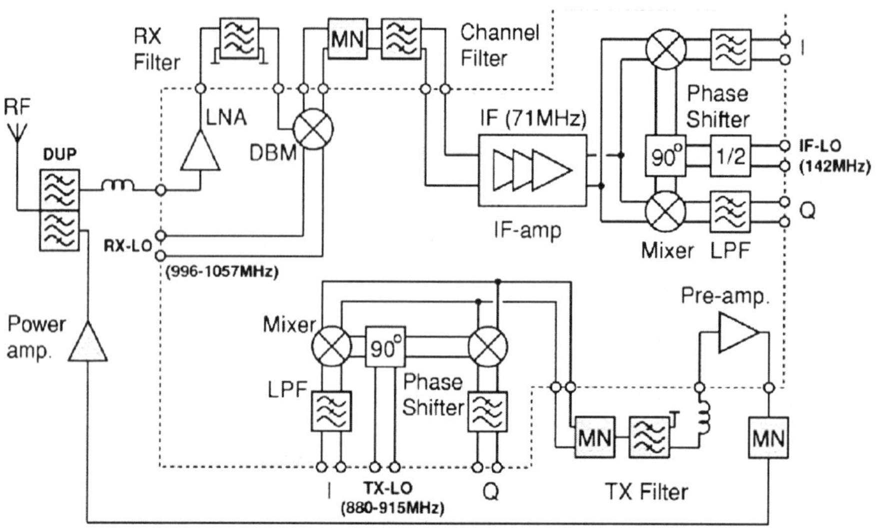

Figure 9.14 CMOS dual-IF GSM transceiver architecture; from Orsatti [266], © 1999 IEEE.

Figure 9.15 summarizes the noise figure, IIP3, and power dissipation of the key components in the receive chain. It can be seen that the attenuation introduced by the duplexer prior to the LNA contributes directly to the system noise figure. Attenuation after the LNA contributes much less due to the high gain of the LNA, which is 15 dB. The LNA is designed with a 1.9 dB NF which is comparable to that achieved in bipolar designs. The LNA is based on the cascode configuration discussed in Chapter 7. Inductor degeneration is used to achieve good linearity and low noise. The inductor is implemented using bondwires to achieve a higher Q than on-chip spiral inductors. A trans-impedance output stage is used to drive the 50-ohm load of the external RX Filter needed for image rejection. While the external RX Filter is lossy, its effect on the system noise figure is suppressed by the gain of the LNA. The highest contribution to noise figure is from the RF mixer, which has a high noise figure of 12.6 dB due to the inherent nature of the mixer circuit as discussed in Chapter 7. A double-balanced active mixer is used. Since the gain of the mixer and LNA add up to 25 dB, the IF AGC amplifier [267] can still inject appreciable noise into the receiver. In this design, the AGC achieves a 3.8-dB noise figure with 60 dB of dynamic range by dividing the amplification into three stages. The noise figure contributes to less than 10% of the total noise figure. The noise figures of the IF mixer and baseband filters do not contribute significantly to the system noise figure due to the high gain from the AGC, LNA, and RF mixer.

While the duplexer degrades the noise performance of the receiver, it improves the linearity of the receiver by rejecting power of unwanted signals. In particular, the duplexer and IF filter lowers the IIP3 requirement by about 25 and 55 dB, respectively.

The power dissipation of the receiver is dominated by the LNA. Recall that to achieve low noise, the LNA must dissipate more power, especially, for a CMOS design because CMOS has a g_m/I ratio that varies inversely with $\sqrt{I}$. To double the transconductance, the bias current must increase by four times. A bipolar technology, on the other hand, has a higher g_m/I ratio that varies inversely to I. Doubling the transconductance requires only a doubling of the bias current. However, since CMOS tends to scale down in feature size quicker than bipolar, the power advantage of bipolar is diminishing. For instance, this CMOS design dissipates about 20 mA in receive mode which is comparable to the 50 mA dissipation reported in [175]. To maintain a fair comparison, it should be noted that the CMOS design does not contain a synthesizer while the bipolar design does include a synthesize and is also based on an older technology.

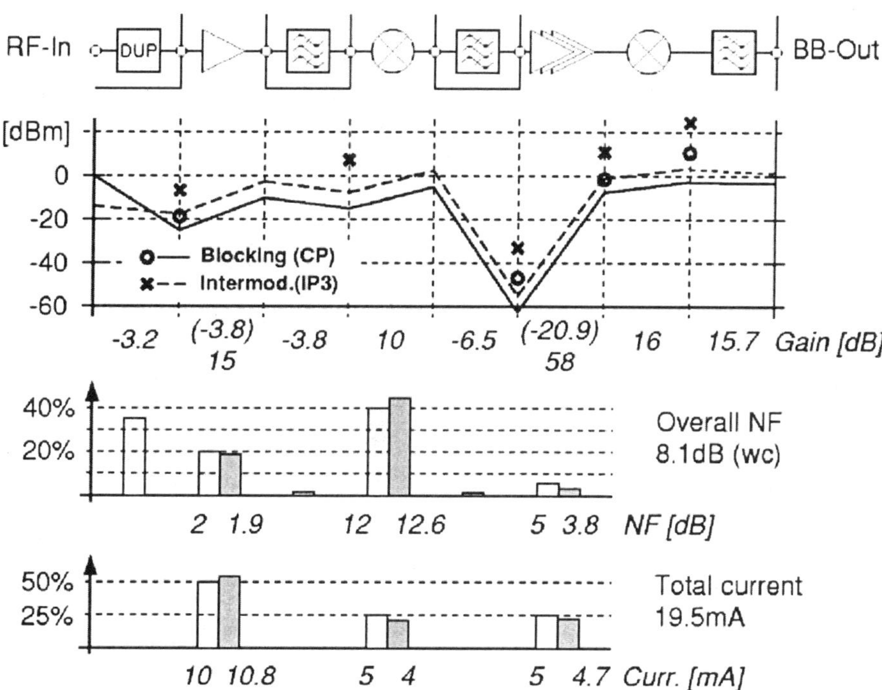

Figure 9.15 Receiver performance. ▇ represents performance actually achieved and ☐ represents performance target for the design; from Orsatti [266], © 1999 IEEE.

For the RF transmitter, the key performance requirements are distortion and out-of-band emission. The main sources of distortion come from compression of the power amplifier, intermodulation products, and phase noise. Linearity of the power amplifier and LO phase noise also determine the amount of out-of-band emission.

Linearity of the power amplifier helps to prevent excessive spectral regrowth. For Class-5 handsets, the maximum transmit power is 29 dBm and would require an off-chip power amplifier. This transceiver incorporates an on-chip 8-dBm preamplifier to drive an external power amplifier. While the output compression point is 5.8 dBm, the continuous phase modulation used in GSM allows transmission without additional output backoff. The transmitted signal resides within the spectral mask shown in Figure 9.9.

The modulator is based on a differential pair with RC and CR phase shifters at the differential inputs. Component matching of the phase shifter determines the phase and amplitude matching of the quadrature modulator. To meet the GSM specification, at least 30-35 dB of IMD rejection is required. The corresponding amplitude and phase matching are 0.3 dB and 3 degrees, which can be satisfied with the simple RC-CR quadrature phase shifter.

Low phase noise helps to minimize spectral leakage and phase jitter. In GSM, phase noise must satisfy the allowable out-of-band emission specified in Table 9.12. The transceiver IC meets the emission requirement as well as the phase jitter requirement of 5 degrees RMS. Specifically, 2-degrees RMS phase jitter has been measured. However, the phase noise requirement can be more difficult to satisfy when a VCO is integrated on-chip.

9.1.4.2 CMOS Low-IF GSM Transceiver

Because of the stringent requirement needed to meet GSM specifications, most transceivers are based on the superheterodyne architecture. The example [174] discussed in this section is also based on the superheterodyne architecture but with an IF frequency that is only a few hundred kHz in contrast to the 71 MHz used in the previous example. The main objective of the work reported in [174] is to eliminate the need for external filters.

Recall from discussions in Chapter 4, images are folded in-band due to the negative frequency components of a real signal. See Figure 9.16a. However, if the signal were complex, then the negative frequency component can be suppressed. For instance, $\cos(\omega_o t)$ has spectral peaks at $\pm\omega_o$ but $e^{j\omega_o t}$ has a spectral peak only at $+\omega_o$. Based on this observation, a low-IF receiver can be constructed using a complex signal representation for the received and RF LO as shown in Figure 9.16b. The LO is centered on $-\omega_{LO}$ whereas the desired signal is centered on $+\omega_o$ and the mirror signal is centered on $\omega_o - 2\omega_{IF}$. The image in the negative frequency have been suppressed with a complex filter that can have relatively relaxed Q, shown as dashed line in Figure 9.16b. After down-conversion, the image is positioned at $-\omega_o + \omega_{LO} = -\omega_{IF}$ as shown in Figure 9.16c. Since the center frequency is at a fairly low frequency, on the order of a few hundred kHz, a low-Q complex filter can be used to suppress the image as shown in Figure 9.16d.

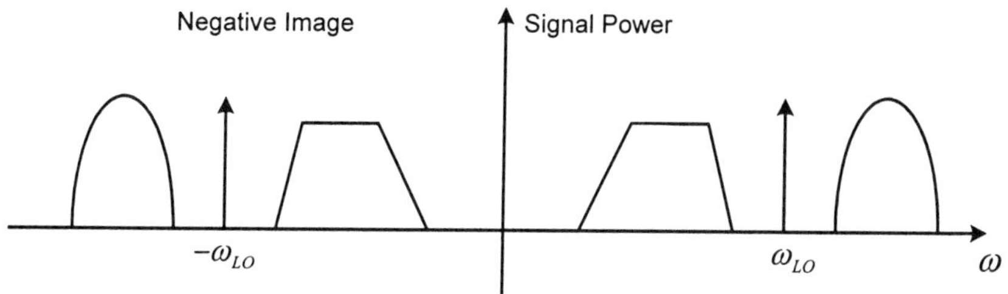

(a) Double sideband signal

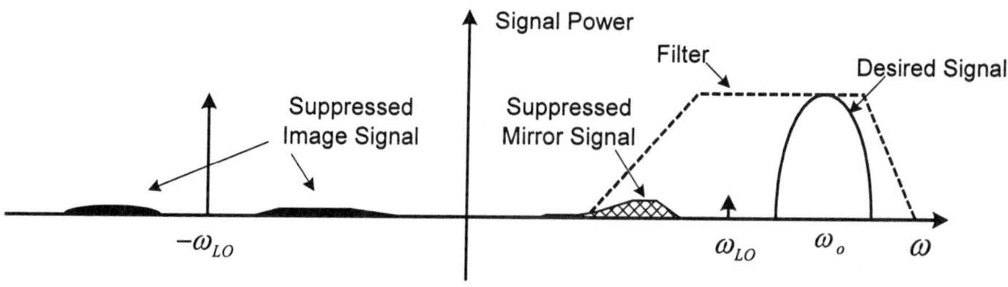

(b) Double sideband to single sideband conversion

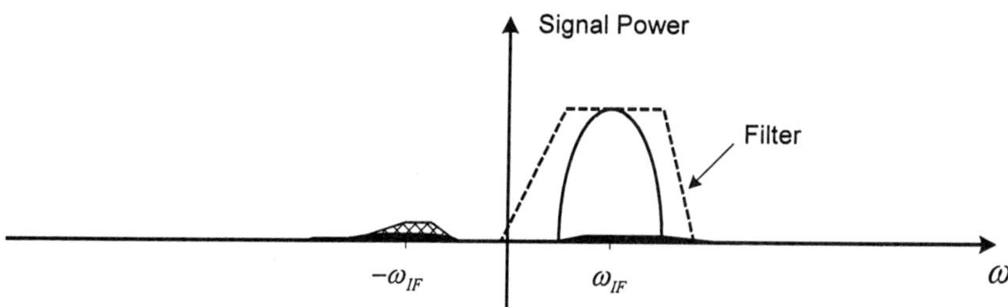

(c) Single sideband down-converted signal with IF carrier at $-\omega_{IF}$

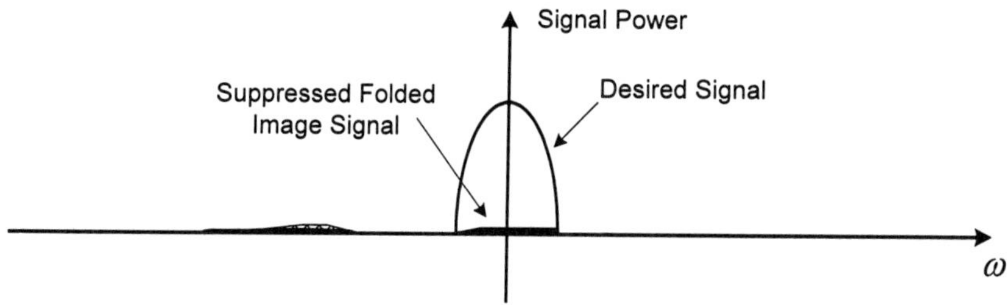

(d) Single sideband down-conversion to DC

Figure 9.16 Frequency spectrum of low-IF conversion.

Complex signals do not exist in a real world. Eventually, the complex signal must be converted to a real signal but before then complex operations may be carried out using quadrature representation as discussed in Chapter 2. When the real part of the complex signal is taken, the image sideband is regenerated in a process often referred to as single sideband (SSB) to double sideband (DSB) conversion. In the low-IF architecture, the RF signal is first converted from DSB to SSB and the SSB signal is then down converted using a quadrature mixer shown in Figure 9.17. The quadrature mixer takes two complex input signals, one representing the RF signal $I_{RF} + jQ_{RF}$ and the other representing the RF LO $\cos(\omega_{RF}t) - j\sin(\omega_{RF}t)$. The mixer generates the product of the two complex signals. Since the product of two complex signals is also complex, the mixer output consists of two outputs, one for the real part and the other for the imaginary part. The complex mixer output is at low IF and has a spectrum shown in Figure 9.16c.

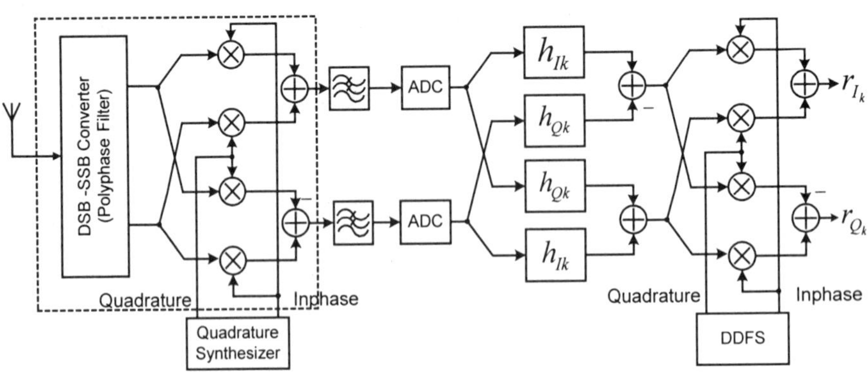

Figure 9.17 Complete low-IF architecture with final complex digital down-conversion.

To select the desired signal, a complex filter could be used, implemented in either analog or digital circuits. Since the low-IF signal is running at a few hundred kHz, it could be efficiently converted to digital signal without an excessively high sampling rate. However, the low-IF signal does impose a high dynamic range requirement due to high power blockers. With greater than 60 dB dynamic range requirement, a 12-14 bit ADC is needed. Such high resolution ADC's are now available at few MHz sampling rates. Therefore, it is reasonable to assume that the final selection and down conversion to baseband can be carried out in the digital domain.

Figure 9.17 shows the complete low-IF architecture with the SSB filter and final down-conversion to DC using digital filters, multipliers, and a quadrature DDFS. Recall from Chapter 2 that complex digital filter can be represented by a complex impulse response, $h_I(n) + jh_Q(n)$, with a corresponding frequency response shown as dashed line in Figure 9.16. The DDFS generates $\cos(\omega_{IF}t) - j\sin(\omega_{IF}t)$ needed to down convert the complex IF signal to baseband. Finally, the real baseband signal can be obtained through SSB-to-DSB conversion, which could be as simple as taking the real part of the quadrature mixer output.

An important observation is that the low-IF architecture shown in Figure 9.17 requires no external filtering to eliminate the image. The architecture relies on SSB generation for adequate suppression of the image frequency components. A prerequisite for adequate image suppression is extremely good phase and amplitude matching in the in-phase and quadrature components. For instance, in the case of a single tone $e^{j\omega_o t}$, a phase offset of ϕ_I and ϕ_Q results in a signal shown below

$$A\cos\left(\omega_o t+\phi_I\right)+jA\sin\left(\omega_o t+\phi_Q\right) \tag{9.14}$$

If phase matching is defined as $\Delta\phi=\phi_Q-\phi_I$, then $e^{j\omega_o t}$ may be represented by

$$0.5Ae^{j\phi_I}\left(1+e^{j\Delta\phi}\right)e^{j\omega_o t}+0.5Ae^{-j\phi_I}\left(1-e^{-j\Delta\phi}\right)e^{-j\omega_o t} \tag{9.15}$$

The second term represents a negative frequency component or image of ω_o. Thus, any phase mismatch regenerates the image sideband. The image-rejection ratio (IRR) is defined as the ratio of the amplitude of the desired sideband to that of the image sideband and can be expressed by

$$IRR=\sqrt{\frac{1+\cos\Delta\phi}{1-\cos\Delta\phi}}. \tag{9.16}$$

Similar derivation shows that IRR due to amplitude mismatch can be expressed as

$$IRR=\frac{2+\dfrac{\Delta A_I+\Delta A_Q}{A}}{\left|\dfrac{\Delta A_I-\Delta A_Q}{A}\right|}, \tag{9.17}$$

where A is the amplitude of the sinusoid, ΔA_I is the amplitude mismatch of the in-phase component, and ΔA_Q is the amplitude mismatch of the quadrature component.

According to (9.16), a 45-dB IRR requires a phase match of 0.6 degrees. Also, if the worst-case amplitude match is assumed whereby $\Delta A_Q=-\Delta A_I$, 0.6% amplitude matching is required in both the I and Q channels to achieve a 45-dB IRR. A simple RC-CR phase splitter achieves a matching of about 3 degrees. To achieve the tight matching needed for SSB down-conversion, polyphase filter [268] is used instead. In this example, a passive polyphase filter is used and is shown in Figure 9.18.

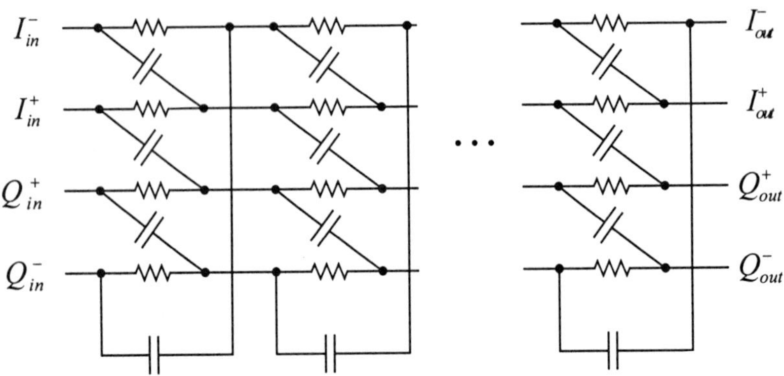

Figure 9.18 A passive polyphase filter.

The polyphase filter shown in Figure 9.18 is less prone to component mismatches as compared to the conventional RC-CR phase splitter [174]. Also, in contrast to the RC-CR phase splitter, the polyphase filter can be designed over a broader bandwidth, making it suitable not only for generating the SSB LO at RF but also for converting the RF carrier from DSB to SSB.

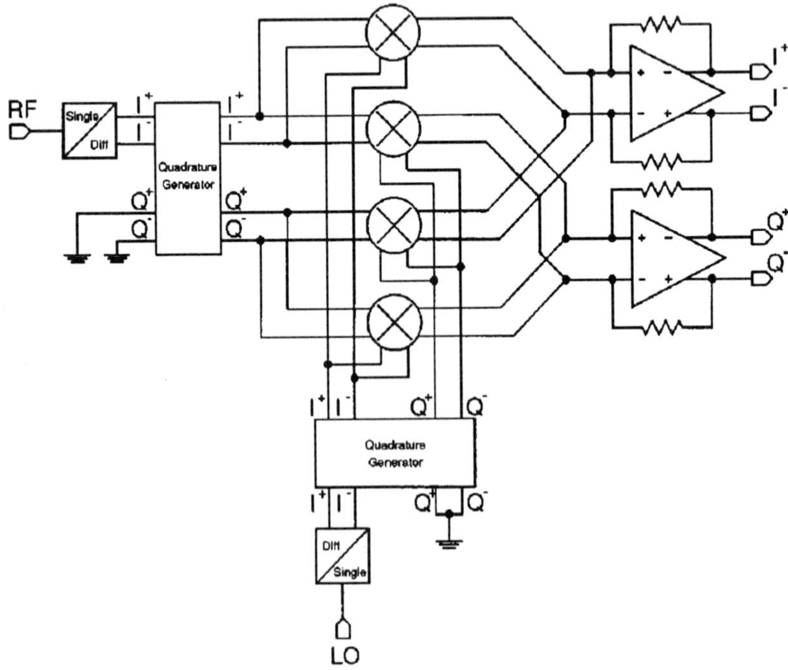

Figure 9.19 Circuit block diagram of the CMOS SSB down-converter; from Crols [174], © 1995 IEEE.

In [174], a SSB down-converter has been implemented in 0.7-μm CMOS that implements the part enclosed by the dashed line in Figure 9.17. A circuit block diagram is shown in Figure 9.19. The RF signal is converted to a differential signal via a single-to-differential conversion block. The differential RF signal is converted to a SSB signal via a passive polyphase filter. Likewise, a passive polyphase filter is used to convert the RF LO to a SSB tone centered at $-\omega_o + \omega_{IF}$. The four mixers effectively perform the complex multiplication of the SSB RF signal with the SSB LO to generate the down-converted complex IF signal. The mixer [192] is based on the linear passive mixer discussed in Chapter 7. Currents from the output of the mixers are summed appropriately via the summing amplifiers whose outputs are voltages representing the real and imaginary part of the IF signal.

The SSB down-converter chip does not provide the entire transceiver function. In particular, it contains no LNA, transmitter, or LO. More recently, a more integrated transceiver chip [269][270] for DCS-1800 has been reported which contains a SSB down converter with LNA, a direct-up conversion transmitter, AGC, and on-chip synthesizer. The performance of the CMOS transceivers discussed is summarized in Table 9.14.

Table 9.14 Summary of CMOS Dual-IF Transceiver IC's

Features	**900 MHz GSM (SSB)** [174]	**1800 MHz DCS (SSB)** [270]	**900 MHz GSM (DSB)** [266]
Technology	0.7 μm	0.25 μm	0.25 μm
Area	6 mm^2	15.4 mm^2	$2.2 \times 2.2\ mm^2$
Power	100 mA @ 5 V	95.5 mA @ 2 V (TX) 80 mA @ 2V (RX)	20 mA @ 2.5V (RX) 55 mA @ 2.5V (TX)
IF	100-300 kHz	100 kHz	71 MHz
Noise Figure	24 dB	8.2 dB (total) 4.7 dB (flicker noise)	8.1 dB (total) 3.2 dB (duplexer)
IIP3	27.9 dBm	-6.2 dBm	-4 dBm
Phase Matching	0.3 degrees	NA	< 3 degree
Amplitude Matching	0.5 dB	NA	< 0.3 dB
Image rejection (RX)	46 dB	32.2 dB	43 dB
Phase Noise	NA	-143.7 dBc/Hz (at 3 MHz offset)	<-138 dBc/Hz (at 3 MHz offset)
RMS Phase Accuracy	NA	NA	1.6 degrees
Image power (TX)	NA	-33 dBc	-31 dBc

The noise figure of the 900 MHz SSB chip is quite high (24 dB) since there is no LNA. Also, the passive polyphase filter, which converts the RF input from DSB to SSB, tends to be noisy due to the resistive network. While the receiver has a poor noise performance, it has a high IIP3 of 27.9 dBm. The good linearity illustrates the trade-off between noise performance and receiver linearity. With on-chip LNA and elimination of the input polyphase filter, the integrated DCS transceiver IC achieves a much lower noise figure of 8.2 dB but with poorer IIP3 of –6.2 dBm. The elimination of the input polyphase filter contributes to a lower image rejection ratio of 32.2 dB as compared with 46 dB achieved with the 900 MHz SSB chip.

When compared to the 900 MHz DSB transceiver, the DCS SSB transceiver shows worse noise and linearity performance since the IF is much closer to DC. Receiver degradation due to high blocking signals and noise that affect DCR is therefore similarly observed in the DCS IC. These degradations, on the other hand, are mitigated in the DSB IC with external filters and a higher IF of 71 MHz. While the SSB transceiver chips do not perform as well in terms of linearity and noise as compared to the more conventional DSB design, they do offer a means to achieve fully integrated solutions by eliminating the need for external filtering. Other SSB transceiver architectures are also possible and are exploited to achieve integrated solutions for GSM/DCS systems [271].

9.1.5 Digital Transceiver

Figure 9.20 shows a general architecture of a digital GSM transceiver with the shaded blacks representing the analog components needed to make a complete handset. These components include the RF front-end, ADC, DAC, crystal, regulators, speaker, and microphone. The RF front-end has been discussed in the previous section. The design of ADC's and DAC's has not been discussed in detail. Interested reader may want to refer to [272] for further reading on this topic. The crystal oscillator provides a clock reference to the digital transceiver as well as the synthesizer in the RF front-end. The power regulators provide stable temperature compensated voltage supplies to the IC's. Typically, the regulators regulate down the 6-9V output from the battery to 2-3 V required by the IC's.

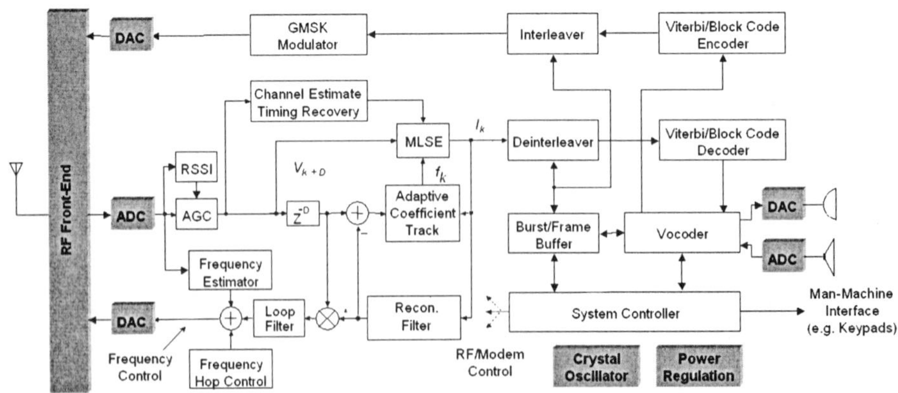

Figure 9.20 Architecture of a digital GSM transceiver.

The digital radio has traditionally been implemented in several chips, one for the analog-digital interface, one for the vocoder, one for the digital transceiver, and a few RF chips. Recently, two-chip solutions have become available [273] for the digital transceiver that implement the vocoder and transceiver on one chip and all the analog interfaces on the other chip. Highly integrated mixed-signal single-chip digital transceivers [274] include the complete digital transceiver functions shown in Figure 9.20 as well as the ADCs and DACs. The RF front-end, antenna, and crystal oscillator are still implemented as external components.

Since the GSM data rate of 270.8 kbps is relatively low, current digital transceivers are implemented with DSP cores accompanied by hardware accelerators for computation intensive functions such as Viterbi decoding. In this section, the transceiver architecture will be described in terms of functional as well as computational requirements, followed by a discussion of how these functions could be implemented.

9.1.5.1 Digital Transmitter

Voice is captured by the microphone and filtered to band limit the signal to 3.4 kHz. The filtered signal is then digitized at a sample rate of 8 kSps. With 13 bits per sample, the digitized source generates a bit stream at 104 kbps, stored in the frame buffers in 20 ms blocks. The vocoder compresses the block of 2080 bits by a factor of 8 using a RPE-LTP coder [259]. Each speech frames consisting of 260 bits is then encoded using unequal error protection in which the most important class Ia bits are encoded with an error-detecting block code for transmission while the least important class II bits are not encoded at all. In between, the class Ib bits are encoded using a rate 1/2 constraints five convolutional code. Figure 9.5 shows the unequal error protection scheme employed by GSM for speech data. Other TDMA channels are encoded differently but in this example the other encoding schemes will not be described. The following discussions on error correction for the speech channel easily apply to the other encoding schemes.

The (53,50) cyclic code used for the class Ia speech data has a generator polynomial shown below [72]

$$g(D) = D^3 + D + 1. \tag{9.18}$$

In Chapter 2, a systematic code generator has been derived based on (2.161) which requires the implementation of finding the remainder of $D^{N-K}u(D)$ divided by $g(D)$. The modulo operation can be implemented with the circuit shown in Figure 9.21. Intuitively, the circuit can be viewed as an infinite impulse response filter with an impulse response of $g^{-1}(D)$ and an input of $D^{N-K}u(D)$. The output response is

$$\frac{D^{N-K}u(D)}{g(D)}. \tag{9.19}$$

Given that a systematic code is generated, only the remainder polynomial is needed since its coefficients correspond to the parity check bits. The remainder polynomial is represented by the register content after *N-K* bits have been shifted into the coder.

Referring to Figure 9.21, with the control signal *s* set low, the first 50 bits are shifted into the linear feedback shift register (LFSR) and also to the output of the encoder to form the first 50 bits of the systematic code. Then, *s* goes high, opening the feedback in the LFSR and switching the LFSR output to the encoder output. In three clock cycles, the parity bits are shifted to the output completing the systematic codeword.

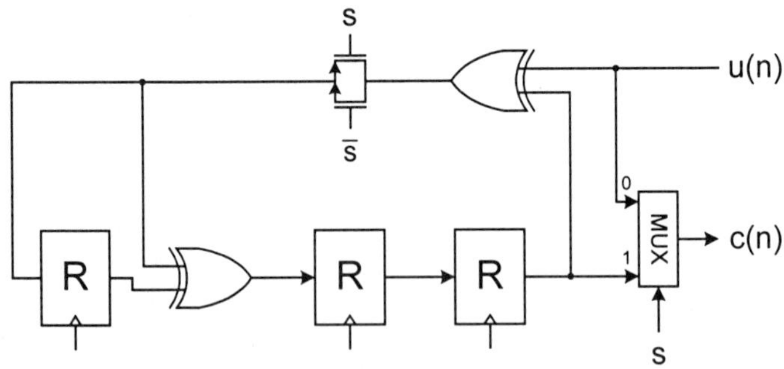

Figure 9.21 GSM (53,50) cyclic encoder for the Class Ia speech data.

Figure 9.22 shows the implementation of the convolutional encoder used to encode the encoded class Ia bits and the class Ib speech data. It has a rate of 1/2 and constraint length 5 with generator polynomials shown in Table 2.9.

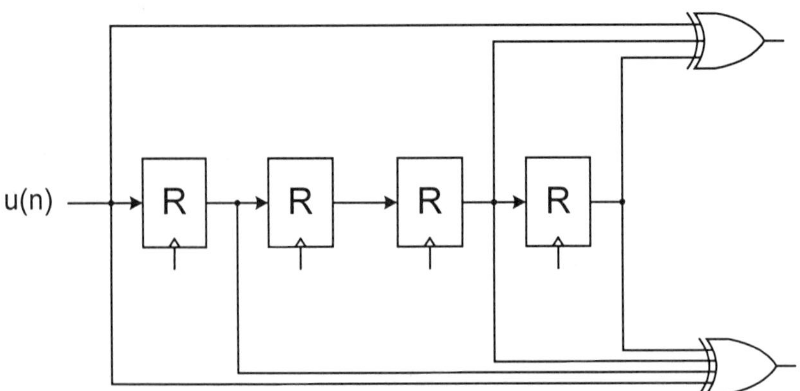

Figure 9.22 GSM rate1/2 constraint length five convolutional encoder.

The coded symbols and the class II bits form a frame of data with 456 bits. Block-diagonal interleaving and inter-burst interleaving are then performed on several of these frames as described in Section 9.1.3.3. The interleaver can be implemented by manipulating data and addresses to data within the frame buffer, which is essentially a block of memory. The memory may be logically partitioned into burst buffers in which TDMA bursts are formed and queued for transmission.

Though it is not shown in Figure 9.20, ciphering is performed after the interleaver and is described in [72]. Ciphering occurs at the coded symbol rate and involves scrambling the coded symbols with a keyed sequence which cannot be easily decoded without prior knowledge of the encryption key. The key processing and management is handled in the subscriber identity module (SIM) and the ciphering is implemented with bit-level logic gates, such as an XOR.

The ciphered data is then sent to a GMSK modulator. GSM uses a modulation index of 0.5 and a BT of 0.3. The corresponding Gaussian filter has an impulse response shown in Figure 9.23 which has a duration of approximately three bits. The filter impulse response peaks at about 100 kHz. If a constant stream of '1's is sent to the modulator, the filter response will rise and flatten out at about half the bit rate or 135.4 kbps. Given a modulation index of 0.5, the maximum frequency deviation is a quarter of the bitrate or 67.7 kHz, same as that of MSK.

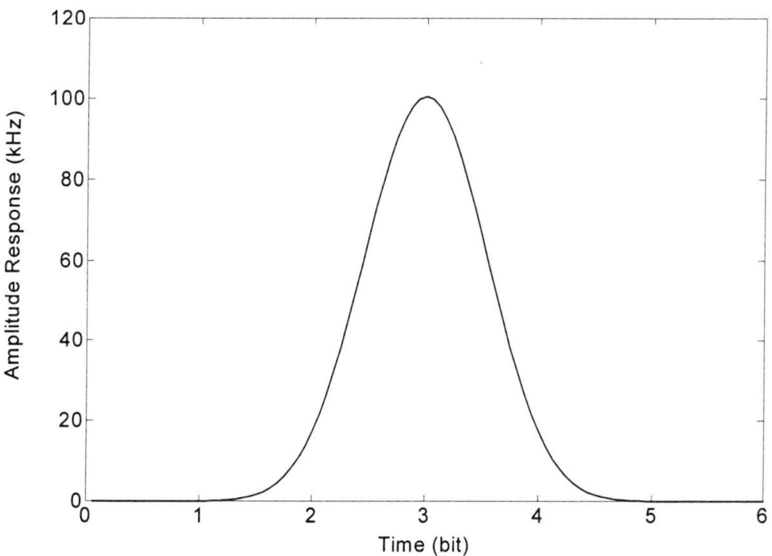

Figure 9.23 Gaussian pulse shape with BT of 0.3 and bitrate of 270.833 kbps.

The modulator can be implemented using a DDFS or a table lookup as shown in Figure 8.37. In the DDFS approach, the scale factor h is set to 0.5, which is a simple shift toward the LSB. The Gaussian filter could be implemented directly using an FIR filter with CSD coefficients to avoid the use of multipliers. The architectures discussed in Chapter 8 can be applied to implement this filter. The number of taps required depends on the over-sampling ratio which should be sufficiently large to minimize distortion in the modulated waveform. In Figure 9.23, the over-sampling ratio is chosen to be sixteen which results in a 48-tap FIR filter. Since only a binary alphabet is used, the symbol ROM can be replaced by a simple hardwired circuit that converts from a unipolar data sequence to NRZ format. The overall sampling rate is sixteen times the data rate or 4.33 MHz.

Alternatively, a table-lookup approach may be used to implement the modulator. In this case, the modulation can be implemented in the phase domain with a potential saving in hardware. Figure 9.24 shows the phase pulse corresponding to the Gaussian filter response shown in Figure 9.23. Note that the phase pulse has an approximate duration of two bits as compared with three bits needed by the associated frequency pulse. The required ROM size then reduces by a factor of two. With a modulation index of 0.5, there are four phase states. According to (8.25), the required ROM size for each of the sine and cosine ROMS becomes

$$K_{ov}\rho 2^L N_{ROM} = 16 \cdot 4 \cdot 2^2 \cdot N_{ROM}.$$

That is a $256 \times N_{ROM}$ ROM is required, where N_{ROM} is the bit resolution used to represent the modulated in-phase and quadrature waveforms.

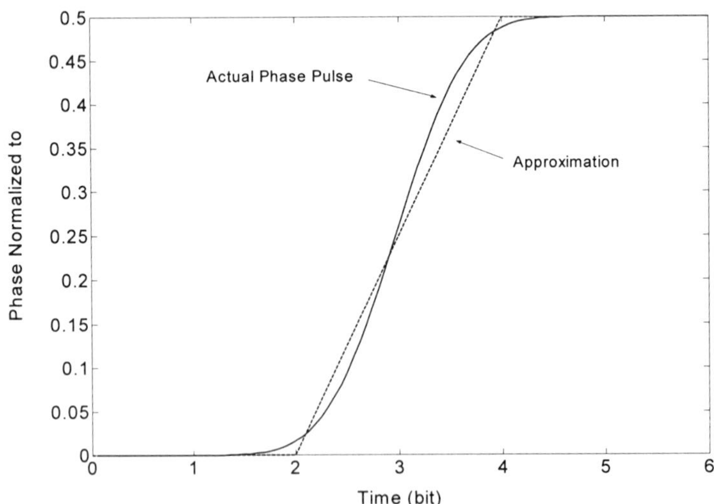

Figure 9.24 Phase pulse for BT = 0.3 and bitrate = 270.833 kbps.

The bit resolution should be chosen for adequate RMS phase error performance, specified to be 5 degrees. Figure 9.25 shows the RMS phase error as a function of the

ROM bit resolution. It appears that four bits is sufficient to meet this requirement. However, since other components in the transmitter in particular the synthesizer contribute significant phase jitters, more bits should be chosen to relax the constraint on the RF section. A reasonable choice would be greater than eight bits above which the RMS phase error becomes less than 0.25 degrees.

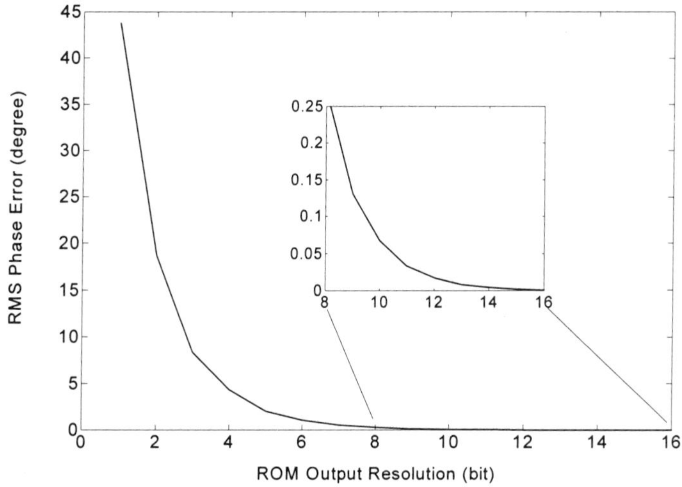

Figure 9.25 RMS phase error as a function of the ROM bit resolution.

With an over-sampling ratio of 16, the ROM must run at 4.33 MHz. However, it is possible to reduce the over-sampling ratio by 2-4 if an interpolator is used at the output of the ROM's.

Table 9.15 summarizes the computation complexity of the main components within the transceiver. Complexity is measured in terms of the number of operations, MOPS, and memory size. The types of operation are bit operation (Bit OP), memory access, addition, and multiplication. The complexity required to process one speech frame is determined for the (53, 50) encoder, Viterbi encoder, interleaver, and modulator. For the modulator, two approaches are considered, one based on DDFS and the other based on ROM table lookup.

For every bit processed, the (53, 50) encoder shown in Figure 9.21 requires three register operations, and two XOR's. Opening and closing the feedback in the LFSR and switching of the 2:1 MUX are done only once for every 50 bits. Therefore, their contributions to computation can be neglected. A total of approximately 250 (i.e. 5×50) operations are required for every speech frame. Although a register can be viewed as a memory element, it is tallied as a bit operation in the (53, 50) encoder since only a single bit is stored. Similarly, for each of the 182 information bits, seven bit operations are needed to implement the Viterbi encoder, resulting in a total of 1274 Bit OP's.

As shown in Figure 9.10, diagonal interleaving requires blocks from the $(n\text{-}1)^{\text{th}}$, n^{th}, and $(n+1)^{\text{th}}$ frames to form the nth interleaved frame. Using an 8-bit word, a 57×8 memory can be used to double buffer half of the 456-b encoded speech frame. Besides the 57×8 buffer, one more 57×8 memory block is used to store an interleaved TDMA burst and a 65×8 to double buffer the original speech frame.

Table 9.15 Transmitter complexity breakdown.

Transmitter Block		Number of operations (kOP/speech frame)				MOPS	Memory Size
		Bit OP	Memory	Add	Multiply		
Speech Buffering			0.033			0.002	65×8
(53,50) Encoder		0.25	0.007			0.17	3×1
Rate ½ Viterbi Encoder		1.27	0.062			2.47	4×1
Interleaver			0.114			0.114	114×8
Modulator (DDFS)	Filter		350.21	342.91	350.21	619.67	47×16
	DDFS		29.18	21.89	21.89	43.33	2×16 512×10
Modulator (Table Lookup)		29.18	21.89	7.30		34.64	512×10

The DDFS-based modulator requires a 48-tap FIR with 47 additions, 48 multiplications, and 47 registers. In this case, the registers are considered memory elements since each register holds a multi-bit signal. Depending on the implementation choice, the multiplication could be implemented with shift-and-add operations when the filter coefficients are represented using CSD. A factor of 16 must be accounted for when computing the number of operations for addition, multiplication, and memory read/write as shown in Table 9.15. The DDFS complexity can be inferred from Figure 8.16. Assuming that 10 bits are required at the output of the phase accumulator, then 256 entries are required in each ROM. Both ROM's then require a total of 512 memory locations. Conservatively, 16 bits are used to represent the frequency pulse amplitude. Moreover, two registers and one adder are required for the phase accumulator. Two additional adders and four XOR's are needed to compute the DDFS output. The XOR's in this case are considered as multiplication operations since each XOR is performing a $1\times N$ multiplication. The number of memory operations corresponds to the two register operations and the two ROM reads per DDFS output sample. For a speech frame, a total of $4\times16\times456=29184$ memory operations need to be processed by the DDFS. The number of other types of operations can be computed based on a similar reasoning.

In contrast to the DDFS-based approach, the table-lookup approach requires much lower complexity with only 4 bit operations for address generation and 512 memory locations to store the in-phase and quadrature components of the transmitted signal. However, if the correlation span is large, the DDFS approach may be more desirable since the complexity of the Gaussian filter grows linearly with L whereas the lookup table size grows exponentially with L.

The required MOPS depends on the schedule used to execute the transmitter operations. Figure 9.26 shows one of many schedules that could be used. To reduce the overall one-way speech delay to about 60 ms, the encoding and interleaving need to be accelerated as shown in Figure 9.26. During each speech frame, the speech data is double buffered, where one of the buffers stores the current vocoder output while the other reads out the previous vocoder frame for processing by the FEC coder. Having two buffers allows the vocoder to generate data without interruption as the previous data is being read from the other buffer for channel encoding. Each buffer exchanges roles every 20 ms. Therefore, storing 260 bits of speech data alone requires 0.002 MOPS which is derived by dividing the corresponding number of memory operations listed in Table 9.15 by 20 ms.

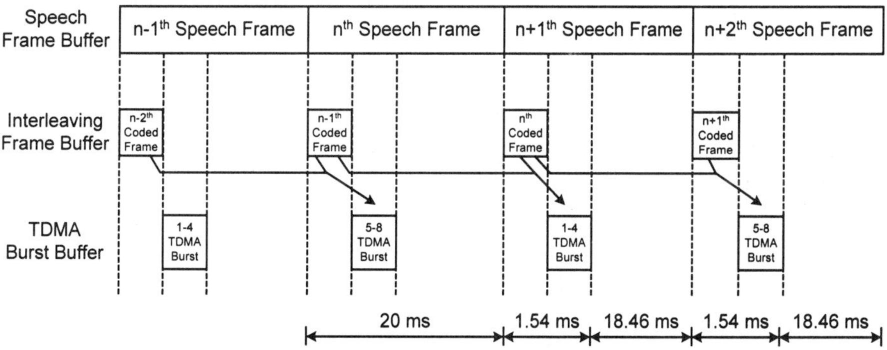

Figure 9.26 Resource scheduling in forming the TDMA burst.

Since half of the previous coded frames are needed for the interleaving as depicted in Figure 9.10, the coded data are double buffered in a similar fashion as for the uncoded speech frame. The interleaved frame is stored in a TDMA burst buffer as shown in Figure 9.26. In the encoding process, bits are read out from the speech frame buffer and encoded by the (53, 50) and Viterbi encoders. The latter half of the encoded bits are written into buffer *A* and the first half of the encoded bits are interleaved with the bits stored in buffer *B*. In the next frame, the roles of buffers *A* and *B* are reversed, implementing the double buffer for half of the encoded speech frame mentioned earlier. The interleaved bits are written into the TDMA burst buffer in the interleaved format. The interleaving process should be done fast enough such that the Interleaved data is stored in the TDMA burst ready for transmission in the next four TDMA frames. Since each TDMA frame occupies 4.615 ms, the amount of time allotted to perform the channel coding and interleaving is approximately 20 – 18.46 = 1.54 ms as shown in Figure 9.26.

If 1.54 ms is allocated to channel coding and interleaving, then 2.76 MOPS is required to read 260 bits from the speech buffer, block encode the 50 class I bits, convolutional encode the class I bits, and interleave the encoded bits. The MOPS estimate assumes that coding and interleaving are processed in a streaming fashion whereby as bits get processed they are immediately passed down to the next processing entity. For example, as bits are read out from the speech buffer they are sent to the encoders which produce bits with a negligible latency. The encoded bits are immediately written to the half-frame double buffer in the interleaved order and as group of interleaved bytes become available they are written into the TDMA burst buffer.

Note the above MOPS estimates assume that each memory access involves the transfer of one byte of data, which ignores the bit operations needed to generate the byte of data. Interleaving, in particular, requires a notable amount of bit operations as discussed in Section 8.7.1. Also, the MOPS rating for bit operations tends to be misleading since a high MOPS rating for bit operations does not imply high computation requirement. For example, a toggle flip-flop operating at 500 MHz translates to 500 MOPS according to the above definition. Therefore, MOPS rating for bit operations must be interpreted carefully on a case-by-case basis.

The buffered TDMA bursts are sent out once every TDMA frame at a burst rate of $270.8\overline{33}$ kbps. The DDFS-based modulator requires significant amount of MOPS, particularly for the Gaussian pulse filter. Assuming an over-sampling ratio of 16, i.e. a sampling rate of $4.\overline{33}$ MHz, the filter requires $143 \times 4.\overline{33} = 619.67\ MOPS$ where the value 143 corresponds to the total number of operations executed per sample and can be derived by taking the total number of operations per speech frame divided by 16*456. Using the data in Table 9.15 in a similar fashion for the DDFS, its complexity can be computed to be $10 \times 4.333 = 43.33\ MOPS$. The total computational complexity then becomes 663 MOPS.

In contrast to the DDFS-based implementation, the table-lookup approach shows substantial saving in computation complexity. While 512 ROM entries are needed to store all the possible transmitted waveforms, only two entries need to be accessed per sample time. Together with the operations required for address generation, a total of eight operations are needed per sample, resulting in a computation requirement of $8 \times 4.33 = 34.64\ MOPS$.

It should be noted that the complexity metrics listed in Table 9.15 depend on implementation. Take for instance, the Gaussian filter. Its filter coefficients could be implemented without any multipliers if CSD representation is used to allow equivalent implementation based only on shifts and adds. A CSD-based implementation can be efficiently implemented in a dedicated architecture, some of which have been discussed in Section 8.2.6. On the other hand, with a programmable DSP, it is generally more efficient to simply use the multiply and accumulate unit, where the coefficients are implemented with real multipliers. Therefore, while informative, these computation metrics must be interpreted carefully with respect to the final implementation choice.

9.1.5.2 Digital Receiver

As shown in Figure 9.20, the down-converted RF signal is digitized with a pair of ADC's assuming quadrature down conversion. A MLSE then demodulates the digital signal to produce an output bit sequence that is placed in the proper order by the de-interleaver. The de-interleaved sequence is decoded by the Viterbi/block code decoder. The decoded data finally reaches the vocoder which reconstructs the speech signal. Since the training sequence resides in the middle of the TDMA burst, temporary frame storage is required to obtain the channel estimates.

The MLSE requires a few auxiliary blocks including an automatic frequency control (AFC) loop, AGC loop, and adaptive coefficient tracking (ACT) loop. The MLSE performance is sensitive to frequency drift and amplitude variations since the path metric is derived from the sum of Euclidean distances between the reconstructed signal and the actual received signal. Any frequency or amplitude error causes a large deviation in the Euclidean distance even for the correct path in the trellis. The AFC loop minimizes the receiver degradation due to frequency offsets by estimating and tracking the offsets due to oscillator drift and Doppler shift. The AGC loop minimizes the amplitude variations in the received signal by tracking its amplitude variations. The ACT loop tracks the time variation of the channel impulse response due to mobility or other time varying effects.

The AFC first estimates the frequency offset which could be relatively large due to inaccuracy of the oscillators, in particular those in the mobile handsets. Typical oscillator frequency stability could be 5-10 ppm or even 30 ppm for low-cost crystals. To estimate the frequency, the receiver tunes to the FCCH channel which is broadcast by the basestation every 10 TDMA frames or 46.15 ms as shown in Figure 9.3. The FCCH channel contains an unmodulated carrier at the maximum frequency deviation of either $f_0 + 0.25R_b$ or $f_0 - 0.25R_b$, corresponding to a data sequence of all zeros or all ones, respectively. The FCCH also serves the important function of allowing the mobile unit to detect the nearest basestation when the mobile unit attempts to enter into the network. By detecting the FCCH channel, the mobile unit discovers the timing structure of the multiframe structure shown in Figure 9.3. It can then synchronize on to the network by using the SCH frame which occurs immediately after the FCCH frame.

At baseband the FCCH channel exhibits a tone frequency at $\pm 0.25R_b$. Realizing this fact, a frequency estimator shown in Figure 8.47 can be used to detect the FCCH channel as well as to estimate its frequency along with any constant frequency offset. The frequency estimator shown in Figure 8.47 utilizes a shift register to introduce a delay of W clock cycles corresponding to the period of the preamble for estimation of frequency offsets in the presence of frequency selective fading. Since GSM does not use a PN code for the FCCH channel, the estimator offers no performance gain against frequency selective fading. However, it can still be used with $W = 1$ for the highest accuracy. The output of the estimator is decimated down to the symbol rate and can be processed to determine its average and variance. If an FCCH channel has been received then the variance output should be close to zero and the average output corresponds to the frequency offset estimate. On the other hand, if an FCCH channel has not been received then the variance would be high. To distinguish the two cases, a comparator can be used with a threshold that could be determined with Monte Carlo simulation.

The frequency estimator performs the FCCH detection and frequency estimation. As shown in Figure 9.20, the estimator output is added to the frequency hop control word and the AFC tracking loop output. The sum is converted to an analog signal applied to the VCO in the RF front-end for frequency correction. The AFC tracking loop output corresponds to the filtered error signal formed by taking the phase difference between the received signal and a reconstructed signal based on the decoded MLSE output. The reconstruction filter uses the estimated channel response as its coefficients. The '*' denotes complex conjugate and the complex multiplication has a cross-coupled structure similar to that shown in Figure 8.47. To time align the data sample at the complex multiplier, the received signal is delayed by *D*, which corresponds to the decoding delay of the Viterbi processor in the MLSE.

Immediately after the FCH frame, a SCH frame is transmitted by the basestation. The proximity of the FCH and SCH channels allows for frame synchronization once the FCH channel has been detected. The channel estimator determines the channel response based on a 64-b training sequence embedded in the SCH channel. The channel estimator consists of a complex matched filter shown in Figure 8.51. Since the GMSK signal used in GSM can be approximated by an offset quadrature amplitude modulation, similar to OQPSK, the complex matched filter provides an efficient way to estimate the channel response as discussed in Section 8.6.2.2. Moreover, the filter coefficients are $\pm 1 \pm j$ and therefore can be implemented without real multipliers.

Since the channel experiences time variation due to Doppler spread, the coefficients also vary with time and should be tracked. The ACT loop shown in Figure 9.20 performs the tracking and consists of an adaptive coefficient track block that employs the LMS algorithm to update the channel coefficients $\{f_k\}$. The LMS algorithm uses the error signal generated by taking the difference between the received signal and the signal derived from the reconstruction filter.

The channel estimation process also allows the receiver to synchronize in time to the received symbols. This is accomplished by finding the area around the peak of the impulse response. In particular, if six tap coefficients are used to model the channel, then the channel estimator would find the six taps with the highest powers that are around the peak. Since the receiver may be sampled at the minimum rate of two times per symbol, the resolution in channel estimation may not be sufficient. Therefore, interpolation may be applied to increase the resolution needed to achieve symbol timing.

The channel estimate and received signal still have a residual frequency offset due to inaccuracy in the frequency estimator. The residual offset is tracked out when the ACT loop and the AFC loop are activated. By selecting the appropriate loop filter coefficients (Section 2.5.2) and LMS step size (Section 2.7.7.3), the time needed to reach steady state can be minimized for a given steady-state error.

The AGC loop adjusts the variable gain amplifier in the RF front-end to maintain the received signal power at a constant level. The feedback loop relies on the RSSI measurement which can be implemented using the circuit shown in Figure 8.67. Long term averaging on the RSSI measurement is used to provide a stable gain control. However, if a sharp amplitude change occurs during the training sequence then the TDMA frame will most likely be lost.

It can be shown that GMSK can be approximated by staggered QAM as shown below

$$\begin{aligned} GMSK\ signal &= \cos(\phi_k) + j\sin(\phi_k) \\ &\approx q_k + jq_{k-D} \end{aligned} \tag{9.20}$$

where q_k is the kth sample of an amplitude pulse shape and D is a time delay [275][276][277]. For MSK, (9.20) offers an exact representation with $D = 0.5T_s$ and q_k being a half cycle cosine pulse shape [277].

With a staggered QAM approximation, the path metric computation simplifies from the one shown in
Figure 8.38b to that shown in Figure 8.50. Also, the time dispersion due to multipath and pulse shaping q_k can be combined into an equivalent response h_k that can be estimated with the channel estimator shown in Figure 8.51. Given the estimated channel coefficients, the branch metric computation unit generates the branch metrics used by the Viterbi processor to determine the optimum received sequence. Numerous architectures are available to implement a Viterbi processor, some of which have been described in Section 8.7.3.

The complexity of a Viterbi decoder depends on the number of states, which in this case, depends on the length of the equivalent channel response L. Most designs typically use L = 6. Intuitively, this choice results from the fact that the pulse shape q_k has approximately the duration of three bit times, which leaves a 4-bit margin for multipath delay spread. The 4-bit margin corresponds to approximately 15 μs that meets most of the delay spreads associated with the GSM channel profiles shown in Table 9.2. A 6-tap response can be equalized using a 32-state MLSE.

The demodulated bits from the MLSE are de-interleaved and then decoded first with the 16-state rate ½ convolutional decoder. If hard decoding is used, then simple bit processing can be used to generate the branch metrics as shown in Figure 8.61a. The branch metrics are fed into the Viterbi decoder that can be implemented with the architectures discussed in Section 8.7.3. The trellis diagram for the decoder is shown in Figure 9.27. The decoded bits for the four transitions associated with states [0000] and [0001] are annotated along the associated state transitions in Figure 9.27. The rest of the transitions can be determined by inspection of the encoder diagram shown in Figure 9.22.

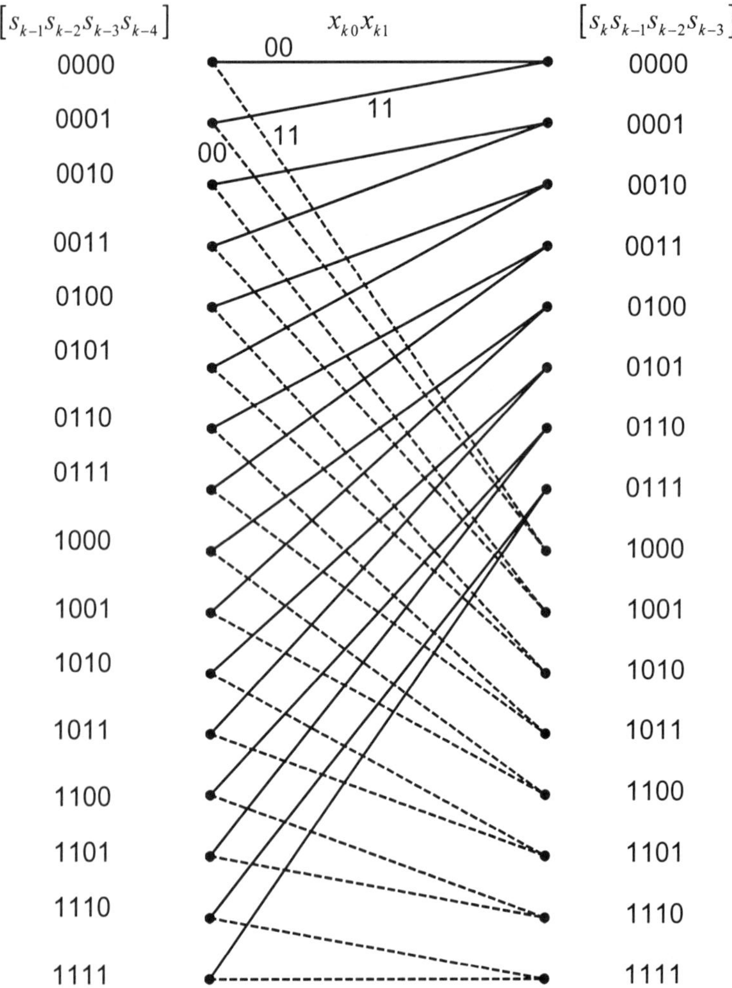

Figure 9.27 Trellis diagram of a 16-state convolutional decoder. Dashed and solid lines indicate an information bit of '1' or '0'.

The parity-check bits of the (53, 50) block code are computed by the decoder shown in Figure 9.28. The decoder has an implementation similar to that of the encoder shown in Figure 9.21 but with an additional parity-check circuit at its output to compare the computed parity-check with the received parity-check bits. If the parity check shows no match, then a frame erasure is declared. If the parity check shows a match, then the decoded frame is passed onto the vocoder to reconstruct the speech signal.

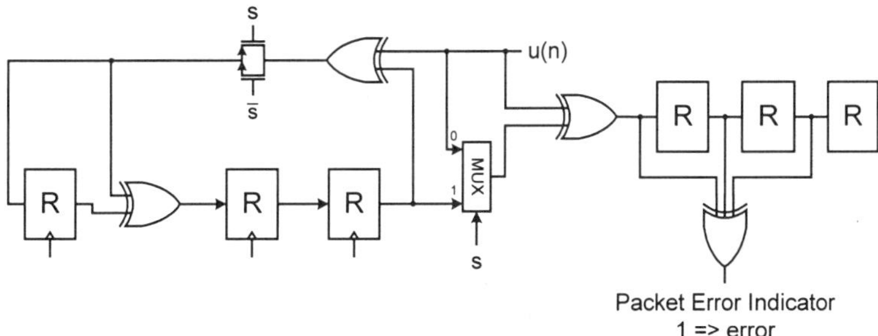

Figure 9.28 Decoder for the (53, 50) error-detecting code.

Table 9.16 summarizes the computation complexity of the digital GSM receiver. The computations are categorized in the same manner as Table 9.15 and can be determined by referring to the associated architecture block diagrams. The 32-state MLSE dominates the complexity of the receiver requiring a total of 81,790 operations and 141.75 MOPS. The supporting tracking loops, channel estimator, and signal reconstruction requires a total of 24,140 operations and 51.17 MOPS. The 7.63 MOPS needed for AFC is not counted since it is not incurred for each TDMA burst but rather it occurs at a lower duty cycle. The convolutional decoding is the second largest in computation complexity, requiring a total of 36,530 operations and 76.59 MOPS. Buffering and de-interleaving require the least computation. In terms of memory requirements, the traceback operation in both the MLSE and convolutional decoder requires the highest memory storage, on the order of 3-4 Kbytes.

Table 9.16 Digital receiver complexity breakdown.

Receiver Block		Number of operations (kOP/speech frame)				MOPS	Memory Size
		Bit OP	Memory	Add	Multiply		
TDMA Burst Buffering			0.16			0.28	156×8
(53,50) Decoder		0.37				3.7	6×1
Rate ½ Convolutional Code	BMU		3.02	9.07		25.35	32×2
	Viterbi Processor		9.13	12.10		44.51	16×8
	Traceback		3.21			6.73	3024×4
De-interleaver			0.11			0.19	171×8
Signal Reconstruction			3.74	4.06	3.74	20.00	20×8
Channel Estimator			0.65	0.68	0.62	20.33	26×8
ACT Loop			2.18	2.18	0.94	9.19	14×8
AFC Loop			1.59	1.25	1.56	7.63	2714×8
AGC Loop			0.17	0.47	0.31	1.65	4×8
MLSE	BMU		14.85	22.27	14.85	90.07	128×8
	Viterbi Processor		11.14	14.85		45.04	32×8
	Traceback		3.83			6.64	3712×5

The MOPS requirement is determined based on the task scheduling shown in Figure 9.29, which is one of many possible schedules that can be used. Due to the interleaving delay, data for the nth frame is not available until after approximately eight TDMA frames. In each TDMA frame, the received data is buffered such that the MLSE can decode the TDMA slot as a complete block. The channel estimate can be done while the data is being stored into the 156×8 buffer which has a wordlength of one byte assuming that the input wordlength is 8-b wide. The stored input is read out and decoded by the MLSE decoder and the results written into the de-interleaver buffer. In the eighth TDMA frame when all 456 bits have been demodulated, the de-interleaver reads out from its buffer and de-interleaves the bits into their original order. It may be possible to eliminate buffers needed for error decoding by streaming the output of the de-interleaver into the error decoder. For error decoding, 0.1 ms is allotted for the (53, 50) error decoder and 0.477 ms for the Viterbi decoding. The decoded outputs may be streamed into the vocoder to generate the reconstructed speech signal.

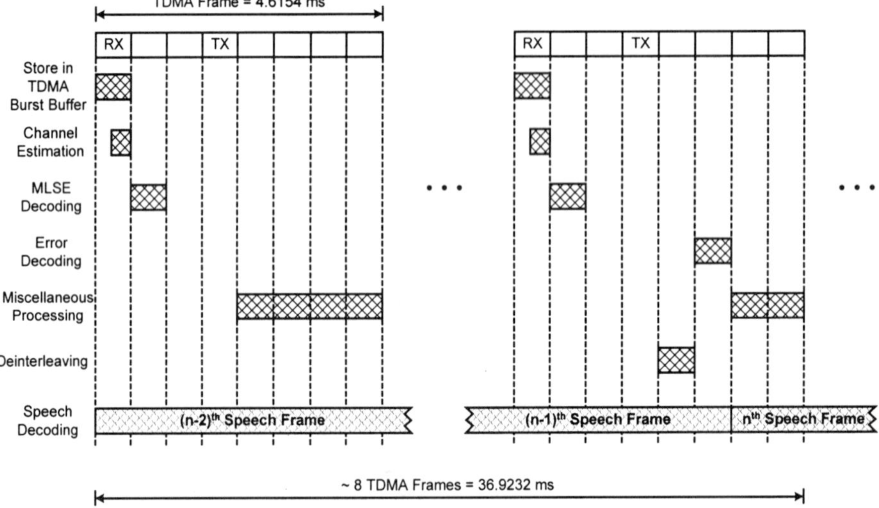

Figure 9.29 Resource scheduling in the reception of a TDMA burst.

It should be noted that the MOPS requirement for the MLSE and Viterbi error decoding can be reduced by 2-4 times if additional time is allocated from other tasks. For instance, the third time slot in each TDMA frame is not used. Therefore, both the second and third time slots may be allocated for MLSE decoding and reducing the MOPS requirement from 141.75 MOPS to 70.88 MOPS. Similarly, the times used for miscellaneous tasks may be borrowed by the more computationally intensive tasks. However, borrowing time from other tasks may not always be possible since during those times the receiver may be performing system maintenance functions such as monitoring the RSSI from neighboring. However, the loading on the computationally intensive tasks can still be balanced with the tasks that require much lower processing power.

9.1.5.3 DSP-based Implementation

The serial and parallel architectures discussed in Section 8.1.5 are the two primary approaches to implement the digital computations required by the digital transceiver. The parallel approach requires higher complexity in terms of hardware resources but meets the high MOPS requirement. Moreover, if the supply voltage is scaled down the parallel approach can potentially achieve a low power solution at a scaled down clock rate as discussed in Section 8.1.7. On the other hand, the serial approach requires fewer hardware resources but at the sacrifice of higher speed operation needed to achieve the same computation power. Usually, voltage cannot be scaled down due to the high-speed requirement. Therefore, it does not provide a means for low power operation.

Two other variants of the transceiver architecture should also be considered. These are the programmable processor and dedicated datapath architecture discussed in Section 8.1.6. Programmable processors provide flexibility to perform multiple functions and/or different parameters of the same architecture; for example, the ability to process different numbers of states for the Viterbi processor. The programmability is obtained by time sharing a processing element (PE) that is shared by sequencing of a series of computation tasks. Depending on the computation, the sequence of tasks can be programmed into memory and executed at run time. The overhead required to program the controls to the PE reduces the energy efficiency as well as the computation efficiency. The Dedicated datapath architecture, on the other hand, uses dedicated hardware resources that cannot be programmed but can achieve higher computation and energy efficiency.

Since the bitrate is relatively low in GSM, most current digital transceivers are implemented with programmable digital signal processors (DSP) for ease of upgradability via software. Dedicated architectures are of course also possible and examples have been discussed in the previous section. In this section, DSP-based approaches are described.

DSP-based designs can be grouped into two general architectures as shown in Figure 9.30. One or more PE's form the central computation engine in the DSP. The PE shares data via the shared buses and memory. The Von Neumann architecture contains one address bus, data bus, and memory space while the Harvard architecture uses separate memories for the data and program. The PE implements operation primitives that form the basis for the instruction set of the DSP. Large complex transceiver functions can be implemented using the DSP by programming it using its instruction set. The instruction decoder in the DSP sequences through the instructions stored in memory and decodes the instructions to execute the corresponding hardware resources in the PE, which might contain multiply-and-accumulate (MAC) units, arithmetic logic units (ALU), shift registers, and temporary registers. A DSP based on the Harvard architecture with a single-MAC PE is shown in Figure 9.31.

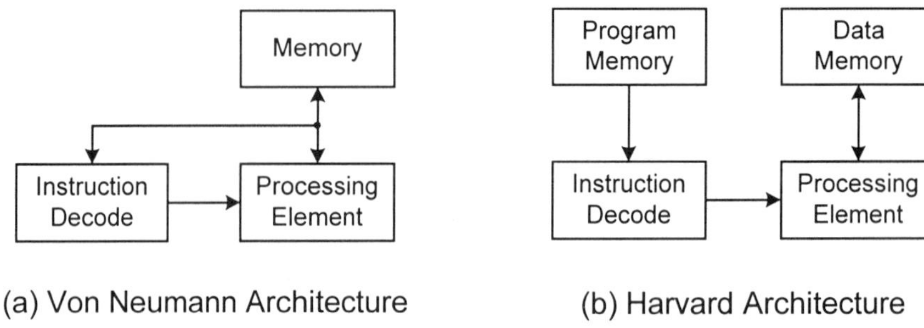

Figure 9.30 DSP architectures.

Table 9.17 summarizes commercial DSP's developed by Texas Instruments since the early 1980's. A clear trend is the move to reduce the instruction cycle needed to perform a given function. To accomplish this, current DSP's employ the following advanced features

- PE's containing parallel hardware resources such as a dual MAC or multiple processor cores (e.g. TMS320C55X and TMS320C602X).
- Hardware accelerators such as ACS units to enable ACS computation in a few instruction cycles (e.g. TMS320C54X)
- Larger bus width from 16 bits to 32 bits that enables fewer instructions to read/write from memory (e.g. TMS320C55X).
- Faster bus and instruction cycle from 10's MHz to 100's MHz.
- Parallel processing to allow execution of multiple PE's (e.g. TMS320C54X and TMS320C602X).

Efficient parallel processing has been the primary focus for current high-performance DSP designs that achieve greater than 1 GOPS. The two basic techniques employed to achieve the high computation power are the single instruction multiple data (SIMD) and the multiple instructions multiple data (MIMD) architectures. The former is most suitable for vector operations such as filtering while the latter is most suitable for less homogeneous functions such as Viterbi processing. TMSC620X is one example of an MIMD processor that uses very long instruction word (VLIW) to allow multiple instructions to be decoded and executed on parallel processing elements in one instruction cycle of 4-5 ns duration.

Although only the TI DSP's have been listed in Table 9.17, the general trend applies to other DSP's. For instance, the DSP16000 series developed by Lucent and Lode DSP core by Atmel employ dual MAC's and achieve similar performance as the TMS320C54X series from TI.

DSP's are generally implemented with the complex instruction set computer (CISC) architecture. Therefore, each instruction represents multiple operations. For instance, two memory reads, two address generations, and one arithmetic operation could all be done in one instruction cycle. Therefore, one MIPS corresponds to several MOPS.

Referring to Table 9.15 and Table 9.16, the receiver clearly dominates the transceiver computation requirement and requires approximately 281 MOPS, excluding speech processing. In contrast, the transmitter requires only approximately 35 MOPS given the ROM-based implementation. Assuming four operations per instruction, the entire transceiver could be implemented in 79 MIPS. A GSM vocoder can be implemented with about 10-20 MIPS. The transceiver then requires approximately a total of 90-100 MIPS, well within the computation power of current programmable technology.

Table 9.17 Texas Instruments DSP summary.

Commercial DSP	CMOS Technology	MIPS	mW/MIPS	General Features
TMS320-1X	NA	5	20-40	• One data/program bus (16 b) • Single MAC • 4 kb on-chip RAM • 24 – 128 kb on-chip ROM
TMS320C2X	NA	10-40	3.6	• Separate data and program busses (16 b) • Single MAC • 72 kb on-chip RAM • 512 kb on-chip ROM
TMS320C54X	0.15-0.25 μm	30-200	0.54 – 1.5	• Eight 16-b busses • Multiply, ALU, dual ACC, ACS HW accelerator • Single/multiple processor cores • 008 - 4Mb on-chip RAM • 0.03 – 0.8 Mb on-chip ROM
TMS320C55X	NA	140-800	0.05	• Multiple 32-b busses • Dual MAC, dual ALU, quad ACC • ACS HW accelerator • Variable 8-48b instructions
TMS320C620X	0.15-0.25 μm	1200-2400	0.5	• 256-b instruction words • Dual MAC and 6 ALU • 8 instructions/cycle • 1-7 Mb on-chip RAM

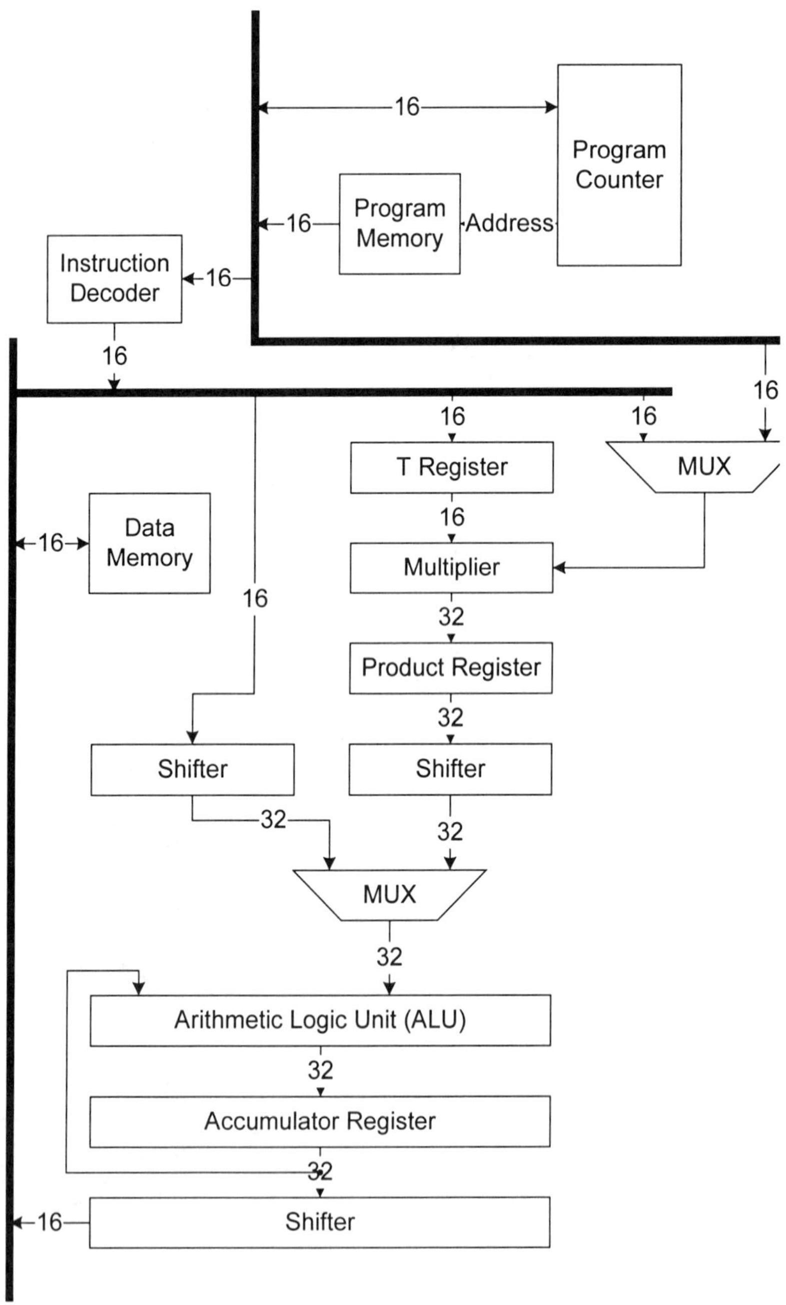

Figure 9.31 DSP architecture with a single-MAC PE.

9.2 Other Examples

The case study on GSM radio IC design just discussed in the previous section serves to illustrate a complete top-down radio design that brings together various system, architecture, and circuit design trade-offs described in Chapter 2 through Chapter 8. In this section, other examples will be discussed to provide further illustrations of integrated radio IC design. However, due to length constraints, these examples will be discussed in less detail. Numerous references are provided for further reading.

9.2.1 Digital Enhanced Cordless Telecommunications

DECT is another European standard that is based on TDMA/FDMA. It has many similarities with GSM but differs in the system application which, in the case of DECT, is targeted primarily for short-range digital cordless and private branch exchange (PBX). Due to the short-range requirement and the local connectivity (i.e. low capacity), DECT has a looser set of radio constraints as compared to GSM. The less stringent requirement also allows DECT to operate at higher data rates without the need for sophisticated equalizations needed in GSM. Table 9.18 lists some of the key features of DECT.

Table 9.18 Summary of DECT system parameters.

System Parameters	Values
Downlink frequency band	1880-1900 MHz
Up-link frequency band	1880-1900 MHz
Duplexing	TDD
Carrier spacing	1728 kHz
Number of time slots	12
Modulation	GMSK (BT = 0.5)
Channel rate	1152 kbps
Speech data rate	32 kbps
Maximum transmit power (mobile)	24 dBm

The total available bandwidth is 20 MHz and since TDD is used the up-link and downlink frequency bands are the same. In contrast to GSM, TDD is used during each time slot of duration 10 ms. The receiver and transmitter each uses half of the time slot. Given that DECT is intended for local connectivity in quasi-stationary and indoor environments, TDD can be used and still achieve adequate performance. The total number channels within the 20 MHz band is 120 which is about ten times lower than GSM. The cordless and PBX applications do not have a high capacity requirement like GSM. A lower capacity requirement loosens other constraints, such as the source data rate and bandwidth efficiency. DECT's source rate is about three times higher than that of GSM and bandwidth efficiency two times lower. GMSK is also used in DECT to enable non-linear amplification and simple discriminator type demodulators that help to reduce the cost of the transceiver.

Commercial DECT chip sets are highly integrated. Most are still based on bipolar technology [278][279][280]. Chip sets based on BiCMOS and SiGe have also been reported [281][283]. DECT transceivers have also been implemented in CMOS. In particular, [284] reports a transceiver based on the wideband IF architecture discussed earlier in Chapter 6.

Figure 9.32 shows the wideband IF architecture. An image reject mixer is used to convert the RF signal down to IF, in this case, 181-197 MHz. Therefore, LO1 tunes around 1.7 GHz and LO2 tunes around 181-197 MHz. The output of the first image reject mixer is lowpass filtered and the filtered output is further down converted to baseband using a complex image reject mixer. The channel select filter following the image reject mixer eliminates signal powers from undesired channels prior to digitizing the received signal with the ADC.

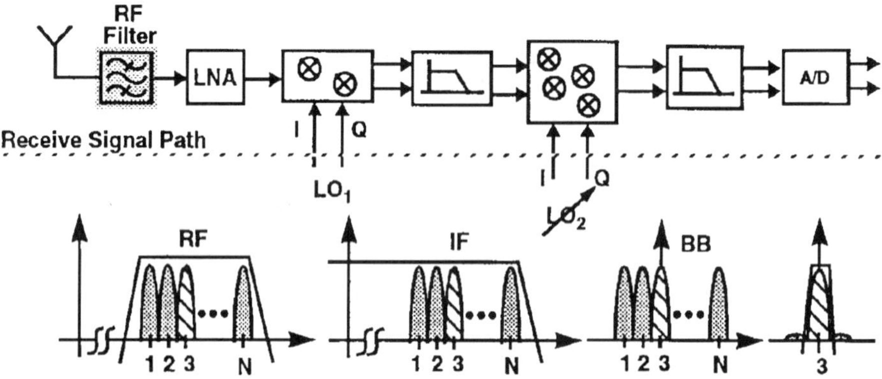

Figure 9.32 Wideband-IF receiver architecture; from Rudell [284], © 1997 IEEE.

The wideband-IF architecture is similar to the single sideband low-IF architecture discussed in 9.1.4.2 and shown in Figure 9.17. The main differences between the two designs are that the IF frequency is relatively higher for the wideband IF architecture and that the IF processing is done in analog rather than digitally. Like the low-IF architecture, the wideband-IF architecture offers immunity to DC offsets and high levels of integration. A single-chip receiver IC has been implemented in 0.6-μm CMOS with performance shown in Table 9.19.

The receiver is fully integrated with the exception of the frequency synthesizer. The sensitivity is 7 dB above that required by DECT. Using (5.40), the NF is approximately 13 dB which is less than the 19 dB required by DECT. The IIP3 and 1-dB input compression point are higher than the –26 dBm and –33 dBm required by DECT, respectively. The short range and low capacity requirement have lowered the linearity requirements in contrast to that of GSM.

Table 9.19 Summary of wideband-IF receiver performance.

Chip Features	Performance
Technology	0.6-μm CMOS
Area	7.5 x 6.5 mm^2
Supply	3.3 V
Power consumption	198 mW
Sensitivity	-90 dBm
IIP3	-7 dBm
1-dB ICP	-24 dBm
Image rejection	-55 dBc

9.2.2 900-MHz ISM-Band Indoor PCS

This section describes a frequency-hop radio transceiver intended for operation in the 900 MHz Industrial Scientific Medical (ISM) band. The FCC allocated three ISM bands for license-free civilian spread-spectrum transmission. These bands comprise of 902-928 MHz, 2.4-2.4835 GHz, and 5.725-5.825 GHz. FCC specified minimal restrictions to encourage broad use and deployment of diverse wireless systems within these bands. To transmit in the ISM bands, the FCC requires spread-spectrum modulation, direct-sequence and/or frequency hop, to minimize the power spectral density of transmissions in the ISM bands. Table 9.20 summarizes the FCC specifications for the 900-MHz ISM band using frequency-hop spread spectrum.

Table 9.20 Summary of 900-MHz ISM-band specification for FHSS.

System Parameters	Values
Frequency band	902-928 MHz
Peak power	1 W
Out-of-band emission	50 dBc
Maximum bandwidth per hop (BW)	500 kHz
Minimum hop spacing	Min(25 kHz, BW)
Minimum hopping set	50 frequencies
Minimum hop rate	2.5 hops/sec

Many of the earlier systems were implemented for the 900 MHz bands. Progressively, more systems are being implemented in the higher frequency bands to obtain more bandwidth as well as to avoid congestion as more systems deploy in these bands. As the 900 MHz band became crowded with digital cordless and WLAN transmissions, more systems are being implemented in the 2.4 GHz band, in particular the IEEE 802.11 WLAN and others that need more bandwidth. One such system claims a bitrate in excess of 64 Mbps using slow frequency-hop in the 2.4 GHz band [285][286][287][288]. This system exploits spatial diversity via digital beamforming to achieve the high bitrate. A CMOS chip-set has been implemented and reported in [289][290][291]. Currently, the wireless industry is marching into the 5 GHz band to develop systems that are also broadband but based mainly on the IEEE 802.11a and HiperLAN WLAN standards, which support bitrates up to 54 Mbps.

This section focuses on one of the earlier systems implemented for the 900-MHz ISM band. The system is intended to implement an indoor personal communications system (PCS), similar to a digital cordless system like DECT. Since the application again is targeted for short-range and low capacity, the system constraints are less stringent than that of GSM. Similar to DECT, this system uses TDD and has a low bandwidth efficiency of 0.5 bps. The poor bandwidth efficiency arises from the use of FSK, which as one might recall from Chapter 2 has a bandwidth efficiency which degrades inversely to the constellation size. The rationale behind using FSK is to achieve long battery life at the sacrifice of bandwidth efficiency. Due to the low capacity requirement, the trade-off of less bandwidth efficiency for more energy efficiency seems reasonable. In fact, for similar reasons, Bluetooth also uses a less bandwidth efficient modulation (Gaussian frequency shift keying) to achieve low power operation.

Frequency hop spread-spectrum has been used instead of direct-sequence to reduce the power in the overall implementation. See discussion in Section 3.4 on the design trade-offs between frequency hop and direct-sequence spread spectrum. The hop rate of up to 160 khops/sec is chosen to be much higher than that specified by the FCC to achieve good performance in the presence of slow fading. In particular, faster hopping allows a shorter interleaver and therefore reduces the latency, which is critical for real-time applications such as voice. The hopping rate corresponds to eight symbols per hop. An 8×40 convolutional interleaver is used to mitigate the effects of slow fading, which is typical of an indoor environment. Note that by combining the interleaver with frequency hopping, frequency diversity is achieved as described in Section 2.7.4.2.

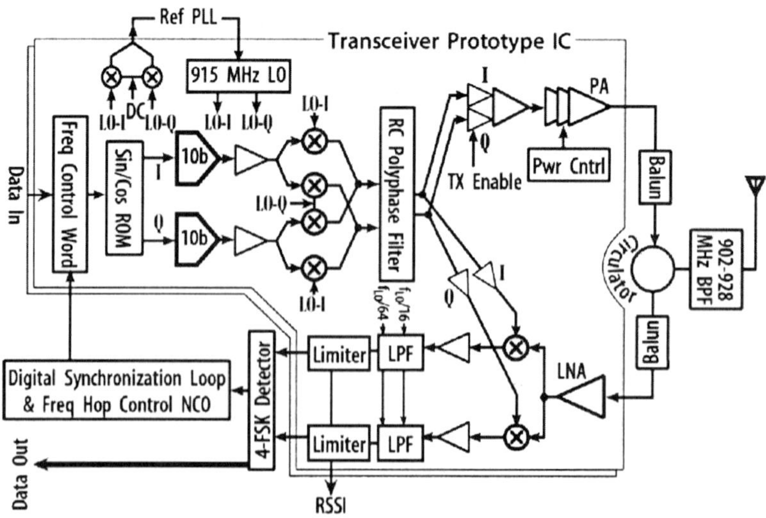

Figure 9.33 Block diagram of the FHSS transceiver; from Rofougaran [294], © 1998 IEEE.

The FH radio transceiver has been implemented in two CMOS chips, an integrated direct-conversion RF transceiver chip [294][296] and a digital baseband chip [296]. Figure 9.33 illustrates the overall transceiver architecture. The shaded region indicates the part that has been integrated onto the RF chip. The 4-FSK detector and synchronization loops have been implemented in the digital chip. The frequency control logic along with other miscellaneous control logic is implemented in FPGA chips. The reason for using the FPGA chips is purely for flexibility during testing. It should be fairly straightforward to integrate the control logic into the same digital chip.

Details of the digital chip are illustrated in more detail in Figure 9.34 as the shaded area. The digital receiver input consists of two I/Q channels from the RF receiver that implements dual antenna diversity based on equal-gain combining. The output of the mixer is filtered by a switched capacitor filter which performs the channel selection. The filtered signal is then limited to a fixed power and quantized to one bit. The limiter essentially implements the function of an AGC combined with a one-bit quantizer. While one-bit quantization could result in high SNR loss, in the case of FSK in which the data modulates the carrier frequency, the loss tends to be less. Measured results of the baseband chip showed a loss of only 1 dB at 0.1% BER. It should be mentioned that increasing the oversampling ratio at the limiter output reduces the loss in SNR. Typically, the receiver inputs are over-sampled by a factor of up to 160 times the baud rate. With a maximum data rate of 160 kbps, the baud rate is 80 kbaud.

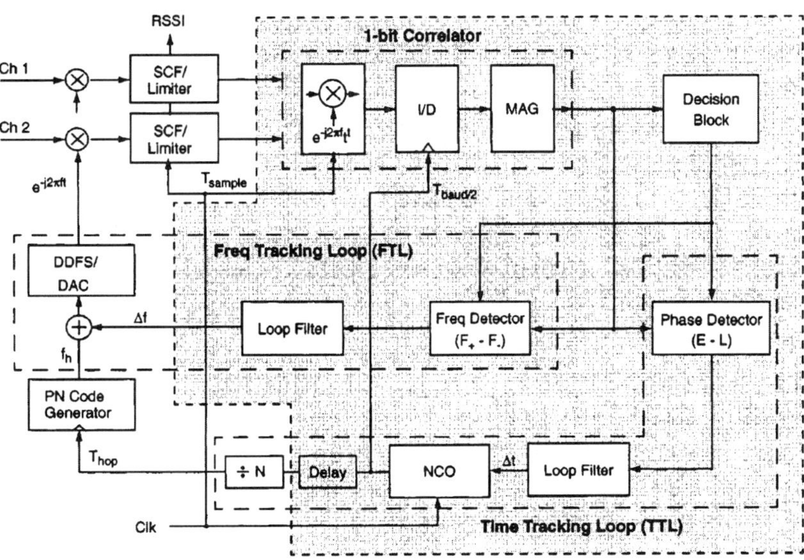

Figure 9.34 Digital baseband receiver architecture of the FHSS system; from Liu [296], © 1996 IEEE.

The one-bit quantized signal is correlated with locally generated tones at $\pm\Delta f$ and $\pm 2\Delta f$ using a complex correlator shown in Figure 8.36. The tone spacing Δf is equal to the baud rate given the non-coherent detection as discussed in Section 2.4.4. The one-bit quantization helps to reduce the hardware significantly in the FSK detector as well as the synchronization loops. For instance, rather than using a real multiplier to correlate the local tones with the received signal, an XOR logic gate can be used instead. Moreover, the locally generated tones, which are also quantized to one bit, can be implemented with an NCO rather than a DDFS. The simplified hardware architecture is shown in Figure 9.35 which can be compared to the one shown in Figure 8.36,

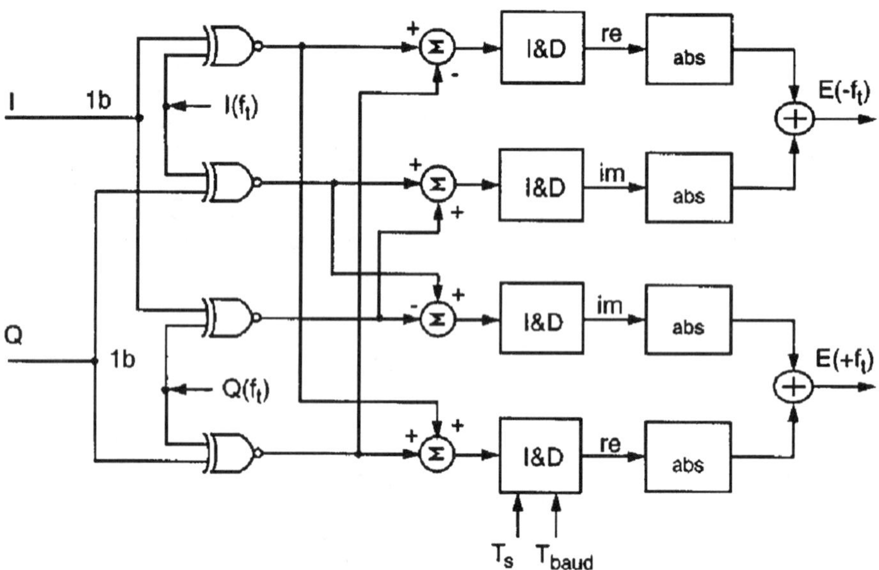

Figure 9.35 Multiplierless quadrature 1-b FSK detector block diagram; from Liu [296], © 1996 IEEE.

The digital chip has been designed with various optimizations to achieve low power, including clock gating, voltage scaling, and efficient use of parallel/serial datapaths. For instance, to reduced hardware complexity, the complex correlator shown in Figure 9.35 has been implemented as a time shared datapath that performs both the early-late gate correlation needed for timing recovery as well as the FSK detection. Furthermore, to minimize the power dissipation, the high-speed operation is performed in a parallel datapath and the lower-speed operations after the integrate and dumps are performed in a time-shared serial datapath as shown in Figure 9.36.

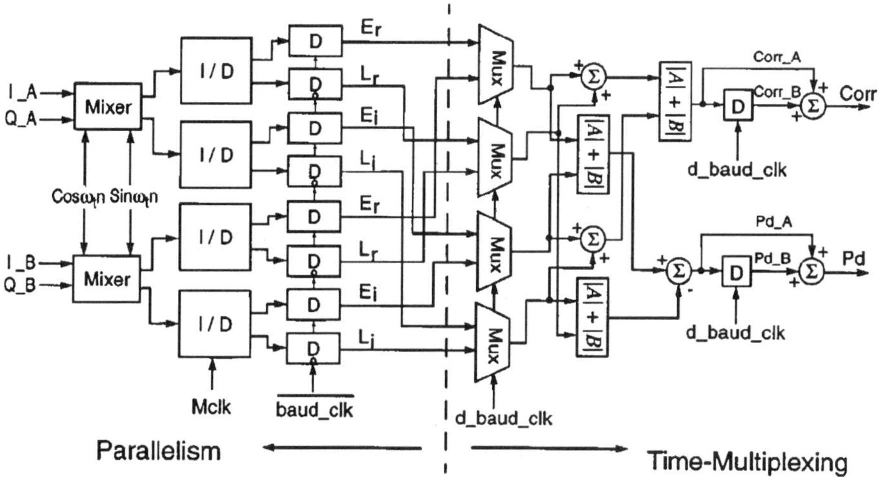

Figure 9.36 Block diagram of a time-shared multiplierless quadrature 1-b FSK detector and early-late correlator with dual-branch equal-gain diversity; from Liu [296], © 1996 IEEE.

The RF transceiver diagram is shown in Figure 9.33. The transceiver implements a direct-conversion architecture for both the receiver and the transmitter, and requires a few external components, such as the RF bandpass filter, balun, circulator, and a reference PLL. A quadrature DDFS with I/Q DAC implements the frequency hopping. By using a single-sideband modulation as shown in Figure 8.35a, the DDFS only has to generate tones from 0 to 13 MHz to cover the entire 902-928 MHz band since the negative frequency components can be generated by changing the sign at the combiner input. In other words, the set of frequencies $f_{LO} \pm f_{hop}$ could be generated centered on the 915 MHz band with a total bandwidth of 26 MHz by choosing a value of 915 MHz for f_{LO} and 0 to 13 MHz for f_{hop}. Data modulation can be easily combined with the hopping sequence by changing the frequency control word as discussed in Section 8.4.2.1. To achieve good spur levels, the DDFS uses a 10-bit output and to provide a frequency resolution of 6 Hz at 100 MHz sampling rate, the phase accumulator has a wordlength of 24 bits. A fine resolution is needed to tracking frequency offsets, especially when the Doppler shift tends to be small in indoor environments.

Perhaps the most critical design requirement for the transmitter is its image rejection performance of –50 dBc specified by the FCC for unlicensed transmission. Since FSK modulation with frequency hopping involves tones that could result in many intermodulation products and images when subject to sampling, and non-linear mixing and amplifications, judicious choice of sampling rate, filtering, and mixer topology is essential to achieve the adequate image suppression. To improve the linearity of the up-conversion path, the transmitter uses a linear passive double-balanced mixer, discussed earlier in Section 7.5.3. Such passive mixers have a conversion loss but high IIP3, which helps to reduce the intermodulation distortion levels.

The DAC outputs contain images at multiples of the sampling rate, which should be chosen to allow adequate suppression from the image-reject filters. See Section 6.1.4. Ideally, an image-reject filter should be used after the DAC's to reject the images at multiples of 100 MHz. However, such filters increase the complexity as well as power dissipation of the transceiver. Instead, since the images are set apart by at least 100 MHz, the RF bandpass filter can be used instead to also serve as the image-reject filter. However, such an approach requires that the up-converted signal at the mixer output contains low-level intermodulation products. Otherwise, the non-linear power amplifier will generate new mixed frequencies induced by the undesired intermodulation products.

There are several methods to alleviate this problem, including decreasing the LO drive to the mixer, use of a linear amplifier with a high output backoff, or inserting a bandpass filter prior to power amplification. However, none of these are power efficient. Instead, the authors implemented a passive polyphase filter to null out the third-order intermodulation products at the output of the mixer as shown in Figure 9.33. For a detailed description, refer to [295]. Finally, to achieve a –50 dBc image rejection, phase matching of 1 degree and amplitude matching of 0.1 dB are necessary as discussed in Section 9.1.4.2. To accomplish this, clock lines of the DDFS are laid out with attention to balancing the routing for the I and Q sections and LC tuned loads are used in a cross-coupled quadrature oscillator.

Since the transmitter implements a direct-conversion architecture, it is prone to LO re-radiation that could occur through various mechanisms described in Section 6.1.1. Here, the non-zero DC offset at the output of the DAC's could modulate the carrier frequency and result in re-radiation at the LO frequency. If not suppressed, the re-radiated signal can desensitize the receiver and result in poorer BER performance. To minimize the LO re-radiation, the effect of the DC offset should be minimized. Instead of using large AC coupling capacitors to reduce the DC offsets, the LO frequency is decreased to 890 MHz while the DDFS output frequency is increased to 38 MHz so that the up-converted RF frequency still centers around the 900-MHz ISM band. The re-radiated LO signal now resides in the transition band of the RF bandpass filter and therefore gets attenuated by approximately 30 dB.

VCO pulling is also a major problem with the transmitter. In this design, the problem is resolved by making the RF PLL bandwidth narrow so that less of the PA signal feeds back to the VCO input. The PA uses a class A/B configuration since FSK is a constant envelop modulation and as long as the zero crossings are preserved, data can still be demodulated at the receiver. Certain amount of spectral regrowth is still observed since the FSK modulation has discontinuous phases at the hop boundaries.

The RF receiver shares the same LO with the transmitter and uses a common-gate configuration with tuned loads for the LNA, whose output is fed to an active double-balanced mixer with passive loads. The active mixer is similar to the one described in Section 7.5.2. The use of active mixers helps to improve the noise performance of the receiver since active mixers have a conversion gain rather than a conversion loss.

The receiver implements the direct-conversion architecture and bears the problems stressed by the high dynamic range requirement and the large DC offsets due to numerous mechanisms discussed in Chapter 6. To address the high dynamic range requirement in only a single stage, multistage filtering is devised to achieve a balanced trade-off between NF and the overall IIP3 as shown in Figure 9.37. In particular, the 6th-order Elliptic channel-select filter is designed using three biquads with a gain distribution that leads to a high IIP3 but with minimal compromises in the NF. For a discussion on the trade-offs between linearity and NF, see Chapter 4. The 6th-order Elliptic filter has a cutoff corresponding to the hop bandwidth of the transmitted signal. The 2nd-order Butterworth prefilter is employed to reduce undesired signals at 14.3 MHz since for practical reasons, the Elliptic filter clock rate is limited to 14.3 MHz. The receiver chip achieves a system NF of 8.6 dB and a total IIP3 of –8.3 dBm.

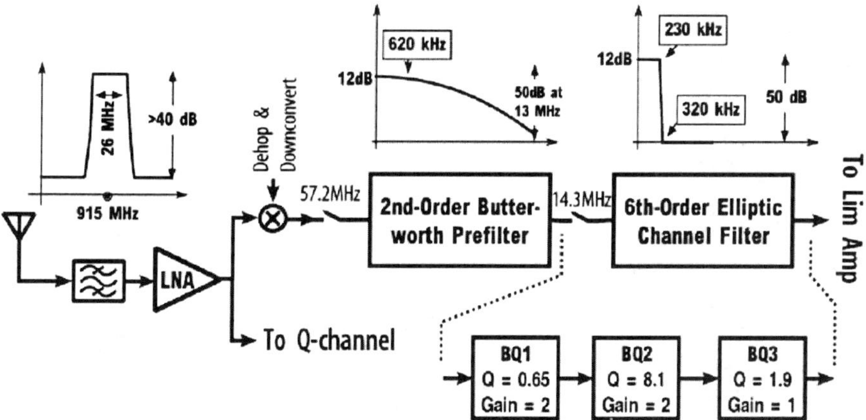

Figure 9.37 Block diagram of the receiver front-end highlighting the multi-stage filtering designed to achieve a suppression of 50 dB for undesired channels while maintaining good linearity and noise performance; from Rofougaran [295], © 1998 IEEE.

Typically, in the receiver an AGC is needed to bring the small received signal level up to a level that can drive an ADC. However, in this receiver since one-bit quantization is used, an AGC and ADC are not necessary. Rather, a limiting amplifier is employed that maintains a constant signal power at its output. Moreover, DC feedback is implemented to cancel out the high DC offsets that occur at the output of the limiter as shown in Figure 9.38. The DC feedback loop results in an IIP2 of +22 dBm. Due to the nature of the modulation, FSK is less sensitive to DC offsets since it has little signal content around DC as shown in Figure 9.39. Therefore, the bandwidth of the feedback loop has less effect on the performance of the demodulator.

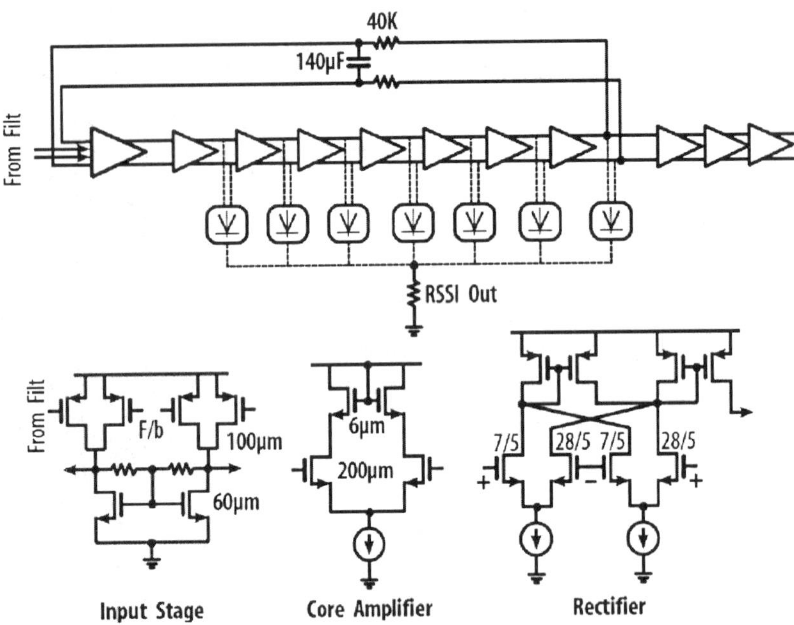

Figure 9.38 Limiting amplifier and logarithmic received signal-strength indicator (RSSI). Individual circuits shown for input stage, seven core amplifiers in cascade, and rectifier; from Rofougaran [295], © 1998 IEEE.

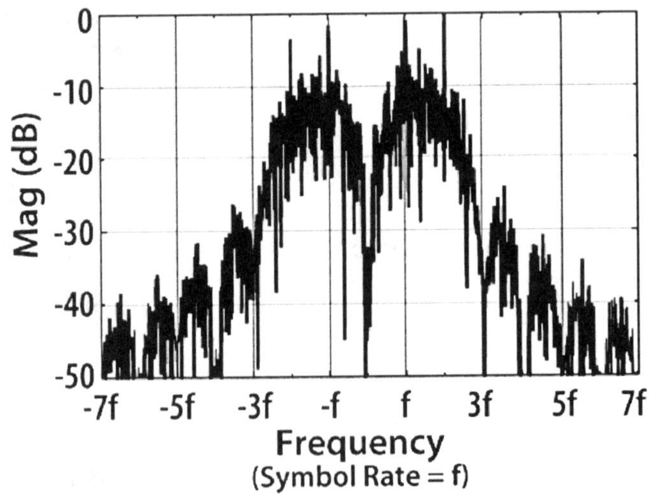

Figure 9.39 Spectrum of 4-FSK modulation; from Rofougaran [295], © 1998 IEEE.

Finally, the RF chip uses twelve square spiral inductors with inductor sizes of 30 nH or 50 nH. A large inductor has poor Q due to the series resistance which increases with the length of the winding and the large capacitor that forms between the winding and the substrate. Due to the low loss substrate, the capacitive coupling and the increased series resistance decrease the obtainable Q. Moreover, the large capacitance tends to cause the inductor to resonate at a lower frequency than the one designed for. See Section 7.3.3 for a review of trade-offs associated with spiral on-chip inductors.

To alleviate the problems associated with on-chip inductors, the authors placed vias around the inductors and performed a post-etching to remove a large portion of the substrate underneath the inductors, effectively eliminating the low loss substrate as well as substantially reducing the capacitance to the substrate [293][295]. The inductors with substrate etched away underneath are shown in Figure 9.40. The performance of the FHSS RF transceiver chip-set is summarized in Table 9.21.

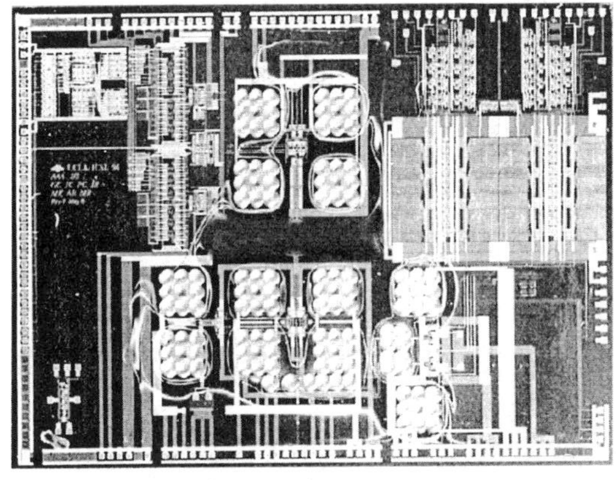

(a)

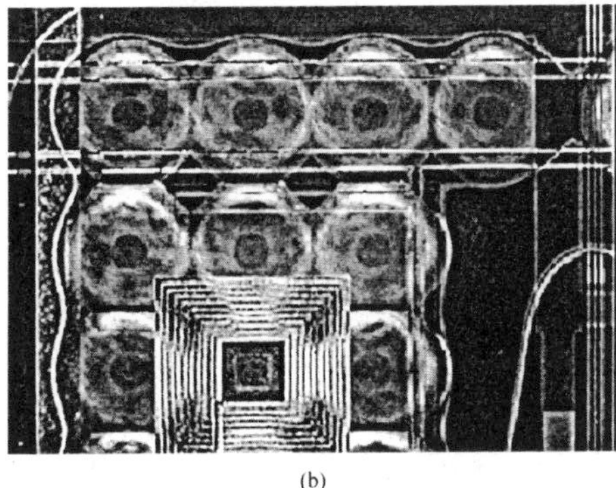

(b)

Figure 9.40 Suspended on-chip spiral inductors: a) die photo after etching of cavities under inductors and b) detailed view of pits under inductor and under long RF interconnects; from Rofougaran [295], © 1998 IEEE.

Table 9.21 Summary of FH transceiver chip-set performance.

	Chip Feature	Performance
General	Technology	1.0 μm CMOS
	Die Area	$10.5 \times 7.3\ mm$ (RF chip)
		$5.4 \times 5.4\ mm$ (Digital chip)
	Supply Voltage	3 V
Transmitter	Signal Bandwidth	320 kHz (4-FSK)
	Hop Rate	Up to 160 khops/sec.
	Data Rate	Up to 160 kbps with 4-FSK
	Symbol Rate	80 kbaud
	Output Power	-17 dBm - 13 dBm
	Current Drain	100 mA from 3 V
	Spurious Levels	-55 dBc @ 3 dBm output
Receiver	Sensitivity	-102.4 dBm @ 0.1 % BER
	Noise Figure	8.6 dB
	IIP2 (High Gain)	22 dBm
	IIP3 (High Gain)	-8.3 dBm
	Input Matching	-10 dB
	Transistor Count	48,000 (Digital chip)
	Current Drain	120 mA from 3 V (RF chip)
		1.5 mA from 3 V @ 10 MHz (Digital chip)

9.2.3 Wideband CDMA

CDMA based on direct-sequence spread spectrum has become increasingly popular due to its potential to achieve high spectral efficiency in terms of bps/Hz/km^2 as discussed in Chapter 5. However, it was not until Qualcomm demonstrated the feasibility of deploying CDMA cellular networks using direct-sequence spread spectrum in the early 90's that CDMA began to grow in popularity. The key contribution by Qualcomm is in applying power control to direct-sequence CDMA, without which the capacity enhancement cannot be achieved.

Qualcomm's CDMA system is based on the IS-95 standard [5], which specifies a relatively low chip rate of 1.22 Mchips/sec. Recently, with the effort to obtain a worldwide standard in third-generation wireless, also known as IMT-2000, a number of proposals were submitted to the ITU based on wideband CDMA (WCDMA). While the goal had been to create a unified standard, ITU ended up accepting multiple competing systems for IMT-2000. These include the Universal Terrestrial Radio Access (UTRA) WCDMA [298][299], UTRA wideband TDMA (WTDMA) [300], cdma2000 [301], TD-SCDMA [302], and Universal Wireless Communications (UWC)-136 [303]. Both UTRA WTDMA and UWC-136 are TDMA based systems, where the former originated from Europe and the latter from the US. UTRA WCDMA, cdma2000, and TD-SCDMA are CDMA based systems developed by Europe, US, and China, respectively.

Of all the 3G systems, those based on WCDMA have the highest potential to dominate the future system, in particular the UTRA WCDMA and cdma2000 standards. The former has received significant support from Europe while the latter has received strong support within the US. Backers of the two standards are wooing supports from others around the world to bolster their positions in this growing market. Table 9.22 summarizes some of the key features of WCDMA.

Table 9.22 Summary of WCDMA system parameters.

System Parameters	Values
Downlink frequency band	2110-2170 MHz
Up-link frequency band	1920-1980 MHz
Duplexing	FDD
Carrier spacing	1.25, 5, 10, 20 MHz
Number of time slots	10 ms
Modulation	Adaptive
Spread-spectrum type	Direct sequence
Chip rate	1.024, 4.096, 8.192, 16.384 Mcps
Data rate	Up to 2 Mbps

Radio IC's targeted for third-generation systems based on WCDMA are just beginning to emerge. For instance, [304] reports a chip set based on BiCMOS and CMOS where the RF and baseband chips are implemented with a 25 GHz 0.3 μm BiCMOS process and the ADC's are implemented with a 0.5 μm CMOS process. The RF chip achieves a –114 dBm sensitivity for a data rate of 128 kbps at a chipping rate of 4.096 Mchips/sec. The NF of the RF chip is 5.1 dB at the maximum voltage gain of 94 dB and the IIP3 is –9.5 dBm. The receiver drains 128 mA from a 2.7 V supply.

Third-generation WCDMA system stresses all aspects of the radio design, in particular on the noise and linearity performance over a wide bandwidth. Thus far, no one has been able to implement a chip-set based only on CMOS due to the high performance requirement of WCDMA. For more relaxed system though the feasibility of using a complete CMOS solution has been demonstrated. In particular, a radio chip-set for a research system designed to allow ubiquitous information access in a pico-cellular indoor environment has been implemented in 0.8 μm CMOS [305].

The system is also referred to as InfoPad [306][307][308] intended to provide wideband information access by mobile units in an indoor environment as shown in Figure 9.41. The system is based on TDD and uses direct-sequence spread spectrum. The link is asymmetric where the uplink carries control information and runs at 32 kbps/user whereas the downlink carries user data rate runs at 1 Mbps/user. The modulation uses DQPSK resulting in a processing gain of 15 dB at a chip rate of 1 Mchips/sec, by definition (3.6). The RF carrier frequency is at 1.056 GHz and the transmission bandwidth is 43 MHz assuming an excess bandwidth of 30%. Since the intended application is for a pico-cellular indoor system, the maximum transmitter power is 0 dBm.

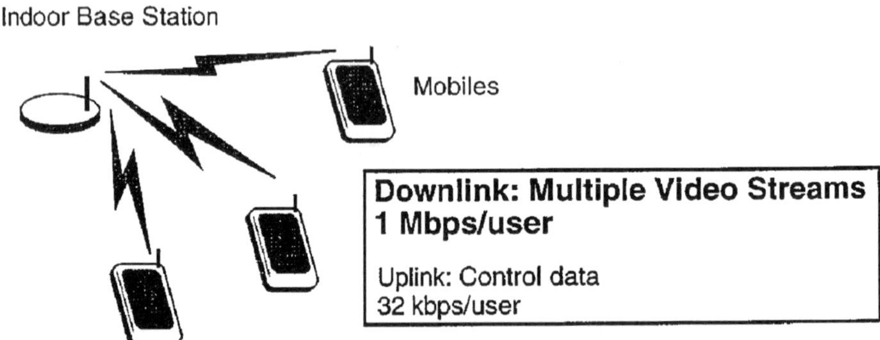

Figure 9.41 The InfoPad system; from Sheng [305], © 1996 IEEE.

The radio chip-set consists of three chips, a digital transmitter, an RF receiver, and a digital receiver. Figure 9.42 shows the digital transmitter which consists of 15 independent DQPSK modulators and Walsh code spreaders. A transmit power control command could be inserted into each spread channel to perform power control. A pilot channel is inserted using the all-zero Walsh code to allow the receiver to synchronize onto the transmitter and to perform channel estimation required for coherent detection as discussed in Section 8.6.1. The composite spread user signal with the pilot code is spread again by a long PN code before being filtered by a pair of I/Q raised-cosine pulse shape filter. The implementation of the DQPSK and Walsh spreading are straightforward. The implementation of the transmit filter follows the approach used in [309] for QAM transmission. The Transmitted symbol is interpolated by four prior to filtering and a polyphase architecture, described in Section 8.2.7.1, is used so that the four sub-filters only have to run at the chip rate. Moreover, IF up conversion is implemented whereby the local oscillator can be generated without complex circuitry, such as a DDFS. Since the spread signal is over-sampled by four, an IF frequency equal to a quarter of the interpolated sampling rate could be used. Digitally, the IF corresponds to repeating sequences of $\{0\ 1\ 0\ -1\}$ and $\{1\ 0\ -1\ 0\}$ for the quadrature and in-phase components, respectively. Because of the zero terms, only two of the sub-filters in each channel are needed. Thus, a hardware saving of two fold can be obtained. The digital IF signal is then up converted to RF in quadrature using an external single-sideband modulator as shown in Figure 9.42.

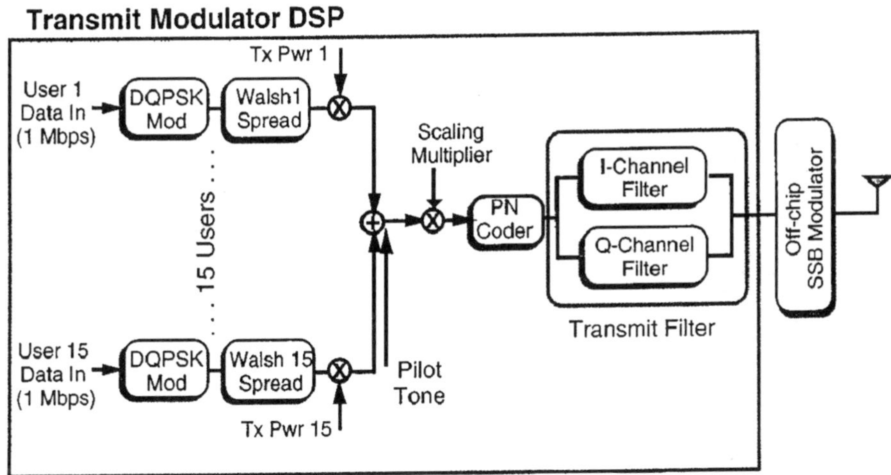

Figure 9.42 Block diagram of the WCDMA transmitter; from Sheng [305], © 1996 IEEE.

The RF receiver chip, shown in Figure 9.43, implements a sub-sampling down-converter discussed in Section 7.5.4. A sub-sampling converter has the advantage in terms of simplicity. However, it faces two major difficulties: 1) the problem of image folding and 2) switch noise injection and phase matching. The first problem arises due to aliasing of images from sub-bands that are multiples of the sub-sampling rate, a direct consequence of not sampling faster than the Nyquist rate. In general, signals within the input bandwidth of the sampling switch will be fold in-band. For instance, assuming 1-GHz of input bandwidth for the sampling switch and a sampling rate of 64 MHz, then approximately sixteen times more noise is generated at the output of the sampling switch, assuming uniform distribution of noise across the 1-GHz band. This can seriously degrade the NF of the system.

To mitigate the image-folding problem, the RF receiver employs two external filters, one placed before the LNA and the other after the LNA to provide rejection of 30-40 dB in the nearest image frequency. Unfortunately, such filtering does not provide adequate adjacent channel interference for all types of systems. In particular, for high performance systems such as GSM where blocker signals could have power as high as 80 dB above the desired signal, the attenuation provided by the two filters cannot suppress the blockers sufficiently to negate the interference power being folded in-band. Moreover, the post LNA filter typically introduces an insertion loss that degrades the NF of the system. See Section 4.2.1. Even with micro-cellular systems, such as DECT, the attenuation may not be sufficient to mitigate the distortion caused by the folding of image bands. Fortunately, for a pico-cellular system, the blocker requirement may not be as severe so that the external filters can provide sufficient suppression of interference to minimize receiver desensitization.

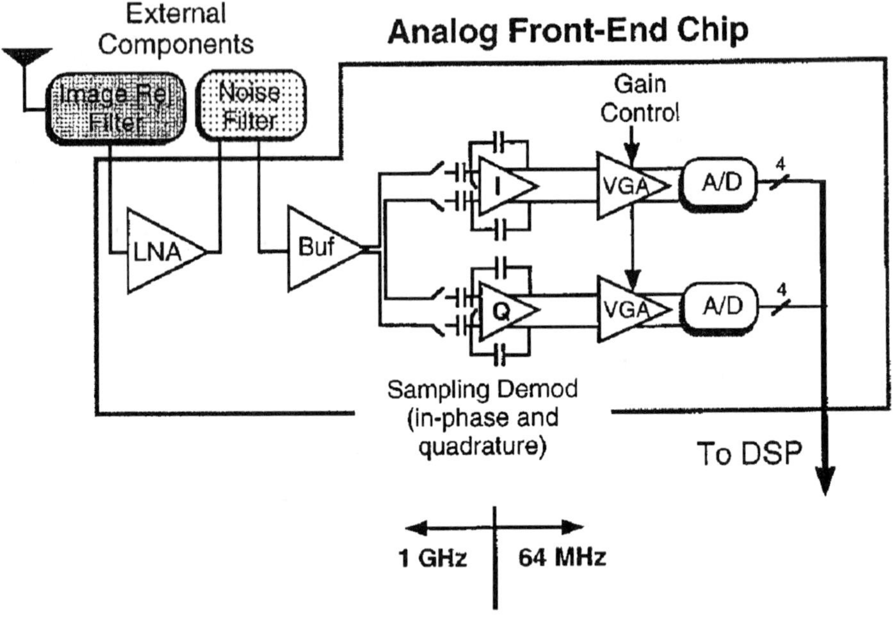

Figure 9.43 Block diagram of the WCDMA RF receiver chip; from Sheng [305], © 1996 IEEE.

The second design challenge associated with the sub-sampling receiver is in designing a "perfect" switch defined as one which can switch instantaneously, noiselessly, and is broadband with perfect I/Q phase matching. None of the above can be achieved in practice. To achieve fast switching typically means large signal drive on the sampling clock that could result in large power dissipation. Furthermore, due to charge stored near the conducting channel and parasitic overlap capacitance, a sample clock transition can cause charge injection that introduces error at the output of the sampling switch. Also, a broadband switch is desired so that the RF input could be tracked and held at the correct voltage. Unfortunately, a broadband sample-and-hold circuit is difficult to implement. Finally, I/Q phase matching requires a careful layout, which tends to be process dependent and prone to parasitic effects. The receiver chip attempts to address the above problems with a track-and-hold circuit shown in Figure 9.44. The differential circuit topology helps to minimize the effect of charge injection and the use of a telescopic cascode amplifier provides a fast tracking response in the track and hold circuit.

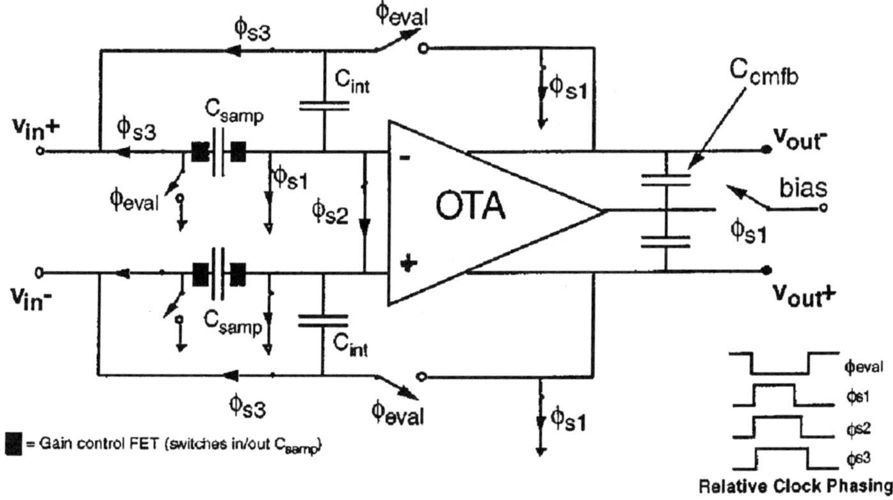

Figure 9.44 The track-and-hold circuit for the sub-sampling mixer; from Sheng [305], © 1996 IEEE.

The track-and-hold outputs are sampled by an ADC, whose outputs are sent to the digital baseband receiver chip shown in Figure 9.45. The receiver architecture is similar to the Rake receiver described in Section 8.6.1 and shown in Figure 8.49. Table 9.23 summarizes the performance of the radio chip-set.

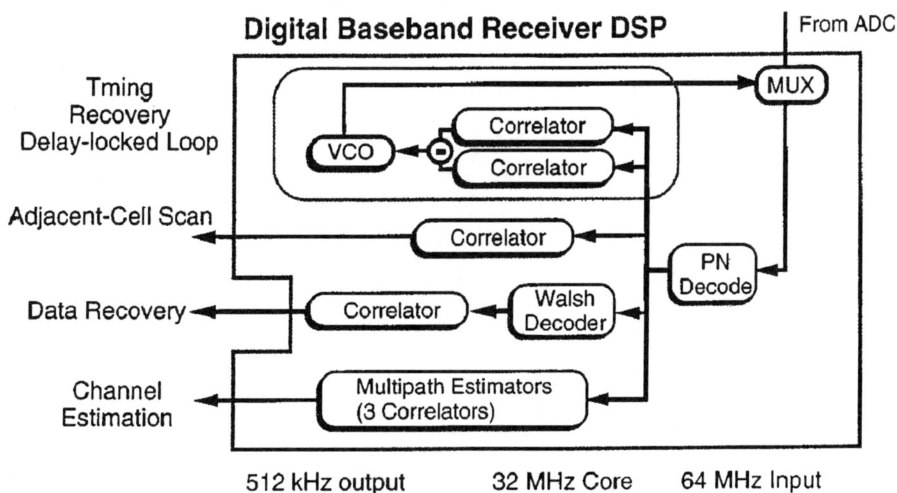

Figure 9.45 Block diagram of the WCDMA digital receiver chip; from Sheng [305], © 1996 IEEE.

Table 9.23 Summary of WCDMA transceiver chip-set performance.

	Chip Feature	Performance
General	Technology	0.8 μm CMOS
	Chip Rate	32 Mchips/sec.
	Data Rate	1 Mbps (up link)
	Center Frequency	1.056 GHz
	Signal Bandwidth	43 MHz
Digital TX	Die Area	9.9 x 10 mm
	Transistor Count	112,000
	Sampling Rate	32 MHz
	Power Dissipation	165 mW
RF Receiver	Die rea	6.2 x 6.2 mm
	Noise Figure	7.2 dB (LNA)
	Power Dissipation	106.8 mW
Digital RX	Die Area	7.5 x 7.4 mm
	Transistor Count	87,000
	Sampling Rate	32 MHz
	Power Dissipation	46.5 mW

9.3 Single-Chip Radio

This book has focused on a system approach to designing a digital radio system on a chip. System trade-offs, design techniques, architectures, and circuits have been discussed and illustrated in several case studies described in the previous section. A single-chip solution offers lower component count and therefore reduces the integration cost and increases the reliability of the final product. A single-chip CMOS radio solution is particularly attractive due to its low fabrication cost, resulting from the high volume production of digital CMOS IC's. Being able to integrate RF and analog circuitry along with the digital transceiver on a standard CMOS process leverages the economy of scale that currently exists for CMOS technology. However, several design challenges exist which have prevented single-chip integration. These challenges are listed below

- Low-Q on-chip components (e.g. less than 10 at 1-2 GHz) due to substrate loss;
- Inability to deliver high transmission power due to short-channel effects, lower g_m/I ratio, and poor thermal isolation;
- Substrate coupling of large-swing digital signals to sensitive RF signals.

The low-Q limitation could be overcome with the use of bondwire inductors which can achieve a Q of 20-30 at 10-20 GHz [310]. However, at lower resonant frequencies, the Q degrades roughly as the square root of the frequency. The poor power handling of CMOS makes it difficult to integrate a power amplifier on chip. Due to the inherent limitations in standard silicon technology this is an extremely difficult problem especially for systems that require linear amplification at high transmit power greater than 20 dBm. For instance, IS-95 CDMA and GSM system have this stringent requirement. IS-95 in particular requires linear amplification at 28 dBm. GSM does not require linear amplification but requires higher transmit power of 29 dBm (Class 5) and 33 dBm (Class 4). Current solutions use an external GaAs power amplifier to achieve the high transmit power level and reliability with extended voltage and thermal stress. However, CMOS solutions for power amplifiers are beginning to emerge for systems that can tolerate non-linear amplification [311]. Integrated PA solutions are also emerging for low tier systems such as digital cordless phones and Bluetooth that do not require high-power transmissions [168]. Finally, substrate coupling of digital signals that have a swing between 2 to 3 V can greatly distort sensitive RF circuitry that operates on signals with a swing of only a few μV.

Note that these issues also hold for bipolar technology. Though bipolar does have the advantage of higher g_m/I ratio and slightly higher substrate isolation due to less doping concentration in the substrate as compared to CMOS which requires higher doping concentration to prevent latch-up.

The above limitations have been major factors in preventing CMOS from replacing bipolar and GaAs technology for RF applications even though serious research and development in RF CMOS have begun since the early 1990's. While single-chip CMOS radio solutions have been a focal point of most wireless IC developments, no commercial solution yet exists for high tier systems such as GSM and CDMA. However, many research prototypes have been developed as discussed in Section 9.2. On the other hand, for low-tier systems, single-chip radios that include both the RF and digital transceivers are emerging. For instance, in [167] a BiCMOS single-chip DECT radio has been reported. As for CMOS, a single-chip solution that integrates both the RF and digital transceiver for spread-spectrum digital cordless telephones has been reported in [168]. This CMOS radio IC is described in more detail below.

9.3.1 CMOS Single-Chip Spread-Spectrum Cordless Radio

The single-chip transceiver [168][312] discussed in this example is targeted for the unlicensed digital spread-spectrum cordless phone in the 900 MHz band. The cordless phone requires a low range of less than 100 m and low capacity. In most cases, only a few phones are active in a local area and therefore the system requires fairly low capacity. The low range and capacity relax many of the RF and baseband parameters in contrast to high-tier systems such as GSM.

The cordless phone system is based on time-division duplex and FDMA with a channel bit rate of 100 kbps. Each frame consists of a transmit and receive slot of 100 ms duration with 16 ms, 8 ms, 8 ms, and 64 ms partitioned for preamble, service, synchronization, and voice traffic. The speech is coded using adaptive differential pulse code modulation (ADPCM) [313]. While the channel rate is 100 kbps per frequency channel, the data is spread with a chip sequence running at 1.5 Mchips/sec. The spreading allows FCC compliance for unlicensed transmission in the 900 MHz ISM band that spans 902-928 MHz and provides minimal degree of robustness against interference.

To achieve single-chip integration of almost the entire transceiver, the direct-conversion architecture is used as shown in Figure 9.46. The transmitter takes the ADPCM coder output and spreads it using a 1.5 Mchips/sec PN code. The spread data directly modulates a VCO to generate a signal at twice the RF frequency. The transmitted waveform is obtained by dividing the VCO output by two and amplifying the result with an on-chip power amplifier. At the receiver, the received signal is amplified by an LNA, down-converted in quadrature, and filtered with several stages of switched-capacitor lowpass filters. DC-offset cancellation is implemented with the lowpass filters to reduce the effect of DC offset inherent in a direct-conversion receiver. The received signal is then adjusted by an AGC followed by another offset cancellation block whose output is demodulated using a discriminator implemented with discrete-time switched-capacitor circuits. The output of the discriminator is then despread and sent to the ADPCM decoder.

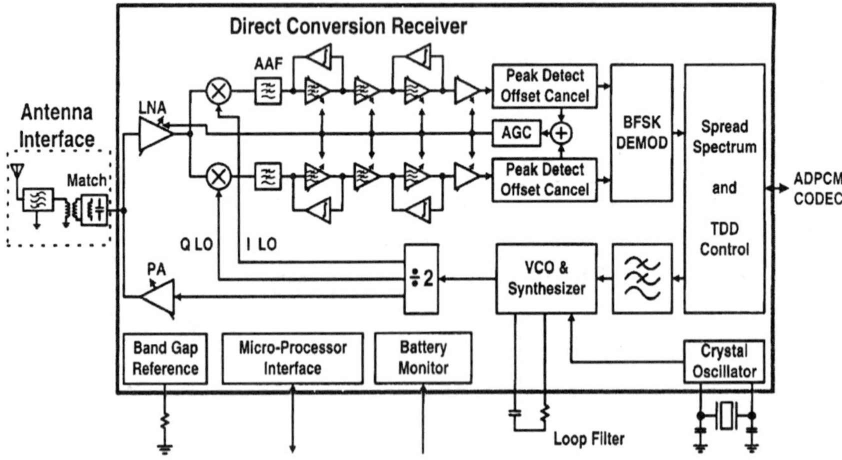

Figure 9.46 Transceiver architecture of the single-chip cordless telephone radio; from Cho [314], © 1999 IEEE.

The single-chip transceiver requires only a few external components, including: an ISM-band filter, a balun, an RF matching network, an RC loop filter for the PLL, a crystal resonator, and a resistor for biasing. It should be noted that while all of the communication functions have been implemented on a single chip, the microcontroller and vocoder have not been integrated. The transceiver is implemented in 0.6 μm CMOS technology.

9.3.1.1 Transmitter

To simplify the transceiver hardware, direct modulation is used to generate the BFSK signal as shown in Figure 9.47. Depending on the maximum frequency deviation or modulation index, the loop filter bandwidth should be sufficiently narrow so that the VCO does not track the modulated data. In this design, the VCO outputs at twice the RF frequency to provide isolation between the RF and VCO control input such that VCO pulling can be avoided. The alternative is to use an I/Q modulator with offset VCO which requires more hardware to implement a dual DAC, an I/Q mixer, and a combiner.

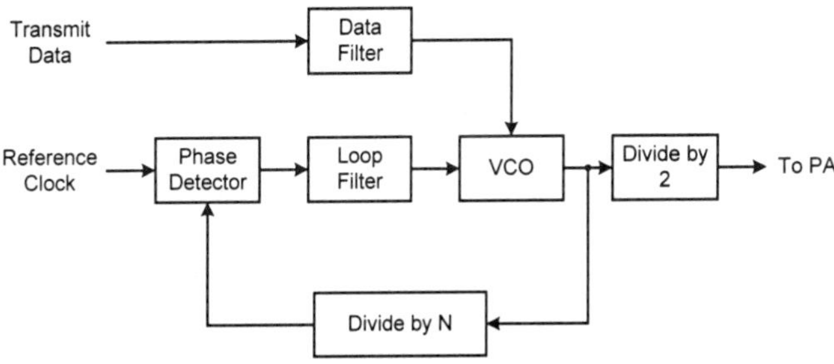

Figure 9.47 Direct modulator architecture.

Direct modulation does complicate the synthesizer design. In particular, the baseband data needs to be fed into the VCO in addition to the frequency control signal. The PLL should also be designed to have a narrow loop bandwidth to prevent the VCO from tracking the modulated data. A narrow loop bandwidth increases the phase noise but suppresses thermal noise. The PLL loop bandwidth should therefore be adjusted appropriately in the receive mode to achieve a balanced trade-off between phase noise and thermal noise. The VCO implemented in this radio is shown in Figure 9.48.

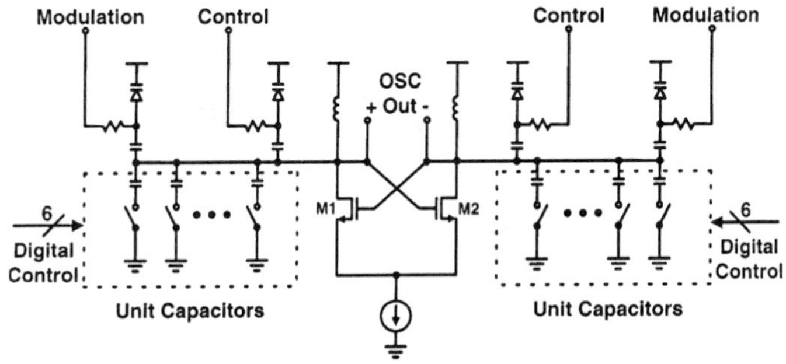

Figure 9.48 VCO circuit diagram; from Cho [314], © 1999 IEEE.

The VCO is based on the LC oscillator discussed in Chapter 4 with two sets of varactors, one for frequency control and one for modulation. Calibration capacitors are included to center the VCO frequency on 1.83 GHz via a six bit digital word.

9.3.1.2 Receiver

The receiver is based on the direct-conversion architecture discussed in Chapter 6. It has the advantage of low hardware complexity but places high dynamic range requirements on the front-end circuits and provides no frequency isolation that leads to LO re-radiation and large DC offsets. For a low-tier system, these problems can be more easily mitigated since there are fewer strong interferers in the channel and the communication range is relatively low. The primary problem then lies in the large DC offsets that can reduce the dynamic range of the receiver, or worse, saturate the amplifiers.

In this radio, the offset cancellation method attenuates offsets in the baseband circuits without sacrificing the receiver bandwidth, in contrast to using an AC coupling capacitor. Feedback integrators cancel offsets in the lowpass baseband filters prior to reception of data whereas the feed-forward loop cancels offset during data reception. The peak detector determines the received signal power as well as the DC offset. Signal power is used to adjust the VGA and offset estimate is used to cancel the DC offset.

The BFSK demodulator shown in Figure 9.46 takes the AGC output and demodulates the data with discrete time processing using switched-capacitor circuits running at 6 MHz clock rate. To simplify hardware, the demodulator uses a discriminator type detection that recovers the data based on the rotation of the signal constellation. Little details have been given in [168][312] for the demodulation, AFC, and clock recovery. Figure 9.49 illustrates one of the many possible ways to implement a demodulator. Most approaches (e.g. [315][316]) require very little hardware. For this transceiver, the digital processing only requires about 8000 standard-cell gates.

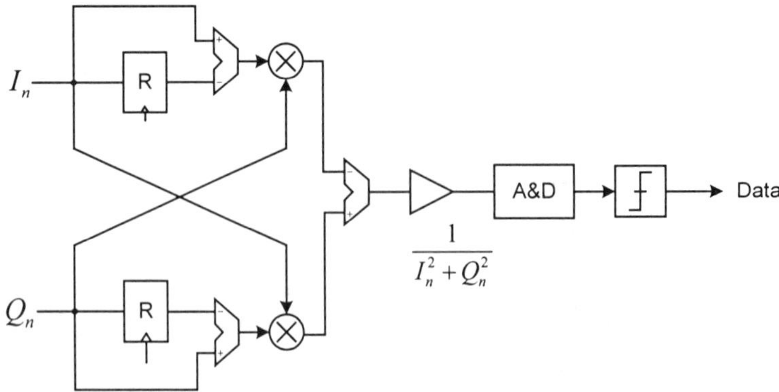

Figure 9.49 Discriminator architecture.

Given the I/Q samples, the discriminator can be implemented using the architecture shown in Figure 9.49. The architecture can be built using digital processing or discrete time analog processing. A simple AFC can be implemented by counting the time between zero crossing during the preamble in the I and Q channels. To eliminate data dependence on the frequency estimate, the preamble usually carries a tone at either the maximum or minimum frequency deviation with no data modulation. For better performance, the discriminator output can be digitized and the frequency directly estimated and cancelled out [315]. A simple implementation of clock recovery can be based on a digital PLL that uses the early-late gate as the phase detector. The demodulated data in this case are chips that are despread in the burst-mode controller using PN-acquisition architectures described in Chapter 8. For lowest complexity a serial acquisition loop can be used.

9.3.1.3 Integration Issues

Typically a T/R switch is required at the front-end to switch the antenna between the transmitter and the receiver. A T/R switch is usually an external component implemented in GaAs that provides fast switching with low on resistance that translates to low loss of 0.5-1 dB. Moreover, the GaAs switch provides fairly good isolation of 30 to 40 dB. In CMOS, a T/R switch tends to have poor performance both in terms of isolation as well as insertion loss. In this design, to achieve high level of integration, the T/R switch is avoided by shorting the transmitter and receiver into a single RF port as shown in Figure 9.46. The transmitter and receiver isolation at the shared RF port is achieved by turning off the transmitter during the receive mode and turning the receiver off during the transmit mode. It is important to ensure that the RF chain being turned off presents high impedance to minimize the effect on input matching at the RF port.

An external PA is generally implemented in GaAs technology because of the requirements on high current and thermal isolation. This holds true for high performance systems such as GSM and IS-95 where the output power is high. In a cordless phone system, the required power does not exceed 20 dBm because of the short communication range. Thus, if desired, the power amplifier can be integrated on chip with care in thermal isolation and de-coupling of the high power signal from the sensitive front-end circuitry. In this design, de-coupling is achieved by employing time-division duplexing to isolate the PA from the receiver. While thermal isolation and stress are important to achieve high reliability and long device lifetime, the authors of this work have not revealed their solutions to address this problem.

Several design techniques have been applied to allow digital circuits to reside on the same chip with sensitive RF circuitry. First, to reduce coupling, CMOS logic gate transitions occur staggered in time (5 ns) relative to the sampling instants of the switched-capacitor circuits. Second, differential circuits are used to reduce coupling that occurs in common mode. Third, differential source-coupled logic is used for the prescalar in the synthesizer to minimize current glitches that couple to RF circuits through the supply and substrate.

Finally, LC loads have been used generously in the front-end circuitry. Due to the lossy substrate in CMOS, a Q of 4 is achieved at 900 MHz with a spiral octagonal shape inductor [312]. For increased Q, the inductor has a hollow center area and three layer metals are used. See Chapter 4 for a review of design trade-offs for on-chip spiral inductors.

9.3.1.4 Performance

The performance of the single-chip cordless radio is summarized in Table 9.24. The die area is 36.2 mm^2 in 0.6 μm CMOS. The small size offers a low cost solution that can vary from \$4 to \$12 depending on the fabricating cost and volume. The transmitter waveform has an accuracy of 1 % with four selectable transmit power of 2 dBm, 8 dBm, 14 dBm, and 20 dBm. Note that the power amplifier current dissipation dominates the total transmitter current consumption at high output power of 14 dBm and 20 dBm. At these power levels, the fraction of output power to total current consumption is 21% and 48%, respectively. In contrast, at 2 dBm and 8 dBm output power, the fraction of output power to total current consumption is 2% and 6%, respectively.

Table 9.24 Performance Summary of the single-chip cordless radio IC [168][312].

	Chip Feature	Performance
General	Technology	0.6 μm double poly triple metal CMOS
	Die Area	$5.4 \times 6.7\ mm^2$
	Supply Voltage	3.3 V
Transmitter	Bandwidth	1.5 MHz
	Modulation Accuracy	±1%
	Modulation index	0.35
	Channel Spacing	750 kHz
	Current Drain	77 mA @ 2 dBm output power 98 mA @ 8 dBm output power 118 mA @ 14 dBm output power 210 mA @ 20 dBm output power
Receiver	Sensitivity	-104 dBm @ 0.1 % BER
	Noise Figure	4.5 dB
	IIP2 (high gain)	22 dBm
	IIP3 (high gain)	-21 dBm
	Selectivity @ -83 dBm received signal power and 0.1 % BER	1 dB co-channel 23 dB 1.5 MHz 57 dB 3.0 MHz 67 dB 4.5 MHz
	Current Drain	159 mA

With a noise figure of 4.5 dB, the receiver has a sensitivity of –104 dBm. At a bitrate of 100 kbps, this implies a E_b / N_0 requirement of 15.5 dB, which is consistent with the use of discriminator for demodulation. The IIP2 of 22 dBm is needed for direct-conversion architecture due to its sensitivity to DC offsets. The IIP3 of –21 dBm is low in contrast to GSM. However, since the cordless system has a much shorter range requirement, the input power fluctuation is not as high. The selectivity in this case is characterized at –83 dBm of input power level, which relaxes the IIP3 requirement.

The radio chip micrograph is shown in Figure 9.50. The layout is fairly compact. The on-chip spiral inductors can be seen in the synthesizer and LNA/PA sections of the die.

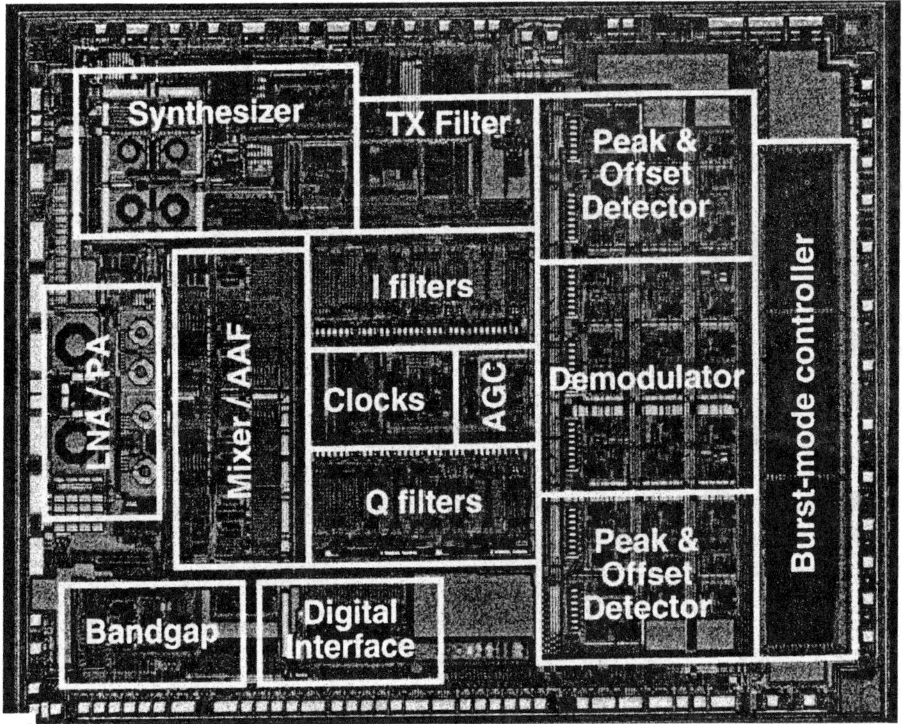

Figure 9.50 Cordless radio chip micrograph; from Cho [314], © 1999 IEEE.

9.4 Concluding Remarks

Traditional design is based largely on chipsets implemented in mixed technology. Specifically, RF power amplifier circuits have been implemented in GaAs, other front-end circuits have been implemented in bipolar, and baseband analog interface and digital processing in CMOS. With rapid decrease in feature size down to 0.1-0.18 μm, CMOS is achieving comparable performance as that of bipolar. Thus, the current trend is to push single-chip integration of radio systems in CMOS. In this chapter, several highly integrated CMOS radio chipsets have been described, including a single-chip spread-spectrum cordless radio. With careful system, architecture, and circuit design, integration of a complete radio system on a chip in CMOS is becoming very promising for the 1-2 GHz range.

While substantial progress has been made to integrating radio systems on a chip, there are still many technology and packaging related issues that cannot be solved directly with circuit techniques and system architecture design. For example, CMOS has a lossy substrate that limits the Q of on-chip inductors to below 10. This in combination of lower g_m / I ratio makes it a higher power solution in comparison with traditional bipolar designs. The high power dissipation of the example described in Section 9.3 indicates the limitation of CMOS in power dissipation.

No one technology is perfect. Bipolar technology is mature and primed for analog circuits but does not offer high level of integration for large digital designs. Digital CMOS, on the other hand, is ideal for large digital designs but is not optimized for analog designs. With lower g_m / I, CMOS requires higher power to provide a given gain and current drive for RF and analog circuits but since CMOS has virtually zero dynamic power dissipation it can result in extremely low power digital circuits. Moreover, due to the doping profile needed to prevent latch-ups in digital circuits, digital CMOS has a higher substrate loss (1-10 $\Omega \cdot cm$) that has limited on-chip inductor Q's to about 10 [317].

GaAs has transition frequency in the 80-100 GHz range and can withstand high voltage and thermal stress. Thus, it is widely used for power amplifier circuits. It also has a high resistivity substrate ($100\, M\Omega \cdot cm$) making it ideal for RF microwave designs. However, GaAs has the disadvantage of inability to support even medium levels of integration. SiGe has finally emerged after many years in the development stage. It offers a transition frequency (100 GHz) comparable to GaAs but with the added advantage of allowing higher levels of integration. Yet another emerging technology is silicon on insulator (SOI) or silicon on sapphire (SOS) which offers low substrate loss such that high-Q on-chip inductors could be integrated along with CMOS circuits. In the race toward a single-chip solution for Bluetooth, many companies are taking a high-risk approach by using these relatively new technologies.

While integrating the radio in a single technology, especially CMOS, has its appeal, many other on-going activities are aimed to leverage the advantages of different technologies for different part of the radio system. The mixed technology solution is integrated on a single substrate. Micro-assembly technologies such as Low-temperature co-fired ceramic (LTCC) [318][319] and multichip module (MCM) [320] [321][322] offer a means to accomplish this. Besides being able to leverage the advantages of different technologies, LTCC and MCM offer a means to embed discrete passives, filters, and resonators into a single substrate along with the IC. An example of this [323] is shown in Figure 9.51. In this case, the substrate is based on high-resistivity silicon. The IC, a GSM transceiver, is flip-chip bonded to the silicon substrate which contains embedded high-Q passives, e.g. an inductor with a Q of 60.

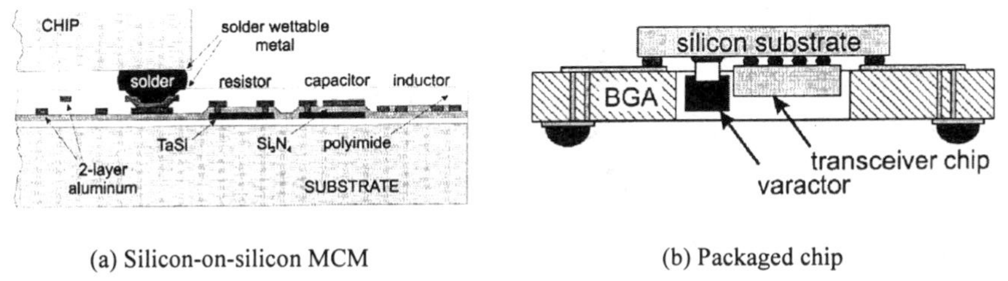

(a) Silicon-on-silicon MCM

(b) Packaged chip

Figure 9.51 Silicon-on-silicon integration of a GSM transceiver; from Davis [323], © 1998 IEEE.

It is not clear which technology approach will offer the best single-chip solution for future radio communications. However, one thing is clear - despite the various difficulties and many design challenges, we are hedging onwards to integrate a complete radio system on a chip. Whether the single-chip radio will be based on CMOS, SiGe, SOI/SOS, or a mixture of different technologies, future solutions will most likely depend on system requirements and economic considerations.

References

[1] T. Lewis, Empire of The Air: The Men Who Made Radio, E. Burlingame Books, New York, 1991.

[2] J. N. Sheth and R. S. Sisodia, "A strategic vision of the wireless industry: communications unbound," International Engineering Consortium, 1997.

[3] L. J. Harte, A. D. Smith, and C. A. Jacobs, IS-136 TDMA Technology, Economics, and Services, Artech House, Boston, 1998.

[4] J. S. Lee and L. E. Miller, CDMA Systems Engineering Handbook, Artech House, Boston, 1998.

[5] TIA/EIA Standard, "Mobile station-base station compatibility standard for wideband spread spectrum cellular systems," TIA/EIA-95-B, March 1999.

[6] J. E. Padgett, C. G. Gunther, and T. Hattori, "Overview of wireless personal communications," IEEE Communications Magazine, Vol. 33, No. 1, pp. 28-41, Jan. 1995.

[7] W. H. Tuttlebee, "Cordless personal communications," IEEE Communications Magazine, Vol. 30, pp. 42-53, December 1992.

[8] S. M. Redl, M. K. Weber, M. W. Oliphant, An Introduction to GSM, Artech House, Boston, 1995.

[9] R. D. Carsello, R. Meiden, S. Allpress, F. O'Brien, J .A. Tarrallo, N. Ziesse, A. Arunachalam, J. M. Costa, E. Berruto, R. C. Kirby, A. Maclatchy, F. Watanabe, and H. Xia, "IMT-2000 standards: radio aspects," IEEE Personal Communications, vol. 4, no. 4, pp. 30-40, August 1997.

[10] T. Ojanpera and R. Prasad, Wideband CDMA for Third Generation Mobile Communications, Artech House, Boston, 1998.

[11] B. P. Crow, L. Widjaja, L. Kim, P. T. Sakai, "IEEE 802.11 wireless local area networks," IEEE Communications Magazine, vol. 35, no. 9, pp. 116-126, September 1997.

[12] Bluetooth SIG, "Specification of the Bluetooth System," Vol. 1, December 1999.

[13] H. Nyquist, "Certain topics in telegraph transmission theory," AIEE Transactions, vol. 47, pp. 617-644, 1928.

[14] S. Benedetto, E. Biglieri, V. Castellani, Digital Transmission Theory, Prentice Hall, New Jersey, 87.

[15] A. D. Whalen, Detection of Signals in Noise, Academic Press, San Diego, 71.

[16] GSM Recommendation 6.10, "GSM full rate speech transcoding", ETSI PT 12, February 1992.

[17] C. E. Shannon, "A mathematical theory of communication," Bell System Technical Journal, vol. 27, part I, pp. 379-423 and part II, pp. 623-656, July 1948.

[18] J. B. Anderson and S. Mohan, Source and Channel Coding, Kluwer Academic Publishers, Boston, 1991.

[19] A. J. Viterbi, CDMA Principles of Spread-Spectrum Communication, Addison Wesley, Reading, 1995.

[20] Brennan, D. G., "Linear diversity combining techniques," Proceedings of IRE, Vol. 47, pp. 1075-1102, June 1959.

[21] J. B. Johnson, "Thermal agitation of electricity in conductors," Phys. Rev., vol. 32, July 1928, pp. 97-109.

[22] H. Nyquist, "Thermal agitation of electric charge in conductors," Phys. Rev., vol. 32, July 1928, pp. 110-113.

[23] J. R. Herman, "Survey of man-made noise," Progress in Radio Science, vol. 1, pp. 315-348, 1971.

[24] H. T. Friis, "A note on a simple transmission formula," Proc. IRE, 34, 1946.

[25] K. Pahlavan and A. H. Levesque, Wireless Information Networks, John Wiley & Sons, New York, 1995.

[26] D. Akerberg, "Properties of a TDMA pico-cellular office communication system," Proc. IEEE GLOBECOM, Florida, pp. 1343-1349, December 1988.

[27] M. Marsan, G. C. Hess, S. S. Gilbert, "Shadowing variability in an urban land mobile environment at 900 MHz," Electronics Letters, vol. 26, pp. 646-8, May 1990.

[28] W. C. Jakes, Microwave Mobile Communications, IEEE Press, New Jersey, 1974.

[29] R. J. C. Bultitude, "Measurement, characterization and modeling of indoor 800/900 MHz radio channels for digital communications," IEEE Communications Magazine, Vol. 25, No. 6, pp. 5-12, June 1987.

[30] P. A. Bello, "Characterization of randomly time-variant linear channels," IEEE Transactions on Communications Systems, CS-11, pp. 360-393, 1963.

[31] T. Kailath, "Sampling models of variable networks," Proc. IRE, Vol. 38, pp. 291-299, 1950.

[32] G. L. Turin, F. D. Clapp, T. L. Johnston, S. B. Fine, D. Lavry, "A Statistical model of urban multipath propagation," IEEE Transactions on Vehicular Technology, VT-21, pp. 1-9, Feb. 1972.

[33] J. C.-I. Chuang, "The effects of time delay spread on portable radio communications with digital modulation," IEEE Journal on Selected Areas in Communications, SAC-5, pp. 879-889, June 1985.

[34] R. A. Monzingo, Introduction to Adaptive Arrays, Wiley, New York, 1980.

[35] A. J. Paulraj and C. B. Papadias, "Space-time processing for wireless communications," IEEE Signal Processing Magazine, Vol. 14, pp. 49-83, November 1997.

[36] G. A. Arredondo, W. H. Chriss, and E. H. Walker, "A Multipath fading simulator for mobile radio," IEEE Transactions on Communications, pp. 1325-1328, November 1973.

[37] D. Parsons, The Mobile Radio Propagation Channel, Pentech Press, London, 1992.

[38] W. C. Y. Lee, Mobile Cellular Telecommunications Systems, McGraw-Hill, New York, 1989.

[39] A. A. M. Saleh and R. A. Valenzuela, "A Statistical model for indoor multipath propagation," IEEE J. Selected Areas in Communications, SAC-5, No. 2, pp. 128-137, Feb. 1987.

[40] T. S. Rappaport, "Indoor radio communication for factories of the future," IEEE Communications Magazine, Vol. 27, No. 5, pp. 15-24, May 1989.

[41] T. S. Rappaport, "Characterization of UHF multipath radio channels in factory buildings," IEEE Transactions on Antenna Propagation, Vol. 37, pp. 1058-1069, August 1989.

[42] D. M. J. Devasirvatham, "Time delay spread and signal level measurements of 850 MHz radio waves in building environments," IEEE Transactions on Antenna Propagation, AP-34, No. 11, pp. 1300-1305, November 1986.

[43] S. Howard and K. Pahlavan, "Doppler spread measurements of the indoor radio channel," IEE Electronics Letters, vol. 26, pp. 107-109, January 1990.

[44] R. B. Ertel, P. Cardieri, K. W. Sowerby, T. S. Rappaport, J. H. Reed, "Overview of spatial channel models for antenna array communication systems," IEEE Personal Communications, vol. 5, no. 1, pp. 10-22, February 1998.

[45] V. Castellani, L. Lo Presti, and M. Pent, "Multilevel DCPSK over real channels," IEEE International Conference on Communications, Minneapolis, June 1974.

[46] S. A. Gronemeyer and A. L. McBride, "MSK and offset QPSK modulation," IEEE Transactions on Communications, vol. 24, no. 8, pp. 809-820, August 1976.

[47] Y. Akaiwa and Y. Nagata, "Highly efficient digital mobile communications with a linear modulation method," IEEE Journal on Selected Areas in Communication, vol. 5, no. 5, pp. 890-895, June 1987.

[48] T. Aulin and C. E. W. Sundberg, "Continuous phase modulation – Part I: Full response signaling," IEEE Transactions on Communications, vol. 29, no. 3, pp. 196-209, March 1981.

[49] T. Aulin, N. Rydbeck, and C. E. W. Sundberg, "Continuous phase modulation – Part II: Partial response signaling," IEEE Transactions on Communications, vol. 29, no. 3, pp. 210-225, March 1981.

[50] G. D. Forney, Jr., "The Viterbi algorithm," IEEE Proceedings, vol. 61, pp. 268-278, March 1973.

[51] S. Pasupathy, "Minimum shift keying a spectrally efficient modulation," IEEE Communications Magazine, vol. 17, pp. 14-22, July 1979.

[52] K. Murota and K. Hirade, "GMSK modulation for digital mobile radio telephony," IEEE Transaction on Communications, COM-29, pp. 1044-1050, July 1981.

[53] J. Li and M. Kavehrad, "OFDM-CDMA systems with nonlinear power amplifier," IEEE Wireless Communications and Networking Conference, vol. 3, pp. 1167-71, September 21-24 1999.

[54] P.-R. Chang and C.-F. Lin, "Design of spread-spectrum multicode CDMA transport architecture for multimedia services," IEEE Journal on Selected Areas in Communications, Vol. 18, pp. 99-111, January 2000.

[55] H. C. Osborne, "A generalized polarity-type Costas loop for tracking MPSK signals," IEEE Transactions on Communications, Vol. 30, pp. 2289-2296, October 1982.

[56] F. M. Gardner, Phaselock Techniques, Wiley Intersciences, New York, 1979.

[57] D. C. Rife and R. R. Boorstyn, "Single-Tone Parameter Estimation from Discrete-Time Observations," IEEE Transactions on Information Theory, Vol. IT-20, No. 5, pp. 591-598, Sept. 1974.

[58] F. Classen, H. Meyr, P. Sehier, "Maximum-Likelihood Open Loop Carrier Synchronizer for Digital Radio," IEEE International Conference on Communications, pp. 493-7, 1993.

[59] F. Classen and H. Meyr, "Two Frequency Estimation Schemes Operating Independently of Timing Information," IEEE GLOBECOM, pp. 1996-2000, 1993.

[60] V. Aggarwal and C. Chien, "A robust low-complexity estimation technique for frequency acquisition over frequency-selective channels," IEEE International Conference on Communications, Vancouver, June 1999.

[61] H. P. Hartmann, "Analysis of a dithering loop for PN code tracking," IEEE Transactions on Aerospace and Electronic Systems, vol. 10, no. 1, pp. 2-9, January 1974.

[62] W. C. Lindsey, Synchronization Systems in Communication and Control, Prentice-Hall, New Jersey, 1971.

[63] P. Elias, "Error-free coding," IRE Transaction on Information Theory, IT-4, pp. 29-37, 1950.

[64] R. E. Blahut, Theory and Practice of Error Control Codes, Addison Wesley, Reading, 1984.

[65] S. Lin and D. J. Costello, Jr., Error Control Coding, Prentice Hall, New Jersey, 1983.

[66] R. C. Bose and D. K. Ray-Chaudhuri, “On a class of error correcting binary group codes,” Information and Control, vol. 3, pp. 68-79, March 1960.

[67] I. S. Reed and G. Solomon, “Polynomial codes over certain finite fields,” J. Soc. Ind. Appl. Math., vol. 8, pp. 300-304, June 1960.

[68] V. K. Bhargava and S. B. Wicker, Reed-Solomon Codes and Their Applications, IEEE Press, New Jersey, 1994.

[69] J. E. Meggitt, “Error correcting codes and their implementation,” IRE Transactions on Information Theory, vol. 7, pp. 232-255, October 1961.

[70] R. W. Hamming, “Error detecting and error correcting codes,” Bell Laboratory System Technical. Journal, vol. 29, pp. 147-160, April 1950.

[71] J. P. Stenbit, “Table of Generators for BCH Codes,” IEEE Trans. Information Theory, vol. IT-10, pp. 390-391, October 1964.

[72] GSM Recommendation 5.03, “Channel Coding”, ETSI PT 12, October 1994.

[73] A. J. Viterbi, “Error bounds for convolutional codes and an asymptotically optimum decoding algorithm,” IEEE Transactions on Information Theory, Vol. IT-15, pp. 177-179, January 1979.

[74] Larsen, K. J., “Short convolutional codes with maximal free distance for rates ½, 1/3, and ¼,” IEEE Trans. Information Theory, vol. IT-19, pp. 371-372, May 1973.

[75] D. G. Daut, J. W. Modestino, and L. D. Wismer, “New short constraint length convolutional code construction for selected rational rates,” IEEE Trans. Information Theory, vol. IT-28, pp. 793-799, September 1982.

[76] R. Steele, Mobile Radio Communications, Pentech Press, IEEE Press, New Jersey, 1992.

[77] E. R. Berlekamp, “On decoding binary Bose-Chaudhuri-Hocquenghem codes,” IEEE Trans. Information Theory, IT-11, pp. 577-580, Oct. 1965.

[78] J. L. Massey, “Shift-register synthesis and BCH decoding,” IEEE Trans. Information Theory, IT-15, pp. 122-127, Jan. 1969.

[79] R. T. Chien, “Cyclic decoding procedure for the Bose-Chaudhuri-Hocquenghem codes,” IEEE Trans. Information Theory, IT-10, pp. 357-363, October 1964.

[80] G. D. Forney, Jr., Concatenated Codes, MIT Press, Cambridge, Mass., 1966.

[81] G. Ungerboeck, “Trellis-coded modulation with redundant signal sets – Parts I and II,” IEEE Communications Magazine, vol. 25, pp. 5-21, Feb. 1987.

[82] E. Biglieri, D. Divsalar, P. J. McLane, M. K. Simon, Introduction to Trellis-Coded Modulation with Applications,” Macmillan, New York, 1991.

[83] G. Ungerboeck, “Channel coding with multilevel/phase signals,” IEEE Trans. Information Theory, vol. IT-28, pp. 55-67, January 1982.

[84] H. V. Poor and G. W. Wornell, Wireless Communications, Prentice Hall, New Jersey, 1998.

[85] C. Leung, “Optimized selection diversity for Rayleigh fading channels,” IEEE Transactions on Communications, COM-30, pp. 554-557, March 1982.

[86] M. Kavehrad and B. Ramamurthi, “Direct-sequence spread-spectrum with DPSK modulation and diversity for indoor wireless communications,” IEEE Transactions on Communications, COM-35, pp. 224-236, February 1987.

[87] R. Price and P. E. Green, Jr., “A Communication technique for multipath channels,” Proc. IRE, vol. 46, pp. 555-570, March 1958.

[88] S. U. H. Qureshi, “Adaptive Equalization,” Proceedings of the IEEE, vol. 73, No 9, September 1985.

[89] C. A. Belfiore and J. H. Park, Jr., “Decision feedback equalization,” Proceedings of the IEEE, vol. 67, pp. 1143-1156, Aug. 1979.

[90] C. D. Forney, Jr., "Maximum likelihood sequence estimation of digital sequences in the presence of intersymbol interference," IEEE Transactions on Information Theory, IT-18, pp. 363-378, May 1972.

[91] J. G. Proakis, Digital Communications, McGraw Hill, New York, 1995.

[92] J. L. Ramsey, "Realization of optimum interleavers," IEEE Transaction on Information Theory, IT-16, pp. 338-345.

[93] G. D. Forney Jr., "Burst correcting codes for the classic bursty channel," IEEE Transaction on Communications Technology, COM-19, pp. 772-781, October 1971.

[94] S. H. Reiger, "Codes for the correction of clustered errors," IRE Transaction on Information Theory, IT-6, pp. 16-21, March 1960.

[95] S. Verdu, "Recent progress in multiuser detection," Advances in Communications and Signal Processing, Springer-Verlag, Berlin, 1989.

[96] G. J. Foschini, "Layered space-time architecture for wireless communication in a fading environment when using multiple antennas," Bell laboratories Technical Journal, vol. 1, no. 2, autumn 1996, pp. 41-59.

[97] P. W. Wolniansky, G. J. Foschini, G. D. Golden, R. A. Valenzuela, "V-BLAST: an architecture for realizing very high data rates over the rich-scattering wireless channel," URSI International Symposium on Signals, Systems, and Electronics, pp. 295-300, September 29 – October 2 1998.

[98] A. Salamasi and K. S. Gilhausen, "On the system design aspects of code division multiple access (CDMA) applied to digital cellular and personal communications networks," 41st IEEE Vehicular Technology Conference, St. Louis, pp. 57-62, May 1991.

[99] A. Naguib, A. R. Calderbank, H. Jafarkhani, N. Seshadri, and V. Tarokh, "Recent progress in space-time coding," Sixth Workshop on Smart Antennas in Wireless Mobile Communications, Stanford University, July 22-23, 1999.

[100] A. C. Dhanantwari, S. Stergiopoulos, W. Phillips, and W. Robertson, "Adaptive beamforming with near-instantaneous convergence for matched filter processing," *Proceedings of the 1996 Canadian Conference on Electrical and Computer Engineering*, vol. 2, pp. 683-6, May 1996.

[101] J. Litva and T. K.-Y. Lo, *Digital Beamforming in Wireless Communications*, Artech House, Boston, 1996.

[102] D. Gerlach and A. Paulraj, "Base station transmitting antenna arrays for multipath environments," *Signal Processing*, vol. 54, pp. 59-73, October 1996.

[103] J.-W. Liang, J.-T. Chen, and A. Paulraj, "A two-stage hybrid approach for CCI/ISI reduction with space-time processing," *IEEE Communications Letters*, vol. 1, pp. 163-5, November 1997.

[104] K. I. Pedersen, P. Mogensen, and J. B. Andersen, "Vector RAKE receiver for third generation systems – an evaluation," *Sixth Workshop on Smart Antennas in Wireless Mobile Communications*, Stanford University, July 22-23, 1999.

[105] S. N. Diggavi and A. Paulraj, "Performance of multisensor adaptive MLSE in fading channels," *IEEE 47th Vehicular Technology Conference*, vol. 3, pp. 2148-52, May 1997.

[106] B. Widrow, P. E. Mantey, L. J. Griffiths, and B.B. Goode, "Adaptive antenna systems," *Proceedings of the IEEE*, vol. 55, no. 12, pp. 2143-2159, December 1967.

[107] B. Widrow and M. E. Hoff, Jr., "Adaptive switching circuits," in *IRE WESCON Conv. Rec.*, part 4, pp. 96-104, Aug. 1960.

[108] C. A. Baird, Jr., "Recursive minimum variance estimation for adaptive sensor arrays," *IEEE International Conference on Cybernetics and Society*, October 9-12, Washington DC, pp. 412-414, 1972.

[109] W. F. Gabriel, "Adaptive arrays – an introduction," *Proceedings of the IEEE*, vol. 64, no. 2, pp. 239-272, February 1976.

[110] D. N. Godard, "Channel equalization using a Kalman filter for fast data transmission," *IBM J. Res. Develop.*, vol. 18, pp. 267-273, May 1974.

[111] I. S. Reed and J. D. Mallett, L. E. Brennan, "Rapid convergence rate in adaptive arrays," *IEEE Transactions on Aerospace Electronic Systems*, vol. 10, no. 6, pp. 853-863, November 1974.

[112] B. Frieflander, "Lattice filters for adaptive processing," *Proceedings of IEEE*, vol. 70, pp. 829-867, August 1982.

[113] N. E. Lay, A. Polydoros, "Per-survivor processing for channel acquisition, data detection, and modulation classification," *28th Asilomar Conference on Signals, Systems and Computers*, Vol. 2, pp. 1169-73, October 31 – November 2 1994.

[114] M. K. Simon, J. K. Omura, R. A. Scholtz, B. K. Levitt, *Spread Spectrum Communications Handbook*, McGraw Hill, 1994.

[115] R. L. Pickholtz and D. L. Schilling and L. B. Milstein, "Theory of spread spectrum communications – a tutorial," *IEEE Transactions on Communications*, COM-30, pp. 855-884, May 1982.

[116] D. J. Torrieri, *Principles of Secure Communication Systems*, Artech House, Massachusetts, 1992.

[117] K. S. Gilhousen, I. M. Jacobs, R. Padovani, L. A. Weaver, Jr., Increased capacity using CDMA for mobile satellite communication," *IEEE Journal on Selected Areas in Communications*, Vol. 8, pp. 503-514, May 1990.

[118] K. S. Gilhousen, I. M. Jacobs, R. Padovani, A. J. Viterbi, L. A. Weaver, Jr., and C. E. Wheatley III, "On the capacity of a cellular CDMA system," *IEEE Transactions on Vehicular Technology*, Vol. 40, pp. 303-312, May 1991.

[119] A. M. Viterbi and A. J. Viterbi, "Erlang capacity of a power controlled CDMA system," *IEEE Journal on Selected Areas in Communications*, Vol. 11, No. 6, pp. 892-900, Aug. 1993.

[120] G. L. Turin, "Introduction to spread spectrum anti-multipath techniques and their application to urban digital radio," *Proceedings of the IEEE*, Vol. 68, No. 3, pp. 328-353, March 1980.

[121] J. T. Harvey, "High-speed M-sequence generation," *Electronic Letters*, Vol. 10, pp. 480-481 Nov. 1974.

[122] R. Gold, "Optimum binary sequences for spread spectrum multiplexing," *IEEE Transactions on Information Theory*, IT-13, pp. 619

[123] R. T. Barghouthi and G. L. Stuber, "Rapid sequence acquisition for DS/CDMA systems employing Kasami sequences," *IEEE Transactions on Communications*, Vol. 42, pp. 1957-1968, Feb.-April 1994.

[124] S. C. Martin and G. S. Sidhu, "Orthogonality and correlation properties of Walsh functions for the design of fast acquisition ranging codes," *IEEE International Symposium on Information Theory*, p. 80, 1976.

[125] H. T. Nicholas III, H. Samueli, "A 150-MHz direct digital frequency synthesizer in 1.25-μm CMOS with –90 dBc spurious performances," *IEEE Journal of Solid-State Circuits,* Vol. 26, pp. 1959-1969, December 1991.

[126] L. K.-L. Lau, "A silicon compiler for direct digital frequency synthesizers," *Masters Thesis, School of Engineering and Applied Sciences, University of California at Los Angeles*, 1991.

[127] L. Tan, E. Roth, G. Yee, H. Samueli, "An 800-MHz quadrature digital synthesizer with ECL-compatible output drivers in 0.8-μm CMOS," *IEEE Journal of Solid-State Circuits,* Vol. 30, pp. 1463-1473, Dec. 1995.

[128] G. Chang, M. K. Ku, A. A. Abidi, and H. Samueli, "A low power agile frequency synthesizer in 1.2-μm CMOS," *ISSCC 1994 Digest of Technical Papers*, Vol. 34, pp. 32-33, Feb. 1994.

[129] A. Polydoros and M. K. Simon, "Generalized serial search code acquisition: the equivalent circular state diagram approach," *IEEE Transactions on Communications*, Vol. 32, pp. 1260-1268, May 1984.

[130] Polydoros and C. L. Weber, "A unified approach to serial search spread spectrum code acquisition," *IEEE Transactions on Communications*, Vol. 32, No. 5, pp. 542-560, May 1984.

[131] D. Avidor, S.-S. Hang, and J. Omura, "A direct-sequence spread spectrum transceiver chip," Proceedings of the Custom Integrated Circuits Conference, pp. 16.4.1-16.4.4, San Diego, California, May 9-12, 1993.

[132] Shibano T., Lizuka K., et. al., "Matched Filter for DS-CDMA of up to 50MChips/s Based on Sampled Analog Signal Processing", *IEEE International Solid-State Circuits Conference Digest of Technical Papers*, p.100-1, Feb. 1997.

[133] C. Deng and C. Chien, "A PN-acquisition ASIC for wireless CDMA systems," *Proceedings of the IEEE Custom Integrated Circuits Conference*, Orlando, Florida, May 21-24, 2000.

[134] R. B. Ward, "Acquisition of pseudonoise signals by sequential estimation," *IEEE Transactions on Communications*, Vol. 13, pp. 475-483, December 1965.

[135] A. Weinberg, "Generalized analysis for the evaluation of search strategy effects on PN acquisition performance," *IEEE Transactions on Communications*, Vol. 31, pp. 37-49, January 1983.

[136] D. M. DiCarlo and C. L. Weber, "Multiple dwell serial search: performance and application to direct sequence code acquisition," *IEEE Transactions on Communications*, Vol. 31, No. 5, pp. 650-659, May 1983.

[137] S. G. Glisic, "Automatic Decision Threshold Level Control in Direct-Sequence Spread Spectrum Systems", *IEEE Transactions on Communications*, Vol. 39, No. 2, Feb. 1991.

[138] B.-Y. Chung, C. Chien, H. Samueli, and R. Jain, "Performance analysis of an all-digital BPSK direct-sequence spread spectrum IF receiver architecture," *IEEE Journal on Selected Areas in Communications*, Vol. 11, pp. 1096-1107, 1993.

[139] B. Razavi, "Design considerations for direct-conversion receivers," *IEEE Transactions on Circuits and Systems II: Analog and Digital Signal Processing*," vol. 44, no. 6, pp. 428-435, June 1997.

[140] A. A. Abidi, "Direct-conversion radio transceivers for digital communications," IEEE Journal of Solid-state Circuits, vol. 30, no. 12, pp. 1399-1410, December 1995.

[141] Thomas H. Lee, *The Design of CMOS Radio-Frequency Integrated Circuits*, Cambridge University Press, Cambridge, 1998.

[142] D. B. Leeson, "A simple model of feedback oscillator noise spectrum," *Proceedings of the IEEE*, vol. 54, pp. 329-330, 1966.

[143] J. L. Plumb and E. R. Chenette, "Flicker noise in transistors," *IEEE Transactions on Electron Devices*, Vol. 10, pp. 304-308, September 1963.

[144] R. C. Jaeger and A. J. Broderson, "Low-frequency noise sources in bipolar junction transistors," *IEEE Transactions on Electron Devices*, Vol. 17, pp. 128-134, February 1970.

[145] N. Gerlich, "On the spatial multiplexing gain of SDMA for wireless local loop access," *Second European Personal Mobile Communications Conference*, September 30 – October 2, 1997, pp. 229-235.

[146] C. B. Papadias and A. Paulraj, "A space-time constant modulus algorithm for SDMA systems," *IEEE 46th Vehicular Technology Conference*, April 28 – May 1, New York, pp. 86-90, 1996.

[147] Paulraj, "Space-time processing for third-generation wireless networks," *First Annual UCSD Conference on Wireless Communications*, San Diego, March 8-10, 1998, pp. 133-137.

[148] J. Paulraj and B. C. Ng, "Space-time modems for wireless personal communications," *IEEE Personal Communications Magazine*, vol. 5, no. 1, pp. 36-48, February 1998.

[149] J. Wu and R. Kohno, "A wireless multimedia CDMA system based on transmission power control," *IEEE Journal on Selected Areas in Communications*, vol. 14, no. 4, pp. 683-691, May 1996.

[150] F. Simpson and J. M. Holtzman, "Direct sequence CDMA power control, interleaving, and coding," *IEEE Journal on Selected Areas in Communications*, vol. 11, no. 7, pp. 1085-1095, September 1993.

[151] H. Takanashi, T. Tanaka, T. Oono, H. Matsuki, and T. Hattori, "High-quality and high-speed wireless multimedia transmission technology for Personal Handy-phone System," *Wireless Personal Communications*, vol. 9, no. 3, Kluwer Academic Publishers, pp. 255-270, May 1999.

[152] GSM Recommendation 6.32, "Voice activity detection," *ETSI/PT 12*, February 1992.

[153] GSM Recommendation 6.31, "Discontinuous transmission for full-rate speech traffic channels", *ETSI PT 12*, October 1992.

[154] GSM Recommendation 3.50, "Transmission planning aspects of the speech service in the GSM PLMN system," *ETSI SMG*, July 1994.

[155] G. Brasche and B. Walke, "Concepts, services, and protocols of the new GSM phase 2+ general packet radio service," *IEEE Communications Magazine*, vol. 35, no. 8, pp. 94-104, August 1997.

[156] A. Furuskar, S. Mazur, and F. Muller, H. Olofsson, "EDGE: enhanced data rates for GSM and TDMA/136 evolution," *IEEE Personal Communications*, vol. 6, no. 3, pp. 56-66, June 1999.

[157] A. K. Salkintzis, "Radio resource management in cellular digital packet data networks," *IEEE Personal Communications*, vol. 6, no. 6, pp. 28-36, December 1999.

[158] M. Sreetharan and R. Kumar, *Cellular Digital Packet Data*, Artech House, Boston, 1996.

[159] T. Nishikawa, et. al., "A 60-MHz 240 mW MPEG-4 Video-Phone LSI with 16Mb Embedded DRAM," *International Solid-State Circuits Conference Slide Supplement*, San Francisco, CA, February 2000, pp. 182-183.

[160] O. K. Al-shaykh, I. Moccagatta, and H. Chen, "JPEG-2000: a new still image compression standard," *32nd Asilomar Conference on Signals, Systems, and Computers*, Pacific Grove, CA, November 1-4 1998, pp. 99-103.

[161] J. R. Boucher, *Voice Teletraffic Systems Engineering*, Artech House, Boston, 1988.

[162] L. Kleinrock, *Queuing Systems Volume 1: Theory*, John Wiley & Sons, New York, 1975.

[163] C. Chien, M. B. Srivastava, R. Jain, P. Lettieri, V. Aggarwal, and R. Sternowski, "Adaptive radio for multimedia wireless links," *IEEE Journal on Selected Areas in Communications*, vol. 17, no. 5, pp. 793-813, May 1999.

[164] A. A. Abidi, "Low-power radio-frequency IC's for portable communications," *Proceedings of the IEEE*, vol. 83, no. 4, pp. 544-569, April 1995.

[165] C. D. Hull, J. Leong, Tham, and R. R. Chu, "A direct-conversion receiver for 900-MHz (ISM band) spread-spectrum digital cordless telephone," IEEE Journal of Solid-State Circuits, vol. 31, no. 12, pp. 1955-1963, December 1996.

[166] D. Haspeslagh, J. Oeuterick, L. Kiss, J. Wenin, A. Vanwelsenaers, C. Enel-Rehel, "BBTRX: a baseband transceiver for a zero IF GSM hand portable station," *Proceedings of the IEEE 1992 Custom Integrated Circuits Conference*, pp. 10.7/1-4, New York, NY, USA, May 3-6, 1992.

[167] F. Op 'tEynde, J. Craninckx, and P. Goetschalckx, "A fully-integrated zero-IF DECT transceiver," *2000 IEEE International Solid-State Circuits Conference Digest of Technical Papers*, pp. 138-139, San Francisco, CA, 7-9 Feb. 2000.

[168] T. Cho, E. Dukatz, M. Mack, D. Macnally, M. Marringa, S. Mehta, C. Nilson, L. Plouvier, and S. Rabii, "A single-chip CMOS direct-conversion transceiver for 900 MHz spread-spectrum digital cordless phones," *IEEE International Solid-state Circuits Conference*, San Francisco, CA, February 1999, pp. 228-229.

[169] H. Darabi, and A. A. Abidi, "An ultra low power single-chip CMOS 900 MHz receiver for wireless paging," Proceedings of the IEEE 1999 Custom Integrated Circuits Conference, pp. 213-216, May 1999, San Diego, California.

[170] A. Parssinen, J. Jussila, J. Ryynanen, L. Sumanen, and I. A. K. Halonen, "A 2-GHz wide-band direct conversion receiver for WCDMA applications," *IEEE Journal of Solid-state Circuits*, vol. 34, no. 12, pp. 1893-1903, December 1999.

[171] T.-P. Liu, E. Westerwick, N. Rohani, and R.-H. Yan, "5GHz CMOS radio transceiver front-end chipset," *2000 IEEE International Solid-State Circuits Conference Digest of Technical Papers*, pp. 320-321, San Francisco, CA, 7-9 Feb. 2000.

[172] J. S. M. Steyaert and J. Crols, "Low-IF topologies for high-performance analog front ends of fully integrated receivers," *IEEE Transactions on Circuits and Systems II: Analog and Digital Signal Processing*, vol. 45, no. 3, pp. 269-282, March 1998.

[173] J. C. Rudell, J.-J. Ou, R. S. Narayanaswami, G. Chien, J. A. Weldon, L. Lin, K.-C. Tsai, L. Tee, K. Khoo, D. Au, T. Robinson, D. Gerna, M. Otsuka, P. R. Gray, "Recent developments in high integration multi-standard CMOS transceivers for personal communication systems," *International Symposium on Low Power Electronics and Design*, Monterey, CA, August 1998, pp. 149-154.

[174] J. Cros and M. Steyaert, "A single-chip 900 MHz CMOS receivers front-end with a high performance low-IF topology," *IEEE J. of Solid-State Circuits*, pp. 1483-1492, December 1995.

[175] T. Stetzler, I. G. Post, J. H. Havens, M. Koyama, "A 2.7-4.5 V single chip GSM transceiver RF integrated circuits," *IEEE Journal of Solid-State Circuits*, vol. 30, no. 12, pp. 1421-1429, December 1995.

[176] H. Samueli, T.-J. Lin, R. Hawley, S. Olafson, "VLSI architectures for a high-speed tunable digital modulator/demodulator/bandpass-filter chip set," *1992 IEEE International Symposium on Circuits and Systems*, San Diego, CA, May 10-13 1992, pp. 1065-1068.

[177] B. Walden, "Analog-digital converter survey & analysis," *IEEE Emerging Technology Conference*, Dallas, April 14, 1999.

[178] K. C. Wang, K. R. Nary, R. B. Nubling, S. M. Beccue, R. L. Pierson, M. F. Chang, P. M. Asbeck, R. T. Huang, "AlGaAs/GaAs heterojunction bipolar transistor technology for multi-giga-samples per second ADCs," *24th European Microwave Conference*, Cannes, France, September 5-8, 1994, vol. 1, pp. 177-184.

[179] C. Chien, R. Jain, E. Cohen, H. Samueli, "A12.7 Mchips/sec All-Digital BPSK Direct-Sequence Spread Spectrum IF Transceiver in 1.2-μm CMOS", *IEEE Journal of Solid-State Circuits*, Vol. 29, No. 12, December 1994.

[180] A. Rofougaran, G. Chang, J. J. Rael, J.-Y.-C. Chang, M. Rofougaran, P. J. Chang, M. Djafari, M.-K. Ku, E. W. Roth, A. A. Abidi, H. Samueli, "A single-chip 900-MHz spread-spectrum wireless transceiver in 1 μm CMOS," *IEEE Journal of Solid-State Circuits*, vol. 33, no. 4, pp. 515-534, April 1998.

[181] W. B. Kuhn, N. K. Yanduru, A. S. Wyszynski, "Q-enhanced LC bandpass filters for integrated wireless applications," IEEE Transactions on Microwave Theory and Techniques, vol. 46, no. 12, *1998 IEEE MTT-S International Microwave Symposium*, Baltimore, MD, June 7-12 1998, pp. 2577-2586.

[182] R. C. Johnson, *Antenna Engineering Handbook*, McGraw Hill, New York, 1993.

[183] P. R. Gray, *Analysis and Design of Analog Integrated Circuits*, Wiley, New York, 1993.

[184] A. Van der Ziel, "Thermal Noise in Field-effect transistors," *Proceedings of the IRE*, vol. 50, pp. 1808-1812, August 1962.

[185] A. Van der Ziel, *Noise*, Prentice Hall, New York, 1954.

[186] K. Laker and W. Sansen, *Design of Analog Integrated Circuits and Systems*, McGraw Hill, New York, 1996.

[187] J. Craninckx and M. S. J. Steyaert, "A 1.8-GHz low-phase noise CMOS VCO using optimized hollow spiral inductors," *IEEE Journal of Solid-state circuits*, vol. 32, no. 5, May 1997, pp. 736-744.

[188] S. Ramo, J. R. Whinnery, T. Van Duzer, *Fields and Waves in Communication Electrons*, Wiley, New York, 1994.

[189] J. Smith, Modern *Communication Circuits*, McGraw Hill, New York, 1986.

[190] D. K. Shaeffer and T. H. Lee, "A 1.5V, 1.5 GHz CMOS low noise amplifier," *IEEE Journal of Solid State Circuits*, May 1997.

[191] P. E. Chadwick, "High performance integrated circuit mixers," *Radio Receivers & Assoc. Syst.*, London, pp. 1-9 1981.

[192] J. Crols, M. S. J. Steyaert, "A 1.5 GHz highly linear CMOS downconversion mixer," *IEEE Journal of Solid-state Circuits*, vol. 30, no. 7, pp. 736-742, July 1995.

[193] P. Y. Chan, A. Rofougaran, K. A. Ahmed, and A. A. Abidi, "A highly linear 1 GHz CMOS downconversion mixer," *European Solid-State Circuits Conference*, Sevilla, Spain, pp. 210-213, 1993.

[194] G. C. Gillette, "The digiphase synthesizer," *23rd Annual Symposium on Frequency Control*, Fort Monmouth, New Jersey, pp. 201-210, 1969.

[195] B. Razavi, *Monolithic Phase-locked Loops and Clock Recovery Circuits: Theory and Design*, IEEE Press, Piscataway, NJ, 1996.

[196] P. V. Brennan, *Phase-locked loops: Principles and Practice*, McGraw Hill, NY, 1996.

[197] B. Razavi and J. J. Sung, "A 6-GHz 68 mW BiCMOS Phase-locked loop," *IEEE Journal of Solid-State Circuits*, vol. 29, no. 12, pp. 1560-1565, Dec. 1994.

[198] S. K. Enam and A. A. Abidi, "NMOS ICs for clock and data regeneration in Gb/s optical fiber receivers," *IEEE Journal of Solid State Circuits*, vol. 27, pp. 1763-1774, December 1992.

[199] M. K. Kazimierczuk, "A new concept of class F tuned power amplifier," *27th Midwest Symposium on Circuits and Systems*, Morgantown, WV, June 11-12 1984, pp. 425-428, 1984.

[200] C. Duvanaud, S. Dietsche, G. Pataut, J. Obregon, High-efficient class F GaAs FET amplifiers operating with very low bias voltages for use in mobile telephones at 1.75

GHz," *IEEE Microwave and Guided Wave Letters*, vol. 3, no. 8, pp. 268-270, August 1993.

[201] A. S. Tanenbaum, *Computer Networks*, 3rd Edition, Prentice Hall, NJ, 1996.

[202] M. Steenstrup, *Routing in Communications Networks*, Prentice Hall, NJ, 1995.

[203] S. Mazor and P. Langstraat, *A Guide to VHDL*, 2nd Edition, Kluwer Academic Press, Boston MA, 1993.

[204] M. G. Arnold, *Verilog Digital Computer Design: Algorithm into Hardware*, Prentice Hall, NJ, 1999.

[205] J. M. Rabaey, *Digital Integrated Circuits: A Design Perspective*, Prentice Hall, NJ, 1996.

[206] T. A. DeMassa, *Digital Integrated Circuits*, Wiley, NY, 1996.

[207] D. Mann and P. Cobb, "When Dhrystone leaves you high and dry," *EDN*, Vol. 43, No. 11, pp. 117-118, 120, 122, 124, 126, 129, May 21 1998.

[208] http://www.ARM.com

[209] T. G. Noll, "Carry-save architectures for high-speed digital signal processing," *Journal of VLSI Signal Processing*, Vol. 3, No. 1-2, pp. 121-140, June 1991.

[210] R. B. Freeman, "Checked carry select adder," *IBM Technical Disclosure Bulletin*, Vol. 13, No. 6, pp. 1504-1505, Nov. 1970.

[211] C. H. Becker and R. E. McKay, "Developing propagate and generate terms in a carry look-ahead adder," *IBM Technical Disclosure Bulletin*, Vol. 13, No. 12, pp. 3619-3622, May 1971.

[212] D. V. Kanani and K. H. O'Keefe, "A note on conditional-sum addition for base-2 systems," *IEEE Transactions on Computers*, Vol. 22, No. 6, p. 626, June 1973.

[213] A. Chandrakasan and R. Brodersen, *Low-Power CMOS Design*, IEEE Press, NJ, 1998.

[214] H. J. M. Veendrick, "Short-circuit dissipation of static CMOS circuitry and its impact on the design of buffer circuits," *IEEE Journal of Solid-State Circuits*, Vol. 19, No. 4, pp. 486-473, August 1984.

[215] A. P. Chandrakasan, "Low-power digital filtering using approximate processing," *IEEE Journal of Solid-State Circuits*, Vol. 31, No. 3, pp. 395-400, March 1996.

[216] G. C. Temes, H. J. Orchard, and M. Jahanbegloo, "Switched-capacitor filter design using the bilinear z-transform," *IEEE Transactions on Circuits and Systems*, Vol. 25, No. 12, pp. 1038-1044, December 1978.

[217] F. Lu, H. Samueli, J. Yuan, and C. Svensson, "A 700-MHz 24-b pipelined accumulator in 1.2-μm CMOS for application as a numerically controlled oscillator," *IEEE Journal of Solid-State Circuits*, Vol. 28, No. 8, pp. 878-886, August 1993.

[218] H. T. Nicholas III, H. Samueli, and B. Kim, "The optimization of direct digital frequency synthesizer performance in the presence of finite wordlength effects," *Proceedings of the 42nd Annual Frequency Control Symposium*, pp. 357-363, Baltimore, MD, June 1-3 1988.

[219] L. K.-L. Lau, R. Jain, H. Samueli, H. T. Nicholas III, and E. G. Cohen, "DDFSGEN: a silicon complier for direct digital frequency synthesizers," *Journal of VLSI Signal Processing*, Vol. 4, No. 2-3, pp. 213-226, May 1992.

[220] L. K. Tan, E. W. Roth, G. E. Yee, and H. Samueli, "An 800-MHz quadrature digital synthesizer with ECL-compatible output drivers in 0.8 μm CMOS," *IEEE Journal of Solid-State Circuits*, Vol. 30, No. 12, pp. 1463-1473, December 1995.

[221] N. Caglio, J.-L. Degouy, D. Meignant, P. Rousseau, and B. Leroux, "An integrated GaAs 1.25 GHz clock frequency FM-CW direct digital synthesizer," *15th Annual GaAs IC Symposium Technical Digest 1993*, pp. 167-170, October 10-13 1993, San Jose, CA.

[222] J. G. Proakis, "*Introduction to Digital Signal Processing*," MacMillan, New York, 1988.

[223] H. Samueli, "An improved search algorithm for the design of multiplierless FIR filters with powers-of-two coefficients," *IEEE Transactions on Circuits and Systems*, Vol. 36, No. 7, pp. 1044-1047, July 1989.

[224] T. Yoshino, R. Jain, P. T. Yang, H. Davis, W. Gass, and A. H. Shah, "A 100-MHz 64-tap FIR digital filter in 0.8-μm BiCMOS gate array," *IEEE Journal of Solid-State Circuits*, Vol. 25, No. 6, pp. 1494-11501, December 1990.

[225] R. Jain, P. T. Yang, B. Y. Chung, C. Chien, L. K. Tan, and T. Yoshino, "FIRGEN: a CAD system for automatic layout generation of high-performance FIR filter," *Proceedings of the IEEE 1990 Custom Integrated Circuits Conference*," pp. 14.6/1-4, May 13-16 1990, Boston, MA.

[226] V. Verma and C. Chien, "A VHDL based functional compiler for optimum architecture generation of FIR filters," *Proceedings of the 1996 IEEE International Symposium on Circuits and Systems*, pp. 564-567, May 12-15 1996, Atlanta, GA.

[227] E. B. Hogenauer, "An economical class of digital filters for decimation and interpolation," *IEEE Transaction on Acoustic, Speech, and Signal Processing*, Vol. 29, pp. 155-162, April 1981.

[228] H. Samueli and T. Lin, "A VLSI architecture for a universal high-speed multirate FIR digital filter with selectable power-of-two decimation/interpolation ratios," *Proceedings of the 1991 IEEE International Conference on Acoustics, Speech, and Signal Processing*, pp. 1813-1916, May 1991.

[229] T. Lin, Architectures and Circuits for High-Performance Multirate Digital Filters with Applications to Tunable Modulator/Demodulator/Bandpass-Filter Systems, *Ph.D. Dissertation, School of Engineering and Applied Sciences, University of California at Los Angeles*, 1993

[230] L. Liu, T. Yoshino, S. Sprouse, and R. Jain, "An interleaved/retimed architecture for the lattice wave digital filter," *IEEE Transactions on Circuits and Systems*, Vol. 38, No. 3, pp. 344-347, March 1991.

[231] L. Liu, R. Jain, and T. Yoshino, "A 23.8 MHz 73-dB attenuation recursive digital filter IC," *Proceedings of the IEEE 1992 Custom Integrated Circuits Conference*, pp. 19.1/1-4, May 3-6 1992, New York, NY.

[232] L. D'Luna, P. Yang, D. Mueller, K. Cameron, H.-C. Liu, D. Gee, F. Lu, R. Hawley, S. Tsubota, C. Reames, and H. Samueli, "A dual-channel QAM/QPSK receiver IC with integrated cable set-top box functionality," *Proceedings of the IEEE 1998 Custom Integrated Circuits Conference*, pp. 351-354, May 11-14 1998, New York, New York.

[233] J. S. Wu, M. L. Liou, H. P. Ma, and T. Z. Chiueh, "A 2.6-V, 44-MHz all-digital QPSK direct-sequence spread-spectrum transceiver IC," *IEEE Journal of Solid-State Circuits*, Vol. 32, No. 10, pp. 1499-1510, October 1997.

[234] J. K. Hinderling, T. Rueth, K. Easton, D. Eagleson, D. Kindred, R. Kerr, and J. Levin, "CDMA mobile station modem ASIC," *IEEE Journal of Solid-State Circuits*, Vol. 28, No. 3, pp. 253-260, March 1993.

[235] B.-E. Jun, D.-J. Park, and Y.-W. Kim, "Convergence analysis of sign-sign LMS algorithm for adaptive filters with correlated Gaussian data," IEEE International Conference on Acoustics, Speech, and Signal Processing, vol. 2, pp. 1380-1383, May 1995, New York.

[236] E. Eweda, "Transient performance degradation of the LMS, RLS, sign, signed regressor, and sign-sign algorithms with data correlation," *IEEE Transactions on Circuits and Systems II: Analog and Digital Signal Processing*, vol. 46, no. 8, pp. 1055-1062, Aug 1999.

[237] F. Lu and H. Samueli, "A 60-Mbd, 480-Mb/s, 256-QAM decision-feedback equalizer in 1.2-μm CMOS," *IEEE Journal of Solid-State Circuits*, Vol. 28, No. 1, pp. 330-338, March 1993.

[238] R. B. Joshi, p. Yang, H. C. Liu, K. Kindsfater, K. Cameron, D. Gee, V. Hung, G. Gorman, T. Shauhyuam, A. Hung, R. Khan, O. Lee, S. Tollefsrud, E. C. Berg, J. Y. Lee, T. Kwan, C. H. Lin, A. Buchwald, C. D. Jones, and H. Samueli, "A 52 Mb/s universal DSL transceiver IC, *IEEE International Solid-State Circuits Conference Digest of Technical Papers*, pp. 250-251, February 15-17, 1999, SF, CA.

[239] L. D'Luna, L. Tan, D. Mueller, J. Laskowski, K. Cameron, J. Y. Lee, D. Gee, J. Monroe, H. M. Law, J. Chang, M. Wakayama, T. Kwan, C. H. Lin, A. Buchwald, T. Kaylani, F. Lu, T. Spieker, R. Hawley, and H. Samueli, "A single chip universal cable set-top box/modem transceiver, *IEEE International Solid-State Circuits Conference Digest of Technical Papers*, pp. 334-335, February 15-17, 1999, SF, CA.

[240] L. K. Tan, J. S. Putnam, F. Lu, L. J. D'Luna, D. W. Mueller, K. R. Kindsfater, K. B. Cameron, R. B. Joshi, R. A. Hawley, and H. Samueli, "A 70-Mbps variable-rate 1024-QAM cable receiver IC with integrated 10-b ADC and FEC decoder," *IEEE Journal of Solid-State Circuits*, Vol. 33, No. 4, pp. 2205-2218. December 1998.

[241] G. Feygin, P. G. Gulak, and P. Chow, "A multiprocessor architecture for Viterbi decoders with linear speedup," *IEEE Transactions on Signal Processing*, Vol. 41, No. 9, pp. 2907-2917, September 1993.

[242] K. Kawazoe, S. Honda, S. Kubota, and S. Kato, "Universal-coding rate scarce-state-transition Viterbi decoder, *IEEE International Communications Conference*, pp. 1583-1587, June 14-18, 1992, Chicago, IL.

[243] B. K. Min and N. Demassieux, "A versatile architecture for VLSI implementation of the Viterbi algorithm," *IEEE International Conference on Acoustics, Speech, and Signal Processing*, pp. 1101-1104, May 14-17, 1991, Toronto, Ontario.

[244] N. J. P. Frenette, P. J. McLane, L. E. Peppard, and F. Cotter, "Implementation of a Viterbi processor for a digital communications system with a time-dispersive channel," *IEEE Journal on Selected Areas in Communications*, Vol. 4, No. 1, pp. 160-167, January 1986.

[245] P. G. Gulak and E. Shwedyk, "VLSI structures for Viterbi receivers: Part I – general theory and applications," *IEEE Journal on Selected Areas in Communications*, Vol. 4, No. 1, pp. 142-154, January 1986.

[246] J. Sparso, H. N. Jorgensen, E. Paaske, S. Pedersen, and T. R. Petersen, "An area-efficient topology for VLSI implementation of Viterbi decoders and other shuffle-exchange type structures," *IEEE Journal of Solid-State Circuits*, Vol. 26, No. 2, pp. 90-96, February 1991.

[247] P. J. Black and T. H. Meng, "140-Mb/s, 32-state, radix-4 Viterbi decoder," *IEEE Journal of Solid-State Circuits*, Vol. 27, No. 12, pp. 1877-1885, December 1992.

[248] I. Kang and A. N. Willson Jr., "Low-power Viterbi decoder for CDMA mobile terminals," *IEEE Journal of Solid-State Circuits*, Vol. 33, No. 3, pp. 473-481, March 1998.

[249] C. B. Shung, H. D. Lin, R. Cypher, P. H. Siegel, and H. K. Thapar, "Area-efficient architectures for the Viterbi algorithm – part I: theory," *IEEE Transactions on Communications*, pp. 636-644, Vol. 41, No. 4, April 1993.

[250] P. J. Black and T. H. Y. Meng, "A 1-Gb/s, four-state, sliding block Viterbi decoder," *IEEE Journal of Solid-State Circuits*, Vol. 32, No. 6, pp. 797-805, June 1997.

[251] G. Fettweis and H. Meyr, "High-speed parallel Viterbi decoding algorithm and VLSI architecture," *IEEE Communications Magazine*, Vol. 29, No. 5, pp. 46-55, May 1991.

[252] C. M. Radar, “Memory management in a Viterbi decoder,” *IEEE Transactions on Communications*, Vol. 29, No. 9, pp. 1399-1401, September 1981.

[253] H. L. Lou, “Implementing the Viterbi algorithm,” *IEEE Signal Processing Magazine*, Vol. 12, No. 5, pp. 42-52, September 1995.

[254] J. L. Dornstetter and D. Verhulst, “Cellular efficiency with slow frequency hopping Analysis of the digital SFH900 mobile system, “*IEEE Journal on Selected Areas in Communications*," Vol. 5, No. 5, pp. 835-848, June 1987.

[255] GSM Recommendation 5.05, “Radio transmission and reception”, *ETSI PT 12*, October 1993.

[256] GSM Recommendation 5.02, “Multiplexing and multiple access on the radio path”, *ETSI PT 12*, December 1995.

[257] GSM Recommendation 5.04, “Modulation”, *ETSI 300 032*, March 1992.

[258] GSM Recommendation 3.50, “Transmission planning aspects of the speech service in the GSM PLMN system,” *ETSI-SMG Technical Specification*, July 1994.

[259] GSM Recommendation 6.10, “GSM full rate speech transcoding”, *ETSI PT 12*, February 1992.

[260] M. R. L. Hodges, S. A. Jensen, P. R. Tattersall, "Laboratory testing of digital cellular radio systems," *British Telecom Technology Journal*, Vol. 8, No. 1, pp. 57-66, January 1990.

[261] R. D'Avella, L. Moreno, M. Sant'Agostino, "An adaptive MLSE receiver for TDMA digital mobile radio," *IEEE Journal on Selected Areas in Communications*, Vol. 7, No. 1, pp. 122-129, January 1989.

[262] G. Weinberger, “The new millennium: wireless technologies for a truly mobile society,” 2000 IEEE *International Solid-State Circuits Conference Digest of Technical Papers,* pp. 20-24, San Francisco, CA, February 7-9, 2000.

[263] J. Seyenhans, A. Vanwelsenaers, J. Wenin, and J. Baro, “An integrated Si bipolar RF transceiver for a zero IF 900 MHz GSM digital mobile radio frontend of a hand portable phone,” *Proceedings of the IEEE 1991 Custom Integrated Circuits Conference*, pp. 7.7/1-4, San Diego, CA, USA, May 12-15, 1991.

[264] F. Marshall, Behbahani, W. Birth, A. Fotowai, T. Fuchs, R. Gaethke, E. Heimeri, Sheng Lee, P. Moore, S. Navid, and E. Saur, “A 2.7 V GSM transceiver ICs with on-chip filtering,” pp. 148-149, *1995 IEEE International Solid-State Circuits Conference. Digest of Technical Papers*, San Francisco, CA, 15-17 Feb. 1995.

[265] T. Yamawaki, M. Kokubo, K. Irie, H. Matsui, K. Hori, T. Endou, H. Hagisawa, T. Furuya, Y. Shimizu, M. Katagishi, and J. R. Hildersley, “A 2.7-V GSM RF transceiver IC,” *IEEE Journal of Solid-State Circuits*, pp. 2089-2096, vol.32, no.12, Dec. 1997.

[266] P. Orsatti, F. Piazza, and Q. Huang, “A 20-mA-receive, 55-mA-transmit, single-chip GSM transceiver in 0.25- µm CMOS,” *IEEE Journal of Solid-State Circuits*, vol.34, no.12, p.1869-80, Dec. 1999.

[267] F. Piazza, P. Orsatti, and Q. Huang, “A 2 mA/3 V 71 MHz IF amplifier in 0.4-µm CMOS programmable over 80 dB range,” *IEEE International Solid-State Circuits Conference Digest of Technical Papers*, pp. 78-79, San Francisco, Feb. 1997.

[268] M. J. Gingell, “Single sideband modulation using sequence asymmetric polyphase networks,” *Electronic Communications*, vol. 48, pp. 21-25, 1973.

[269] M. Steyaert, M. Borremans, J. Janssens, B. de Muer, I. Itoh, J. Craninckx, J. Crols, E. Morifuji, S. Momose, and W. Sansen, “A single-chip CMOS transceiver for DCS-1800 wireless communications, *1998 IEEE International Solid-State Circuits Conference Digest of Technical Papers*, pp. 48-49, San Francisco, CA, 5-7 Feb. 1998.

[270] J. Janssens, M. Steyaert, B. de Muer, M. Borremans, and I. Itoh, "A 2V CMOS cellular transceiver front-end," *2000 IEEE International Solid-State Circuits Conference Digest of Technical Papers*, pp. 142-143, San Francisco, CA, 7-9 Feb. 2000.

[271] S. Wu and B. Razavi, "A 900 MHz/1.8 GHz CMOS receiver for dual band applications," *IEEE Journal of Solid-State Circuits*, vol. 33, pp. 2178-2185, Dec. 1998.

[272] R. J. van de Plassche, *Integrated Analog-to-Digital and Digital-to-Analog Converters*, Kluwer Academic Press, Boston, MA, 1984.

[273] www.philips.com

[274] www.infineon.com/products/wireless/cellular

[275] M. K. Simon, "A generalization of minimum shift keying (MSK)-type signaling based upon input data symbol pulse shaping," *IEEE Transactions on Communications*, vol. 24, pp. 845-856, August 1976.

[276] P. A. Laurent, "Exact and approximate construction of digital phase modulations by superposition of amplitude modulated pulses," *IEEE Transactions on Communications*, vol. 34, pp. 150-160, February 1986.

[277] M. Luise and U. Mengali, "A new interpretation of the average matched filter for MSK-type receivers," *IEEE Transactions on Communications*, vol. 39, no. 1, January 1991.

[278] J. Fenk, "Highly integrated RF-IC's for GSM and DECT systems-A status review," *IEEE Transactions on Microwave Theory and Techniques*, pp. 2531-2539, vol.45, no. 12, Dec. 1997.

[279] P. T. M. Van Zeijl, N. Van Erven, F. Risseeuw, and M. Van Roosmalen, "An RF ASIC for DECT TDMA Applications," *Proceedings of the 23rd European Solid-State Circuits Conference*, pp. 144-147, Southampton, UK, 16-18 Sept. 1997.

[280] G. L. Puma, K. Hadjizada, S. van Waasen, C. Grewing, P. Schrader, W. Geppert, A. Hanke, M. Seth, and S. Heinen, "A RF Transceiver for digital wireless communication in a 25 GHz Si Bipolar Technology," *2000 IEEE International Solid-State Circuits Conference Digest of Technical Papers*, pp. 144-145, San Francisco, CA, 7-9 Feb. 2000.

[281] M. Bopp, M. Alles, M. Arens, D. Eichel, S. Gerlach, R. Gotzfried, F. Gruson, M. Kocks, G. Krimmer, R. Reimann, B. Roos, M. Siegle, and J. Zieschang, "A DECT transceiver chip set using SiGe technology," 1999 *IEEE International Solid-State Circuits Conference Digest of Technical Papers*, pp. 68-69, San Francisco, CA, USA, 15-17 Feb. 1999.

[282] S. Atkinson, A. Shah, and J. Strange, "A single chip radio transceiver for DECT," *Proceedings of 8th International Symposium on Personal, Indoor and Mobile Radio Communications*, vol. 3, pp. 840-843, Helsinki, Finland, 1-4 Sept. 1997.

[283] G. C. Dawe, J. M. Mourant, and A. P. Brokaw, "A 2.7 V DECT RF transceiver with integrated VCO. (1997 IEEE International Solid-State Circuits Conference Digest of Technical Papers, pp. 308-309, San Francisco, CA, USA, 6-8 Feb. 1997.

[284] J. C. Rudell, J. J. Ou, T. Cho, G. Chien, F. Brianti, J. A. Weldon, and P. R. Gray, "A 1.9-GHz wide-band IF double conversion CMOS receiver for cordless telephone applications," *IEEE Journal of Solid-State Circuits*, vol.32, (no.12), pp. 2071-2088, Dec. 1997.

[285] J. Y. Lee, H. C. Liu, J. S. Putnam, K. Kindsfater, and H. Samueli, "A 42 MB/s multi-channel digital adaptive beamforming AQM demodulator for wireless applications," *Proceedings of the IEEE 1997 Custom Integrated Circuits Conference*, pp. 323-326, May 5-8, 1997, Santa Clara, CA.

[286] J. Y. Lee, H. C. Liu, and H. Samueli, "Performance analysis of synchronization errors on digital adaptive beamforming for high bit-rate QAM receiver," *1997 IEEE 47th Vehicular Technology Conference*, pp. 1064-1068, May 4-7, 1997, Phoenix, AZ.

[287] J. Y. Lee and H. Samueli, "Adaptive beamforming techniques for a frequency-hopped QAM receiver," *1996 IEEE 46th Vehicular Technology Conference*, pp. 1700-1704, April 28 – May 1, 1996, NY, NY.

[288] J.-Y. Lee and H. Samueli, "Adaptive antenna arrays and equalization techniques for high bit-rate QAM receivers," *1996 5th International Conference on Universal Personal Communications*, pp. 1009-1013, Sept. 29 – Oct. 2, 1996, NY, NY.

[289] Jind-Yeh Lee, Huan-Chang Liu, and H. Samueli, "A digital adaptive beamforming QAM demodulator IC for high bit-rate wireless communications," *IEEE Journal of Solid-State Circuits*, vol.33, no.3, pp. 367-377, March 1998.

[290] J. S. Putnam and H. Samueli, "A 4-channel diversity QAM receiver for broadband wireless communications," *IEEE International Solid-State Circuits Conference Digest of Technical Papers*, pp. 338-339, February 15-17, SF, CA.

[291] F. Behbahani, J. Leete, W. Tan, Y. Kishigami, A. Karimi-Sanjaani, A. Roithmeier, K. Hoshino, and A. Abidi, "An adaptive 2.4GHz Low-IF receiver in 0.6 μm CMOS for wideband wireless LAN," 2000 *IEEE International Solid-State Circuits Conference Digest of Technical Papers*, pp. 146-147, San Francisco, CA, 7-9 Feb. 2000.

[292] A. A. Abidi, "Direct-conversion radio transceivers for digital Communications," *IEEE Journal of Solid-State Circuits*, vol.30, no.12, pp.1399-1410, Dec. 1995.

[293] A. Rofougaran, J. Y. C. Chang, M. Rofougaran, A. A. Abidi, "A 1 GHz CMOS RF front-end IC for a direct-conversion wireless receiver," *IEEE Journal of Solid-State Circuits*, vol.31, no.7, pp. 880-889, July 1996.

[294] A. Rofougaran, G. Chang, J. J. Rael, J. Y. C. Chang, M. Rofougaran, J. P. Chang, M. Djafari, M. K. Ku, E. W. Roth, A. A. Abidi, A.A., H. Samueli, "A single-chip 900-MHz spread-spectrum wireless transceiver in 1-μm CMOS. I. Architecture and transmitter design," *IEEE Journal of Solid-State Circuits*, vol.33, no.4, pp. 515-534, April 1998.

[295] A. Rofougaran, G. Chang, J. J. Rael, J. Y. C. Chang, M. Rofougaran, P. J. Chang, M. Djafari, J. Min, E. W. Roth, A. A. Abidi, H. Samueli, "A single-chip 900-MHz spread-spectrum wireless transceiver in 1-μm CMOS. II. Receiver design," *IEEE Journal of Solid-State Circuits*, vol.33, no.4, pp.535-547, April 1998.

[296] Huan-Chang Liu, J. S. Min, H. Samueli, "A low-power baseband receiver IC for frequency-hopped spread spectrum communications," *IEEE Journal of Solid-State Circuits*, vol.31, no.3, pp.384-394, March 1996.

[297] S. Verdu, "Minimum probability of error for asynchronous Gaussian multiple access," *IEEE Transactions on Information Theory*, vol. 32, no. 1, pp. 85-96, January 1986.

[298] S. Akhtar and D. Zeghlache, "Capacity evaluation of the UTRA WCDMA interface," *IEEE 50th Vehicular Technology Conference*, pp. 914-918, Amsterdam, Netherlands, September 19-22, 1999.

[299] P. M. Mistry, "Third generation cellular (3G): W-CDMA and TD-CDMA," *Northcon/98 Conference Proceedings*, Seattle, WA, Oct. 21-23, 1998.

[300] K. Koutsopoulos, G. Lyberopoulos, J. Markoulidaki, R. Menolascino, and A. Rolando, "On the dimensioning and efficiency of the UTRA air-interface modes (W-CDMA and TD-CDMA)," *CSELT Technical Reports*, vol. 27, no. 6, pp. 831-843, December 1999.

[301] T. Takenaka, H. Tamura, T. Maruyama, S. Aizawa, and A. Nishiumi, "Development of cdma2000 RTT-based experimental system for IMT-2000," IEEE 51st Vehicular Technology Conference, pp. 133-137, Tokyo, Japan, May 15-18, 2000.

[302] S. Li and J. Li, "TD-SCDMA standardization and prototype development," *Fifth Asia-Pacific Conference on Communications and Fourth Optoelectronics and Communications Conference*, vol. 2, Beijing, China, Oct. 18-22, 1999.

[303] N. R. Sollenberger, N. Seshadri, and R. Cox, "The evolution of IS-136 TDMA for third-generation wireless services," *IEEE Personal Communications*, vol. 6, no. 3, pp. 8-18, June 1999.

[304] Parssinen, J. Jussila, J. Ryynanen, L. Sumanen, and K. A. I. Halonen, "A 2-GHz wide-band direct conversion receiver for WCDMA Applications," *IEEE Journal of Solid-State Circuits*, pp. 1893-903, vol.34, no.12, Dec. 1999.

[305] S. Sheng, L. Lynn, J. Peroulas, K. Stone, I. O'Donnell, and R. Brodersen, "A low-power CMOS chipset for spread spectrum communications," *1996 IEEE International Solid-State Circuits Conference. Digest of Technical Papers*, pp. 346-347, San Francisco, CA, USA, 8-10 Feb. 1996.

[306] T. E. Truman, T. Pering, R. Doering, and R. W. Brodersen, "The InfoPad multimedia terminal: a portable device for wireless information access," *IEEE Transactions on Computers*, vol.47, no.10, pp. 1073-1087, Oct. 1998.

[307] S. Narayanaswamy, S. Seshan, E. Amir, E. Brewer, R. W. Brodersen, F. Burghardt, A. Burstein, Yuan-Chi Chang, A. Fox, J. M. Gilbert, R. Han, R. H. Katz, A. C. Long, D. G. Messerschmitt, and J. M. Rabaey, "A low-power, lightweight unit to provide ubiquitous information access application and network support for InfoPad," *IEEE Personal Communications*, vol.3, no.2, pp. 4-17, April 1996.

[308] B. C. Richards and R. W. Brodersen, "InfoPad: The design of a portable multimedia terminal," *Proceedings 2nd International Workshop on Mobile Multimedia Communications*, pp. S2/1/1-6, Bristol, UK, 11-13 April 1995.

[309] B. Wong and H. Samueli, "A 200 MHz all-digital QAM modulator and demodulator in 1.2 mm CMOS for digital radio applications," *IEEE Journal of Solid-State Circuits*, vol. 26, no. 12, pp. 1970-1980, December 1991.

[310] S.-J. Kim, Y.-G. Lee, S.-K. Yun, and H.-Y. Lee, "Novel high-Q bondwire inductors for RF and microwave monolithic integrated circuits," *1999 IEEE International Microwave Symposium Digest*, vol. 4, pp. 1621-1624, Anaheim, CA, 13-19 June 1999.

[311] K.-C. Tsai and P. R. Gray, "A 1.9-GHz, 1-W CMOS class-E power amplifier for wireless communications," *IEEE Journal of Solid State Circuits*, vol. 34, no. 7, pp. 962-970, July 1999.

[312] T. Cho, E. Dukatz, M. Mack, D. MacNally, M. Marringa, S. Mehta, C. Nilson, L. Plouvier, S. Rabii, and S. Luna, "A single chip CMOS direct conversion transceiver for 900 MHz spread spectrum digital cordless phones," *Workshop on CMOS Wireless Transceiver Designs*, Taipei, Taiwan, June 14, 1999.

[313] M. H. Sherif, D. O. Bowker, G. Bertocci, B. A. Orford, G. A. Mariano, "Overview of CCITT embedded ADPCM algorithm," *IEEE International Conference on Communications*, Atlanta, GA, USA, pp.1014-18, April 1990.

[314] T. Cho, E. Dukatz, M. Mack, D. Macnally, M. Marringa, S. Mehta, C. Nilson, L. Plouvier, S. Rabii, S, "A single-chip CMOS direct-conversion transceiver for 900 MHz spread-spectrum digital cordless phones," *1999 IEEE International Solid-State Circuits Conference Slides Supplement*, pp.228-229, San Francisco, CA, USA, 15-17 Feb. 1999.

[315] K. Matsuyama, S. Yoshioka, M. Shimizu, N. Aoki, and Y. Tozawa, "A burst GFSK-modem for wireless LAN systems," *1995 IEEE International Symposium on Personal, Indoor and Mobile Radio Communications*, pp. 198-202, Toronto, Canada, September 27-29, 1995.

[316] H. Suzuki, K. Momma, and Y. Yamao, "Digital portable transceiver using GMSK modem and ADM codec," *IEEE Journal on Selected Areas in Communications*, vol. 2, no. 4, pp. 604-610, July 1984.

[317] M. Park, S. Lee, H. K. Yu, and K. S. Nam, "Optimization of high Q CMOS-compatible microwave inductors using silicon CMOS technology," *1997 IEEE Radio Frequency*

Integrated Circuits Symposium Digest of Technical Papers, pp. 181-184, Denver, CO, June 8-11, 1997.

[318] S. Scrantom, G. Gravier, T. Valentine, D. Pehlke, and B. Schiffer, "Manufacture of embedded integrated passive components into low temperature co-fired ceramic systems," Proceedings of the *1998 International Symposium on Microelectronics*, vol. 3582, pp. 459-466, San Diego, CA, November 1-4, 1998.

[319] S. Consolazio, K. Nguyen, D. Biscan, K. Vu, A. Ferek, and A. Ramos, "Low temperature cofired ceramic (LTCC) for wireless applications," *1999 IEEE MTT-S International Topical Symposium on Technologies for Wireless Applications*, pp. 201-205, Vancouver, Canada, February 21-24, 1999.

[320] R. G. Arnold and D. J. Pedder, "Microwave characterization of microstrip lines and spiral inductors in MCM-D technology," *IEEE Transactions on Components, Hybrids, and Manufacturing Technology*, vol. 15, no. 6, pp. 1038-1045, December 1992.

[321] R. Goyal and V. K. Tripathi, "MCM design methodology for portable wireless communication systems design," *1996 International conference on Multichip Modules*, pp. 230-233, Denver, CO, April 17-19, 1996.

[322] R. C. Frye, R. P. Smith, R. R. Kola, Y. L. Low, Y. M. Lau, and L. K. Tai, "Silicon MCM-D with integrated passive components fore RF wireless applications," *The 34th ISHM-Nordic Annual Conference*, pp. 138-142, Oslo, Norway, September 21-24, 1997.

[323] P. Davis, P. Smith, E. Campbell, J. Lin, K. Gross, G. Bath, Y. Low, M. Lau, Y. Degani, J. Gregus, R. Frye, and K. Tai, "Silicon-on-silicon integration of a GSM transceiver with VCO resonator," *1998 IEEE International Solid-State Circuits Conference Digest of Technical Papers*, pp. 248-249, San Francisco, CA, USA, February 5-7, 1998.

Index